Springer Series in
SOLID-STATE SCIENCES 94

Springer
*Berlin*
*Heidelberg*
*New York*
*Barcelona*
*Hong Kong*
*London*
*Milan*
*Paris*
*Singapore*
*Tokyo*

**Physics and Astronomy** ONLINE LIBRARY

http://www.springer.de/phys/

Springer Series in

# SOLID-STATE SCIENCES

---

*Series Editors:*

M. Cardona P. Fulde K. von Klitzing R. Merlin H.-J. Queisser H. Störmer

The Springer Series in Solid-State Sciences consists of fundamental scientific books prepared by leading researchers in the field. They strive to communicate, in a systematic and comprehensive way, the basic principles as well as new developments in theoretical and experimental solid-state physics.

126 **Physical Properties of Quasicrystals**
Editor: Z.M. Stadnik

127 **Positron Annihilation in Semiconductors**
Defect Studies
By R. Krause-Rehberg and H.S. Leipner

128 **Magneto-Optics**
Editors: S. Sugano and N. Kojima

129 **Computational Materials Science**
From Ab Initio to Monte Carlo Methods
By K. Ohno, K. Esfarjani, and Y. Kawazoe

130 **Contact, Adhesion and Rupture of Elastic Solids**
By D. Maugis

131 **Field Theories for Low-Dimensional Condensed Matter Systems**
Spin Systems and Strongly Correlated Electrons
By G. Morandi, P. Sodano, A. Tagliacozzo, and V. Tognetti

Series homepage – http://www.springer.de/phys/books/sss/

---

Volumes 1–125 are listed at the end of the book.

N. Tsuda K. Nasu
A. Fujimori K. Siratori

# Electronic Conduction in Oxides

Second, Revised and Enlarged Edition

With 199 Figures and 13 Tables

Springer

Professor Dr. Nobuo Tsuda
Department of Applied Physics, Faculty of Science, Science University of Tokyo
1–3 Kagurazaka, Shinjuku-ku,
Tokyo 162–8601, Japan
e-mail: tsuda@ap.kagu.sut.ac.jp

Professor Dr. Keiichiro Nasu
Division of Solid State Theory, Institute of Materials Structure Science, High Energy Accelerator Research Organization
1–1 Oho, Tsukuba, Ibaraki 305–0801, Japan
e-mail: knasu@nasuws4.kek.jp

Professor Dr. Atsushi Fujimori
Department of Complexity Science and Engineering,
The University of Tokyo
7–3–1 Hongo, Bunkyo-ku,
Tokyo 113–0033, Japan
e-mail: fujimori@phys.s.u-tokyo.ac.jp

Dr. Kiiti Siratori
Tiba-ken, Narasino-si, Tudanuma 5–2–14,
275–0016 Japan
e-mail: siratori@jca.apc.org



ISSN 0171-1873
ISBN 3-540-66956-6 2nd Edition Springer-Verlag Berlin Heidelberg New York
ISBN 3-540-52637-4 1st Edition Springer-Verlag Berlin Heidelberg New York

Library of Congress Cataloging-in-Publication Data
Denki dendosei sankabutsu. English.
Electronic conduction in oxides / N. Tsuda ... [et al.].– 2nd rev. and enl. ed. p. cm. – (Springer series in solid-state sciences, ISSN 0171-1873 ; 94) Includes bibliographical references and index. ISBN 3540669566 (alk. paper) 1. Energy-band theory of solids. 2. Free electron theory of metals. 3. Oxides-Electric properties. 4. Electric conductivity. I. Tsuda, N. (Nobuo), 1936- II. Title. III. Series.
QC176.8.E4 D3813 2000 530.4'12–dc2l 00-041925



Springer-Verlag Berlin Heidelberg New York
a member of BertelsmannSpringer Science+Business Media GmbH

Printed in Germany



Typesetting: Data conversion by A. Duhm
Cover concept: eStudio Calamar Steinen
Cover production: *design & production* GmbH, Heidelberg

SPIN: 10744868 57/3141/tr - 5 4 3 2 1 0 – Printed on acid-free paper

# Preface to the Second Edition

This is a revised version of the first edition published in 1991. At the same time, this is a revised version of the Syokabo edition, which was written in Japanese and published in 1993 as a revised version of the original edition published in 1983.

Compared with the first edition, the following revisions have been made: a new chapter on electron–electron interaction has been prepared by a new co-author A. Fujimori. The substances in the previous Chap. 4 have been changed, and $La_{1-x}Sr_xMnO_3$ substituted for $V_2O_3$, which has been extensively reviewed in *Reviews of Modern Physics*, Vol. 70, p. 1039 (1998) by M. Imada, A. Fujimori and Y. Tokura. Section 4.6, NiO, was rewritten by A. Fujimori. The other chapters have also been revised by each author to accommodate new developments which have appeared since the publication of the first edition.

As a result, the references have been increased from 1088 to 1293, and 83 figures are new or improved.

The authors are: N. Tsuda for Chaps. 1, 2 and Sects. 5.1–5.5 and 5.9, K. Nasu for Chap. 2, A. Fujimori for Chap. 3 and Sects. 5.6 and 5.7, and K. Siratori for Sect. 5.8.

The authors would like to express their gratitude to many authors and publishers for allowing them to reproduce their diagrams, and to H.K.V. Lotsch and C.E. Ascheron for their encouragement to complete this book. The authors are indebted to S. Lyle for having improved the English, and to Mrs. P. Treiber for her cooperation in the production process. Especially, they are indebted to Ms. A. Duhm for all her efforts to complete this book. Thanks are also to S. Makiya of Syokabo for his cooperation. Without their help, this book could not have been accomplished.

Finally, we acknowledge, with thanks, the authors of all the papers referred to in this book.

Tokyo, Tsukuba  
Tokyo, Narasino, July 2000

*N. Tsuda, K. Nasu*  
*A. Fujimori, K. Siratori*

# Preface to the First Edition

This book is a revised and up-dated translation of *Denki Dendōsei Sankabutsu* (Electronic Conduction in Oxides) published by Shokabo in Tokyo in 1983 as the second volume in the Material Science Series, which was edited for postgraduate students by T. Suzuki, S. Chikazumi, and S. Nakajima.

Since the publication of the first edition, we have witnessed the historic discovery of high-$T_c$ superconductors by J.G. Bednorz and K.A. Müller. The Shokabo edition has thus been thoroughly revised to accommodate the recent developments, and K. Nasu joined as the fourth author.

The book is compiled as follows: After a short introductory chapter, Chap. 2, written by Tsuda, is devoted to a brief review of transport phenomena and electronic states in oxides. In Chap. 3, electron–phonon and electron–electron interaction are treated theoretically by Nasu and Yanase. Nasu discusses the present status of theoretical studies of the electron–phonon interaction in solids and Yanase explains the electron correlation. Chapter 4 treats the physics of various representative oxides in detail. Sections 4.1–4.5 and 4.10 were written by Tsuda and Sects. 4.6–4.9 by Siratori. This chapter is intended not as an exhaustive review of the properties of each oxide, but rather as an illustration of the concepts which have developed out of the research into transport phenomena in conductive oxides. Many of these concepts are due to N.F. Mott. At the end of Chap. 4, the properties of high-$T_c$ oxides are reviewed by Tsuda. The reader is kindly asked to forgive the inevitable omission of certain important works in this field.

The authors would like to express their gratitude to H.K.V. Lotsch for his encouragement to complete this book and to A.M. Lahee for improving the manuscript. One of the authors (KS) is indebted to D. Ihle for his valuable comments on Sect. 4.8.

Finally, we acknowledge with thanks the authors of all the papers referred to in this book.

Tokyo, Okazaki | *N. Tsuda, K. Nasu*
Sakai, Toyonaka, October 1990 | *A. Yanase, K. Siratori*

# Contents


**1. Introduction** .......... 1

**2. Introduction to Electronic States in Oxides and an Overview of Transport Properties** .......... 5

2.1 Atoms in a Ligand Field .......... 5

2.2 Electronic Energy Bands .......... 10

2.3 Electron–Electron Interaction .......... 10

2.3.1 Direct Interaction .......... 10

2.3.2 Indirect Interaction .......... 11

2.4 Electron–Phonon Interaction .......... 11

2.4.1 The Adiabatic Approximation .......... 11

2.4.2 The Fröhlich Model, the Deformation Potential and the Simple Metal .......... 12

2.4.3 Polarons .......... 13

2.5 Randomness .......... 15

2.5.1 Anderson Localization .......... 15

2.5.2 Variable Range Hopping .......... 15

2.6 The Seebeck Coefficient and Hall Mobility .......... 16

2.7 Magnetic Susceptibility .......... 17

2.8 Metal–Insulator Transition (MIT) .......... 19

2.9 Good Conductors .......... 38

2.9.1 The NaCl Structure .......... 40

2.9.2 The Corundum Structure .......... 44

2.9.3 The Rutile Structure .......... 45

2.9.4 The Perovskite Structure .......... 47

2.9.5 The $K_2NiF_4$ Structure .......... 52

2.9.6 $ReO_3$ and $M_xWO_3$ .......... 52

2.9.7 Pyrochlores $A_2B_2O_{7-x}$ .......... 53

2.9.8 Spinels .......... 54

2.9.9 Low-Dimensional Oxides .......... 55

**3. Theories for Many-Electron Systems with Strong Electron–Phonon and Interelectron Coulombic Interactions** ... 57
3.1 Single-Body Problems in Strongly Coupled Electron–Phonon Systems ... 60
3.1.1 Electrons, Phonons and Their Couplings ... 60
3.1.2 Weak Coupling and Large Polarons ... 61
3.1.3 Strong Coupling, Self-Trapping, Broken Symmetry and Dimensionality ... 62
3.1.4 Dynamics of Self-Trapping ... 65
3.2 Two-Body Problems in Strongly Coupled Electron–Phonon Systems ... 67
3.2.1 Bipolarons ... 67
3.2.2 Charge Separation of Self-Trapped Exciton ... 69
3.3 Excitons and Solitons in One-Dimensional Charge Density Wave States ... 71
3.3.1 Phase Diagram of the Ground State ... 74
3.3.2 Nonlinear Lattice Relaxation and Proliferations of Excitons in One-Dimensional CDW ... 76
3.3.3 One-Dimensional Extended Peierls–Hubbard Model ... 78
3.3.4 Relaxation in One-Dimensional CDW ... 80
3.4 Direct and Indirect Excitons in Three-Dimensional CDW State ... 83
3.4.1 Three-Dimensional Extended Peierls–Hubbard Model for $BaBiO_3$ ... 85
3.4.2 Direct and Indirect Excitons ... 85
3.5 Competition Between Superconductivity and CDW State ... 87
3.5.1 The Many-Polaron System ... 88
3.5.2 Phase Diagram ... 90
3.6 Superconducting Transition Temperatures of Strongly Coupled Electron–Phonon Systems ... 94
3.6.1 Expected Behaviour of $T_c$ ... 94
3.6.2 Interpolating Theory for $T_c$ by CPA ... 95
3.7 Many-Electron System Coupling Strongly with Nonlinear Phonons ... 98
3.7.1 BCS Limit, Nonlinear Phonons and Isotope Effects ... 98
3.7.2 Anharmonic Peierls–Hubbard Model ... 100
3.7.3 Anharmonicity and Metal–Insulator (CDW, SDW) Transitions ... 101
3.7.4 Isotope Effects and Anharmonicity by the BCS Theory 104
3.7.5 Migdal–Eliashberg Theory ... 105
3.8 Non-Grassmann Path Integral Theory for Long-Range Coulomb Repulsion ... 107

3.8.1 Quadratic Form for Long-Range Coulomb Interaction . 108
3.8.2 Path-Integral for Both Short- and Long-Range Parts . . 109
3.8.3 One-Body Green's Function Free from Grassmann Algebra . . . 110
3.8.4 Time-Dependent Bloch–De Dominicis Theorem . . . 113
3.8.5 Light Absorption Spectrum of the SDW State . . . 116

**4. Electron–Electron Interaction and Electron Correlation** . . . 119
4.1 Introduction . . . 119
4.2 Microscopic Models of Interacting Electrons . . . 120
4.3 One-Electron Theories and Electron Correlation . . . 123
4.3.1 Hartree–Fock Approximation . . . 124
4.3.2 Local Density Approximation . . . 126
4.3.3 Electron Correlation Effects . . . 126
4.4 Electronic Structure of Transition-Metal Ions . . . 127
4.4.1 Hartree–Fock Scheme . . . 127
4.4.2 Ligand-Field Theory . . . 128
4.4.3 $d$ Bands and Carrier Doping in Mott Insulators . . . 131
4.5 Hybridization Between $d$ and $p$ Electrons . . . 133
4.5.1 Mott–Hubbard Type and Charge-Transfer Type . . . 133
4.5.2 Configuration-Interaction Theory . . . 136
4.6 Magnetic Interactions . . . 140
4.6.1 Superexchange Interaction . . . 140
4.6.2 Local Moment in Metals . . . 142
4.7 Correlated Metals . . . 144
4.7.1 Metal–Insulator Transition . . . 144
4.7.2 Hubbard Model . . . 145
4.7.3 Fermi-Liquid Properties . . . 146
4.7.4 Long-Range Coulomb Interaction . . . 150
4.7.5 Mixed Valence States . . . 151
4.7.6 Anderson Localization . . . 152

**5. Representative Conducting Oxides** . . . 157
5.1 $ReO_3$: The Most Conductive $d\epsilon$ Conductor . . . 158
5.1.1 Crystal Structure . . . 158
5.1.2 Electronic Properties . . . 160
5.2 $SnO_2$ and $TiO_2$: Oxide Semiconductors . . . 168
5.2.1 Electronic Energy Band Structure of $SnO_2$ . . . 169
5.2.2 Electrical Conductivity of $SnO_2$ . . . 170
5.2.3 Optical Properties of $SnO_2$ . . . 175
5.2.4 $TiO_2$ . . . 177
5.3 $LiTi_2O_4$ and $LiV_2O_4$: Weak-Coupling Superconductor and Temperature-Dependent Magnetism . . . 181

5.3.1 Crystal Structure ... 182
5.3.2 Electronic Properties ... 183
5.3.3 Superconducting Properties ... 185
5.3.4 Insulating Properties: Nonzero Density of States ... 186
5.3.5 $LiV_2O_4$-$ZnV_2O_4$ ... 190
5.4 $WO_3$ and $M_xWO_3$: Large Polarons ... 191
5.4.1 Structure ... 192
5.4.2 Electronic Properties in the Insulating Range and the Metal–Insulator Transition ... 193
5.4.3 Superconductivity and Screening of the Electron–Phonon Interaction ... 198
5.5 $M_xV_2O_5$ and $M_xMoO_3$: Low-Dimensional Oxides ... 199
5.5.1 Crystal Structure of $\beta$-$Na_xV_2O_5$ ... 199
5.5.2 Electronic Properties of Na-Vanadium Bronze ... 200
5.5.3 Magnetic Properties ... 205
5.5.4 Specific Heat ... 206
5.5.5 EPR and NMR in $Na_xV_2O_5$ ... 207
5.5.6 Molybdenum Bronzes ... 209
5.6 NiO: Origin of the Band Gap and Hole Conduction ... 213
5.6.1 Optical and Magnetic Properties ... 213
5.6.2 Transport Properties ... 216
5.6.3 Electronic Structure ... 221
5.6.4 Electronic Structure of Acceptor Level ... 225
5.6.5 Band Theory of Mott Insulators ... 227
5.7 Perovskite-Type Mn Oxides: Magnetoresistance ... 230
5.7.1 Crystal Structure ... 230
5.7.2 Ferromagnetic Metal–Paramagnetic Insulator Transition ... 231
5.7.3 Electronic Structure ... 234
5.7.4 Charge and Orbital Ordering ... 236
5.7.5 Polaron Effects ... 237
5.8 $Fe_3O_4$, Verwey Transition ... 243
5.8.1 Phase Diagram of the Iron–Oxygen System ... 244
5.8.2 The Spinel Structure ... 246
5.8.3 Verwey's Model: Order–Disorder Transformation of $Fe^{2+}$ and $Fe^{3+}$ ... 248
5.8.4 Comment by Anderson: Frustration on the $B$ Lattice ... 250
5.8.5 Transport Phenomena and the Fluctuation of Charge ... 252
5.8.6 Band Structure ... 259
5.8.7 Fluctuating Local Lattice Distortion and Electron–Phonon Coupling ... 263
5.8.8 Itinerant Versus Localized Character of Carriers ... 268
5.9 High-$T_c$ Superconductors ... 270
5.9.1 $d\gamma$ Conductors ... 271

5.9.2 $La_2CuO_4$ . . . . . . . . 272
5.9.3 $La_{2-x}M_xCuO_4$ . . . . . . . . 276
5.9.4 $YBa_2Cu_3O_{7-x}$ . . . . . . . . 288
5.9.5 $Bi_2Sr_2CaCu_2O_8$ . . . . . . . . 305

**References** . . . . . . . . 321

**Index** . . . . . . . . 361

# 1. Introduction

The surface of the earth is almost entirely composed of oxides. Over the centuries much effort has been expended to reduce these oxides to metals such as aluminum, copper, and iron. Metals can carry an electric current and are ductile, whereas oxides have generally been considered to be insulating and brittle. They remain brittle even at room temperature, except for a certain specially fabricated zirconium oxide. However, as far as their electric properties are concerned, there are actually many good conductors and in fact mercury–copper oxide shows superconductivity at temperatures as high as 160 K. Thus oxides cover the entire range of conductivity from insulators, through semiconductors and metallic conductors, to superconductors. Nevertheless they have not been technologically exploited in the same way as silicon, copper, and $Nb_3Sn$. Their use has been largely confined to applications such as insulating materials. One reason is that it is very difficult to keep the oxygen content at the desired level. A second reason is their brittleness, and a third is the crystalline transformation. In Si, a typical dimension of a carrier orbital is 30 Å and the orbitals overlap for doping concentration above about $4 \times 10^{18}$cm$^{-3}$. In transition element oxides, however, an orbital is typically 1 Å in size and overlap only occurs at high doping concentrations of about $10^{22}$cm$^{-3}$. At such high doping levels it is difficult to maintain the lattice structure of the host material and a different crystal structure often appears to accommodate the dopants. In this case, the concept of doping is not appropriate and we classify oxides according to their crystal structure. The electrical properties change drastically from one structure to another and their modification is usually limited within each structure. These phase changes may be seen as a disadvantage compared with Si, but at the same time they can be a great advantage because an unexpected variety of properties are found to develop over the various crystal structures.

Compared with a simple metal, the electrical properties of oxides show certain characteristic features. One is the metal–insulator transition, in which at a certain temperature or pressure, an insulator turns to a metal. The word transition is usually used even when this change occurs at a certain composition. This phenomenon has attracted much attention and was the most popular research theme before high-temperature superconductivity exploded onto the scene. The mechanism of the transition is not simple and the phenomenon itself seems to become more and more complex as research progresses. It is a result of many electron–electron and/or electron–phonon interactions.

Another characteristic feature of oxide conductivity is a temperature dependence of the resistivity that is stronger than $T^1$. This is mostly due to scattering by optical phonons but for certain transition-element oxides, such as cuprates, electron–electron scattering is not negligible. In transition metal oxides, the conduction bands are mostly formed with oxygen $2p$- and metal $d$-orbitals. The $s$-electrons enter into deeper bonding orbitals. The oxygen orbitals are usually located lower in energy, which leads to the observed ionicity. At the same time, the directionality of the $p$- and $d$-orbitals, and the strong Coulomb interaction in the $d$-orbital cloud manifest more directly in the transport phenomena, whereas in the simple metals, $s$-orbitals are the main constituent of the conduction band. The stronger atomic-electron-like character of the conduction electrons induces a complex many-body aspect in the oxides.

It is a difficult task to describe an electronic state in a quasiperiodic potential. When a potential is nearly periodic, any aperiodicity is taken into account as a scattering of the Bloch electrons. However when the aperiodicity becomes significant, there is no itinerant state extending over the entire region and the orbital is more or less localized over a small number of atomic distances. There thus appears an intermediately extended state. The aperiodicity in oxides is due either to a static random distribution of ions or to a dynamical interaction among the hopping electrons.

When the electrons are localized, the characteristic feature of the electron system appears in the form of magnetic or dielectric properties. Oxides are frequently utilized as ferromagnets or ferroelectric materials. For such applications, the conductivity can be detrimental to the desired properties, and ways to suppress the conductivity have been extensively studied. Historically, it was exactly such studies of hopping conduction that were the forerunner of the present much wider interest in oxide conductivity.

Since the solid consists of electrons and phonons, the interaction among these greatly influences the electronic states and, to a lesser extent, the phonon states. A carrier state is rather well described as a combined state of an electron and phonons. Such a state is called a polaron. This is not a good name in fact, since it gives the impression of a localized state. At lower temperatures such a carrier falls into a more stable two-carrier state, which is the superconducting state. In some cases, localized two-electron states appear, which are known as bipolarons.

Recent study of high-temperature superconductivity has elucidated various anomalous electronic properties for cuprates. The anomalous properties cannot be explained simply by the electron–phonon interaction, and the importance of the electron–electron interaction has been recognized not only for describing the insulating properties but also for describing the metallic properties. Extremely numerous studies have been accumulated for cuprates all over the world, and they are pushing up the level of precision in solid state physics.

In this book, the various properties of the oxides will be treated in four chapters. The reader is expected to have a knowledge of solid state physics to the level of the famous textbooks "Introduction to Solid State Physics" by Kittel [1] and also "Solid-State Physics" by Ibach and Lüth [2].

In the next chapter, various basic concepts necessary to understand the transport properties in oxides will be reviewed and we will endeavor to convey their physical meaning in a very concise way. This chapter will serve as a guide to important basic works and the reader is expected to consult the original papers for certain details and derivatives. In this chapter, too, the conductive oxides will be tabulated except for the high-$T_c$ oxides which will appear in Chap. 5. The conductive oxides will be classified into $d\epsilon$, $d\gamma$ and $s$–$p$ conductors. Their properties will be reviewed according to their crystal structures.

In the third chapter, the electron–phonon interaction will be explained theoretically. The reason for preparing this section is that we consider it very important to review this field since it will facilitate the study of high-$T_c$ superconductivity.

In the fourth chapter, the electron–electron interaction is explained. This has been a most interested theoretical problem concerning the metal–insulator transition and high-$T_c$ superconductivity. However, it is not yet fully clarified how the traditional one-electron description of the transport property of conductive oxides is to be modified by the strong electron–electron interaction, and this many-body problem of metallic conductivity is now being discussed by many researchers.

In the fifth chapter, various transport properties will be explained with reference to representative oxides. (At the beginning of Chap. 5, it will be explained just how representative they are. Some of them in the first edition have been deleted to reduce the total number of pages.) This chapter comprises the main part of this book. It is not our intention to introduce all the various oxides to the reader but rather to introduce the various concepts that have been developed in the field of oxide conductors. It is from this viewpoint that a number of oxides have been selected for explicit mention.

Since almost all the relevant phenomena are the consequence of many-body effects, it is impossible to treat them in a unified quantitative way, and various models have been proposed to describe in detail a certain aspect of the complicated many-body interactions. Some phenomena have been analyzed repeatedly using various models and from various viewpoints and yet without definite conclusions. There are many such cases. Therefore one should not consider that all the phenomena have been clarified. The interpretations presented in Chap. 5 are not always conclusions agreed upon by all the researchers in this field.

The representative oxides discussed in Chap. 5 will be classified into two classes. In one class, the electron–phonon interaction plays an essential role, and in the other, it is the electron–electron interaction that is dominant. The first five materials from $ReO_3$ to $M_xV_2O_5$ (Sects. 5.1–5.5) might be considered as the phonon materials whereas the subsequent three materials from NiO to $Fe_3O_4$ (Sects. 5.6–5.8) could be classed as electron materials to a first approximation. In the last section (Sect. 5.9) the high-$T_c$ Cu oxides will be treated in detail. In these materials both the electron and the phonon interactions will be important. Since this topic presents new and still unresolved problems, our intention is simply to introduce various experimental results with equal weight; we avoid attempts to systematize the results in terms of mechanisms.

From a technological point of view, ionic transport is also an important conductivity problem in oxides. However, it is beyond the scope of this book and we refer the interested reader to the literature [3].

# 2. Introduction to Electronic States in Oxides and an Overview of Transport Properties

## 2.1 Atoms in a Ligand Field

The basic structural units of transition element oxides are an $MO_6$ octahedron and an $MO_4$ tetrahedron which are shown in Fig. 2.1. The metal-$d$ orbitals ($M_d$) overlap with the ligand $2p$ orbitals ($O_{2p}$) forming molecular orbitals.

In the following, it will be seen how the atomic orbitals are rearranged in the clusters.

Firstly, ligand oxygen ions are approximated by negative point charges [4], and these point charges influence the electrostatic potential of the central $d$ electrons. In case of the octahedron, the potential is

$$
\begin{aligned}
V(\boldsymbol{r}) &= \sum_{i=1}^{6} \frac{Ze^2}{4\pi\epsilon_0 |\boldsymbol{r} - \boldsymbol{R}_i|} \\
&= \frac{1}{4\pi\epsilon_0} \left\{ \frac{6Ze^2}{a} \right. \\
&\qquad \left. + \frac{7Ze^2}{2a^5} r^4 \sqrt{\frac{4\pi}{9}} \left( Y_{40}(\theta,\phi) + \sqrt{\frac{5}{14}} (Y_{44} + Y_{4-4}) \right) + \dots \right\}, \qquad (2.1)
\end{aligned}
$$

where $\boldsymbol{R}_i$ is the position vector of the ligand point charge. $Ze$ is the ligand valence and $r$, $\theta$, and $\phi$ are the polar coordinates of the central $d$ electron. $Y_{lm}$ are the spherical harmonics. Under this ligand field, five degenerate $3d$ orbitals, $\phi_{322}, \phi_{321}, \phi_{320}, \phi_{32\bar{1}}$ and $\phi_{32\bar{2}}$ split into two groups. The atomic functions arc

$$\phi_{32m} = R_{32}(r)\Theta_{2m}(\theta)\Phi_m(\phi) \,, \qquad (2.2)$$

$$\Theta_{20}(\theta) = \frac{\sqrt{5}}{2\sqrt{2}} (3\cos^2\theta - 1) \,, \quad \Theta_{2\pm1}(\theta) = \mp\frac{\sqrt{15}}{2} \sin\theta\cos\theta \,, \qquad (2.3)$$

$$\Theta_{2\pm2}(\theta) = \frac{\sqrt{15}}{4} \sin^2\theta \,, \quad \Phi_m(\phi) = \frac{1}{\sqrt{2\pi}} \exp(im\phi) \,. \qquad (2.4)$$

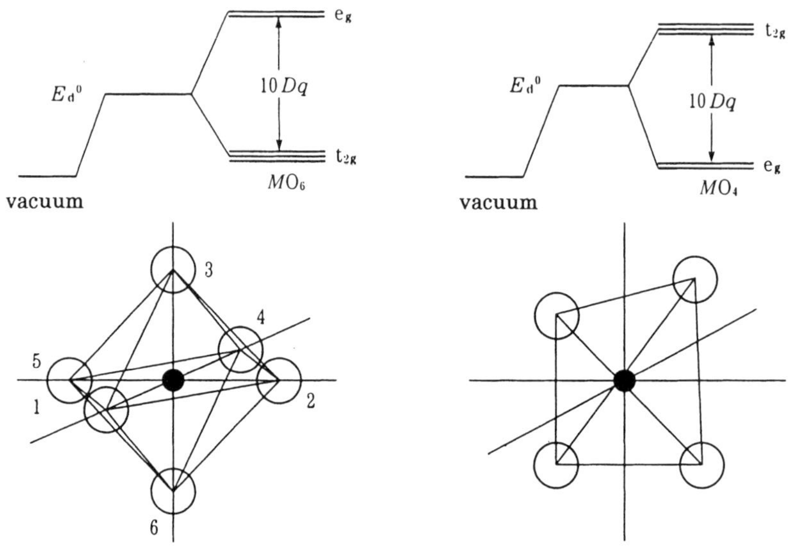


**Fig. 2.1.** Energy level splittings in the octahedral and the tetrahedral positions. Atomic $d$ states split into $E_g$ and $T_{2g}$ states

$R_{32}(r)$ is the radial function. The $5 \times 5$ matrix of $V(\boldsymbol{r})$ is diagonalized as follows [4]:

$$
\begin{array}{c}
\phi_{322}, \phi_{321}, \phi_{320}, \phi_{32\bar{1}}, \phi_{32\bar{2}} \\
\left( \quad\quad V(\boldsymbol{r}) \quad\quad \right)
\end{array}
\rightarrow
\begin{array}{c}
\phi_1,\ \phi_2,\ \phi_3,\ \phi_4,\ \phi_5 \\
\begin{pmatrix}
E_{e_g} & & & & \\
& E_{e_g} & & & \\
& & E_{t_{2g}} & & \\
& & & E_{t_{2g}} & \\
& & & & E_{t_{2g}}
\end{pmatrix}
\end{array},
$$

where

$$
\begin{aligned}
\phi_1 &= \phi_{320}, \phi_2 = \frac{1}{\sqrt{2}}(\phi_{322} + \phi_{32\bar{2}}) \\
\phi_3 &= \phi_{321}, \phi_4 = \phi_{32\bar{1}}, \phi_5 = \frac{1}{\sqrt{2}}(\phi_{322} - \phi_{32\bar{2}})
\end{aligned}
\tag{2.5}
$$

and

$$
E_{e_g} - E_{t_{2g}} = 10\,Dq = \frac{1.67 Ze^2}{4\pi\epsilon_0} \frac{1}{a^5} \int_0^\infty r^4 R_{32}^2(r) r^2 \mathrm{d}r\,. \tag{2.6}
$$

The wavefunctions are recombined as follows and are named $d\epsilon$ and $d\gamma$ orbitals, respectively:

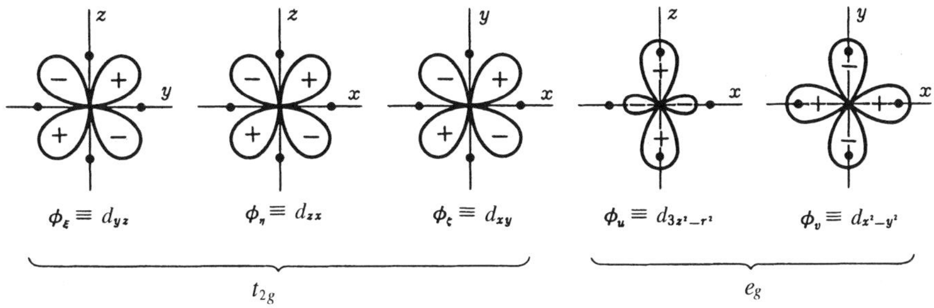


**Fig. 2.2.** Schematic representation of $d\epsilon(T_{2g})$ and $d\gamma(E_g)$ orbitals. Anions are shown by small solid circles [4]

$$d\epsilon \begin{cases} d_{xy} = \sqrt{\frac{15}{4\pi}} \frac{R_{32}(r)}{r^2} xy = -\frac{\mathrm{i}}{\sqrt{2}} \left(\phi_{322} - \phi_{32\bar{2}}\right) , \\ d_{yz} = \sqrt{\frac{15}{4\pi}} \frac{R_{32}(r)}{r^2} yz = \frac{\mathrm{i}}{\sqrt{2}} \left(\phi_{321} + \phi_{32\bar{1}}\right) , \\ d_{zx} = \sqrt{\frac{15}{4\pi}} \frac{R_{32}(r)}{r^2} zx = -\frac{1}{\sqrt{2}} \left(\phi_{321} - \phi_{32\bar{1}}\right) , \end{cases} \tag{2.7}$$

$$d\gamma \begin{cases} d_{x^2-y^2} = \sqrt{\frac{15}{16\pi}} \frac{R_{32}(r)}{r^2} \left(x^2 - y^2\right) = \frac{1}{\sqrt{2}} \left(\phi_{322} + \phi_{32\bar{2}}\right) , \\ d_{3z^2-r^2} = \sqrt{\frac{5}{16\pi}} \frac{R_{32}(r)}{r^2} \left(3z^2 - r^2\right) = \phi_{320} . \end{cases} \tag{2.8}$$

The former triply degenerate orbitals have the $T_{2g}(t_{2g})$ symmetry and the latter doubly degenerate orbitals $E_g(e_g)$ symmetry. In the octahedron, the $d\gamma$ orbitals extend toward the negative ions whereas the $d\epsilon$ orbitals deflect from those directions as shown in Fig. 2.2. Therefore the former have higher energy than the latter. In the tetrahedron, the order of the energy levels is reversed because the $d\epsilon$ orbitals extend toward the anions in this case. This point charge model, however, is not realistic because the oxygen $2p$ orbitals also extend widely in the same space.

As $d$ electrons come nearer to oxygen ions, the point charge model ceases to be a good approximation and the spatial distribution of $O_{2p}$ electrons should be taken into account. In fact, the $M_{3d}$ orbitals overlap with $O_{2p}$ orbitals, and their mixing occurs when the $d$ orbitals and the ligand $p$ orbitals have the same symmetry. With cubic symmetry, the symmetry is represented by 10 irreducible representations designated as

$$A_{1g}, A_{1u}, A_{2g}, A_{2u}, E_g, E_u, T_{1g}, T_{1u}, T_{2g}, \text{and } T_{2u}$$

by Mulliken's symbols and

$$\Gamma_1, \Gamma_1', \Gamma_2, \Gamma_2', \Gamma_{12}, \Gamma_{12}', \Gamma_{15}', \Gamma_{15}, \Gamma_{25}', \text{and } \Gamma_{25}$$

in that order by Bouckaert–Smolchowski–Wigner's symbol [6]. The $g$ and $u$ stand for even and odd symmetry under the inversion $(x, y, z \to -x, -y, -z)$.

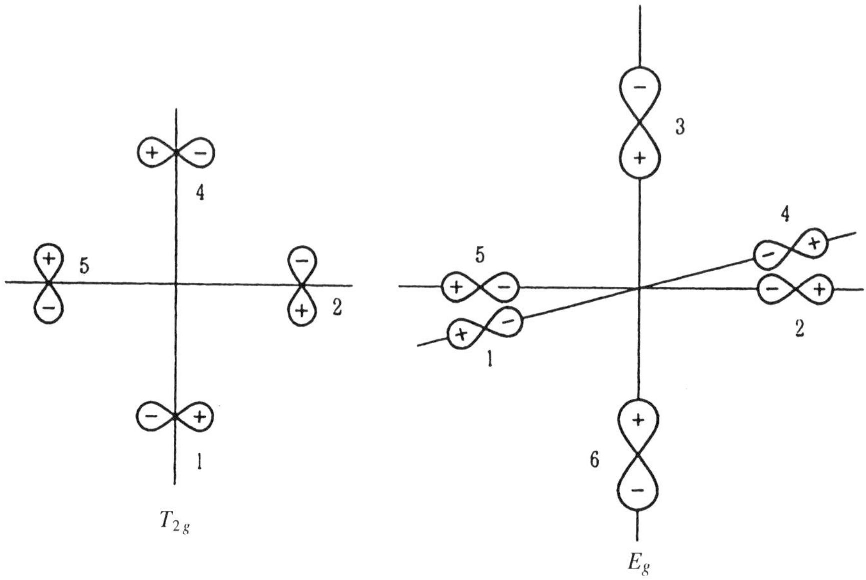


**Fig. 2.3.** Schematic representation of $O_{2p}$ orbitals of $T_{2g}$ and $E_g$ symmmetry in an $MO_6$ cluster [4]

Examples of basis functions of $E_g(\Gamma_{12})$ and $T_{2g}(\Gamma'_{25})$ have already been given by (2.7–2.8). The other basis functions in the form of polynomials are as follows [6].

$$
\begin{aligned}
&\Gamma_1 : 1\,,\\
&\Gamma'_1 : xyz(x^4(y^2 - z^2) + y^4(z^2 - x^2) + z^4(x^2 - y^2))\,,\\
&\Gamma_2 : x^4(y^2 - z^2) + y^4(z^2 - x^2) + z^4(x^2 - y^2)\,,\\
&\Gamma'_2 : xyz\,,\\
&\Gamma'_{12} : xyz(3x^2 - r^2), 3xyz(y^2 - z^2)\,,\\
&\Gamma_{15} : x, y, z\,,\\
&\Gamma'_{15} : yz(y^2 - z^2), zx(z^2 - x^2), xy(x^2 - y^2)\,,\\
&\Gamma_{25} : x(y^2 - z^2), y(z^2 - x^2), z(x^2 - y^2)\,.
\end{aligned}
$$

The $6 \times 3$ ligand $p$ functions are also rearranged into those symmetry types around the origin. In Fig. 2.3, ligand orbitals of $E_g$ and $T_{2g}$ symmetry are shown schematically. They are [4]

$$\psi_{3z^2-r^2} = \frac{1}{\sqrt{12}}\left(-2\phi_{3z} + 2\phi_{6z} + \phi_{1x} - \phi_{4x} + \phi_{2y} - \phi_{5y}\right)\,, \tag{2.9}$$

$$\psi_{xy} = \frac{1}{2}\left(\phi_{1y} - \phi_{4y} + \phi_{2x} - \phi_{5x}\right)\,. \tag{2.10}$$

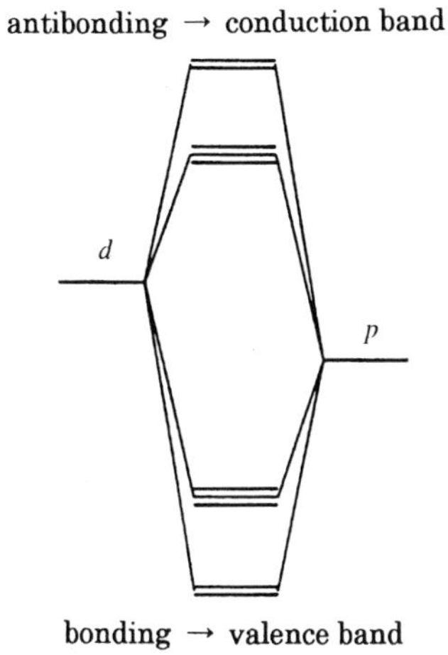


**Fig. 2.4.** Formation of bonding and antibonding orbitals in the octahedral positions. Level order of the antibonding states is the same as that in Fig. 2.1. Splittings of the initial $d$ and $p$ states due to ionicity have been neglected

The $p$ function $\phi_{3z}$ is $\sqrt{3/4\pi}R_{21}(r)z/r$ and is centered at point 3 in Fig. 2.3. The others are identified by comparing with Fig. 2.3. $d_{xy}$ orbitals mix with $\phi_{xy}$ and $d_{3z^2-r^2}$ mix with $\phi_{3z^2-r^2}$.

Figure 2.4 shows the energy diagram for the octahedron schematically. The atomic orbitals mix and divide into the upper antibonding and the lower bonding orbitals [8]. Usually $O_{2p}$ levels are lower than $M_{3d}$ levels and correspondingly the bonding orbitals are mainly composed of $O_{2p}$ orbitals whereas the antibonding orbitals are mainly composed of $M_{3d}$ orbitals. In the recent high-$T_c$ copper oxides, the $O_{2p}$ levels are considered to be exceptionally high. As seen from Fig. 2.4, Fig. 2.1 is the upper half of this diagram and the level order does not change in either model.

For the tetragonal symmetry, the irreducible representations and the polynomials are [4, 7]

$$
\begin{aligned}
A_{1g} &: 1 \text{ or } 3z^2 - r^2 \,, \\
A_{2g} &: xy(x^2 - y^2) \,, \\
B_{1g} &: x^2 - y^2 \,, \\
B_{2g} &: xy \,, \\
E_g &: zx, yz \,, \\
A_{1u} &: xyz(x^2 - y^2) \,, \\
A_{2u} &: z \,, \\
B_{1u} &: xyz \,, \\
B_{2u} &: z(x^2 - y^2) \,, \\
E_u &: x, y \,.
\end{aligned}
$$

## 2.2 Electronic Energy Bands

A lattice is composed by periodically arranging a basic unit. For instance the $ReO_3$ structure of Fig. 5.1 is composed by connecting the $MO_6$ octahedra at their corners into a simple cubic lattice. In this case, the molecular orbitals are connected throughout the whole lattice, with the antibonding orbitals forming the conduction band and the bonding orbitals forming the valence band.

In the oxides, the orbitals are rather well approximated by linearly combining the atomic orbitals with a phase factor which satisfies the periodic boundary condition as

$$\psi_{\boldsymbol{k},l,m}(\boldsymbol{r}) = \frac{1}{\sqrt{N}} \sum_j \exp\left[\mathrm{i}\boldsymbol{k}\cdot(\boldsymbol{R}_j + \boldsymbol{d}_l)\right] \phi_m(\boldsymbol{r} - \boldsymbol{R}_j - \boldsymbol{d}_l)\,, \tag{2.11}$$

where $\phi_m$ is a wavefunction of an atomic electron centered at $\boldsymbol{R}_j + \boldsymbol{d}_l$ where $\boldsymbol{R}_j$ is the position vector of the $j$-th unit cell and $\boldsymbol{d}_l$ is the position vector of the $l$-th atom in that cell. This representation of the orbitals is called the LCAO or tight binding approximation [6, 9]. The function is of course the Bloch function [10]. Using these functions as bases, the Hamiltonian is diagonalized for each wavevector $\boldsymbol{k}$ and the eigenenergies and eigenfunctions are obtained.

When the amplitude of the $d$ function is much larger than the others, the band is called a $d$ band. In a typical transition element oxide, the conduction band is a $d$ band and the valence band is a $p$ band. As $ReO_3$ is a rare crystal whose band structure is rather well represented analytically [11, 12], it will be explained in detail in Sect. 5.1.

## 2.3 Electron–Electron Interaction

### 2.3.1 Direct Interaction

The band picture of electrons fails when various interactions, which were neglected in the band calculation, become larger than the band width. They are: non-periodic potentials, electron–phonon interactions and parts of electron–electron interactions. The electron–electron interaction is taken into account as an averaged interaction in the usual band calculation which is useful for $s$–$p$ electron oxides. However, such a mean field approximation is sometimes not applicable to $d$-electron oxides. For instance, the band calculation predicts a Fermi surface for certain insulating materials. Since the electron–electron interaction is important to understand the electronic conduction phenomena, it is explained in Chap. 4.

### 2.3.2 Indirect Interaction

When electrons interact indirectly by exchanging phonons, a superconducting state may occur. It is often supposed that high-$T_c$ superconductivity is a result of exchange of antiferromagnetic spin fluctuations (Sect. 5.9). When the fluctuation of a localized magnetic moment is exchanged between conduction electrons, the Kondo effect appears [13, 14].

Kondo treated the scattering of electrons by a localized magnetic moment in a higher order approximation and found that the scattering rate increases logarithmically as temperature decreases when the exchange interaction between the carrier and the localized moment is antiferromagnetic [13]. Hamann expressed the resistivity as [15]

$$\rho_{\mathrm{K}} = \frac{1}{2} R_0 \left\{ 1 - \frac{\ln(T/T_{\mathrm{K}})}{\left[ (\ln(T/T_{\mathrm{K}}))^2 + \pi^2 S(S+1) \right]^{1/2}} \right\}, \tag{2.12}$$

$$T_{\mathrm{K}} = \frac{D}{k_{\mathrm{B}}} \exp\left( -\frac{1}{|J|\rho} \right), \tag{2.13}$$

where $T_{\mathrm{K}}$ is the Kondo temperature which is not a critical temperature, and $R_0$ is proportional to the concentration of magnetic impurities. $J$ is the exchange constant between the carrier and the localized moment, and $2D$ is the band width of the carriers with the density of states $\rho$. Due to the mixing of localized and itinerant orbitals, the localized magnetic moment is effectively reduced, and a higher temperature magnetic susceptibility ($7T_{\mathrm{K}} < T < 100\,T_{\mathrm{K}}$) is approximately expressed by [14]

$$\chi = \frac{g^2 {\mu_{\mathrm{B}}}^2 S(S+1)}{3.66 k_{\mathrm{B}}(T + 4.5T_{\mathrm{K}})} . \tag{2.14}$$

The Kondo effect is also observed for oxides. For instance $T_{\mathrm{K}}$ is 48 K for $CaRu_{0.9}Mn_{0.1}O_3$ [16].

## 2.4 Electron–Phonon Interaction

### 2.4.1 The Adiabatic Approximation

Energy band calculation is carried out by putting each ion at a fixed lattice point. When the ions deviate from their fixed positions, the electron state thus determined will become an excited state in the new environment and a transition to a new state will occur. The adiabatic approximation or Born–Oppenheimer approximation, assumes that the electron can follow the motion of the ions instantaneously. Due to this assumption there occurs an error of order of $\sqrt{m/M}$ in the electronic energy and $m/M$ in the phonon energy, where $m$ is the electron mass and $M$ the ion mass [17,18]. In fact, the electron cannot follow the ionic motion instantaneously and this assumption is not appropriate for describing transport phenomena. The electron–phonon interaction causes temperature-dependent resistivity, superconductivity, phonon softening, polaron formation and so on.

### 2.4.2 The Fröhlich Model, the Deformation Potential and the Simple Metal

In a simple metal, lattice vibrations are usually described by the acoustic modes because in most cases only one atom exists in the primitive cell. However in oxides, there are at least two atoms in the cell and optical phonons appear.

*Frölich* approximated the interaction between the electron and the optical phonon by a point charge model. The interaction is given by (2.15) [19],

$$\begin{aligned} H_{\mathrm{e-p}} &= \sum_{\boldsymbol{q}} V_{\boldsymbol{q}} [b_{\boldsymbol{q}}^{\dagger} e^{-\mathrm{i}\boldsymbol{q}\cdot\boldsymbol{r}} - b_{\boldsymbol{q}} e^{\mathrm{i}\boldsymbol{q}\cdot\boldsymbol{r}}] \,, \\ V_{\boldsymbol{q}} &= \frac{\mathrm{i}}{|\boldsymbol{q}|} \sqrt{\frac{\hbar\omega_{\mathrm{l}} e^2}{2V\epsilon_0} \left( \frac{1}{\epsilon_{r\infty}} - \frac{1}{\epsilon_{rs}} \right)} \,, \end{aligned} \tag{2.15}$$

where $\omega_{\mathrm{l}}$ is the frequency of the longitudinal optical phonon. $V$ is the volume of a crystal. $\epsilon_0\epsilon_{r\infty}$ and $\epsilon_0\epsilon_{rs}$ are the dielectric constants at the high and low frequency limit and $b_{\boldsymbol{q}}^{\dagger}$ and $b_{\boldsymbol{q}}$ are the phonon creation and annihilation operators, respectively. In the weak interaction limit, *Howarth* and *Sondheimer* obtained expressions for the conductivity as (5.26, 5.27, 5.29) [20].

In Si there are two atoms in a primitive cell but there is no electric dipole moment. However, the energy levels of the electrons are shifted by the modulation of atomic distance by longitudinal acoustic phonons. *Bardeen* and *Shockley* assumed that the shift is proportional to the dilation and the interaction Hamiltonian is given by

$$\begin{aligned} H_{\mathrm{e-p}} &= E_1 \mathrm{div} \boldsymbol{u}(\boldsymbol{r}) \\ &= E_1 \sum_{\boldsymbol{q}} \sqrt{\hbar/2N_0 M\omega(\boldsymbol{q})} \mathrm{i}(\boldsymbol{q}\cdot\boldsymbol{e}) [b_{\boldsymbol{q}} e^{\mathrm{i}\boldsymbol{q}\cdot\boldsymbol{r}} - b_{\boldsymbol{q}}^{\dagger} e^{-\mathrm{i}\boldsymbol{q}\cdot\boldsymbol{r}}] \,, \end{aligned} \tag{2.16}$$

where $\boldsymbol{u}(\boldsymbol{r})$ is the displacement of an atom with mass $M$ and number density $N_0$. $\omega$ is the phonon frequency. $\boldsymbol{e}$ is a unit vector along the displacement. $E_1$ is a proportionality constant and represents a deformation potential [21]. When the scattering of electrons occurs due to this interaction, the mobility is given by (5.24).

For a simple metal, the interaction of electrons with longitudinal acoustic phonons gives the following temperature dependence of the resistivity [22,23]:

$$\rho = \frac{1}{\sigma} = \frac{3\pi N}{e^2 n^2} \frac{m^* 2E_1^2}{M\hbar k_B \theta_D} \left(\frac{T}{\theta_D}\right)^5 J_5\left(\frac{\theta_D}{T}\right) ,$$
$$J_5(x) = \int_0^x \frac{e^z z^5}{(e^z - 1)^2} dz$$
$$= \frac{x^4}{4} - \frac{x^6}{72} + \cdots (x \ll 1)$$
$$= 124.4 (x \gg 1) , \tag{2.17}$$

where $T$ is the temperature, and $\theta_D$ is the Debye temperature, $n$ being the carrier density and $m^*$ being the band mass of the electron.

The long-wavelength transverse phonons couple weakly with electrons but at shorter wavelength they interact with electrons as strongly as the longitudinal phonons [24].

### 2.4.3 Polarons

When the electron-phonon interaction is small, the dispersion relation $E(\boldsymbol{k})$ of a band electron is modified and is expressed by

$$E = -\alpha\hbar\omega_0 + \frac{\hbar^2 \boldsymbol{k}^2}{2m^*}\left(1 - \frac{\alpha}{6}\right) + \cdots , \tag{2.18}$$

where $m^*$ is the band mass of an electron and $\alpha$ is

$$\alpha = \frac{e^2}{8\pi\epsilon_0\hbar}\sqrt{\frac{2m^*}{\hbar\omega}}\left(\frac{1}{\epsilon_{r\infty}} - \frac{1}{\epsilon_{rs}}\right) , \tag{2.19}$$

using the Fröhlich Hamiltonian [25]. The effective mass of an electron is given by

$$m_p = \frac{m^*}{1 - \alpha/6} , \tag{2.20}$$

which is not applicable when $\alpha > 1$. *Lee* et al. gave an expression

$$E = -\alpha\hbar\omega_0 + \frac{\hbar^2 \boldsymbol{k}^2}{2m^*(1 + \alpha/6)} + \cdots , \tag{2.21}$$

which is considered to be correct for $\alpha < 5$ [26].

Thus due to the electron–phonon interaction, the electron becomes heavier by the factor $(1+\alpha/6)$ and this state is called a large polaron. It is an itinerant electron dressing phonons.

The density of states at the Fermi surface is enhanced by the electron–phonon interaction since the energy of the electrons is modified as (2.21). Usually the enhancement is expressed as

$$N^*(E) = N(E)(1+\lambda)\,, \tag{2.22}$$

where $\lambda$ is the dimensionless electron–phonon coupling constant [22, 27] and is defined for a simple metal as

$$\lambda = 2\int_0^{\omega_0} \frac{\mathrm{d}\omega\alpha^2 F(\omega)}{\omega}\,, \tag{2.23}$$

$$\alpha^2 F(\omega) = \frac{\int \frac{\mathrm{d}^2 k}{v_\mathrm{F}} \int \frac{\mathrm{d}^2 k'}{(2\pi)^3 \hbar v'_\mathrm{F}} \sum_\nu |g_{kk'\nu}|^2 \delta(\omega - \omega_{k-k',\nu})}{\int \frac{\mathrm{d}^2 k}{v_\mathrm{F}}} \tag{2.24}$$

$$g_{kk'\nu} = \sqrt{\frac{\hbar}{2MN\omega_{k-k',\nu}}} M_{kk'\nu}\,,$$

and

$$M_{kk'\nu} = \int \phi_k^*(\boldsymbol{e}_{k-k',\nu}\cdot\nabla V)\phi_{k'}\mathrm{d}\boldsymbol{r}\,,$$

where $\nu$ specifies the phonon mode and the integral is over the Fermi surface. $M$ is a mass of the atom with number density $N$, $\boldsymbol{e}$ is a unit vector along the polarization, $\omega_0$ is the maximum phonon frequency and $V$ is the crystalline potential. $\phi_k$ is the wavefunction of an electron. $\lambda$ is expressed using the averaged quantities as

$$\lambda = \frac{N(E_\mathrm{F})\langle M^2_{kk'\nu}\rangle}{M\langle\omega^2\rangle}\,, \tag{2.25}$$

where $N(E_F)$ is the band structure density of states at the Fermi surface [27]. $\omega$ is the phonon frequency. Thus the electronic specific heat of a metal is also enhanced by the factor $(1+\lambda)$. The paramagnetic susceptibility of the conduction electrons is not enhanced: the population difference between up and down spin electrons is not affected by the correction since the enhancement adheres to the Fermi level.

The dc electrical resistivity due to the electron–phonon interaction is calculated using the Bloch–Boltzmann equation in a relaxation rate approximation [28, 29]. It is given by [in CGS, $\times(4\pi\epsilon_0)^{-1}$ in SI]

$$\rho = \frac{(4\pi)^2}{\omega_\mathrm{p}^2}\int_0^{\omega_0} \frac{\left(\frac{\hbar\omega}{k_\mathrm{B}T}\right)\alpha^2 F(\omega)_\mathrm{trans}}{\left[\exp\left(\frac{\hbar\omega}{k_\mathrm{B}T}\right)-1\right]\left[1-\exp\left(-\frac{\hbar\omega}{k_\mathrm{B}T}\right)\right]}\mathrm{d}\omega\,, \tag{2.26}$$

$$\rho \to \frac{8\pi^2 k_\mathrm{B}\lambda_\mathrm{trans}}{\hbar\omega_\mathrm{p}^2}T\,, \quad (k_\mathrm{B}T \gg \hbar\omega)$$

where suffix "trans" indicates that the integrand of $\alpha^2 F$ is multiplied by a factor $[v_x(k)^2 - v_x(k)v_x(k')]/\langle v_x^2\rangle$, $v_x$ being the Fermi velocity along the electric field. $\lambda_\mathrm{trans}$ is defined for $\alpha^2 F_\mathrm{trans}$.

It must be noted that the maximum phonon energy is as large as 1000 K for oxides, and the assumption of $\hbar\omega \ll k_{\mathrm{B}}T$ in (2.26) is usually not fulfilled.

When the interaction becomes larger, electrons push negative ions farther away, and the overlap between neighbouring orbitals becomes smaller until finally the electrons fall into localized states. This is called the small polaron. According to *Holstein*, the overlap integral $J$ between two electron orbitals is narrowed to

$$J \to J \exp(-S)\,,$$

where $S$ increases with temperature [30]. At higher temperatures than approximately half the Debye temperature, hopping conduction occurs and the mobility is given by (5.37) [31]. At lower temperatures, tunneling conduction occurs and the conductivity becomes larger as temperature is lowered [30].

The above polarization concerns phonons. *Kasuya* and *Yanase* have also considered the polarization of magnons around a spin and have called such a quasi-particle a magnetic polaron [32].

## 2.5 Randomness

### 2.5.1 Anderson Localization

*Anderson* pointed out that when the periodic potential is perturbed, a localized state occurs at some critical randomness [33]. At this point all the waves are reflected or trapped. This localization is not limited to one atom but may extend over many atoms. *Mott* argued that the upper and lower edges of a band would suffer most from randomness and so the localized states would appear there [34]. *Thouless* et al. obtained such a density of states by a computer simulation [35, 36]. The energy $E_{\mathrm{c}}$ which separates the non-localized from the localized state is called the mobility edge.

*Lee* and *Ramakrishnan* reviewed the effects of localization and electron-electron interaction in a disordered system [37]. According to them, the two effects are additive in a metallic state, and for the three-dimensional system $\Delta\sigma_{\mathrm{loc}} \propto T^{p/2}$ and $\Delta\sigma_{\mathrm{int}} \propto T^{1/2}$.

For the two-dimensional system, $\log T$ dependence appears in both cases. It is said that a two-dimensional metal cannot exist at low temperatures and it turns into an insulator [38, 39, 41].

### 2.5.2 Variable Range Hopping

*Mott* proposed a conduction mechanism for a system with random energy levels assuming that the one electron state extends coherently in a local region which is larger than one ion [42–44]. As will be explained in Sect. 5.5, the conductivity is expressed by (5.60) and it reflects the dimension of the conduction network. In oxides, examples are rare as yet but as it is quite a general conduction mechanism, many examples will soon be accumulated. *Efros* calculated the density of states for a lattice which is randomly occupied by localized electrons whose concentration is half that of the lattice point and interact with the long-range Coulomb force. The density of states at the Fermi level diminishes, and the low excitations give the conductivity which is proportional to $\exp[-(T_0/T)^{1/2}]$ for the three-dimensional lattice [45].

## 2.6 The Seebeck Coefficient and Hall Mobility

The Seebeck coefficient, $S$, reflects the electron energy spectrum weighted by the conductivity as will be seen from the following expression for itinerant electrons [22]:

$$S = -\frac{k_\mathrm{B}}{e}\frac{\int(-\partial f/\partial E)N(E)(v^2\tau)_E[(E-E_\mathrm{F})/k_\mathrm{B}T]\mathrm{d}E}{\int(-\partial f/\partial E)N(E)(v^2\tau)_E\mathrm{d}E}, \tag{2.27}$$

where $f$ is the Fermi distribution function, $N(E)$ is the density of states, $E_\mathrm{F}$ is the Fermi energy, $v$ is the velocity of the carrier and $\tau$ is the relaxation time of the momentum distribution.

For a simple metal with a single parabolic band, $S$ is given by

$$S \approx -\frac{k_\mathrm{B}}{e}\frac{\pi^2}{3}\left(\frac{3}{2}+\frac{\tau_1}{\tau_0}\right)\frac{k_\mathrm{B}T}{E_\mathrm{F}}, \tag{2.28}$$

where $\tau$ has been expanded as

$$\tau = \tau_0 + (E-E_\mathrm{F})\tau_1/E_\mathrm{F}\,.$$

$S$ is proportional to $T$ and is much reduced by the factor $k_\mathrm{B}T/E_\mathrm{F}$. When $S$ is as small as a few μV/K, the phonon drag effect cannot be neglected [46].

For a broad band semiconductor, $S$ is given by

$$S = -\frac{k_\mathrm{B}}{e}\left(\frac{E_\mathrm{e}-E_\mathrm{F}}{k_\mathrm{B}T}+\frac{5}{2}+r\right) \quad \text{(electron)}\,,$$

$$S = \frac{k_\mathrm{B}}{e}\left(\frac{E_\mathrm{F}-E_e}{k_\mathrm{B}T}+\frac{5}{2}+r\right) \quad \text{(hole)}\,, \tag{2.29}$$

where $\tau$ for an electron is

$$\tau = \tau_0(T)(E-E_\mathrm{e})^r$$

and $E_\mathrm{e}$ is the band edge energy. $S$ is negative for an electron and positive for a hole.

*Brenig* et al. have obtained an expression for a variable range hopping system as will be explained in Sect. 5.3.4 [43]. *Friedman* and *Pollak* also obtained the $T^{-1/4}$ dependence for a random system whose density of states is of exponential type. According to them, $S$ becomes constant at higher temperature [47]. *Wood* and *Emin* calculated an $(a+bT)$ type Seebeck coefficient for electrons hopping among nonequivalent sites, where $T$ is temperature [48].

For small polaron hopping conduction, $S$ is given by

$$S = \frac{k_\mathrm{B}}{e} \ln \left( \frac{n}{N-n} \right) , \tag{2.30}$$

where $N$ and $n$ are densities of hopping sites and carriers, respectively [31,49]. Here currents from other conduction mechanisms such as band conduction must be negligibly small. However, such a pure hopping conduction mechanism will be scarce in the oxides.

When there are two conduction paths whose conductivities are $\sigma_1$ and $\sigma_2$, the overall $S$ will be given by

$$S = \frac{(\sigma_1 S_1 + \sigma_2 S_2)}{(\sigma_1 + \sigma_2)} , \tag{2.31}$$

where $S_1$ and $S_2$ are the respective $S$ values.

The Seebeck coefficient reflects an energy spectrum of electrons weighted by the conductivity, so it depends on the state of the specimen and may not be intrinsic for sintered materials due to the presence of grain boundaries. However as the conductivity is the normalized conductivity, $S$ is not so sensitive as the conductivity itself and it sometimes reveals valuable information even from sintered materials where conductivity is not an intrinsic property.

The Hall mobility, $\mu_\mathrm{H} = R_\mathrm{H}\sigma$, where $R_\mathrm{H}$ is the Hall coefficient and $\sigma$ is the conductivity, is equal to the drift mobility, $\mu_\mathrm{D} = \sigma/ne$, when the itinerant nature of the electrons is well defined, the band is isotropic and the relaxation time of electron momentum is energy independent. However, for the hopping regime, the conduction path is limited geometrically and $\mu_\mathrm{H}$ is sensitive to the form of the electron wavefunction, the geometrical structure of the hopping sites, etc. and its value and even its sign may differ greatly from $\mu_\mathrm{D}$ [50–55].

## 2.7 Magnetic Susceptibility

Electron spins become antiparallel when they occupy the same orbital due to Fermi statistics. When two orbitals are available, the Coulomb repulsion often becomes smaller for a parallel spin configuration than for an antiparallel one. This appears as the Heisenberg ferromagnetism of a free electron gas or Hund's rule of atomic electrons. In the insulating oxides, antiferromagnetism is often observed, and as the magnetic ions are separated by oxygen ions, the interaction is called a superexchange interaction. This occurs as follows. The localized orbital will be a mixed orbital of metal $d$ and oxygen $2p$ orbitals. The neighboring orbital affects the localized state through virtual transfer and lowers its energy. The energy gain will be

$$E_{\mathrm{ex}} = \frac{1}{4S^2} \sum_{ij} \frac{2t^2}{U} \boldsymbol{S}_i \cdot \boldsymbol{S}_j , \qquad (2.32)$$

where $t$ is the transfer integral between the neighboring orbital and $U$ is the Coulomb repulsion between two electrons on the same orbital [56]. Due to the Fermi statistics, spins have antiparallel alignment when the same orbital is occupied. When conduction electrons exist, the magnetic interaction between localized magnetic moments occurs through the conduction electrons. If they are broad-band electrons, the sign of the interaction oscillates depending on the distance between the localized moments and the carrier density, since the spin density oscillates in the band [57]. It is a RKKY (Ruderman, Kittel, Kasuya, Yoshida) interaction or an $s$–$d$ interaction. However, in the usual transition metal oxides, the application of the $s-d$ interaction is limited since the conduction band is a narrow $d$ band. The conducting $d$ electrons may be magnetically polarized as in the atomic electrons and the ferromagnetic interaction will appear between the localized moments as in $La_{1-x}Sr_xMnO_3$. It is called the double exchange interaction. The spin arrangement is not confined to a ferromagnetic one depending on the carrier concentration [58, 59]. Under certain conditions, the carriers may be segregated from the region of antiparallel spin arrays instead of aligning the localized moments parallel [60].

*Nagaoka* considered the magnetism of the $s$ band in the Hubbard scheme and showed that the ferromagnetism appears in the sc, bcc, fcc, and hcp structures when $0 < n - N \ll N$ and when the repulsive potential is stronger than a certain critical value. $n$ and $N$ are the electron and atomic densities [61].

For a free electron gas, the magnetic susceptibility does not depend on temperature to a first approximation and is called Pauli paramagnetism. It is proportional to the density of states at the Fermi level and is enhanced by electron correlations (Sect. 4.7). The intra-atomic Coulomb repulsion tends to reduce the number density of doubly occupied atoms and thus enhance the susceptibility [62,63]. *Sernelius* and *Berggren* showed that correlation enhances the susceptibility and the specific heat of heavily doped semiconductors near the metal–insulator transition [64]. *Moriya* and *Hasegawa* showed that the susceptibility of the paramagnetic metal may be expressed by a Curie–Weiss type relation depending on $U/t$ where $U$ is the intra-atomic Coulomb interaction and $t$ is the transfer integral [65] (Sect. 4.7).

Temperature-independent paramagnetism (TIP) also appears for electrons in a closed atomic shell when the energy separation between the ground and an excited state is much larger than the thermal energy $k_{\mathrm{B}}T$. This is called Van Vleck paramagnetism [66,67]. The orbital diamagnetism of atomic electrons is also temperature-independent [66, 68].

For interacting localized moments, the susceptibility usually follows the Curie–Weiss law.

$$\chi = \frac{C}{T - \theta} ,$$

$$C = \frac{N\mu_0\mu_B^2 \sum_i S_i(S_i+1)\omega_i g_i^2}{3k_B} = \frac{N\mu_0\mu_B^2 p_{\text{eff}}^2}{3k_B} , \tag{2.33}$$

where $w_i$ is a number of ions with spin $S_i$ in one molecule, $g_i$ is the $g$ factor of ion $i$, and $N$ is the density of molecules.

For a two-dimensional substance, where a long-range ordering is to be absent, almost temperature independent paramagnetism appears for the antiferromagnetic interaction between magnetic moments. This is explained in Sect. 5.9.

For conducting oxides, a Pauli-like susceptibility is often a good indicator of the itineracy of electrons. In Table 2.1, magnetic susceptibility is briefly mentioned. However, it must be noted that TIP does not always mean itinerant band electrons. When electron spins are in an antiparallel alignment, they may show TIP under certain conditions. Also a Curie–Weiss like susceptibility does not always mean localized moments as shown by *Moriya* and *Hasegawa* [65].

## 2.8 Metal–Insulator Transition (MIT)

In a free electron gas, the bare Coulomb potential is screened in a static limit as

$$\frac{e}{4\pi\epsilon_0 r} \to \frac{e}{4\pi\epsilon_0 r} e^{-\kappa_s r} . \tag{2.34}$$

For a degenerate gas, $\kappa_s$ is given in the Thomas-Fermi approximation as

$$\kappa_s = 1.98 n^{1/6} a_B^{-1/2} = e\sqrt{N(E_F/\epsilon_0)} , \tag{2.35}$$

where $n$ is the carrier density, $N(E_F)$ is the density of states at the Fermi level, and $a_B$ is the Bohr radius [66]. In this reduced potential, a bound state appears when $\kappa_s$ is smaller than $1.19/a_B$, that is, when the carrier concentration is smaller than $0.046 a_B^{-3}$ [69–71].

*Mott* pointed out that when electrons fall into bound states, their screening effect on other potentials is reduced and such bound states will appear cooperatively at a certain concentration of electrons. This transition to the insulating state is called the Mott transition. The critical concentration was calculated under various conditions [72,73]. *Edwards* and *Sienko* applied this idea to a variety of substances and found a universal relation between the effective Bohr radius and the critical concentration as shown in Fig. 4.14 [74]. In most MITs, a crystal lattice transforms into a lower symmetry type at the insulating state and it is not clear whether it is a Mott transition. However, all types of MITs occur when the carrier concentration is low and the poor

**Table 2.1.** Electronically conductive oxides. (Degenerate semiconductors are not cited except for few examples. High $T_c$ oxides will be shown in Sect. 5.9)

| Oxides | Specific resistivity at 300 K ($\Omega m$) | Magnetic susceptibility | References |
|---|---|---|---|
| *NaCl* | | | |
| TiO | $3 \times 10^{-6}, \mathrm{sc}T_c = 2.3\,\mathrm{K}$ | Pauli para. | [84–89] |
| VO | $2 \times 10^{-5}$ | Curie-Weiss like | [85–88] |
| NbO | $1.3 \times 10^{-7}$ | | [84, 90–93] |
| LaO | $2.7 \times 10^{-6}$ | Temp. indep. susceptibility | [94] |
| NdO | $7 \times 10^{-7}$ | Temp. indep. susceptibility | [94] |
| SmO | $2 \times 10^{-6}$ | Temp. indep. susceptibility | [94] |
| $EuO_{1-x}$ | $1 - 10^{-7}$, M-I, | Ferromag., Curie-Weiss above $T_c$ | [95] |
| $LiVO_2$ | 1, M?-I at about 473 K, layered NaCl | Not Pauli-para. | [96–98] |
| *Spinel* | | | |
| $LiTi_2O_4$ | $2 \times 10^{-6}(T \sim T_c)$, $\mathrm{sc}T_c = 13.7\,\mathrm{K}$ | Pauli para. | [99–104] |
| $LiM_xTi_{2-x}O_4$ ($M$ = Li, Al, Cr) | M-I (0.12, Li; 0.33, Al; 0.33-0.4, Cr) | | [105] |
| $Li_{1-x}M_xTi_2O_4$ ($M$ = Mn, Mg) | | | [106] |
| Mg | $x < 0.2$, metallic, sc, | Pauli para | |
| Mn | M-I, $x = 0.6, \mathrm{sc}T_c = 3.6\,\mathrm{K}$ ($x = 0.2$) | Curie-Weiss | |

**Table 2.1.** (continued)

| Oxides | Specific resistivity at 300 K ($\Omega m$) | Magnetic susceptibility | References |
|---|---|---|---|
| $LiV_2O_4$ | $8 \times 10^{-6}$ | Curie-Weiss | [107–109] |
| $Fe_30_4$ | $4 \times 10^{-5}$ M(?)-I at 124 K | Ferrimag., $T_c \sim 850$ K | [84, 108] |
| $ZnGa_2O_{4-\delta}$ | $3 \times 10^{-4}$, $d\rho/dT \sim 0$ (reduced oxide) | | [110] |
| *Perovskite-$ReO_3$* | | | |
| $ReO_3$ | $1 \times 10^{-7}$ | Pauli para. | [111] |
| $M_xReO_3$ | | Pauli para. | [111, 112] |
| $M_xWO_3$ | sc $T_c = 7.7$ K ($M$ = Rb, hexagonal) | Pauli para. | [112–115] |
| $A_2P_8W_{32}O_{112}$ | Sintered, (perovskite related) | Pauli para. | [116] |
| $A$ = K | $6.7 \times 10^{-6}$ | | |
| $A$ = Rb | $6.2 \times 10^{-6}$ | | |
| $A$ = Tl | $7.9 \times 10^{-6}$ | | |
| $Na_xTa_yW_{1-y}O_3$ | $10^{-5.2}(x = 0.64)$, M-I $(x - y = 0.18)$ Anderson localization? no structure change | | [117, 118] |
| $M_xNbO_3$ | | | |
| $M$ = Er | $4 \times 10^{-6}(x = 0.76)$ metallic $(0.65 < x < 0.92)$ | | [119, 120] |
| $Na_{1-x}Sr_xNbO_3$ | Anderson localization? | | [121] |
| $RTiO_3$ | | | |

**Table 2.1.** (continued)

| Oxides | Specific resistivity at 300 K ($\Omega m$) | Magnetic susceptibility | References |
|---|---|---|---|
| $R = \mathrm{La}$ | $2\times10^{-5}$ M?-I (125 K), insulator metallic upon oxidization | Giant Pauli para ($T > T_N = 125$ K), canted antiferro. | [122–125] |
| $R = \mathrm{Ce}$ | M?-I (100–65 K) | $T_N = 116$ K, complex antiferromag., antiferromag. metal? | [122, 123] |
| $Sr_{1-x}La_xTiO_3$ | Metallic ($x \geq 0.1$), enhancement of effective mass near $x = 1$ | | [126] |
| $La_{2/3+x}TiO_{3-\delta}$ | MIT at 0.33 electron per molecule | | [127] |
| $Ln_{2/3+x}TiO_{3-\delta}$ (Ln = Ce, Nd) | MIT around 200–250 K | Deviation from Curie–Weiss | [128] |
| $Ca_{n+1}Ti_nO_{3n+1-\delta}$ | Layer structure | Pauli para.-like | |
| ($n = 2, 3, \infty$) | metallic by reduction, gradual MIT | | [129] |
| $CaVO_3$ | $4 \times 10^{-3}$ | Pauli para. | [130] |
| $O_{3.00}$ metallic | | | |
| $O_{3.05}$ insulating | | | [131] |
| $SrVO_3$ | $3 \times 10^{-7}$, single | Weak ferromag. < 85 K, neither Curie–Weiss nor Pauli para. above 85 K paramagnetic, $Sr_8V_8O_{25}$ is ferromagnetic | [130, 132, 133] |
| $Sr_{1-x}Eu_xVO_3$ | MIT at $x = 0.75$ | | [134] |

**Table 2.1.** (continued)

| Oxides | Specific resistivity at 300 K ($\Omega m$) | Magnetic susceptibility | References |
|---|---|---|---|
| $Nd_{1-x}Ca_xVO_3$ | Metallic ($0.32 < x < 1$) | | [135] |
| $La_{1-x}K_xVO_3$ | MIT at $x = 0.12$ | | [136] |
| $La_{1-x}Sr_xVO_3$ | $x > 0.23$ metallic, $x < 0.23$ semicond., Anderson localization | Antiferro., Curie–Weiss like | [69, 137, 138] |
| Sr→Ba | Activation energy = 0.008 eV ($x = 0.2$) | | [138] |
| $La_{1-x}M_xVO_3$ ($M$ =Li, Rb, Cs) | MIT | | [139] |
| $Sr_{n+1}V_nO_{3n+1-y}$ ($n = 1, 2, 3, \ldots$) | Metallic ($n \geq 2$) $4 \times 10^{-4} (n = 2)$ $2 \times 10^{-5}$ $(n = 3), 10^{-5}(\infty)$ | Pauli para. like | [140–143] |
| $Gd_{1-x}Sr_xVO_3$ | $3.2 \times 10^{-4} (x = 0.5)$, $d\rho/dT = 0$ ($x$ =0.4–0.45) | | [138] |
| $CaCrO_3$ | $7 \times 10^{-6}$ single | Canted antiferro. below 325 K, metallic antiferro. | [144, 145] |
| $SrCrO_3$ | $4 \times 10^{-6}$ | Pauli para. | [146] |
| $La_xCa_{1-x}MnO_{2.97}$ | $10^{-3}$ ($0.04 < x < 0.4$, 500 K) M-I (400–450 K), change in the sign of $d\rho/dT$, no structure change | Curie–Weiss | [147] |

**Table 2.1.** (continued)

| Oxides | Specific resistivity at 300 K ($\Omega m$) | Magnetic susceptibility | References |
|---|---|---|---|
| $La_{0.72}Ca_{0.25}MnO_y$ | $2\times10^{-3}$, film, MIT (change in the sign of $d\rho/dT$), giant negative magnetoresistance | | [148] |
| $La_{0.67}Ca_{0.33}MnO_y$ | $1 \times 10^{-3}$, film MIT, giant negative isotropic magnetoresistance | | [149] |
| $(La_{1-x}A_x)_{2/3}Ca_{1/3}MnO_3$ ($A$ = Y or Tb) | Giant magnetoresistance, shift in $T_{\rho_{\max}}$ from $T_c$ | Ferromag., magnetic polaron associated with a lattice distortion | [150] |
| $La_{1-x}Sr_xMnO_3$ | $10^{-4}$ | $0.2 < x < 0.4$ ferromag. while metallic | [151–153] |
| $La_{1-x}(Sr,Pb)_xMnO_{3+\delta}$ | | Ferromag. while metallic | [154, 155] |
| $La_{2/3}Ba_{1/3}MnO_y$ | $8 \times 10^{-5}$, film, giant negative magnetoresistance | Ferromagnetic-like | [156] |
| $Pr_{1-x}Ba_xMnO_3$ | MIT around 120–160 K ($x = 0.4$) | | [157] |
| $Nd_{0.7}Sr_{0.3}MnO_3$ | $5 \times 10^{+4}\Omega$, film, electric field dependent resistance, reduction above $T$ of $\rho_{max}$ | | [158] |
| $Nd_{0.5}Pb_{0.5}MnO_3$ | $3 \times 10^{-3}$, MIT ($d\rho/dT$ changes sign), magnetic polaron hopping motion | $T_c = 184\,K$ | [159] |

**Table 2.1.** (continued)

| Oxides | Specific resistivity at 300 K ($\Omega m$) | Magnetic susceptibility | References |
|---|---|---|---|
| $CaFeO_3$ | 0.3, $d\rho/dT < 0, < 115$ K, $d\rho/dT > 0, > 115$ K sintered | Antiferromag. $< 115$ K | [160] |
| $SrFeO_3$ | $2 \times 10^{-5}$ | Helical $< 134$ K high spin | [161–164] |
| $SrCoO_3$ | $10^{-4}, d\rho/dT <> 0$, sintered | Ferromag. $< 212$ K | [162] |
| $YCoO_3$ | Metallic above 1200 K | | [165] |
| $ErCoO_3$ | Metallic above 1200 K | | [165] |
| $LaCoO_3$ | $d\rho/dT > 0, > 700$ K, | Coexistence of high and low spin | [166–168] |
| | $d\rho/dT < 0, < 600$ K, gradual M-I 520–720 K | Thermally induced high spin state | |
| $La_{1-x}Ca_xCoO_3$ $(0.3 \leq x \leq 0.6)$ | Metallic, $p$ type | | [169] |
| $LaNi_{1-x}M_xO_3$ | | | |
| ($M$ =Cr, Mn, Fe, Co) | MIT | | [170] |
| $La_{1-x}Sr_xCoO_3$ | $d\rho/dT > 0$, $x > 0.25$ sintered, $4 \times 10^{-6}$ at $x = 0.5$ | Ferromag., $x > 0.05$, Curie–Weiss at higher temp. | [171–173] |
| $LaNi_{1-x}Co_xO_3$ | Metallic ($x \leq 0.65$), electron–electron interaction, $\sigma(T)$; power law $T^{0.3-0.4}$ below 2 K | | [174] |
| $LaCr_{1-x}Ni_xO_3$ | MIT at $x = 0.79$ | | [175] |
| $LaNiO_3$ | $10^{-5}$, sintered, | Enhanced Pauli para. | [176–180] |
| $Nd_{1-x}La_xNiO_{3-\delta}$ | Metallic ($x \geq 0.5$) | | [181] |

**Table 2.1.** (continued)

| Oxides | Specific resistivity at 300 K ($\Omega m$) | Magnetic susceptibility | References |
|---|---|---|---|
| $R$NiO$_3$ ($R$ =Pr, Nd, Sm, Eu) | MIT, 135 K (Pr), 201 K (Nd), 403 K (Sm), 480 K (Eu) | $T_N$, 135 K (Pr), 195 K (Nd) 225 K (Sm), 205 K (Eu) | [182, 183] |
| $NdNiO_3$ | Sintered, $1.5 \times 10^{-4}$ MIT at 130 K | $\chi$ constant above 100 K | [184] |
| $LuNiO_3$ | | Antiferromag. < 90 K | [185] |
| $LaCuO_3$ | Sintered, $3 \times 10^{-2}$ | Pauli para. | [179, 186] |
| $La_4BaCu_5O_{13-\delta}$ | Metallic but not sc > 5 K | | [187–190] |
| $La_5SrCu_6O_{15}$ | Metallic but not sc > 5 K | | [189] |
| $La_{6.4}Sr_{1.6}Cu_8O_{20}$ | MIT | | [191] |
| $La_2SrCu_2O_{6.0}$ | Metallic at least above 50 K | | [189] |
| $La_{1-x}Sr_xCuO_{3-\delta}$ | Metallic ($0.2 \leq x \leq 0.25$) | | [192] |
| $EuNbO_3$ | $5 \times 10^{-5}$ | | [193] |
| $Nb_{12}O_{29}$ (perovskite related) | $4 \times 10^{-5}$, sintered | $T_N = 12$ K, metallic and antiferromag. | [194] |
| $CaRuO_3$ | Single, $2.5 \times 10^{-6}$ | Curie–Weiss, non magnetic ordering, antiferromag. interaction, nearly ferromag. metal? | [195–200] |
| $SrRuO_3$ | Single, $2.8 \times 10^{-6}$ | Ferromag. < 160 K, Curie–Weiss | [195–197, 201, 202] |
| $BaRuO_3$ | Metallic, nine layered hexagonal perovskite | | [196] |

**Table 2.1.** (continued)

| Oxides | Specific resistivity at 300 K ($\Omega m$) | Magnetic susceptibility | References |
|---|---|---|---|
| $Ca_{1-x}Na_xRuO_3$ | Metallic | $\chi$ shows a peak at 70 K ($x = 0.05$) | [203] |
| $Ca_{1-x}Mg_xRuO_3$ | Semiconducting | | |
| ($x = 0.1, 0.3, 0.5$) | | | [204] |
| $Ca_{1-x}Sr_xRuO_3$ | Metallic | Ferromag. ($x \geq 0.4$) | [204–207] |
| $La_{0.5}Na_{0.5}RuO_3$ | Metallic | Curie–Weiss | [204] |
| $La_{1-x}Ca_xRuO_3$ | | | [208] |
| $La_{1-x}Sr_xRuO_3$ | | | [209] |
| $Ca_{1-x}(La, Y)_x$ | Metallic ($x, y \ll 1$) | | [210] |
| $Ru_{1-y}Mn_yO_3$ | | | |
| $LaRu_{1-x}Fe_xO_3$ | MIT at $x = 0.1$ | | [211] |
| $Ba_{1-x}K_xRuO_3$ | Metallic | | [212] |
| $Ba_{5/6}Sr_{1/6}RuO_3$ | Metallic | Antiferromag. | [196] |
| $La_{1-x}M_xRhO_3$ | $10^{-3} \sim 1$ | | |
| ($M$ =Ca, Sr, Ba) | $p$-type semiconductor ($x = 0$), metallic by doping Sr, Ca | Curie–Weiss, low spin | [213, 214] |
| $SrIrO_3$ | $4 \times 10^{-5}$, sintered, | Pauli para. | [201] |
| $BaIrO_3$ | Poor metal | Ferromag.? | [215] |
| $BaPbO_3$ | $3 \times 10^{-6}, d\rho/dT > 0$, single, sc$T_c$ = 0.4 K, carrier 0.013/molecule | | [216, 217] |

**Table 2.1.** (continued)

| Oxides | Specific resistivity at 300 K ($\Omega m$) | Magnetic susceptibility | References |
|---|---|---|---|
| $Ba_{1-x}Sr_xPbO_{3-y}$ | $1.5 \times 10^{-5}$, sintered, metallic ($x < 0.5$) | | [218] |
| $BaPb_{1-x}Bi_xO_3$ | sc$T_c = 13\,K$ | | [219] |
| $Ba_{1-x}K_xBiO_3$ | sc$T_c = 22\,K$ metallic and superconducting at $x \approx 0.35$ | | [220–224] |
| $BaPb_{0.75}Sb_{0.25}O_3$ | sc$T_c = 3.5\,K$ | | [225] |
| $Sr(Pb_{1-x}Sb_x)O_{3-\delta}$ | $x \leq 0.2$, metallic | Pauli para. | [226] |
| $Sr(Pb_{1-x}Bi_x)O_{3-\delta}$ | $x \leq 0.05$, metallic | | [226] |
| $Ba(Pb_{1-x}Sb_x)O_{3-\delta}$ | $x \leq 0.3$, metallic | Pauli para. | [226] |
| $M MoO_3$ | | | [227–230] |
| $M = Ca$ | $10^{-5}$, | Pauli para. | |
| $M = Sr$ | $10^{-6}$, | Pauli para. | |
| $M = Ba$ | $2 \times 10^{-5}$, | Pauli para. | |
| $SrMoO_{2.6}N_{0.4}$ | Metallic | | [231] |
| $SrV_{1-x}Mo_xO_3$ | Metallic | | [232] |
| $(Ba,Ca,Sr)TiO_{3-x}$ | $x = 0$, insulator, $SrTiO_{3-x}$ sc | | [233–236] |
| $KSbO_3$ | | | |
| $La_4Re_6O_{19}$ | $1.4 \times 10^{-4}$ | | [237, 238] |
| $La_4Ru_6O_{19}$ | $1.4 \times 10^{-4}$ | | [237] |
| $Bi_3Ru_3O_{11}$ | $2.9 \times 10^{-5}$ | | [237] |

**Table 2.1.** (continued)

| Oxides | Specific resistivity at 300 K ($\Omega m$) | Magnetic susceptibility | References |
|---|---|---|---|
| *Corundum* | | | |
| $V_2O_3$ | $10^{-5}$, M-I at 168 K, another M-I at higher temp. when doped with Cr, heavy fermion? | Antiferro. at lower temp. | [239, 240] |
| $Ti_2O_3$ | $9 \times 10^{-5}$ broad M-I around 660 K? | | [240–242] |
| $Rh_2O_3$ | $4 \times 10^{-2}$, $d\rho/dT < 0$, semimetal | Pauli para. | [243] |
| *Rutile-$MoO_2$ (including Magnéli phase)* | | | |
| $VO_2$ | $5 \times 10^{-6}$ (370 K), metallic $> 340$ K | | [78, 84, 244–247] |
| $CrO_2$ | $3 \times 10^{-6}$ | Ferromag. $< 392$ K | [84, 250] |
| $NbO_2$ | M-I? at 1070 K along *c*-axis | | [251, 252] |
| $MoO_2$ | $2 \times 10^{-6}$ | Pauli para. | [84, 253, 254] |
| $WO_2$ | $3 \times 10^{-5}$ | Pauli para. | [253] |
| $ReO_2$ | $10^{-6}$, two phases | Pauli para. | [253] |
| $RuO_2$ | $2 \times 10^{-7}$ | Pauli para. | [255–261] |
| $RhO_2$ | $< 10^{-6}$ | Pauli para. | [253, 262] |
| $OsO_2$ | $6 \times 10^{-7}$ | Pauli para. | [253, 257] |
| $IrO_2$ | $5 \times 10^{-7}$ | Pauli para. | [253, 255–259] |
| $PtO_2$ | $6 \times 10^{-6}$, semimetal? | Pauli para. | [152, 253, 262] |

**Table 2.1.** (continued)

| Oxides | Specific resistivity at 300 K ($\Omega m$) | Magnetic susceptibility | References |
|---|---|---|---|
| $V_3O_5$ | M-I | One-dimensional spin order? | [263–265] |
| $V_nO_{2n-1}$ $(4 \leq n \leq 8)$ | M-I ($n = 4, 5, 6, 8$) | Antiferromag.? mixed CDW-SDW? metallic and antiferro? ($n = 7$) | [244, 266–275] |
| $Ti_3O_5$ | $10^{-4}$ metallic $> 460$ K | | [276] |
| $Ti_nO_{2n-1}(n = 4 - 9)$ | $10^{-5}$, M?-I, 8?, 9?, $3 \times 10^{-4}$ at 300 K ($n = 8$), bipolaronic conduction? ($n = 4$), semiconductor- semiconductor transition ($n = 7$, 120 K) | | [276–278] |
| $SnO_{2-x}$ | Semiconductor $< 2 \times 10^{-4}$ | | [279] |
| $Na_xTiO_2$ | | | [280] |
| $K_2NiF_4$ | | | |
| $La_2NiO_4$ ($n = 1$ in $La_{n+1}$ $Ni_nO_{3n+1}$) | Stoichiometric, insulator, energy gap 4 eV, nearly stoichiometric, MIT at 650 K | Curie–Weiss, canted antiferromag. $T_N = 230$ K | [281–286] |
| $La_3Ni_2O_7$ | Metallic | | [180] |
| $La_4Ni_3O_{10}$ | Metallic | | [180] |
| $La_{2-x}Sr_xNiO_4$ | $x > 0.8$, Metallic non sc. down to 30 mK, | | [286–289] |
| $Nd_2NiO_4$ | $d\rho/dT > 0$, $> 500$ K, | Curie–Weiss | [281] |
| $La_2CuO_4$ | $10^{-3}$, phase separation | $T_N = 317$ K | [283, 290–292] |

**Table 2.1.** (continued)

| Oxides | Specific resistivity at 300 K ($\Omega m$) | Magnetic susceptibility | References |
|---|---|---|---|
| $La_{2-x}Ln_xCuO_4$ (Pr, Nd) | $1 - 0.1$, $d\rho/dT \sim 0$ | | [293] |
| $Sr_2RuO_4$ | $10^{-6}$ in plane, sc$T_c = 0.93$ K | | [294–298] |
| $Sr_2Ir_{1-x}Ru_xO_4$ | MIT at $x = 0.7$ | Weak ferromag. insulator ($x = 0$) | [299] |
| $Sr_2RhO_4$ | $10^{-2}$, $d\rho/dT > 0$ ($< 150$ K), $d\rho/dT < 0$ ($> 150$ K), non sc. down to 50 mK | $\chi_{max}$ at 240 K | [297, 300] |
| $Sr_{2-x}La_xRhO_4$ | $0 < x \leq 0.15$, Metallic | | [301] |
| *Pyrochlore* | | | |
| $M_2V_2O_{7-x}$ | | | [302] |
| $M$ = Tm | 10, semicon. | | |
| $M$ = Lu | Semicon. | $T_c = 80$ K, ferromag. | |
| $M_2Mn_2O_{1-x}$ | | | |
| $M$ = Tl | 0.1, $d\rho/dT > 0$ in the magn. range | $T_c = 117$ K, ferromag. | [302] |
| $M_2Mo_2O_{7-x}$ | | | [302] |
| $M$ = Y | Insulator? | Spin glass? | |
| $M$ = Nd | Metal, M-I? | Ferromag. | |
| $M$ = Sm | Metal, M-I? | Ferromag. | |

**Table 2.1.** (continued)

| Oxides | Specific resistivity at 300 K ($\Omega m$) | Magnetic susceptibility | References |
|---|---|---|---|
| $M = \mathrm{Gd}$ | Metal?, M-I? | Ferromag. | |
| $M = \mathrm{Tb}$ | Insulator? | | |
| $M = \mathrm{La}$ | Not pyrochlore, two-dimensional | | [303] |
| $Pb_2Tc_2 0_{7-x}$ | | Pauli para. | [302] |
| $M_2Ru_2O_{7-x}$ | | | [302] |
| $M = \mathrm{Tl}$ | M-I? | | |
| $M = \mathrm{Pb}$ | $5 \times 10^{-6}$, trans. to perovskite | Pauli para. | |
| $M = \mathrm{Bi}$ | $7 \times 10^{-6}$ (Bi-Gd: M-I) | Pauli para. | |
| $M = \mathrm{Lu}$ | $10^{-5}$ | | |
| $M = \mathrm{Y}$ | Insulator, correlation induced insulator? | | |
| $M_2Rh_2O_{7-x}$ | | | [302] |
| $M = \mathrm{Tl}$ | Metallic | | |
| $M = \mathrm{Pb}$ | Semicon. | | |
| $M = \mathrm{Bi}$ | $3 \times 10^{-5}$, trans. to perovskite | Pauli para. | |
| $M_2Re_2O_{7-x}$ | | | [302] |
| $M = \mathrm{Pb}$ | | Pauli para. | |
| $M = \mathrm{Cd}$ | $2 \times 10^{-5}$ | | |

**Table 2.1.** (continued)

| Oxides | Specific resistivity at 300 K ($\Omega m$) | Magnetic susceptibility | References |
|---|---|---|---|
| $M_2Os_2O_{7-x}$ | | | [302] |
| $M = Tl$ | $2 \times 10^{-6}$ | | |
| $M = Pb$ | $4 \times 10^{-6}$ | Pauli para. | |
| $M = Ln$ | $10^{-4 \sim -5}$ | | |
| $M_2Ir_2O_{7-x}$ | | | [302] |
| $M = Tl$ | $2 \times 10^{-5}$ | | |
| $M = Pb$ | $2 \times 10^{-6}$ | Pauli para. | |
| $M = Bi$ | $2 \times 10^{-5}$ | Pauli para. | |
| $M = Lu$ | $4 \times 10^{-5}$ | | |
| $M_2Pt_2O_{7-x}$ | | | [302] |
| $M = Pb$ | Semicon. | | |
| $M = Tl$ | Semicon. | | |
| *Others* | | | |
| $Tl_2O_{3-x}$ | $9 \times 10^{-7}$ single | | [304–306] |
| $TlO_{1-x}F$ | $4 \times 10^{-6}$ single | | [304–306] |
| $Zn_3In_{2-x}Sn_xO_6$ | Transparent metal | | [307] |
| $Li_2RuO_3$ | Metallic | Curie–Weiss | [308] |
| $Cu_6O_8MCl$ | | | |
| ($M$ = Sc, In) | $1.5 \times 10^{-4}$ (Sc), $8 \times 10^{-5}$ (In) | Constant $\chi$ except for a peak at low temp. | [309] |

**Table 2.1.** (continued)

| Oxides | Specific resistivity at 300 K ($\Omega m$) | Magnetic susceptibility | References |
|---|---|---|---|
| $Ag_6O_8X$ | sc $T_c = 1.04$ K ($X$=$NO_3$) | | [310] |
| | sc $T_c = 0.3$ K (F) | | |
| | sc $T_c = 0.15$ K ($BF_4$) | | |
| $NaPd_3O_4$ | $2 \times 10^{-5}$, sintered | | [311, 312] |
| $Dy_xPd_3O_4$ | $2.6 \times 10^{-7}$, sintered | | [313] |
| $Dy_xPt_3O_4$ | $1.9 \times 10^{-7}$, sintered | | [313] |
| *Low-dimensional substances* | | | |
| $Pd_{1-x}Li_xO$ | Sintered, $x > 0.01$ metallic, $10^{-5}$ ($x = 0.08$), non-sc. down to 2.2 K | Pauli para. | [314] |
| $M_xV_2O_5$ | | | [315] |
| $M$ = Cu | $2 \times 10^{-6}$, metallic? ($x = 0.55$), bipolaron, one-dimensional | Curie–Weiss | |
| $M$ = Na | $10^{-4}$, semiconductor, bipolaron, one-dimensional | Curie–Weiss, spin glass | |
| *Molybdenum bronzes* | | | [315] |
| $Li_{0.9}Mo_6O_{17}$ | Insulator?– superconductor transition, $d\rho/dT < 0$ below 24 K, $T_c = 1.9$ K, one-dimensional | | |

**Table 2.1.** (continued)

| Oxides | Specific resistivity at 300 K ($\Omega m$) | Magnetic susceptibility | References |
|---|---|---|---|
| $La_2Mo_2O_7$ | $d\rho/dT < 0$ below 125 K, one-dimensional | | |
| $M_{0.3}MoO_3$ | | | |
| ($M$ = K, Rb, Tl) | $4 \times 10^{-6}$ (K), incom. CDW at 180 K, com. at 90 K, one-dimensional | | |
| $K_{0.33}MoO_3$ | Insulator | | |
| $M_{0.9}Mo_6O_{17}$ | | | |
| ($M$ = Na, K, Tl) | Com. CDW 120 K (K), 80 K (Na), 120 K (Tl), two-dimensional | | |
| $Mo_nO_{3n-1}$ | Two-dimensional | | |
| $n = 4$ ($\eta$ and $\gamma$) | Incom. CDW at 109 K and 100 K, respectively | | |
| $n = 8$ | Incom. CDW at 315 K | | |
| $n = 9$ | Incom. CDW at 500 K | | |
| $n = 10$ | Incom. CDW at 610 K | | |
| $Mo_{17}O_{47}$ | $4 \times 10^{-5}$, M-I at 110 K, | Antiferro? < 110 K | |

screening of the Coulomb interaction may well have some effect on the transition. In oxides, $V_2O_3$:Cr, $Li_{1+x}Ti_{2-x}O_4$, $Li_{1-x}Zn_xV_2O_4$, $La_{1-x}Ca_xMnO_{2.97}$, $Na_xTa_yW_{1-y}O_3$, and $Pr(Nd)NiO_3$ are known to transform without a change in the symmetry. In the first oxide, the transition appears at 380 K (Cr 0.51%) and the higher-temperature phase is a non-metallic one. In the following four oxides, the metallic phase appears at certain compositions (Sects. 5.3 and 5.4). The last $Pr(Nd)NiO_3$ are discussed in Sect. 2.9.4.

As seen from Table 2.1, there are many oxides which have a metal-insulator transition. In most of them, a structural change occurs at the transition temperature or the transition composition. Phenomenologically an outstanding feature is a change in the separation between cations. In the insulating phase, the separation usually splits into a wider and a narrower one. *Goodenough* tried to explain various transitions by introducing a concept of critical separation [75].

In a one-dimensional substance, the electron–phonon interaction brings about a sharp transition which doubles the period $a$ to $2a$ as shown in Fig. 2.5. *Peierls* pointed out that this always occurs in a half filled one-dimensional metal and the metal turns to an insulator because an energy gap appears at the Fermi level due to the doubling of the unit cell as shown in Fig. 2.5. This is called the Peierls instability, and it is due to a coupling of a special phonon, whose wavevector is $2k_F$, with an electron near the Fermi level [66, 69]. The

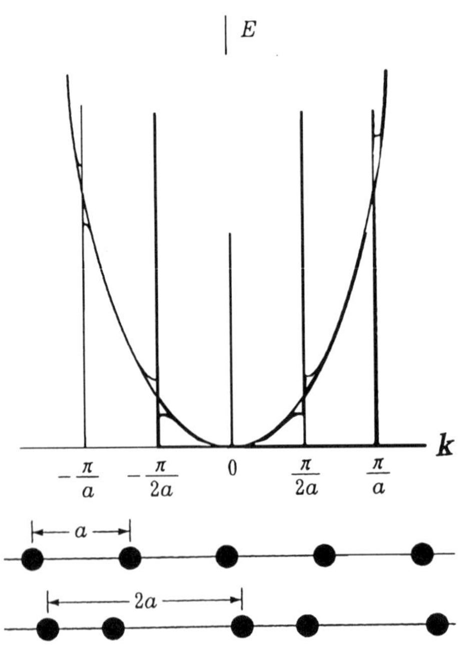


**Fig. 2.5.** Peierls instability in a linear chain. Doubling of the unit cell in real space produces a zone boundary at $\pi/2a$ where $a$ is the initial atomic spacing

coupling phonons are limited because in a one-dimensional structure energy and momentum can only be conserved with such phonons in the first order process. In many metal–insulator transitions, metal ions often become paired in the insulator phase, for instance, $Ti_4O_7$ and $VO_2$ [76,77]. Such pairing has been considered to be the important mechanism of the transition, at least when the electron correlation is small and the band picture is suitable, as it is for Li oxides. However when the electron correlation is large and the band picture no longer holds, a gap is not always introduced by such a pairing. Thus, $VO_2$ is considered to become insulating due to electron correlation and not due to the pairing [78,79].

In a three-dimensional lattice, the conservation of energy and momentum is fulfilled for various phonons, and the pairing becomes obscure. In the second-order process, a lattice modulation with $\boldsymbol{q}_0$ can occur if the top of the valence band and the bottom of the vacant conduction band is connected by this $\boldsymbol{q}_0$. In this case conduction electrons work to prevent such a coupling and the symmetry of the lattice of the metallic oxides is higher than that of insulating oxides. This is seen by comparing the lattices of $ReO_3$ and $WO_3$ (Sects. 5.1 and 5.4).

Sometimes antiferromagnetic ordering occurs below the metal–insulator transition. *Slater* pointed out that when the ordering doubles the size of a unit cell, the Brillouin zone becomes half that of the paramagnetic state and there a gap may open up in the middle of the conduction band as in the Peierls transition. The spin Peierls transition is believed to occur for $CuGeO_3$ [80,81].

*Schlenker* et al. found that the metallic molybdenum oxides of one or two-dimensional character become insulators at lower temperatures due to the formation of charge density wave (CDW) [82]. When some of the wavevectors of the Fermi surface electrons differ by the same wavevector $\boldsymbol{q}$ due to the particular shape of the Fermi surface, the charge distribution becomes unstable to a perturbation with that $\boldsymbol{q}$ and a CDW with that $\boldsymbol{q}$ is formed as will be explained in Sect. 5.5. Such a nesting of the Fermi surface may appear in a lower-dimensional substance.

With a random potential, localized states may appear near the band edge. When the carrier density decreases and the Fermi level crosses the mobility edge, the metal will change to an insulator and the conduction will be by variable range hopping. This will be explained in Sect. 5.4.

When the electron–electron interaction is strong, the electrons repel each other, and a substance, which is anticipated to be a conductor by the band calculation, is often found to be an insulator. This insulating property is explained in Chap. 4.

Phenomenologically, most MITs can be analyzed by a percolation model. This model assumes a random distribution of localized orbitals and when such orbitals span a crystal from one end to the other, it becomes metallic. However at that time a dimension of the localized orbital becomes a parameter and a mechanism of the localization must be clarified with the ideas above. The percolation model will be explained in Sect. 5.4.

...itly, *Imada* et al. have comprehensively reviewed the phenomena of metal to insulator transition in [83] in which the role of electron–electron interaction for the transition is emphasized.

## 2.9 Good Conductors

Oxides have been considered to be representative insulators. However, recently many good conductive oxides have been noticed. When $\mathrm{d}\rho/\mathrm{d}T$ is positive and $\rho$ is smaller than about $10^{-4}\Omega$m, the substance is usually said to be metallic, but this is not a good definition of metallic conductivity. For instance the conductivity of an amorphous metal is considered to become smaller without any lower limit as the randomness increases. However, applying the aforementioned criterion, we have selected out the metallic oxides shown in Table 2.1. It must be noted that this is not an exhaustive illustration of metallic oxides and cannot be a comprehensive database. Table 2.1 also cites briefly other characteristic features but these are very concise descriptions, and the reader should consult the original paper if one is of particular interest.

Here we describe briefly the difference between a typical metal such as iron and copper and the metallic oxides.

The first difference is in the ionicity. In a monoatomic metal the ionicity is zero and in the metallic alloys it is very small. However in a metallic oxide, the ionicity is large and even infrared absorption is observed.

The second difference is in the lower *s* character in the conduction electron of the oxide. Usually the *s* electrons of transition elements are transferred to the oxygen ions to occupy the bonding orbitals and the remaining *d* electrons mainly enter the conduction band, whereas, in a simple metal, they both constitute the conduction band.

The third difference is in the distribution of conduction electrons in the material. In a simple metal the conduction electrons occupy the whole space in the material. In oxides, they occupy only part of the material and sometimes the averaged density has no meaning.

The fourth difference is in the carrier density. In oxides, it covers a wide range and in some cases may be much smaller than in metals.

The fifth difference is in the variety of crystal structures. Even a low-dimensional structure is possible in oxides.

The sixth difference is in the impurity. In oxides, the main impurity is usually not other elements but an excess or insufficiency in the constituent element itself which occurs as a deviation from stoichiometric composition.

As seen from Table 2.1, there are many good conductors among oxides and many more will be found in the near future. Phenomenologically they are classified into the following groups.

1. Metallic conductors in which a deviation from the stoichiometric composition is essential to have metallic conductivity. $SnO_2$, $In_2O_3$, $Tl_2O_3$, TlOF, $SrTiO_3$, EuO, etc.
2. Normal metallic conductors which show metallic conductivity in their stoichiometric compositions. Their $d\rho/dT$ are positive in the whole temperature range and show no discontinuous variation. They show Pauli paramagnetism. $ReO_3$, $M_xReO_3$, TiO, NbO, $LaTiO_3$, $LaNiO_3$, $LaCuO_3$, $MMoO_3$, $SrIrO_3$, $SrCrO_3$, $Pb_2M_2O_7$, $LiTi_2O_4$, $RuO_2$, $OsO_2$, $IrO_2$, $MoO_2$, $WO_2$, $RhO_2$, $PtO_2$, etc.
3. Anomalous metallic conductors whose $d\rho/dT$ is not always positive and sometimes a discontinuity appears in the resistivity. The magnetic susceptibility depends on temperature. The metal–insulator transition occurs at certain temperatures in: $V_2O_3$, $Fe_3O_4$, $VO_2$, $V_nO_{2n-1}$ $(n = 4, 5, 6, 8)$, $Ti_2O_3$, $Ti_nO_{2n-1}$ $(n = 3, 4, 5, 6)$, $EuO_{1-x}$, etc. The metal–insulator transition appears at a certain composition in: $Li_{1+x}Ti_{2-x}O_4$, $M_xWO_3$, $Zn_{1-x}Li_xV_2O_4$, $La_{1-x}Sr_xMnO_3$, $BaPb_{1-x}Bi_xO_3$, $Bi_{2-x}Gd_xRu_2O_7$, etc. Anomalous magnetic behavior appears in: VO, $CrO_2$, $SrVO_3$, $CaCrO_3$, $CaFeO_3$, $SrFeO_3$, $SrCoO_3$, $LaCoO_3$, $LuNiO_3$, $CaRuO_3$, $SrRuO_3$, $La_2NiO_4$, $Nd_2NiO_4$, etc.
4. There are some insulators for which the band calculation predicts metallic conductivity. This suggests the importance of electron–phonon and electron–electron interactions, which are not correctly included in the mean field approximation, to determine transport properties: NiO, CoO, $La_2CuO_4$, etc.
5. One- or two-dimensional conductors which have quite an anisotropic conductivity where a particular interaction is expected to be intensified due to special geometrical characteristics, as in the Peierls transition. A disappearance of long-range order in one dimension and sometimes in two dimensions will affect the transport properties. Recent high-$T_c$ superconductors may be classified into this class: $M_xV_2O_5$, $M_xMoO_3$, $M_{0.9}Mo_6O_{17}$, $LiVO_2$, etc.

Another classification is the $d\epsilon$, $d\gamma$, and $sp$ conductors. They are classified according to the orbital nature of electrons at the Fermi level. The mixing of $O_{2p}$ orbitals is of course not negligible. The most numerous in Table 2.1 are the $d\epsilon$ conductors. The $sp$ and $d\gamma$ conductor classes contain only a few materials, as listed below.

- $s$–$p$ conductors: $SnO_{2-x}$, $BaPb_{1-x}Bi_xO_3$, $Ba_{1-x}K_xBiO_3$, etc.
- $d\gamma$ conductors: $SrFeO_3$, $YCoO_3$, $ErCoO_3$, $LaCoO_3$, $LaNiO_3$, $Ln_2NiO_4$, $Li_xPd_{1-x}O$, $Na_{1-x}Ca_xPd_3O_4$, $PtO_2$?, $Na_xPt_3O_4$?, high-$T_c$ cuprates, etc.

We have classified oxides as $d\gamma$ conductors where there are more than six $d$ electrons per ion. These are the oxides of Ni, Cu, Pd, Ag, and Pt. It must be noted that the high-$T_c$ superconductors are the main members of this group. Many researchers consider that the high-$T_c$ superconductors are the $O_{2p}$ conductors. Even so the mixing with the $d$ orbitals is not negligible.

In Chap. 5, we will explain in detail representative oxides selected from these oxides. At the beginning of Chap. 5, we will explain briefly how representative they are.

Here a more general survey of the metallic oxides according to their crystal structure will be given briefly.

### 2.9.1 The NaCl Structure

The NaCl structure is composed of interpenetrating fcc lattices of each ion. TiO, VO, NbO, and EuO have this structure. EuO is a ferromagnetic insulator and when reduced it exhibits metallic conductivity in the ferromagnetic temperature range. TiO, VO, and NbO are intrinsic conductors. However, the transport and magnetic properties of TiO and VO differ greatly. As shown in Fig. 2.6 a, the resistivity of $TiO_{1.01}$ has a positive temperature coefficient, $d\rho/dT > 0$, while $VO_{1.02}$ has a negative one as shown in Fig. 2.6 b [85–88]. The oxygen content $x$ changes from 0.7 to 1.25 in $TiO_x$ and from 0.8 to 1.3 in $VO_x$. Even when $x = 1.0$, Ti sites may be vacant randomly. The vacancy content depends on the preparation conditions and in the usual compound it may be almost 14% [85–88]. The stoichiometric zero vacancy oxide was prepared under the high pressure of 77 kbar (7.7 Gpa) at 1923 K [316]. It exhibits superconductivity at $T_c = 2.3$ K [317], while the one with vacancies

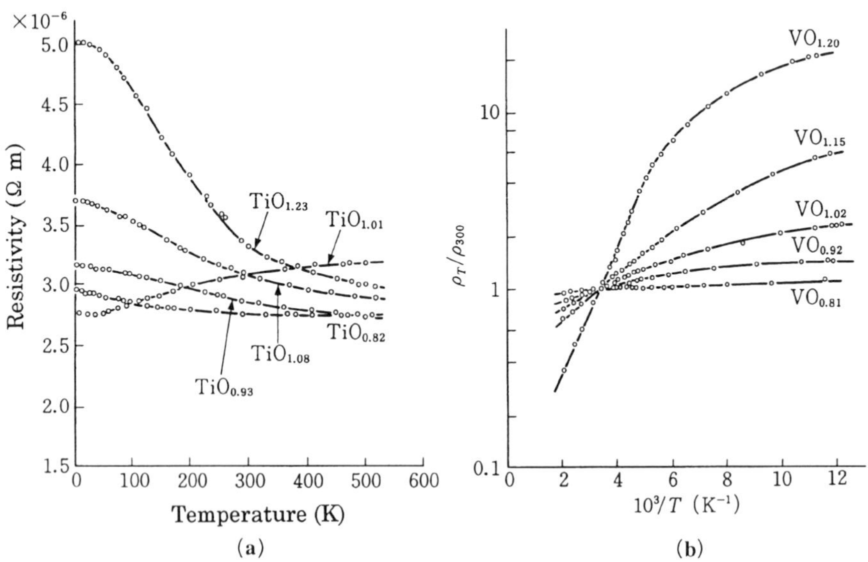


**Fig. 2.6.** Electrical resistivity of $TiO_x$ (**a**) and normalized resistivity of $VO_x$ (**b**) [85, 86]. (With permission from Japan Institute of Metals)

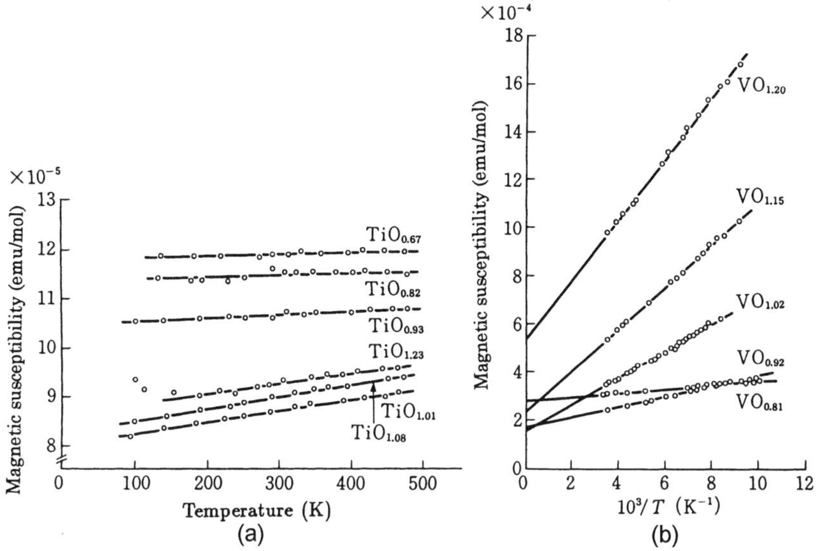


**Fig. 2.7.** Paramagnetic susceptibility of $TiO_x$ (**a**) and $VO_x$ (**b**) [85, 86]. (With permission from Japan Institute of Metals)

has a $T_c$ at 0.7 K. VO also contains about 15% vacancies under the usual preparation conditions. A random distribution of large numbers of vacancies is characteristic of these systems.

The magnetic properties also differ greatly between the two. The susceptibility of $TiO_x$ is nearly temperature-independent as shown in Fig. 2.7, while that of $VO_x$ becomes larger as temperature is lowered [85–88]. It is expressed as [318]

$$\chi = \chi_0 + \frac{C}{T - \theta} . \tag{2.36}$$

From $C = N\mu_0\mu^2/3k_B$, $\mu$ is $0.65\mu_B$ for $x = 1.02$. $N$ is the vanadium number density. ($\theta = 0$ in Fig. 2.7.) $\chi_0$ is larger than that of TiO by 1.6–5 times. Such a Curie or Curie–Weiss susceptibility suggests that the $3d$ electrons partially localize in $VO_x$. The conductivity of $VO_x$ also suggests that the conduction electrons are not simple band electrons. *Neckel* et al. have calculated the band structure of stoichiometric TiO and VO [319]. If the simple band picture is valid, both materials should have similar properties. The results suggest that the one-electron mean field theory is not applicable to VO. Similar phenomena are also observed for other Ti and VO oxides and there seems to be a boundary between Ti and V beyond which localization begins to occur.

This tendency is seen more clearly in Fig. 2.8, in which some results of band calculations of $MO$ ($M$ = Ca → Ni) by *Mattheiss* are shown [320].

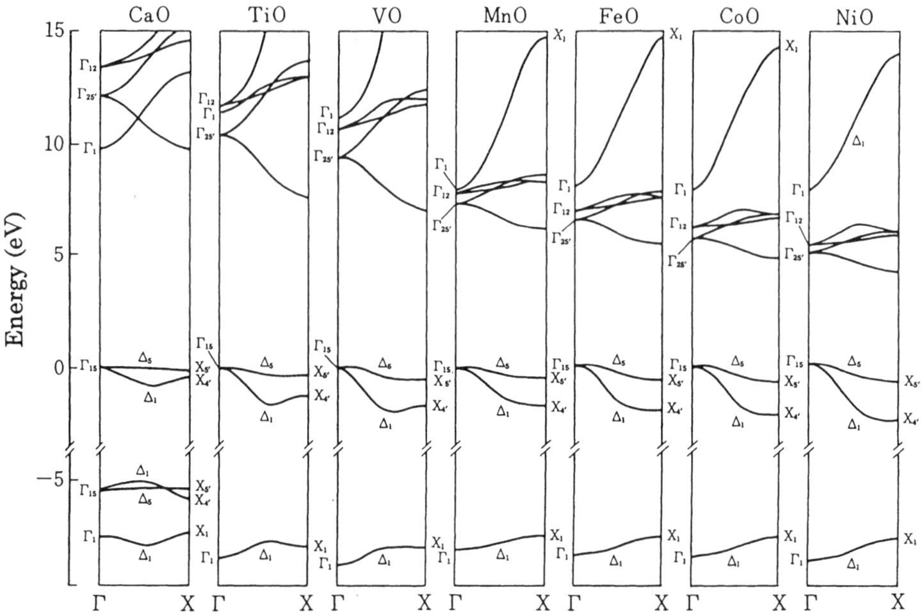


**Fig. 2.8.** Energy band structures of NaCl type mono-oxides calculated by *Mattheiss* [320]. Bands in the positive energy range are conduction bands. Note that there is no gap in the conduction band. $\boldsymbol{k}$ is (0,0,0) at $\Gamma$ and $(\pm\pi/a,0,0)$ or $(0,\pm\pi/a,0)$ or $(0,0,\pm\pi/a)$ at $X$ where $a$ is the cubic-cell dimension. (With permission from The American Physical Society)

As seen from Fig. 2.8 all the $M$Os except for CaO, in which the conduction band is empty, should be metallic. In reality, TiO and VO are metallic and the others are insulators. This result has been believed to show that the mean field theory cannot describe the excited states correctly, at least for MnO, FeO, CoO, and NiO. These are magnetic insulators, and the insulating property is believed to be due to the electron–electron interaction. However, as explained in Sect. 5.6, a small gap can be calculated out even in the one-electron approximation, and the situation is not so simple even though the electron–electron interaction is important.

In Table 2.2, the ionic radii for the four-fold and six-fold coordinations are shown [321]. They are the radii in a high-spin state, HS. In HS, electrons occupy the orbitals as in an isolated atom following the Hund rule. For instance five electrons enter as $(d\epsilon)^3(d\gamma)^2$ and $S = 5/2$. They can also occupy the orbitals as $(d\epsilon)^5$ and $S = 1/2$. When the crystal field splitting is large, such a state may be a ground state and it is called a low-spin state, LS. In Table 2.2 LS radii are not shown to avoid complexity but they are often observed in real crystals. As shown in Table 2.2, the radius becomes smaller as the nuclear charge becomes larger. The $3d$ electrons are then more tightly bound and the intra-ionic density becomes larger. Due to the strong Coulomb repulsion

**Table 2.2.** Periodic System of the Elements

Element
Atomic wt. (most stable)
Ionic radius at a four-(IV)
six- coordinated site (VI) in a high spin State/A (valency)

| | | | | | | | | | | | | | | | | | | |
|---|---|---|---|---|---|---|---|---|---|---|---|---|---|---|---|---|---|---|
| | **H**<br>1.008 | | | | | | | | | | | | | | | | | **He**<br>4.003 |
| IV<br>VI | **Li**<br>6.94<br>0.59(1)<br>0.74(1) | **Be**<br>9.01<br>0.27(2) | | | | | | | | | | | **B**<br>10.81<br>0.12(3) | **C**<br>12.01 | **N**<br>14.01 | **O**<br>15.999<br>1.38(-2)<br>1.40(-2) | **F**<br>19.0<br>1.31(-1)<br>1.33(-1) | **Ne**<br>20.18 |
| IV<br>VI | **Na**<br>22.99<br>0.99(1)<br>1.02(1) | **Mg**<br>24.31<br>0.49(2)<br>0.72(2) | | | | | | | | | | | **Al**<br>26.98<br>0.39(3)<br>0.53(3) | **Si**<br>28.09<br>0.26(4)<br>0.40(4) | **P**<br>30.97<br>0.17(5) | **S**<br>32.07<br>0.12(6)<br>1.84(-2) | **Cl**<br>35.45<br>1.81(-1) | **Ar**<br>39.95 |
| IV<br>IV<br>VI<br>VI | **K**<br>39.10<br>1.38(1) | **Ca**<br>40.08<br>1.00(2) | **Sc**<br>44.96<br>0.73(3) | **Ti**<br>47.88<br>0.86(2)<br>0.67(3) | **V**<br>50.94<br>0.79(2)<br>0.64(3) | **Cr**<br>52.00<br>0.82(2)<br>0.615(3) | **Mn**<br>54.94<br>0.82(2)<br>0.65(3) | **Fe**<br>55.85<br>0.63(2)<br>0.49(3)<br>0.77(2)<br>0.645(3) | **Co**<br>58.93<br>0.735(2)<br>0.61(3) | **Ni**<br>58.69<br>0.70(2)<br>0.60(3) | **Cu**<br>63.55<br>0.62(2)<br>0.73(2) | **Zn**<br>65.39<br>0.60(2)<br>0.745(2) | **Ga**<br>69.72<br>0.47(3)<br>0.62(3) | **Ge**<br>72.61<br>0.40(4)<br>0.54(4) | **As**<br>74.92<br>0.335(4)<br>0.50(5) | **Se**<br>78.96<br>0.29(6) | **Br**<br>79.90<br>1.96(-1) | **Kr**<br>83.80 |
| IV<br>IV<br>VI<br>VI | **Rb**<br>85.47<br>1.49(1) | **Sr**<br>87.62<br>1.16(2) | **Y**<br>88.91<br>0.892(3) | **Zr**<br>91.22<br>0.72(4) | **Nb**<br>92.91<br>0.32(5)<br>0.71(2)<br>0.64(5) | **Mo**<br>95.94<br>0.42(6)<br>0.67(3)<br>0.60(6) | **Tc**<br>(98)<br>0.64(4)<br>0.56(7) | **Ru**<br>101.1<br>0.68(3)<br>0.62(4) | **Rh**<br>102.9<br>0.665(3)<br>0.615(4) | **Pd**<br>106.4<br>0.64(2)<br>0.86(2)<br>0.62(4) | **Ag**<br>107.9<br>1.02(1)<br>0.65(3)<br>1.15(1) | **Cd**<br>112.4<br>0.84(2)<br>0.95(2) | **In**<br>114.8<br>0.79(3) | **Sn**<br>118.7<br>0.69(4) | **Sb**<br>121.8<br>0.77(3)<br>0.61(5) | **Te**<br>127.6 | **I**<br>126.9<br>0.95(5)<br>2.20(-1) | **X**<br>131.3 |
| IV<br>IV<br>VI<br>VI | **Cs**<br>132.9<br>1.70(1) | **Ba**<br>137.3<br>1.36(2) | **La**<br>138.9<br>1.06(3) | **Hf**<br>**178.5**<br>**0.71(4)** | **Ta**<br>180.9<br>0.67(3)<br>0.64(5) | **W**<br>183.9<br>0.41(6)<br>0.65(4)<br>0.58(6) | **Re**<br>186.2<br>0.40(7)<br>0.63(4)<br>0.57(7) | **Os**<br>190.2<br>0.63(4) | **Ir**<br>192.2<br>0.73(3)<br>0.63(4) | **Pt**<br>195.1<br>0.63(4) | **Au**<br>197.0<br>0.70(3) | **Hg**<br>200.6<br>0.96(2)<br>1.02(2) | **Tl**<br>204.4<br>1.50(1)<br>0.88(3) | **Pb**<br>207.2<br>0.94(2)<br>1.18(2)<br>0.775(4) | **Bi**<br>209.0<br>1.02(3) | **Po**<br>(209) | **At**<br>(210) | **Rn**<br>(222) |
| | **Fr**<br>(223) | **RA**<br>(226) | **Ac**<br>(227) | | | | | | | | | | | | | | | |
| | | | | | **Ce**<br>140.1<br>1.034(3)<br>0.80(4) | **Pr**<br>140.9<br>1.013(3)<br>0.78(4) | **Nd**<br>144.2<br>0.995(3) | **Pm**<br>(145)<br>0.979(3) | **Sm**<br>150.4<br>0.964(3) | **Eu**<br>152.0<br>1.17(2)<br>0.95(3) | **Gd**<br>157.3<br>0.983(3) | **Tb**<br>158.9<br>0.923(3)<br>0.76(4) | **Dy**<br>162.5<br>0.908(3) | **Ho**<br>164.9<br>0.894(3) | **Er**<br>167.3<br>0.881(3) | **Tm**<br>168.9<br>0.869(3) | **Yb**<br>173.0<br>0.858(3) | **Lu**<br>175.0<br>0.848(3) |
| | | | | | **Th**<br>232.0 | **Pa**<br>**231.0** | **U**<br>238.0 | **Np**<br>(237) | **Pu**<br>(244) | **Am**<br>(243) | **Cm**<br>(247) | **Bk**<br>(247) | **Cf**<br>(251) | **Es**<br>(252) | **Fm**<br>(257) | **Md**<br>(258) | **No**<br>(259) | **Lr**<br>(260) |

*Source:* R. D.Shannon, C. T.Prewitt: Acta Crystallogr. **25 B**, 925 (1969)

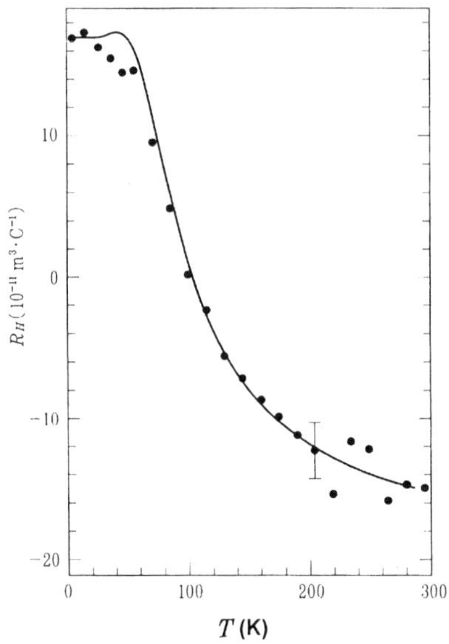


**Fig. 2.9.** Hall coefficient of NbO. The *solid line* is a theoretical fit to the observed results [91] (With permission from The American Physical Society)

among electrons, the mean field theory fails to describe correctly the excited states which are important to characterize the physical properties [322]. At the same time, the electron-phonon interaction will become stronger as the localized character is intensified. Finally the radius tends to increase at the end of the $3d$ period. In this sense, Ti and V correspond to Cu and Ni, and a metal-insulator transition is frequently observed for oxides of these elements as explained in Sect. 2.9.4.

Compared with $3d$ orbitals, $4d$ and $5d$ orbitals extend more widely and the band picture often holds good, though there are many cases where the electron–electron interaction cannot be neglected.

*Schluz* et al. made a band calculation for NbO in which oxygen vacancies distribute periodically [91]. There are three Fermi sheets, and the calculated Hall coefficient explains the observed large temperature dependence as shown in Fig. 2.9. When there are holes and electrons whose mobilities depend differently on temperature, the Hall coefficient may change even its sign at a certain temperature [8]. The calculated electrical resistivity is proportional to temperature above 80 K and fits the experimental result.

### 2.9.2 The Corundum Structure

Corundum is a mineral of $Al_2O_3$. Red ruby and blue sapphire are the gems of $Al_2O_3$ containing Fe and Cr, respectively. $Ti_2O_3$ and $V_2O_3$ have this structure and they show a metal–insulator transition at certain temperatures. A lot of work has been done on $V_2O_3$ and it is the most representative MIT oxide. Recently, *Imada* et al. have reviewed the MIT for various substances including $V_2O_3$ [83]. Therefore, we refer the reader to their excellent review.

### 2.9.3 The Rutile Structure

Rutile is a mineral of $TiO_2$ and is the representative semiconductor in which the electrons have low mobility. On the other hand the isostructural $SnO_2$ is a representative broad-band high-mobility semiconductor. Therefore, $TiO_2$ and $SnO_2$ are compared in detail in Sect. 5.2. $TiO_2$ is also interesting as a useful photo-catalyst.

In this series, there are many 4*d* and 5*d* oxides which show metallic conductivity and Pauli paramagnetism as shown in Fig. 2.10. The experimental Fermi surfaces of $RuO_2$ and $OsO_2$ are consistent with the calculaions by *Mattheiss* [323]. As will be explained in Sects. 2.9.4 and 2.9.7, almost all kinds of magnetic and transport properties which are characteristic of oxides, appear in the oxides of Ru and it is a very interesting element. $RuO_2$ is widely used in the printing boards of electronic circuits.

$Ti_nO_{2n-1}$ ($n = 4-9$) and $V_nO_{2n-1}$ ($n = 4-8$) constitute the homologous series made from $TiO_2$ and $VO_2$, respectively, by subtracting oxygens. In them the oxygen vacancies congregate at a plane and the plane is subtracted periodically. Such structures were extensively studied by Magnéli and they are frequently called the Magnéli phase. Magnéli phases of W and Mo are

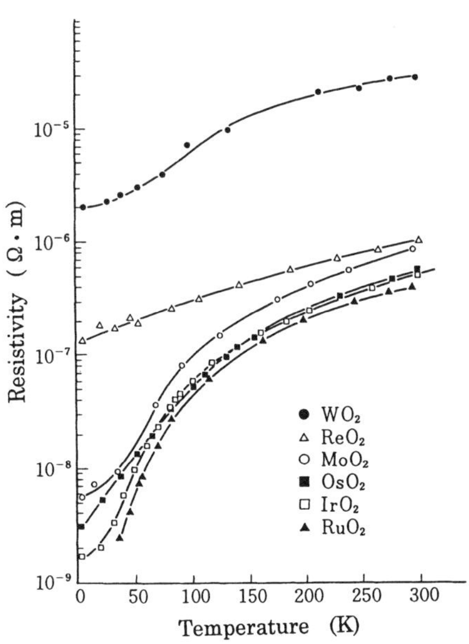


**Fig. 2.10.** Resistivity of rutile type oxides [253]. (With permission from The American Chemical Society)

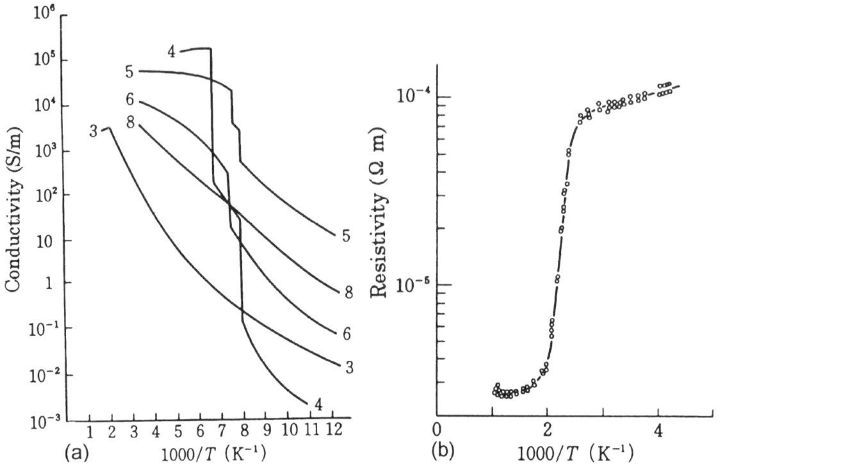


**Fig. 2.11.** Electrical conductivity of $Ti_nO_{2n-1}$. (**a**) $n$ is designated beside each curve [276]. (**b**) $n = 2$ [242]. (With permission from The American Physical Society)

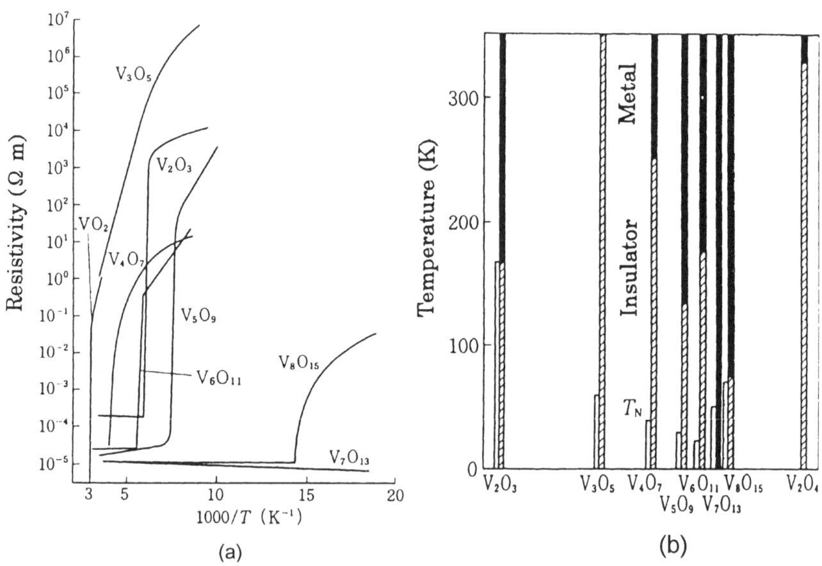


**Fig. 2.12.** Metal–insulator transition in $V_nO_{2n-1}$ [266]. In (**a**) resistivity is plotted and in (**b**) the phase diagram is shown. $T_N$ is the Néel temperature. (With permission from The American Institute of Physics)

well known and will be explained in Chap. 5. The Magnéli phases of Ti and V have a metal–insulator transition as shown in Figs. 2.11 and 2.12 [266,276].

$CrO_2$ is a ferromagnetic metal with Curie temperature at 329 K. The conductivity is shown in Fig. 2.13 [250]. The magnetic moment is $2\mu_B$/molecule and the conduction electrons are believed to align magnetic moments parallel

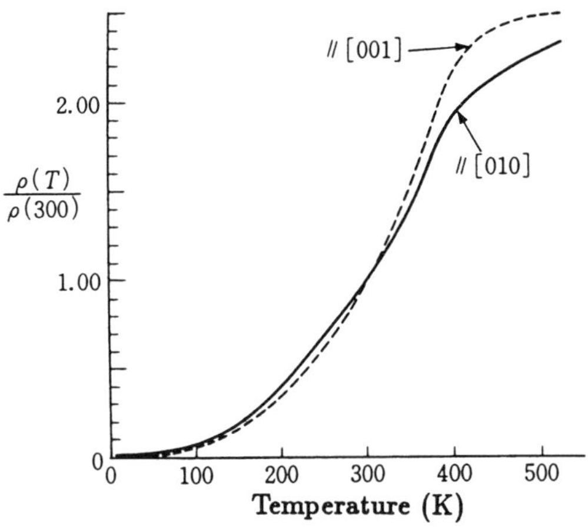


**Fig. 2.13.** Normalized resistivity of $CrO_2$. $\rho(300) = (2.5 \pm 0.5) \times 10^6 \Omega\,m$ [250]

to one another [152]. $MoO_2$ is a $4d$ oxide with a similar electronic configuration to $CrO_2$ but it is Pauli paramagnetic and shows no ferromagnetism.

$VO_2$ shows a MIT at 340 K. In the metallic phase, Pauli paramagnetism appears and below the transition temperature no long-range magnetic ordering occurs. In the insulating phase, pairing occurs between vanadium ions along the $c$ axis. The separation is 0.288 nm above the transition temperature and 0.265 nm and 0.312 nm below. The V ions are also slightly displaced perpendicular to the $c$ axis [84]. It was once believed to be simply a Peierls transition but electron interaction is also important as was explained in Sect. 2.8. A simple Peierls transition assumes that a simple band picture holds good above the transition temperature [78, 84, 244–249].

### 2.9.4 The Perovskite Structure

Figure 2.14 shows the cubic perovskite structure. It is cubic but many perovskites distort a little from this structure at room temperature, and the metal–oxygen–metal bond angle differs from 180° giving large effects on various electronic properties. In this series, there are many useful oxides such as ferroelectric $BaTiO_3$, the piezoelectric PZT ($PbZrO_3$-$PbTiO_3$), the PTC (positive temperature coefficient) resistor, and high-temperature heating elements such as $La_{1-x}Sr_xCrO_3$. Moreover this is a parent material of the high-$T_c$ superconductors. Recently a giant magnetoresistance of $La_{1-x}Sr_xMnO_3$ attracted much attention [143, 149, 150, 156, 324, 325].

There are many magnetic substances among the metallic perovskites and they have been extensively reviewed by *Goodenough* [152].

The band structure of $SrTiO_3$ was calculated by *Mattheiss* as shown in Fig. 2.15 [326, 327]. It is an insulator at the stoichiometric composition but

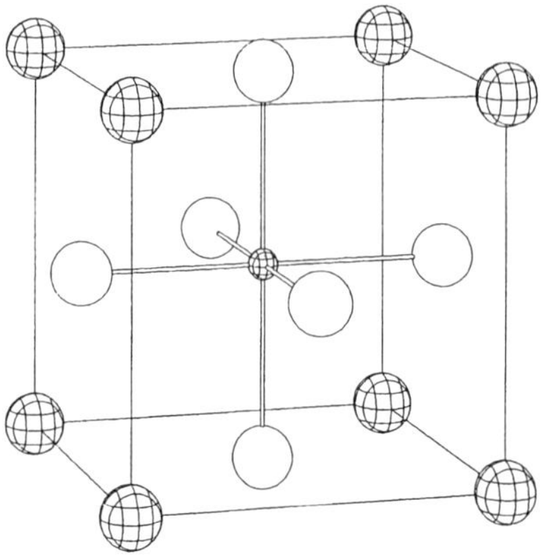

**Fig. 2.14.** The crystal structure of cubic perovskite $ABO_3$. The large $A$ ion is represented by a larger netted sphere in twelve coordination. A smaller netted sphere is a six coordinated smaller $B$ ion. (After A. Yanase)

upon reduction a large conductivity appears and it even shows superconductivity below 0.28 K. The carrier concentration is as low as $10^{25}$ to $10^{26}\mathrm{m}^{-3}$ and $T_c$ depends on the carrier concentration [328]. The band structure is quite similar to that of $ReO_3$ except that the $5s$ orbitals of Sr have a higher energy and Sr behaves mainly as a donor. The band structure of $ReO_3$ will be explained in detail in Sect. 5.1.

The metal–insulator transition is observed for $Ca_{n+1}Ti_nO_{3n+1-x}$. For instance, $CaTiO_{2.991}$ is metallic above 61 K [129].

$LaTiO_3$ is an insulator and shows a canted antiferromagnetism below 150 K. Upon oxidation it becomes metallic but shows no superconductivity [125]. Since the tolerance factor $t = (r_{La} + r_O)/\sqrt{2}(r_{Ti} + r_O)$, where $r$ is an ionic radius, is smaller than 1, the $TiO_6$ octahedra tilt each other as in $GdFeO_3$ [329] and the band width is expected to become narrower [124].

$SrVO_3$ and $CaVO_3$ are metallic and Pauli paramagnetic. Sometimes a trace of ferromagnetic $SrVO_{3.25}$ is included in the former [130, 132, 133, 140].

$CaCrO_3$ shows canted antiferromagnetism below 325 K, and is metallic in the entire temperature range [139, 140]. Thus this is a rare example in which antiferromagnetism and metallic conductivity coexist. $SrCrO_3$ is also metallic but is Pauli paramagnetic [141].

$La_{1-x}Sr_xMnO_3$ is ferromagnetic in the range $0.2 < x < 0.4$ and at the same time it has a metallic conductivity of the order of $10^4\ \mathrm{Sm}^{-1}$ [151–153]. The end members are antiferromagnetic insulators [330, 331]. It is believed that the conduction electrons align the localized moments to produce the ferromagnetism. This is called a double exchange interaction [57–59]. A large magnetoresistance was discovered for $La_{1-x}M_xMnO_3$, and a special magnetic polaron, which associates local distortion of the lattice, has been proposed for explaining the temperature dependence of the resistivity [143, 149, 150,

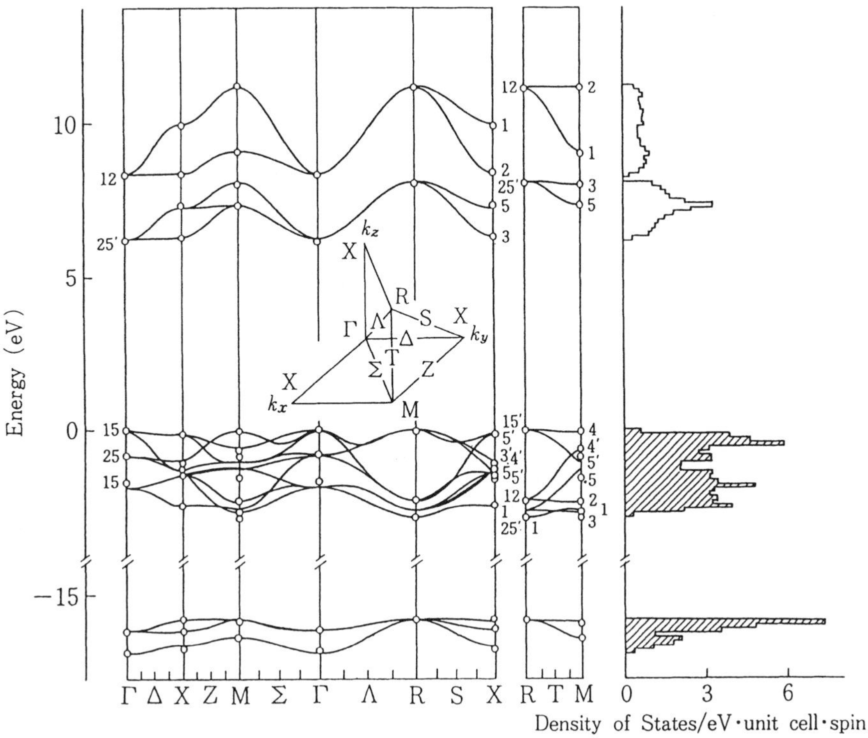

**Fig. 2.15.** Energy band structure of $SrTiO_3$ by *Mattheiss* [326]. Contributing $Ti_{3d}$ and $O_{2p}$ orbitals are shown in Table 5.1 after *Kahn* and *Leyendecker* [502]. The symbols in $\boldsymbol{k}$ space are shown in the *inset.* (With permission from The American Physical Society)

156, 324, 325]. The study of $La_{1-x}Ca_xMnO_3$ has brought about the idea of valence fluctuation [332]. $Mn^{3+}$ and $Mn^{4+}$ states fluctuate with frequency $\nu$. Nowadays, the term is applied to $f$ electron compounds in which the band width is less than 0.1 eV ($\nu < 10^{14}\ s^{-1}$) such as SmS which turns metallic at a pressure of 6 kbar.

$SrFeO_3$ shows an antiferromagnetic-like susceptibility below 134 K and has a helical spin structure. The magnetic moment is $2.7\mu_B$ at 77 K and the effective Bohr magneton number is 5.8 from the paramagnetic susceptibility. There is no change in $d\rho/dT$ at $T_N$, and $d\rho/dT$ is nearly zero below and above $T_N$ [161–163]. $SrFeO_3$ is a $d\gamma$ conductor [164].

$CaFeO_3$ is antiferromagnetic below 115 K and it is an insulator below $T_N$. Valence splitting occurs as $2Fe^{4+} \rightarrow Fe^{3+} + Fe^{5+}$. It is metallic above $T_N$ and the effective Bohr magneton number is 5.6 [160, 333].

$SrCoO_3$ is ferromagnetic below 212 K. The magnetic moment is $1.14\mu_B$ at 77 K and the effective Bohr magneton number is 3.4 [162].

$LaCoO_3$ shows a gradual transition from an insulator to a metal as the temperature is raised between 600 K and 700 K. Both low-spin and high-spin states are thought to coexist [166, 167]. *Asai* et al. claim that a lower-temperature LS state changes into a HS state at about 90 K and not at the MIT temperature [334].

$LaNiO_3$ is metallic. A weakly temperature-dependent paramagnetic susceptibility gives an antiferromagnetic Weiss temperature of 1780 K [177, 178, 335].

$LuNiO_3$ shows Curie-Weiss paramagnetism above 90 K and antiferromagnetism below $T_N$ [185]. In both La and $LuNiO_3$, $Ni^{3+}$ is in the low-spin state, i.e., $(3d)^7 \to (d\epsilon)^6(d\gamma)^1$ and not in a high-spin state $(d\epsilon)^5(d\gamma)^2$. Mixing with $O_{2p}$ is of course not negligible.

$RNiO_3$ shows an MIT at 135 K ($R$ = Pr), 201 K (Nd), 403 K (Nd), and 480 K (Sm) [182, 183], while antiferromagnetic ordering of Ni spins ($S = 1/2$, LS) appears below 135 K, 201 K, 225 K and 205 K, respectively. Below the MIT the lattice expands without a change in symmetry (Pr, Nd) [182].

$SrRuO_3$ is ferromagnetic below 160 K. The saturation magnetic moment is larger than $0.8\mu_B$ and neutron diffraction reveals that it is $1.4\mu_B$ [195, 196]. The high temperature Curie–Weiss susceptibility yields a Weiss temperature of 161 K and an effective Bohr magneton number of 2.6 [196]. It has been suggested that the ferromagnetism is due to itinerant electrons [195]. *Moriya* et al. have considered the magnetism of narrow band electrons allowing for arbitrary spin and charge density fluctuations [65, 336]. They showed that the Curie–Weiss law may be valid even for an itinerant system and the saturation moment may be smaller than the effective magnetic moment deduced from the paramagnetic susceptibility. In the case of a localized moment, it is a matter of course that $S$ is smaller than $\sqrt{S(S+1)}$. The conductivity of a single crystal $SrRuO_3$ is metallic and the resistivity decreases rapidly below $T_c$ as is observed for metallic ferromagnets [197].

$CaRuO_3$ is also metallic. The single crystal resistivity increases gradually but is not proportional to temperature [197]. The higher-temperature susceptibility follows a Curie–Weiss law, with an antiferromagnetic Weiss temperature of 125 K and an effective Bohr magneton number of 3 [196]. Long-range magnetic ordering was not observed down to 4.2 K in the Mössbauer spectrum [198, 337], and there is no change in the single crystal resistivity at 110 K [197]. The Seebeck coefficient of a mixed compound $Ca_{1-x}Sr_xRuO_3$ indicates a hole conduction, and its temperature dependence shows a large bend at about 160 K irrespective of its magnetic property [338]. *Kiyama* et al. claim that $CaRuO_3$ is a nearly ferromagnetic (not antiferromagnetic) metal from the analysis of the temperature dependence of specific heat [200].

$BaRuO_3$ does not have the perovskite structure, but a face sharing layer structure and shows high conductivity. The susceptibility shows a broad maximum at 430 K [196].

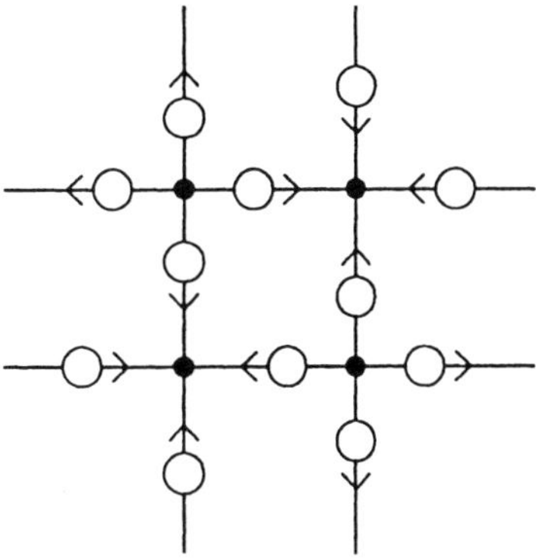

**Fig. 2.16.** Schematic representation of a breathing mode

$BaMoO_3$, $SrMoO_3$, and $CaMoO_3$ show normal metallic conductivity down to 2.5 K [139,228,230,339]. The susceptibility is Pauli paramagnetic [139,227].

$BaPb_{1-x}Bi_xO_3$ ($0.06 < x < 0.35$) was found to be superconducting below about 12 K [219]. As the transition width is narrowest at $x = 0.25$, there is a possibility that the superconducting range is narrower than the parenthesized tetragonal range [340, 341]. $BaPbO_3$ is also superconducting below 0.4 K [217]. The carrier density is as small as $0.7 \times 10^{27}$ $m^{-3}$, that is, 0.13/molecule [216], similar to $SrTiO_{3-x}$. $BaBiO_3$ is not metallic, with an energy gap of 0.2 eV [219]. *Mattheiss* et al. analyzed the characteristic features of the calculated band structures of high-$T_c$ cuprates, and made a guiding principle for searching for a high-$T_c$ superconductor. Following this principle they arrived at $Ba_{1-x}K_xBiO_3$ (BKBO) which is superconducting below 22 K [220]. The conduction band of $BaPbO_3$ is the broad $6s$–$2p$ mixed band and the metallic character is due to the lowering of the $6s$ band by the relativistic effect [342]. The hypothetical cubic $BaBiO_3$ must be metallic from the calculation but is monoclinic and the distortion opens up an energy gap. This distortion is considered to be due to a frozen vibrational mode in which the $BiO_6$ octahedron expands and shrinks as if it were breathing. In this mode the modulation of the overlap between the ligand $O_{2p}$ and $Bi_{6s}$ orbitals becomes maximized, as shown in Fig. 2.16. In monoclinic $BaBiO_3$, the expanded and reduced octahedra appear alternatively like a frozen breathing mode [343]. This system once attracted attention as a possible onsite bipolaron system [344]. From the phonon structures in the tunneling conductance of BKBO the electron–phonon interaction parameter $\lambda$ was determined to be 1.2–1.3, which gives a $T_c$ of 19–20 K. [221, 345–349]. Therefore, superconductivity of BKBO is due to the electron–phonon interaction. Some theoretical studies also support this view [350–352].

### 2.9.5 The $K_2NiF_4$ Structure

$La_2CuO_4$ is the parent oxide of the high-$T_c$ superconductors and will be discussed in Sect. 5.9. The physical properties are very sensitive to the composition.

The properties of $La_2NiO_4$ are also sensitive to the composition. Ceramics prepared in air are believed to contain 3–15% $Ni^{3+}$, and the stoichiometric oxide is synthesized under a controlled reducing atmosphere [284, 285]. The deviation from stoichiometry occurs not only in the oxygen content but also in the lanthanum content [285]. The nearly stoichiometric $La_2NiO_4$ showed a metal to insulator transition at about 650 K. At lower temperatures, the activation energy for the conduction in the basal plane was about 0.06 eV. Along the $c$ axis, the resistivity was three orders of magnitude larger and the oxide was insulating [282, 284]. According to *Eisaki* et al. $La_{2-x}Sr_xNiO_4$ is metallic for $x \geq 0.8$ and the stoichiometric $La_2NiO_4$ is an insulator with an energy gap of 4 eV [286, 287]. In general, $La_2NiO_4$ is one of the homologous oxides, $La_{n+1}Ni_nO_{3n+1}$ ($n$ = 1–3) [180]. The oxides with $n$ = 2 and 3 are metallic [180].

The magnetic susceptibility of the nearly stoichiometric $La_2NiO_4$ was almost temperature independent between 80 and 300 K and was $1.3 \times 10^{-3}$ emu/mol. Judging from the susceptibility peak of the 0.5% $Ni^{3+}$ oxide, canted antiferromagnetic order seems to have occurred below 204 K with a weak ferromagnetic moment along the $c$ axis [283]. The susceptibility is very sensitive to the composition, and was expressed by the sum of the constant term and the Curie–Weiss term for the 0.5% oxide. The ferromagnetic Weiss temperature was 199 K and the effective Bohr magneton number $2\sqrt{S(S+1)}$ was 0.3 when the magnetic field was applied perpendicular to the basal plane [284]. Such composition-dependent canted antiferromagnetism or weak ferromagnetism also appears in another layered oxide $YFe_2O_4$ [353], and in $La_2CuO_4$ [354].

A Mössbauer study of $La_2NiO_4$ showed that $T_N$ was lower than 77 K [355] but this may be due to the deviation of the composition from stoichiometry [284]. In insulating $ZnV_2O_4$ spinel, a slight deviation from stoichiometry also seems to suppress antiferromagnetism [109]. In the metal–insulator spinel $Li_xZn_{1-x}V_2O_4$, the strength of antiferromagnetic interaction is greatly reduced by mobile electrons [109].

$Sr_2RuO_4$ and $Sr_2RhO_4$ also crystallize into this structure, but they are $d\epsilon$ conductors and their superconducting critical temperatures are below 1 K [294–297, 300]. Superconductivity of $Sr_2RuO_4$ is believed to be $p$-wave triplet ($S = 1$) superconductivity [1301].

### 2.9.6 $ReO_3$ and $M_xWO_3$

$ReO_3$ has the simplest structure of the metal oxides. The octahedra are arranged in a simple corner-sharing cubic lattice as shown in Fig. 5.1. It has the highest electrical conductivity among the normal conductive oxides and is a representative broad band conductor. (A more conductive oxide is $CoPdO_2$ in which, however, a metallic layer is composed of purely Pd atoms [1302,1303].) The color is purple red, and the band structure is expressed analytically with considerable accuracy. $ReO_3$ will be explained in detail in Sect. 5.1.

The tungsten bronzes, $M_xWO_3$, are prepared by doping $M$ = H, alkali metals, alkaline earths, Cu, Ag, In, Tl, Sn, Pb, and rare earths into $WO_3$ which has a distorted $ReO_3$ structure. As $x$ can be increased up to 1, the doping density is much higher than in Si. The high density doping is one of the characteristic features of oxide conductors compared with usual semiconductors. The lattice structure changes with this high-density dopant and this complicates the phenomena.

At a certain value of $x$ a metal–insulator transition occurs accompanied by a structural transformation. $Na_xWO_3$ is metallic when $x$ is larger than 0.25 and cubic in the metallic range. It has the vacant perovskite structure while metallic $K_xWO_3$ is hexagonal. The third element distributes randomly in the lattice and electrons move under the random potential. Thus $M_xWO_3$ may be considered as a metal perturbed by the randomness.

The metallic bronzes show superconductivity. For instance the $T_c$ of hexagonal $Rb_xWO_3$ is 4.35 K for $x = 0.2$. It is hexagonal in the range $0.2 < x < 0.33$. $T_c$ is sensitive to the preparation method and there is a report that it has been enhanced to 7.7 K by etching it with acid [113]. Screening of the Coulomb interaction, and then the transition temperature, depend sensitively on the carrier density because the conduction band of $Na_xWO_3$ has two-dimensional character even though the lattice is three-dimensional.

Depending on the carrier density the color changes from blue to red. For instance $WO_3$ is pale green while $H_xWO_3$ is blue. When $WO_3$ is biased negatively in a proton atmosphere such as in a liquid or solid electrolyte, protons diffuse into the bulk and it turns blue. Because of this property it is used as an electrochromic device.

$M_xWO_3$ will be explained in detail in Sect. 5.4 as a degenerate large-polaron material.

$M_xMoO_3$ has a layered structure and electronically it is one-dimensional and will be discussed with other low-dimensional Mo bronzes in Sect. 5.5.

### 2.9.7 Pyrochlores $A_2B_2O_{7-x}$

In the pyrochlore structure $A$ is a large ion such as Y, a rare earth, Tl, In, Pb, or Bi, and $B$ is a small ion such as Ti [356], V [357–360], Cr [361], Mn [361], Nb [362], Mo [363–368], Zr [369], Tc [370], Hf [369], Re [370, 371], Ru [199, 370, 372–377], Rh [271, 373, 374, 377], Pd [369], Os [373, 374, 378], Ir [370, 373, 374, 376, 377], Pt [369, 373], Si [379], Ge [369], Sn [356], and Ga-Sb [356].

The larger $A$ ion has eight-coordination and the smaller $B$ ion has the octahedral six-coordination as shown in Fig. 2.17. The $BO_6$ octahedra share

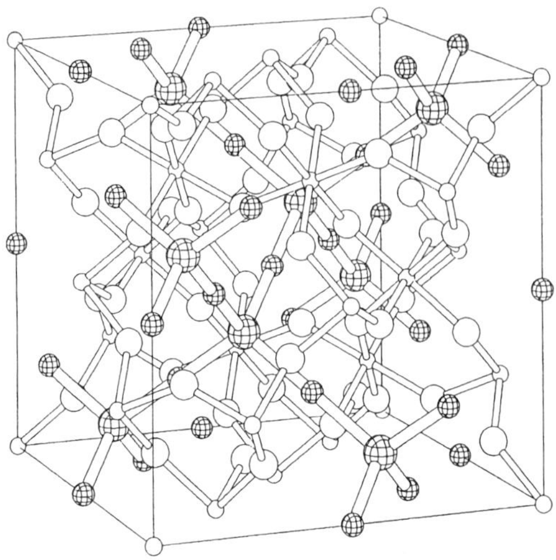

**Fig. 2.17.** A crystal structure of pyrochlores $A_2B_2O_6^1O_1^2$ [377]. Smaller netted sphere: $A$ ion. Smaller open sphere: $B$ ion. Larger open sphere: $O^1$. Larger netted sphere: $O^2$ (after A. Yanase)

corners and make a three-dimensional network. $O^2$ at the center of the tetrahedron of $A$ ions is easily removed [370], and some pyrochlores are synthesized under a high pressure oxygen atmosphere.

$Pb_2Ru_2O_{7-x}$ transforms to an orthorhombic perovskite near 90 kbar at 1673 K [372]. Similar to the perovskites, the pyrochlores show a variety of properties which originate from the narrow energy band. For instance, $Lu_2V_2O_7$ is a semiconducting ferromagnet below 80 K [359]. $Tl_2Mn_2O_7$ is ferromagnetic below 117 K and $d\rho/dT$ is positive in the magnetic region while $d\rho/dT$ is negative in the paramagnetic region. The resistivity of the sintered material is 0.1Ωm at 300 K [361]. $Y_2Mo_2O_7$ behaves magnetically like a spin glass [364]. $Y_2Ru_2O_7$ is not metallic while Ru oxides of Pb and Bi are metallic. The insulating property of $Y_2Ru_2O_7$ is attributed to correlation induced localization [199]. Perovskites of Ru also show a variety of transport and magnetic properties as explained in Sect. 2.9.4. Following *Cox* et al., the density of states at the Fermi level decreases through the series $Pb_2Ru_2O_7$, $Bi_2Ru_2O_7$, $SrRuO_3$, $CaRuO_3$, and is zero for $Y_2Ru_2O_7$ [199]. There is a composition induced metal–insulator transition in $Bi_{2-x}Gd_xRu_2O_7$ [375]. Finally, $Cd_2Nb_2O_7$ is ferroelectric [362].

### 2.9.8 Spinels

Spinel is the name of the mineral $MgAl_2O_4$, with cations in both tetrahedral and octahedral sites. The nearest neighbours of the octahedral cation are also mutually nearest neighbors as shown in Fig. 5.84. There are many ferrimagnetic oxides of this structure such as $NiFe_2O_4$, $CoFe_2O_4$, and $MnFe_2O_4$. They are called ferrites and are widely used in high frequency electronic circuits where they are important for their very high resistivity. There are very few conducting spinels and, in fact, only $LiTi_2O_4$, $LiV_2O_4$, and $Fe_3O_4$ are metallic. The first is superconducting below 13 K, which was the highest transition temperature in oxides until the Cu oxides appeared. $Li_{1+x}Ti_{2-x}O_4$ is an insulator when $x$ is larger than about 0.15. The superconductor-to-insulator transition then occurred at He temperature. This transition may be a good example to examine what will occur when large polarons accumulate and become degenerate. $LiV_2O_4$ shows a Curie–Weiss-type paramagnetism but no long-range magnetic ordering down to 4.2 K. The antiferromagnetic superexchange interaction may be weakened by the conduction electrons. They will be explained in Sect. 5.3.

$Fe_3O_4$ shows metallic conductivity above 120 K but it is not simple metallic conductivity because $d\rho/dT$ is not positive. A molecular polaron has been proposed. Below 120 K, a charge ordering occurs and the oxide becomes an insulator. The transition is known as the Verwey transition. $Fe_3O_4$ may be the most strongly correlated metallic oxide and will be explained in detail in Sect. 5.8.

### 2.9.9 Low-Dimensional Oxides

$\beta$–$M_xV_2O_5$ is a quasi-one-dimensional oxide in which it is supposed that small polarons are densely accumulated and bipolarons are formed. Recently found spin ladder oxides have properties similar to $\beta$–$M_xV_2O_5$.

$A_{0.3}MoO_3$ is a quasi-one-dimensional metallic oxide with a charge density wave below about 180 K where it becomes an insulator. $Li_{0.9}Mo_6O_{17}$ is also a quasi-one-dimensional metal and shows an insulator(?)–superconductor transition at 1.9 K.

$M_{0.9}Mo_6O_{17}$ ($M$ = K, Na, Tl) and $Mo_nO_{3n-1}$ are two-dimensional metals with charge density waves at lower temperatures where they are insulating.

High-$T_c$ Cu-oxides are quasi-two-dimensional substances with a strong electron–phonon and an electron–electron interaction.

How the low-dimensionality affects the transport properties will be described in detail in Sects. 5.5 and 5.9.

# 3. Theories for Many-Electron Systems with Strong Electron–Phonon and Interelectron Coulombic Interactions

The main objective of solid state theory is to clarify the microscopic origins of various macroscopic properties of solids, such as electric, magnetic and optical properties, which change from one crystal to the other. The difference of these macroscopic properties between crystals mainly originates from the difference of the microscopic nature of electrons in each crystal. In particular, it mainly comes from the outer electrons, which are in the outermost orbitals of the atoms or molecules constituting each crystal. Consequently, one of the most important problems in solid state theory is to find the relation between the microscopic nature of the outermost electrons and the macroscopic properties of solids.

Generally speaking, the microscopic behavior of the outermost electrons is governed by the following three competing elements; the itineracy of an electron, the interelectron Coulombic repulsion and the electron–phonon (e–ph) coupling. The first element is a quantum characteristic, and means that even if an electron is confined within an outermost orbital of an atom, it can tunnel from this orbital to neighboring ones, as illustrated by Fig. 3.1 a. In other words, the electron wants to move as "freely" as possible, if it is alone. The strength of the itineracy can be denoted by the resonance transfer energy between two neighboring orbitals ($\equiv T$), whose reciprocal is the time required for the tunneling.

This free motion of an electron is greatly modulated in the many-electron system, since the Coulombic repulsive energy increases, when two electrons happen to be in the same orbital. Thus, the single occupancy of an orbital becomes more probable than a double one, resulting in the occurrence of localized spins and magnetism. The strength of this repulsion is denoted by the energy difference ($\equiv U$) between the double occupancy and the single one, as illustrated by Fig. 3.1 b. If it is finite, the electrons want to be as "independent" of each other as possible.

Through the e–ph coupling an electron can self-induce a local lattice distortion so as to lower the energy level of an orbital, as illustrated by Fig. 3.1 c, and it is then trapped in it. This is the so called self-trapped (ST) state [381, 382]. Since this lowered orbital can also trap another electron, the e–ph coupling results in an attraction between electrons. The strength of this coupling is denoted by the energy lowering ($\equiv S$) of a singly occupied orbital

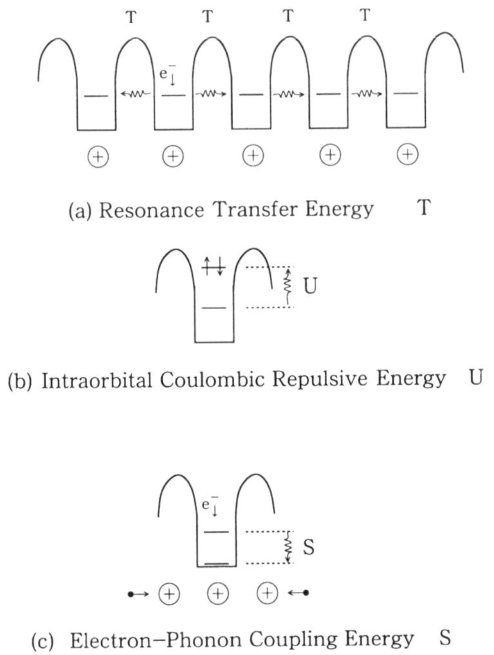


**Fig. 3.1.** (**a**) The resonance transfer energy $T$, (**b**) the intraorbital Coulombic repulsive energy $U$, and (**c**) the e–ph coupling energy $S$. The atomic core seen from an electron is a positive charge, and the *solid line* denotes the potential given by this core. The *small arrow* denotes the spin of the electron

and if it is finite, the electrons want to "associate" with each other as far as possible. These three competing natures of the electrons: the freedom $T$, the independence $U$ and the association $S$, will govern the state of a many-electron system, and result in a variety of macroscopic properties. They will almost always contribute with equal weight, although there will be many cases where one or two elements are hidden.

In the case of single-body problems, such as an electron, a hole or an exciton in a deformable lattice, the main objective is to clarify the phenomena resulting from the competition between $T$ and $S$ [381,382]. On the other hand, the problems related to transitions from normal metallic states to charge density wave (CDW) states [383] are just the $T \leftrightarrow S$ competitions in many-electron systems. The various magnetic properties of solids are mainly related to the $T \leftrightarrow U$ competition [384]. The $S \leftrightarrow U$ competition in many-electron systems is just the competition between the CDW states and magnetic states such as spin density wave (SDW) states [385].

The transitions from normal metallic states to superconducting (SC) states [386] are also related to the $T \leftrightarrow S$ competition in many-electron systems. However, in this case, the nature of the $T \leftrightarrow S$ competition is quite different from the previous one, since the attraction mediated by phonons is quite different from the Coulombic repulsion not only in its sign, but also in its retarded nature. Quantum electrodynamics (QED) tells us that the inter-electron Coulombic repulsion is mediated by photons, and because of their infinite velocity, this interaction has no retardation. In the case of phonons, however, the energy quantum ($\equiv \omega$) is finite, and hence, according to the change of the ratio between $T$ and $\omega$, the nature of the attraction is also greatly changed. In the small limit of $\omega$, $T \gg \omega$, the phonons cannot follow the motions of the electrons at all, but can give only a static potential due to a frozen local lattice distortion. This is the adiabatic case. In this case, the electron will be trapped in the local lattice distortion, since it can follow the motions of phonons completely. On the other hand, in the opposite limit of large $\omega$, $T \ll \omega$, the phonons can follow the motions of the electrons instantaneously, resulting in an attraction with no retardation, just like the Coulombic force. This case can be called the inverse adiabatic limit.

However, situations existing in most solids do not correspond to these extreme limits, but are in intermediate cases where both the adiabatic nature and the inverse adiabatic nature coexist to a certain extent, with their relative importance changing according to the $T \leftrightarrow \omega$ competition. The adiabatic nature appears in the self-trapping [381] or the CDW transition [383], because in these cases we have a frozen lattice distortion. On the other hand, the inverse adiabatic nature is the origin of superconductivity [386], because in this case, we do not have frozen lattice distortion, but rather distortions which move together with the electrons. Thus, we can see that the $T \leftrightarrow \omega$ competition is nothing else but a competition between the CDW state and the SC state in many-electron systems.

As mentioned before, one of the most important problems in magnetism, is to clarify the nature of the $T \leftrightarrow U$ competition. Various theoretical studies have already been devoted to cover almost all regions of $T$ and $U$ [387–390]. As for e–ph coupling in many-electron systems, we often regard it as weak enough to be taken into account within perturbation theory, such as in the Migdal approximation. A typical example is the conventional BCS theory [386], wherein $S/T$ is always assumed to be small. Even the so-called strong coupling theory [391] for superconductivity has not been concerned with very strong cases where $S$ is much larger than $T$.

However, according to recent progress in theoretical and experimental studies on transition metal oxides, and low-dimensional metals and semiconductors, it has become clear that there are many kinds of materials where $T, S$ and $U$ are almost of the same order. In traditional insulators, such as alkali halides, rare gas solids and molecular crystals [382], it is already well known that $S$ can become much greater than $T$.

In this chapter, we will review recent theoretical studies on the $S \leftrightarrow T$ competition including, in parts, the $S \leftrightarrow U$ competition. We will start from one- and two-body problems in insulating solids, and in later sections, will proceed to many-body problems, such as CDWs and superconductivity. In any case, we will focus our attention mainly on the nature of the competition between the four elements $T, U, S$ and $\omega$, from a unified point of view.

The last two sections will be devoted to contemporary and advanced problems. One is the nonlinearity of the phonons, and the second is the path integral method for the long range Coulombic repulsion in the context of the linear e–ph interaction.

## 3.1 Single-Body Problems in Strongly Coupled Electron–Phonon Systems

In this section, we will review various studies related to the $T \leftrightarrow S$ competition in single-body problems, such as an electron, a hole or an exciton coupling strongly with phonons. Practically, these problems are concerned with the optical properties of insulating crystals, such as alkali halides, rare gas solids and molecular crystals.

### 3.1.1 Electrons, Phonons and Their Couplings

Generally speaking, the electron–phonon coupled system can be described by the following Hamiltonian ($\equiv H_1$), ($\hbar = 1$)

$$H_1 = \sum_{\boldsymbol{k},\sigma} E(\boldsymbol{k}) \eta^\dagger_{\boldsymbol{k},\sigma} \eta_{\boldsymbol{k},\sigma} - \sum_{\boldsymbol{k},\boldsymbol{q},\sigma} \sqrt{S_{\boldsymbol{q}} \omega_{\boldsymbol{q}} / 2N} (\xi^\dagger_{\boldsymbol{q}} + \xi_{-\boldsymbol{q}}) \eta^\dagger_{\boldsymbol{k}+\boldsymbol{q}/2,\sigma} \eta_{\boldsymbol{k}-\boldsymbol{q}/2,\sigma} + \sum_{\boldsymbol{q}} \omega_{\boldsymbol{q}} \xi^\dagger_{\boldsymbol{q}} \xi_{\boldsymbol{q}} \,. \tag{3.1}$$

Here, $E(\boldsymbol{k})$ is the energy of an electron with wave vector $\boldsymbol{k}$, and $\eta^\dagger_{\boldsymbol{k},\sigma}$ is its creation operator with spin $\sigma (\equiv \alpha, \beta)$. $N$ is the total numher of atoms in the crystal. $S_{\boldsymbol{q}}$ is the coupling energy between an electron and a phonon with wave vector $\boldsymbol{q}$, and with energy $\omega_{\boldsymbol{q}}$. $\xi^\dagger_{\boldsymbol{q}}$ is its creation operator. In the case of a simple cubic crystal, $E(\boldsymbol{k})$ is given as

$$E(\boldsymbol{k}) = -2T \left[\cos(k_x) + \cos(k_y) + \cos(k_z)\right] \,, \tag{3.2}$$

where $T$ is the aforementioned resonance transfer energy between two neighbouring orbitals, and $k_x, k_y$ and $k_z$ are the Cartesian components of $\boldsymbol{k}$. In this case, the full-energy bandwidth is $12T$, and is the measure of the strength of the itineracy. Typical examples for strong itineracy are the $s$ electrons in alkali metals, and conduction electrons in semiconductors such as Si and Ge. In these cases, the width becomes 10 eV or more. A typical example for weak itineracy is the $d$ electron in transition metal compounds, where the width is about 2 eV or so. Another example of weak itineracy is the holes in alkali halides or rare gas solids, where the width is less than 1 eV. In the case of molecular solids, on the other hand, the width is always a few times 0.1 eV.

There are various phonon modes in a crystal, and a typical example is the longitudinal optical mode in ionic solids, where cations and anions move against each other so as to induce a dipole field. In this case, $\omega_{\boldsymbol{q}}$ is finite and independent of $\boldsymbol{q}$ in the region of small $\boldsymbol{q}$: $|\boldsymbol{q}| \ll 1$ (where the lattice constant is the unit of length). The e–ph interaction of this mode is usually called Fröhlich type [392]: $S_{\boldsymbol{q}} \propto |\boldsymbol{q}|^{-2}$ where electrons feel a dipole field of phonons, so that it is a long-range interaction.

Another typical mode is the longitudinal acoustic one with linear dispersion: $\omega_{\boldsymbol{q}} \propto c|\boldsymbol{q}|$, where $c$ is the sound velocity. The e–ph interaction in this case is usually called a deformation-potential type, where electrons feel the change of atomic core potential due to local dilatations induced by phonons. $S_{\boldsymbol{q}}$ in this case is independent of $\boldsymbol{q}$: $S_{\boldsymbol{q}} \to S$ and denotes a short-ranged interaction as mentioned at the beginning of this chapter. In the case of molecular solids, we not only have intermolecular vibrations of the acoustic type, but also intramolecular ones, which have almost no dispersion. Its e–ph interaction, however, is of short range, like the acoustic mode.

The energy of a phonon changes from crystal to crystal, but roughly speaking, it is in the region from 0.01 eV to 0.1 eV. In the case of monovalent inorganic ionic crystals, it is about 0.03 eV, while in the bivalent case it becomes up to 0.1 eV. On the other hand, in the cases of C-C and C-H stretching modes in organic molecular crystals they become about 0.15 eV and 0.35 eV, respectively.

### 3.1.2 Weak Coupling and Large Polarons

Let us look at the nature of an electron which couples weakly with intramolecular phonons, $S \ll T$, since this case is the most simple and standard. Within second order perturbation theory, we get a new energy ($\equiv E'(\boldsymbol{k})$) of an electron with a phonon cloud around itself. This quasiparticle is usually called a polaron, and is given as

$$\begin{aligned} E'(\boldsymbol{k}) &= E(\boldsymbol{k}) + \Sigma(\boldsymbol{k}) , \\ \Sigma(\boldsymbol{k}) &= -\sum_{\boldsymbol{q}} N^{-1} \frac{S\omega}{2[\omega + E(\boldsymbol{k}+\boldsymbol{q}) - E(\boldsymbol{k})]} , \end{aligned} \tag{3.3}$$

where $\Sigma(\boldsymbol{k})$ is the self-energy of an electron due to the phonon cloud. In the small region of $\boldsymbol{k}$, $|\boldsymbol{k}| \ll 1$, we can use the following expansion form for $E(\boldsymbol{k})$,

$$E(\boldsymbol{k}) \to -6T + Tk^2 , \tag{3.4}$$

and the self-energy can also be expanded as

$$\Sigma(\boldsymbol{k}) = -\sum_{\boldsymbol{q}} N^{-1} \frac{S\omega}{2(\omega + Tq^2)} - \sum_{\boldsymbol{q}} N^{-1} \frac{2S\omega(T\boldsymbol{q}\cdot\boldsymbol{k})^2}{(\omega + Tq^2)^3} . \tag{3.5}$$

The first term of $\Sigma(\boldsymbol{k})$ is the so-called rest energy, and gives the energy lowering due to the lattice distortion cloud or the phonon cloud around the electron. The second one is the mass-correction term. The effective mass of the polaron becomes heavier than the bare electron, because it must carry the phonon cloud whenever it moves. This mass enhancement is easily seen from the mass ratio ($\equiv x$) which is given as,

$$x = \left. \frac{\partial^2 E(k)/\partial k^2}{\partial^2 E'(k)/\partial k^2} \right|_{k=0}, \tag{3.6}$$

since this $x$ becomes bigger than unity.

In (3.5), $\boldsymbol{q}$ denotes the wave vector of each phonon mode that makes up the total phonon cloud, and the denominator $(\omega + Tq^2)$ gives the relative importance between the modes with different $\boldsymbol{q}$'s. Thus, we can determine the maximum ($\equiv q_{\max}$) of $|\boldsymbol{q}|$ that can contribute to the phonon cloud as, $q_{\max} \approx (\omega/T)^{1/2}$, and its reciprocal, $q_{\max}^{-1}$, gives the radius of this cloud. It is usually called the polaron radius, and in the case of large $T, T \gg \omega$, this radius becomes much longer than the lattice constant. For this reason, we call this type of polaron a large polaron.

Although the above-mentioned results are only concerned with intramolecular phonons, the main conclusions are essentially the same for any other modes.

### 3.1.3 Strong Coupling, Self-Trapping, Broken Symmetry and Dimensionality

Let us proceed to the strong region of e–ph coupling, $S \gg T$. In the previous section, we have seen that the mass enhancement occurs because of e–ph coupling. Even in that case, however, the lowest state of the system is still a plane wave state, with translational symmetry. On the other hand, in the case of the strong coupling limit, the mass enhancement becomes large enough to make an electron immobile. In other words, the electron is trapped by the self-induced local lattice distortion with no translational symmetry. This is usually called self-trapping. Generally speaking, broken symmetry can occur only in many-body systems. In the strong coupling case, however, multi-quanta of phonons are involved in the electronic process, and this causes broken symmetry even though it is a one-electron system.

In order to study this problem together with the weak coupling case, $S \ll T$, from a unified point of view, we restrict ourselves to the adiabatic limit, $(S, T) \gg \omega$. In this case, as mentioned before, the phonon can be treated as a classical distortion of the lattice, and the following localized representation of $H_1$ becomes more convenient:

$$H_1 = -T \sum_{\langle l,l' \rangle} \sum_{\sigma} \eta_{l\sigma}^{\dagger} \eta_{l'\sigma} - \sqrt{\frac{S\omega}{2}} \sum_{l\sigma} (\xi_l^{\dagger} + \xi_l) \eta_{l\sigma}^{\dagger} \eta_{l\sigma} + \omega \sum_{l} \xi_l^{\dagger} \xi_l , \tag{3.7}$$

$$\xi_l^\dagger \equiv \sum_q N^{-1/2} \xi_q^\dagger \mathrm{e}^{i\boldsymbol{q}\cdot\boldsymbol{l}}\,, \quad \eta_{l\sigma}^\dagger \equiv \sum_{\boldsymbol{k}} N^{-1/2} \eta_{\boldsymbol{k}\sigma}^\dagger \mathrm{e}^{i\boldsymbol{k}\cdot\boldsymbol{l}}\,, \tag{3.8}$$

where $\xi_l$ and $\eta_{l\sigma}$ denote the annihilation operators of the phonon and the electron localized at a lattice site $\boldsymbol{l}$. $\sum\langle \boldsymbol{l}\boldsymbol{l}'\rangle$ in (3.7) means that the summation over $\boldsymbol{l}$ and $\boldsymbol{l}'$ should be taken under the condition that they are nearest neighbours of the simple cubic lattice. Introducing a real coordinate of the phonon ($\equiv Q_l$),

$$Q_l \equiv \sqrt{\frac{\omega}{2S}}(\xi_l^\dagger + \xi_l)\,, \tag{3.9}$$

we can rewrite (3.7) as

$$\begin{aligned} H_1 = &-T \sum_{\langle \boldsymbol{l},\boldsymbol{l}'\rangle} \sum_\sigma \eta_{l\sigma}^\dagger \eta_{l'\sigma} - S \sum_{\boldsymbol{l},\sigma} Q_l \eta_{l\sigma}^\dagger \eta_{l\sigma} \\ &+ S \sum_{\boldsymbol{l}} \frac{Q_l^2}{2} - \frac{\omega^2}{2S} \sum_{\boldsymbol{l}} \frac{\partial^2}{\partial Q_l^2}\,, \end{aligned} \tag{3.10}$$

where the last term is the kinetic energy of the phonon, and can be neglected within the adiabatic approximation. What we have to calculate here is the adiabatic energy of the system, and we can do it using the following simple trial function $|\bar{\beta}\rangle$,

$$|\bar{\beta}\rangle \equiv \tanh^{3/2}\left(\frac{\bar{\beta}}{2}\right) \sum_{\boldsymbol{l}} \exp\left\{-\frac{\bar{\beta}\,(|l_x| + |l_y| + |l_z|)}{2}\right\} \eta_{l\alpha}^\dagger |0\rangle\,. \tag{3.11}$$

Here, $\bar{\beta}$ is a variational parameter that denotes the reciprocal localization radius of an electron, $l_x$, $l_y$ and $l_z$ are the Cartesian components of $\boldsymbol{l}$, and $|0\rangle$ denotes the true electron vacuum. When $\bar{\beta} \to 0$, it gives a wave-like state with translational symmetry, while if $\bar{\beta}$ is finite, it gives a self-trapped (ST) state. What we investigate now is the relative stability between these two states. The expectation value of $H_1$ is given as

$$\begin{aligned} \langle \bar{\beta}|H_1|\bar{\beta}\rangle = &-T \sum_{\langle \boldsymbol{l},\boldsymbol{l}'\rangle} \langle \bar{\beta}|\eta_{l\alpha}^\dagger \eta_{l'\alpha}|\bar{\beta}\rangle \\ &- S \sum_{\boldsymbol{l}} Q_l \langle \bar{\beta}|\eta_{l\alpha}^\dagger \eta_{l\alpha}|\bar{\beta}\rangle + S \sum_{\boldsymbol{l},\sigma} \frac{Q_l^2}{2}\,, \end{aligned} \tag{3.12}$$

and, according to Feynman's theorem,

$$\frac{\partial}{\partial Q_l} \langle \bar{\beta}|H_1|\bar{\beta}\rangle = 0\,, \tag{3.13}$$

we get a balance equation between the charge density and the lattice displacement as

$$\langle \bar{\beta}|\eta_{l\alpha}^\dagger \eta_{l\alpha}|\bar{\beta}\rangle = Q_l\,. \tag{3.14}$$

Using this relation, we can eliminate $Q_l$ from (3.12), and rewrite $\langle\bar{\beta}|H_1|\bar{\beta}\rangle$ in terms of $\bar{\beta}$ only. For example, the first term is easily shown to become $-6T\text{sech}(\bar{\beta}/2)$. According to this simple principle, *Shinozuka* et al. [393] have calculated $\langle\bar{\beta}|H_1|\bar{\beta}\rangle$ as a function of $\bar{\beta}$, and the results are shown in Fig. 3.2 a. From this figure, we can see that when $S/12T < 0.9$, the state with $\bar{\beta} = 0$ is the lowest one. Hereafter we call this state the "free" or F state. On the other hand, when $S/12T > 0.9$, the state with a finite $\bar{\beta}$ becomes more stable than the F state, and this is called the ST state. From Fig. 3.2 a, we can also see that the F and the ST states are separated by an adiabatic potential energy barrier. The main origin of this barrier is the increase of the kinetic energy coming from the first term of (3.12).

*Shinozuka* et al. have also calculated the two and one-dimensional cases, replacing the simple cubic lattice by a square lattice and a chain lattice [393]. These results are shown in Figs. 3.2 b and c. The square lattice is almost the same as that of three dimensions, except that the critical value of $S/8T$ decreases to about 0.5. In the case of the chain crystal, however, we always have the ST state with a finite $\bar{\beta}$. The reason is very simple. In the one-dimensional case, we can easily prove that the binding energy, coming from the second term of (3.12), is proportional to $\bar{\beta}$ when it is small. On the other hand, the increase of the kinetic energy coming from the first term is proportional to $\bar{\beta}^2$. Thus, the lowest state always has a nonzero value of $\bar{\beta}$.

Since these calculations are within the adiabatic limit, the nonadiabatic effect of phonons on these $T \leftrightarrow S$ competitions should also be clarified. The most intuitive way to take this nonadiabatic effect into account together with the recovery of translational symmetry, is to make a Bloch state by using the aforementioned ST state as a Wannier function. This method was used

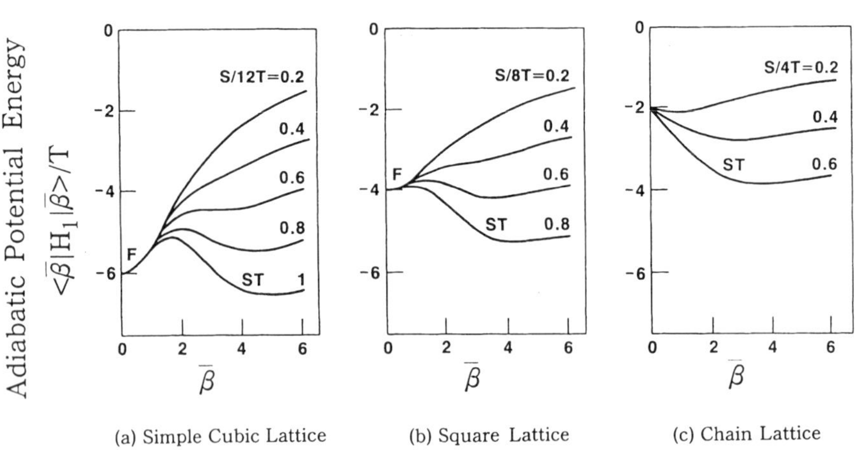


**Fig. 3.2.** The adiabatic potential energy $\langle\bar{\beta}|H_1|\bar{\beta}\rangle$ as a function of $\bar{\beta}$ [393]. (**a**) The three-dimensional cubic lattice, (**b**) the two-dimensional square lattice, and (**c**) the one-dimensional chain lattice. F and ST denote the F and the ST states (With permission from The Physical Society of Japan)

by *Toyozawa* [394] and *Cho* [395]. The other method is to use Feynman's path integral formalism, as done by *Sumi* [396] and *Matsuura* [397]. Their conclusions, however, are essentially the same as that of the adiabatic case. The $x$ defined in (3.6) is found to increase suddenly from 1 to $10^2$ or more, as $S/T$ exceeds a certain critical value, provided $(S,T) \gg \omega$.

Incidentally, the polaron effect due to the Fröhlich type interaction has also been studied in detail. In this case, however, $x$ does not increase suddenly but only gradually as the e–ph coupling increases, because of its long-range nature.

### 3.1.4 Dynamics of Self-Trapping

We have thus seen that the lowest state is clearly divided into two types: the ST state and the F state. This theorem can also be applied not only to electrons, but also holes and excitons created by photons in insulating solids. In the case of excitons, the difference between the two states clearly appears in its light emission spectrum. The emission spectrum from the ST state becomes a very broad Gaussian curve, because the exciton emits multi-quanta of phonons together with a photon, when it is radiatively annihilated. On the other hand, the emission spectrum from the F state becomes a narrow Lorentz curve.

As far as experiments are concerned, a large number of studies have been devoted to measure the emission spectral shape of an exciton in various insulators, and most of them are nicely classified into the aforementioned two kinds. For example, alkali halides, rare gas solids, AgCl, MgO and the molecular crystal of pyrene belong to the ST type, while CdS and AgBr belong to the F type [381,382].

Among them, the ST type emission near the critical situation: $S/12T \geq 0.9$ is very interesting, since in this case we have two locally stable states of the exciton: the F and the ST states, separated by a high potential barrier as shown in Fig. 3.2 a. Immediately after the light excitation, the exciton is in the F state, since according to the Franck–Condon principle, only the electronic part of the system is excited, and the phonon part is still fixed at the ground state. After the lattice relaxation is completed, however, the exciton will go down to the ST state. Thus, during relaxation, at low temperatures it must tunnel through the barrier.

This situation can be easily understood if we assume that the local lattice distortion occurs only in the central site $\boldsymbol{l} = 0$ in (3.12). In this case, $\langle \bar{\beta}|H_1|\bar{\beta}\rangle$ becomes

$$\langle \bar{\beta}|H_1|\bar{\beta}\rangle = -T \sum_{\langle \boldsymbol{l},\boldsymbol{l}'\rangle} \langle \bar{\beta}|\eta^{\dagger}_{\boldsymbol{l}\alpha}\eta_{\boldsymbol{l}'\alpha}|\bar{\beta}\rangle - SQ_0\langle \bar{\beta}|\eta^{\dagger}_{0\alpha}\eta_{0\alpha}|\bar{\beta}\rangle + \frac{SQ_0^2}{2}\,. \tag{3.15}$$

Minimizing the adiabatic energy with respect to $\bar{\beta}$ for each fixed value of $Q_0$, we get an adiabatic potential curve as function of $Q_0$, as shown in

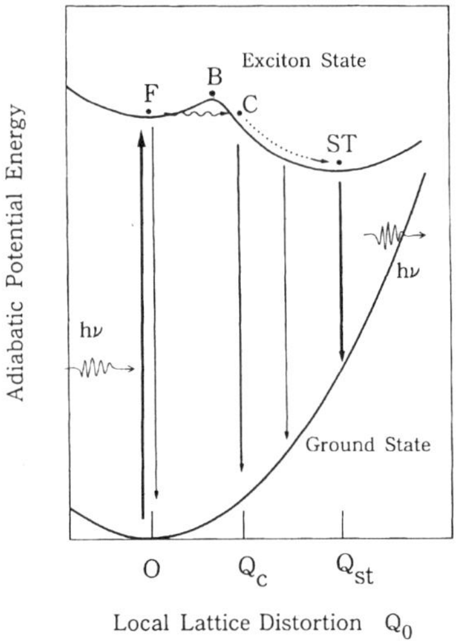


**Fig. 3.3.** The adiabatic potential energy $\langle\bar{\beta}|H_1|\bar{\beta}\rangle$ as functions of $Q_0$. F (at $Q_0 = 0$) is the F state, ST (at $Q_0 = Q_{st}$) is the ST state, B is the top of the barrier, and C (at $Q_0 = Q_C$) is the exit of the tunnel. $h\nu$ and $h\nu'$ denote the exciting and the emitted photons

Fig. 3.3. In this figure, the ground state means the state with no exciton and its adiabatic energy comes only from the elastic energy of the lattice; from the last term of (3.15). On the potential surface of the exciton state, we have two minima: the F and the ST states. The F state is a local minimum at $Q_0 = 0$, while the ST state is a global minimum at $Q_0 = Q_{st}$. These two states are separated by a barrier denoted by B. When this system is excited by light, the transition occurs without changing $Q_0$ according to the Franck-Condon principle mentioned above. Thus, the final state of the optical transition is the F state. However, during the relaxation, it must tunnel the barrier ($F \to B \to C$). After this tunneling, the exciton will slide down rapidly along the potential suface, until it reaches the bottom of the ST state ($C \to ST$), just as if it were trapped in an ant lion pit.

This double-well structure of the potential suface results in the coexistence of the F and ST emissions. The F emission has almost no Stokes shift, while the ST emission has a large Stokes shift as seen from Fig. 3.3. Moreover, we will be able to observe a transient emission during the relaxation, mainly from around the exit of the tunnel ($Q_0 = Q_C$). Such a coexistence of F and ST components was first observed in alkali iodides by *Kuusmann* et al. [398]. Recently, however, *Matsui* et al. [399] have succeeded in observing the time-resolved emisson from the molecular crystal of pyrene, as shown in Fig. 3.4. In this figure, the time written at the right hand side of the spectrum is the time difference between excitation and detection, and the two arrows correspond to the F and ST emissions. According to this result, the F emission rises up

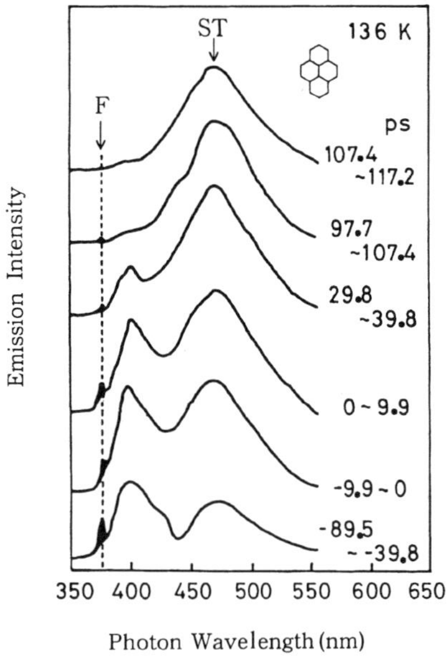


**Fig. 3.4.** The time resolved emission spectra of the molecular crystal of pyrene [399]. The time written at the right hand side of the spectrum is the time difference between the excitation and the detection, and the two *arrows* correspond to the F and ST emissions. (With permission from Elsevier)

quickly and decays within about 100 ps, and after that, the ST emission rises slowly and gradually becomes predominant. In this figure we can also see an emission coming from the exit region, which is between the F region and the ST region. Thus, the dynamic process of self-trapping has been observed experimentally.

In connection with this result, *Nasu* et al. [400], and *Rashba* [401] have undertaken theoretical studies for the rate of this tunneling. *Nasu* has derived a new theory, wherein the dynamics of the shrinkage of the exciton wave function from F to ST is calculated, including the nonadiabatic breakage of the translational symmetry.

## 3.2 Two-Body Problems in Strongly Coupled Electron–Phonon Systems

### 3.2.1 Bipolarons

Let us proceed to two-electron systems which couple strongly with phonons in crystals. As originally proposed by *Vinetskii* [402], the phonon mediated attraction can overcome the interelectron Coulombic repulsion, provided $S$ is sufficiently strong. In this case, two polarons can make a bound state in the space of their relative motion. This is called a bipolaron. When the two constituent electrons have opposite spins to each other, this quasiparticle is usually called a singlet bipolaron. To study these problems theoretically, let us introduce a Hamiltonian ($\equiv H_2$) which has both the e–ph coupling and the interelectron Coulombic repulsion:

$$H_2 = -T \sum_{\langle l,l' \rangle} \sum_{\sigma} \eta_{l\sigma}^{\dagger} \eta_{l'\sigma} - \sqrt{\frac{S\omega}{2}} \sum_{l\sigma} (\xi_l^{\dagger} + \xi_l) n_{l\sigma}$$
$$+ \omega \sum_{l} \xi_l^{\dagger} \xi_l + U \sum_{l} n_{l\alpha} n_{l\beta} , \tag{3.16}$$

where $n_{l\sigma} = \eta_{l\sigma}^{\dagger}\eta_{l\sigma}$, and $U$ is the intra-orbital (or intra-site) Coulombic repulsion mentioned above.

In the strong coupling, $S \gg T$, and adiabatic limits, $S \gg \omega$, we can make use of the form given in (3.10), and can easily calculate the total energy of two electrons localized on the same site. It becomes $-(2S - U)$, while the total energy of a distant pair of polarons is $-S$. Thus, the condition for singlet bipolaron formation is just given as

$$S > U . \tag{3.17}$$

If a large number of such singlet bipolarons are present in a crystal, their center of mass motion can contribute to electrical conductivity, while it does not contribute to the magnetic susceptibility. Hence, these two basic macroscopic quantities become uncorrelated. In the case of unpaired polarons, these two basic macroscopic quantities always have some correlation with each other, since the charge and spin are not separated. On the contrary, when the resistivity changes independently of the susceptibility or of the other magnetic properties, it is a strong indication of the presence of singlet bipolarons.

Experimental studies on $Ti_4O_7$ by *Schlenker* [403] are mainly based on this point of view. The resistivity and the susceptibility of this crystal are shown in Fig. 3.5, as a function of temperature ($\equiv T_{\text{tem}}$). The conduction electrons of this crystal are in the $3d$ orbitals of $Ti^{3+}$, and at high temperatures ($T_{\text{tem}} > 150\,\text{K}$), it is a paramagnetic metallic state. At low temperatures ($T_{\text{tem}} < 130\,\text{K}$), on the other hand, it is a kind of CDW state, wherein neighbouring two $Ti^{3+}$'s make a pair, and this pair spatially orders. The presence of this inter-pair order can be confirmed by X-ray structure analysis. What is interesting is the electronic state realized in the intermediate phase ($130\,\text{K} < T_{\text{tem}} < 150\,\text{K}$). As seen clearly from Fig. 3.5, the sudden increase of the resistivity at $T_{\text{tem}} \approx 130\,\text{K}$ occurs without any change in susceptibility, as the temperature is lowered. This is quite different from the transition at 150 K, where both quantities change suddenly. From these results, the intermediate phase is confirmed to be a bipolaron phase, wherein two $Ti^{3+}$s make a pair without any spatial order among the pairs. Since the spin has already been removed by this pairing below 150 K, the transition at 130 K gives no change in the susceptibility.

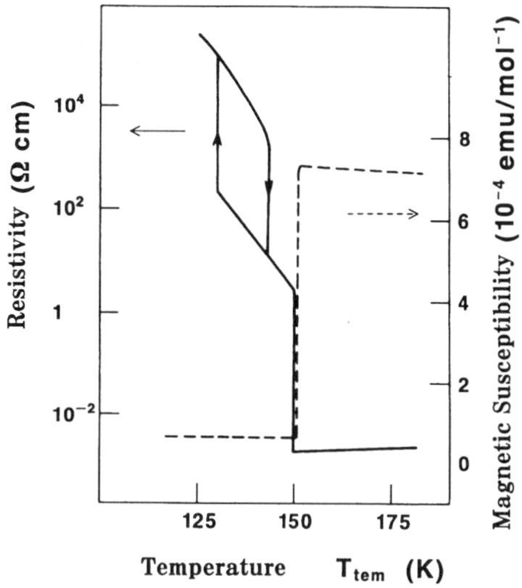


**Fig. 3.5.** The resistivity and the magnetic susceptibility of $Ti_4O_7$ as functions of the temperature [403]. The *solid line* is the resistivity, and the *dashed line* is the magnetic susceptibility. (With permission from Plenum)

In $Na_{0.33}V_2O_5$, *Onoda* [404] has also found the presence of a singlet bipolaron phase with no long range order. He has also found the presence of a collective motion between bipolarons which enhances the conductivity. In conducting polymers such as polypyrrole, *Scott* [405] has also found the presence of singlet bipolarons. The polypyrrole itself is an insulator, but on doping it becomes conductive. What is interesting is the change of the conductivity and that of the ESR signal intensity as a function of the dopant concentration. When the concentration is small both these quantities increase. However, as the concentration increases further, the intensity of the ESR signal begins to decrease, while the conductivity still continues to increase. These results mean that singlet bipolarons are formed in the high doping region.

### 3.2.2 Charge Separation of Self-Trapped Exciton

Let us proceed to the other two-body system composed of a conduction-band electron and a valence-band hole both coupling strongly with phonons in an insulator. This is an excited state created by a photon. In this case, the Coulombic interaction between these two particles is always attractive. As for the phonon mediated interaction, however, there will be various cases as indicated by *Sumi* [406]. We get an attraction, if both the electron and the hole equally expand (or contract) the surrounding crystal lattice through their coupling with phonons. However, we get a repulsion, if the electron expands (or contracts) the surrounding crystal lattice, while the hole contracts (or expands) it. The later case is most interesting since, contrary to the bipolaron, the Coulombic attraction and the phonon mediated repulsion compete with each other. To study this competition theoretically, let us take the following Hamiltonian ($\equiv H_2'$),

$$
\begin{aligned}
H_2' = & -T \sum_{\langle l,l' \rangle} \sum_{\sigma} \eta_{l\sigma}^{+} \eta_{l'\sigma} - T' \sum_{\langle l,l' \rangle} \sum_{\sigma} \zeta_{l\sigma}^{+} \zeta_{l'\sigma} \\
& - \sqrt{\frac{\omega}{2}} \sum_{l,\sigma} (\xi_l^{+} + \xi_l) \left( \sqrt{S} \eta_{l\sigma}^{+} \eta_{l\sigma} - \sqrt{S'} \zeta_{l\sigma}^{+} \zeta_{l\sigma} \right) + \omega \sum_{l} \xi_l^{+} \xi_l \\
& + \sum_{l,l'} \sum_{\sigma\sigma'} W(l - l', \sigma, \sigma') \eta_{l\sigma}^{+} \eta_{l\sigma} \zeta_{l'\sigma'}^{+} \zeta_{l'\sigma'} \, .
\end{aligned}
\tag{3.18}
$$

Here, $T'$ denotes the transfer energy of the hole, and $\zeta_{l\sigma}^{+}$ is its creation operator, while $S'$ is its coupling constant. $W(l - l', \sigma, \sigma')$ denotes both the Coulombic attraction and the exchange repulsion between the electron and the hole. The Coulombic attraction, which is independent of spins $\sigma$ and $\sigma'$, is given as

$$
W(l - l') \equiv \begin{cases} \frac{-e^2}{\epsilon |l - l'|}, & l \neq l' \\ \frac{-e^2}{0.5\epsilon}, & l = l' \end{cases} , \quad \epsilon \rightarrow \text{dielectric constant} \, , \tag{3.19}
$$

while the exchange repulsion is assumed to be $J$ and acts only for the singlet electron–hole pair at $l = l'$. In (3.18), we can see the signs of the two coupling constants are opposite to each other. Consequently, the direction of the lattice distortion which stabilizes the electron is opposite to that of the hole, as mentioned before.

Using this $H_2'$, *Nagasaka* and *Nasu* [407] have calculated the adiabatic potential energies of the singlet and the triplet ST excitons, taking two localized variational wave functions centered at $\boldsymbol{r}_{\mathrm{e}}$ and $\boldsymbol{r}_{\mathrm{h}}$ for the electron and the hole, respectively. Their results are shown in Fig. 3.6, and we can see two minima; the point O and the point C. The point O is the ordinary ST exciton wherein the hole and the electron are localized at the same site, as shown schematically in the lower inset of Fig. 3.6. The direction of the lattice distortion in this case is exclusively determined only by the strongest coupling among S and S′, while the point C is the charge separated ST exciton wherein the hole and the electron are distant from each other, and only loosely trapped by the Coulombic attraction. The direction of the lattice distortion around the electron is opposite to that around the hole, and hence it has no parity, as schematically shown in the upper inset of Fig. 3.6.

This charge separation of the ST exciton is closely related to the ST excitons in alkali halides. According to *Song* [408] and *Kanno* [409], the ST excitons in alkali halides are not the ordinary type but the charge separated type. In alkali halides, the hole is in the $\sigma$-bond of the $p$ orbitals of the halogen ions, and it stabilizes when the halogen sublattice around the hole contracts. The electron on the other hand wants to expand the surrounding

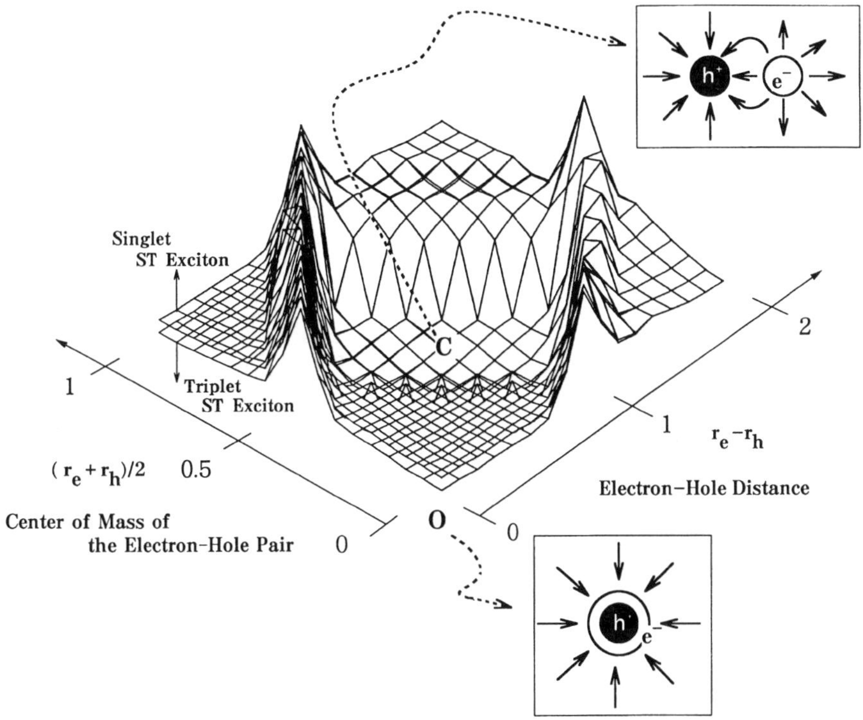


**Fig. 3.6.** The adiabatic potential surfaces of the singlet and the triplet ST excitons [407]. The lattice is simple cubic, and the electron and the hole centered at $\boldsymbol{r}_{\mathrm{e}}$ and $\boldsymbol{r}_{\mathrm{h}}$ can move only along its [100] direction. $12T = 4\,\mathrm{eV}$, $12T' = 2\,\mathrm{eV}$, $S' = 4\,\mathrm{eV}$, $S = 3\,\mathrm{eV}$, $J = 0.5\,\mathrm{eV}$, $\epsilon = 2.19$, and the lattice constant is 3.147 Å. The *arrows* in the two *insets* show the directions of the lattice distortion

halogen sublattice, since it repels the negative ions electrostatically. Thus charge separation occurs. For this problem, various experimental studies are being carried out further by *Tanimura* [410]. *Nagasaka* and *Nasu* [411] have recently derived a new theory which treats the aforementioned characteristics of the alkali halides in detail.

## 3.3 Excitons and Solitons in One-Dimensional Charge Density Wave States

Let us proceed to the problem of quasi one-dimensional many-electron systems, and see the nature of the competition between $T$, $S$ and $U$. These problems are closely related to low-dimensional materials. Various kinds of new chain-like crystals have been designed and synthesized from organic molecules, hydrocarbons and transition metal compounds, in the past thirty years. As far as the electronic properties are concerned, the simplest case among them is such that the constituent atoms or molecules of these chain crystals have one unpaired electron in their outermost orbitals. As mentioned above, the outermost orbitals can make a one-dimensional energy band via $T$, and the lower half of this band is equally occupied by up and down spin electrons. This state is usually called a Pauli paramagnetic metal.

As is well known, however, the one-dimensional metallic state at the absolute zero temperature is always unstable against both e–ph coupling and Coulombic repulsion. In the case of the e–ph coupling alone, Peierls type lattice distortion with a double period occurs, raising or lowering the orbital energy alternately along the chain, and opening up an energy gap at the Fermi level [383]. Since a pair of electrons with opposite spins tends to occupy a lower energy orbital (or lattice site), the metal becomes a charge transfer type insulator, which has a CDW with a double period, as shown in Fig. 3.7 a. In the case of the latter interaction alone, on the other hand, the electrons with opposite spins tend to occupy the lattice sites alternately along the chain, so as to reduce the repulsive energy, resulting in the opening up of an energy gap at the Fermi level of the aforementioned energyband. Thus, the metal becomes an antiferromagnetic insulator [384], which has a SDW with the double period of the original crystal lattice, as shown in Fig. 3.7 b.

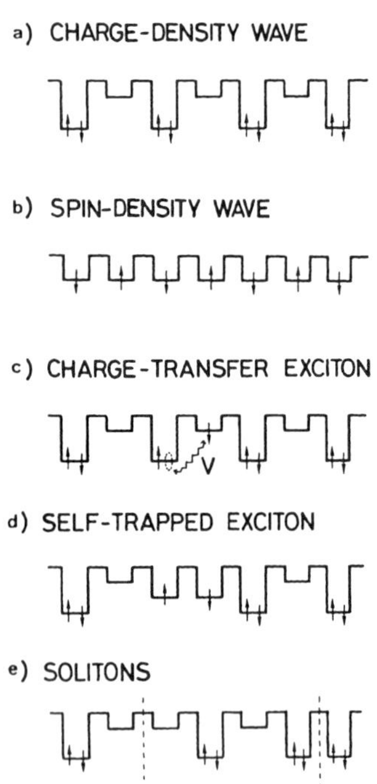


**Fig. 3.7.** Schematic potential energy for the $d_{z^2}$ electrons. (**a**) CDW state, (**b**) SDW state, (**c**) CT exciton, (**d**) STE, (**e**) solitons. *Small arrows* correspond to the electron with up or down spin

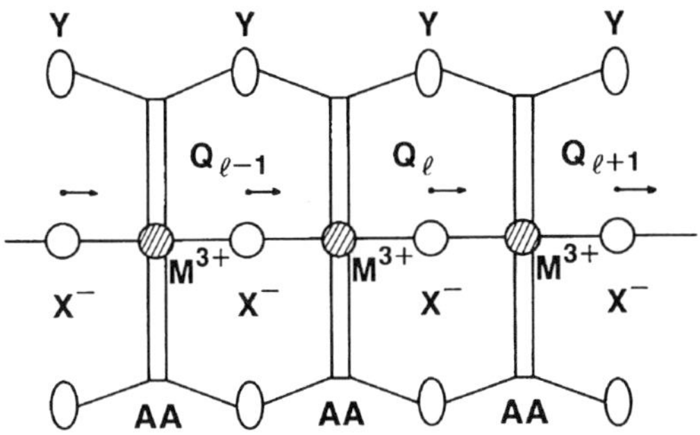


**Fig. 3.8.** The schematic structure of HMMC. $M$ = Pt, Pd, or Ni. $X$ = Cl, Br or I. AA and Y denote a ligand and a counter ion, respectively

Thus the nature of the ground state of a one-dimensional many-electron system is already well known, insofar as only one of the two interactions exists. In real systems, however, these two interactions always coexist, and their relative importance varies from one material to another depending on the constituents, the bonds and the environments. For this reason, we will see in this section, firstly, the nature of the competition between $T$, $U$ and $S$ in the ground state. Secondly, we will see the nature of nonlinear excited states such as excitons and solitons in the CDW state, since they are characteristics of the system with strong e–ph coupling.

One of the typical examples of a quasi one-dimensional CDW is the halogen bridged mixed valence metal complex (HMMC) [412]. It is composed of transition metal ions $M^{3+}$ ($= \mathrm{Pt}^{3+}$, $\mathrm{Pd}^{3+}$, $\mathrm{Ni}^{3+}$) bridged by halogen ions $X^-$ ($= \mathrm{Cl}^-$, $\mathrm{Br}^-$, $\mathrm{I}^-$) as shown schematically in Fig. 3.8. $M^{3+}$ has an unpaired electron in its $d_{z^2}$ orbital, ($z$ is parallel to the chain), and this orbital makes an energy band through the hybridization between the $d_{z^2}$ orbital and the $p_z$ orbital of $X^-$. Thus, this system is expected to be a metal with a half-filled energy band of electrons. However, the displacement of $X^-$ modulates the energy levels of $d_{z^2}$, since $X^-$ has a negative charge, and causes a strong coupling between the electrons and this type of phonons. Because of this e–ph coupling, the aforementioned metallic state becomes unstable, and the charge transfer occurs between two neighbouring $M$s so as to give the following mixed-valence state:

$$M^{(3-\delta)} --- X^- - M^{(3+\delta)} - X^- --- M^{(3-\delta)} --- X^- -$$

$$M^{(3+\delta)} - X^- --- M^{(3-\delta)} --- X^- - M^{(3+\delta)} - X^- ---$$

$$0 \leq \delta \leq 1\,,$$

wherein the $X^-$ is displaced towards the $M^{(3+\delta)}$ in order to stabilize the charge transfer. $\delta$ denotes the degree of this charge transfer. This is nothing but a CDW state with twice the period of the original lattice, and the metal becomes an insulator. As a matter of fact, this is a broken symmetry, and our ground state is doubly degenerate, according to the two possibilities for the charge transfer, as written above.

In the real structure of HMMC, the metallic atom $M$ is coordinated by a ligand ($\equiv AA$) as shown schematically in Fig. 3.8, where $AA$ is a planar organic molecule such as ethylenediamine, ethylamine or cyclohexanediamine. We also have a counter ion ($\equiv Y$) between the chains, where $Y$ is $ClO^{4-}$ or $BF^{4-}$. These counter ions contribute to stabilize the chain structure of this material, and also cause weak inter-chain interactions. At present, we have various kinds of crystals with this structure, according to various combinations of $M, X, AA$ and $Y$. The total number of crystals is about 200 [412–416].

In this CDW state, the displacement of $X^-$ raises and lowers the energy levels of the $d_{z^2}$ orbitals alternately along the chain, just as shown in Fig. 3.7 a, and two electrons with opposite spins tend to occupy the lower energy site (for example, the odd-number site). Most crystals belonging to this HMMC series have the above-mentioned CDW structure. Very recently, however, *Torium* [417] has succeeded in synthesizing a crystal using Ni and Br, and found that the Br atom is exactly in the middle of the two Ni atoms. This is not a CDW state, but a SDW state (Fig. 3.7 b), and its appearance is due to the large $U$ of the $3d_{z^2}$ orbital of Ni [418], which excludes double occupancy. Thus, this family is one of the best candidates to study the competition between $T, U$ and $S$, and, in this section, we will be mainly concerned with it.

Another typical example of an insulator with a Peierls lattice distortion is polyacetylene. However, we will not be concerned with this problem, since many studies on it have already been published [419, 420].

### 3.3.1 Phase Diagram of the Ground State

As a model for HMMC, *Nasu* [385, 421] has studied the phase diagram of the one-dimensional many-electron system with the Hamiltonian $H_2$ given by (3.16). In the present case, $T$ corresponds to the effective resonance transfer energy between neighbouring two $d_{z^2}$ orbitals, $l$ corresponds to the lattice sites of $M$ in this chain crystal, and the phonon coordinate corresponds to the breathing motion of $X^-$. $S$ corresponds to the e–ph coupling of this mode, and $U$ corresponds to the intra-orbital Coulombic repulsion of $d_{z^2}$. Within the mean field theory, we can replace the Hubbard term by its average:

$$U \sum_l n_{l\alpha} n_{l\beta} \to U \sum_l \{ n_{l\alpha} \langle n_{l\beta} \rangle + n_{l\beta} \langle n_{l\alpha} \rangle - \langle n_{l\alpha} \rangle \langle n_{l\beta} \rangle \} , \tag{3.20}$$

where $\langle n_{l\sigma} \rangle$ is the average of $n_{l\sigma}$, and we assume for this average that

$$\langle n_{l\sigma} \rangle = (-1)^l \delta n_\sigma + \frac{1}{2} . \tag{3.21}$$

Here $\delta n_\sigma$ is the amplitude of the density wave, which we should determine self-consistently. When $\delta n_\alpha = \delta n_\beta$, it is a CDW, whereas when $\delta n_\alpha = -\delta n_\beta$,

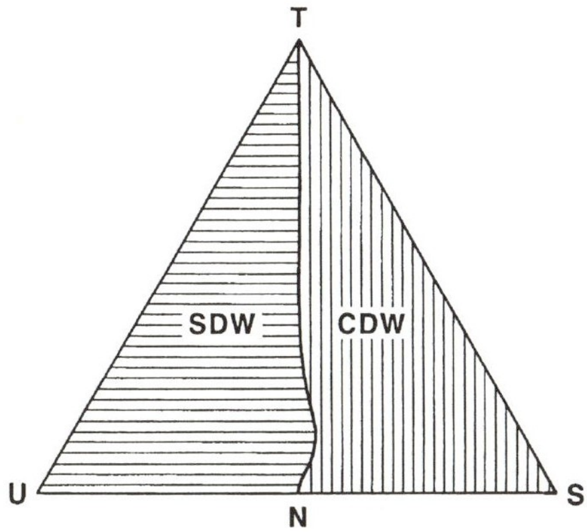


**Fig. 3.9.** The phase diagram on the $T$–$U$–$S$ triangle [421]. The ratios of the three parameters are denoted by the lengths of the perpendiculars from a given point to three axes (With permission from The Physical Society of Japan)

it is an SDW. As for the phonons, we can also assume within the adiabatic approximation that

$$Q_l = (-1)^l Q - 1\,, \tag{3.22}$$

where $Q$ is the amplitude of the Peierls distortion, and 1 denotes the uniform distortion. The resultant phase diagram is shown in Fig. 3.9. It is a triangular phase diagram spanned by $T$, $U$ and $S$, and it is shown to be divided into two parts: SDWs and CDWs, by the line $U = S$. The metallic state appears only at the $T$ vertex.This phase diagram is almost the same as that for the bipolaron case. When $T = 0$, the energy of the two electrons at neighbouring sites is $-S$, while the energy of two electrons at the same site is $-2\,S{+}U$. Thus the boundary in the small $T$ region, $(U, S) \gg T$, is given by $U = S$. According to the proof by *Nasu* et al. [421], however, the boundary is unaltered even in the large $T$ region, if we restrict ourselves to mean field theory. Thus, in the left half of the phase diagram, we have SDWs with $\delta n_\alpha = -\delta n_\beta$ and $Q = 0$, while in the right half, we have CDWs with $\delta n_\alpha = \delta n_\beta$ and $Q \neq 0$.

*Nasu* [385, 421] has also reinforced this mean field result, by including the quantum fluctuation of electrons within the random phase approximation (RPA), leaving the state of phonons still within the adiabatic approximation. This result shows that the SDW region erodes the CDW region given by the mean field theory (Fig. 3.9). This change in the ground state is mainly due to the difference of the lowest excited state above the mean field theory ground state. In the CDW, the lowest Franck–Condon excitation is the charge transfer exciton where one of the electrons in a lower energy site is transferred to a neighbouring high energy site (Fig. 3.7 c). Its energy is almost the same as the energy gap of the CDW. On the other hand, the lowest excitation in the SDW is the magnon, which is just above the ground state, since it has no charge transfer. The ground-state energy gain due to the virtual excitation of these states results in the erosion mentioned above. *Takimoto* et al. [422] have also calculated the phase diagram numerically using a finite ring. This result was almost the same as that of *Nasu* et al. [385, 421].

### 3.3.2 Nonlinear Lattice Relaxation and Proliferations of Excitons in One-Dimensional CDW

The problems related with the nonlinear lattice relaxation process of an exciton in CDW states have been studied very extensively in recent years, from both theoretical and experimental points of view.

In general, the exciton created by a photon in an insulator interacts with the lattice vibrations (or phonons), and this interaction brings about various lattice relaxation processes. In the case of an ordinary insulator, however, the total number of excitons is kept unchanged during this relaxation, except when excitons are created up to a very high density by intense laser lights. Thus, the relaxation of low density excitons in ordinary insulators can be called a "linear lattice relaxation".

On the other hand, the CDW state is a special insulator that is brought about from the metallic state through the Peierls transition. Its ground state therefore has a degeneracy or a multistability. In this multistable situation, an exciton created by a photon from one of the ground states, proliferates during the lattice relaxation process, and finally relaxes down to a collective excited state which encompasses more than two ground states. The word "nonlinear relaxation" is introduced to emphasize this characteristic, in comparison with ordinary insulators.

The purpose of the present section is to study this nonlinearity by taking one-dimensional CDW states as typical examples.

In Fig. 3.10, we have shown the relevant adiabatic energies quite schematically as a function of an order parameter ($\equiv Q$), which denotes the displace-

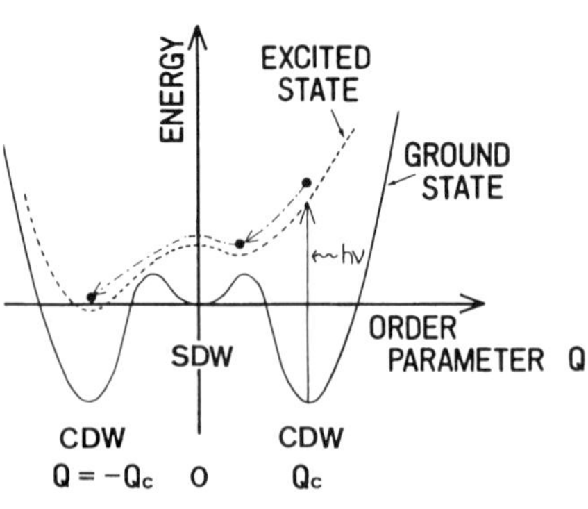


**Fig. 3.10.** Schematical potential energy surfaces of ground and excited state as a function of the CDW-type order parameter $Q$

ment of $X^-$ from the middle point of the adjacent two $M$s. It is nothing but the Peierls type lattice distortion. The solid line denotes the ground state, and the dashed line denotes the excited one. The meaning of the abscissa $Q$ in the case of the excited state is rather conceptional, expressing the local lattice distortion around the exciton (Fig. 3.7 c, d).

As shown in Fig. 3.10, there are three energy minimum points in the ground state. Two minima where $Q$ is equal to $Q_c$ or $-Q_c$ correspond to the CDW states. They have different phases of the Peierls distortion of $X^-$, as shown before. When the electron–electron Coulombic repulsion is taken into account, we get another energy minimum at $Q = 0$, in addition to these CDW states. In this configuration, the up and the down spins are arranged alternately along the metallic sites $M$s to make this repulsion weak, as shown in Fig. 3.7 b. This is the SDW, and the relative stabiltity between this state and the aforementioned CDW one is determined by the competition between the e–ph interaction and the Coulombic repulsion [385]. In the case of the HMMCs, it has been confirmed by various experiments that the ground state is the CDW when $M$ is Pt or Pd, while it is the SDW when $M$ is Ni [418].

Referring to Figs. 3.7 and 3.10, let us briefly discuss the nonlinear lattice relaxation processes of excitons.

The HMMC has a strong light absorption band in the visible region as shown in Fig. 3.11, which corresponds to the charge-transfer (CT) excitation of an electron from the occupied $d_{z^2}$ orbital to the vacant ones, as illustrated in Fig. 3.7 c [423, 424]. Thus, an electron and a hole are created, and they attract with each other through the inter-orbital Coulombic force. Such a state is usually called a CT exciton [424]. In Fig. 3.10, this type photoexcitation from one of the CDW ground states is represented by a vertical arrow. *Iwano* and *Nasu* [424] have calculated the spectral shape of this transition, using the extended Peierls–Hubbard model which will be explained later in detail. Their method is the mean field theory reinforced by the single configuration interaction (CI), which is enough to describe the exciton effect. As for the lattice fluctuation, they have used the classical Monte Carlo method. Once this exciton has been created, it will relax down through the potential curve of the excited state, and induces a local lattice distortion around it.

This photoexcitation is nothing but a backward charge transfer, and the electron number per orbital becomes almost equal in this local region (Fig. 3.7 c), and the Peierls distortion has therefore lost the reason for its presence. Hence, it tends to disappear, as seen from Fig. 3.7 d. Thus, the exciton self-induces a local lattice distortion, and it is trapped in it. This is usually called a self-trapped exciton (STE). However, since this STE appears near the point with $Q = 0$ as shown in Fig. 3.10, the SDW type spin order is also expected to appear in this localized state. In that sense, the STE appearing here has a strong collective nature, being quite different from the one in the ordinary insulators.

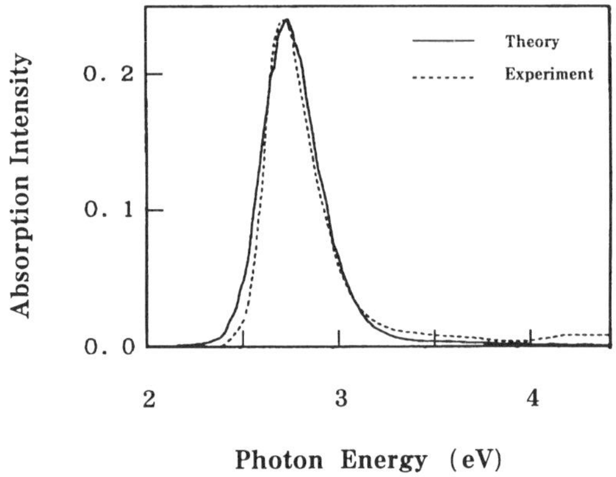


**Fig. 3.11.** The light absorption spectra of the charge-transfer exciton in HMMC [423, 424]. The *dashed line* corresponds to experiment, $M$ = Pt and $X$ = Cl. The *solid line* is a theoretical calculation from (3.32) (With permission from The Physical Society of Japan)

As seen from Fig. 3.10, the STE is expected to undergo further relaxations. In the CDW, as is well known, we have a low-lying excited state with a collective nature. That is, one phase of the ground state can appear locally in the other phase of the ground state, at the expense of creating boundaries between two phases.

This boundary is usually called a soliton, and the exciton is expected finally to relax down to the state with a pair of solitons, as shown in Fig. 3.7 e. In contrast to the initial single charge transfer occurring from the CDW ground state, a large number of charges have now been transferred to make this pair. That is, the exciton in the CDW proliferates through the lattice relaxation process, and it finally relaxes down to the collective excited state, in which a great deal of excitons or electron–hole pairs have condensed. This is nothing but the nonlinear lattice relaxation.

### 3.3.3 One-Dimensional Extended Peierls–Hubbard Model

In order to clarify the aforementioned nonlinearity in more detail, let us study the following quasi one-dimensional extended Peierls–Hubbard model ($\equiv H$),

$$H = -T \sum_{l,l',\sigma} \left( \eta_{l\sigma}^{\dagger} \eta_{l'\sigma} + \text{h.c.} \right) + S \sum_{l} \frac{Q_l^2}{2}$$

$$+ S \sum_{l,l',\sigma} (Q_l - Q_{l'}) \, n_{l\sigma} + U \sum_{l} n_{l\alpha} n_{l\beta} + V \sum_{l,l',\sigma} n_{l\sigma} n_{l'\sigma'} \,, \tag{3.23}$$

where $l'$ denotes a nearest neighboring site of $l$ shown in Fig. 3.8, and $V$ denotes the inter-site Coulombic repulsive energy.

Let us now explain our method to calculate the ground and excited states. In the case of the ground state, we use mean field theory for interelectron interactions, and the adiabatic approximation for phonons. Within mean field theory, we can approximate $H$ by the mean field Hamiltonian ($\equiv H_{\mathrm{HF}}$), in which $n_{l\sigma}$ and $(\eta^+_{l+1\sigma}\eta_{l\sigma})$ are replaced by their averages so that

$$n_{l\sigma} \to \langle n_{l\sigma}\rangle\,, \quad (\eta^+_{l+1\sigma}\eta_{l\sigma}) \to \langle \eta^+_{l+1\sigma}\eta_{l\sigma}\rangle\,. \tag{3.24}$$

These $\langle n_{l\sigma}\rangle$ and $\langle \eta^+_{l+1\sigma}\eta_{l\sigma}\rangle$ are unknown parameters to be determined self-consistently at a later stage. By diagonalizing this $H_{\mathrm{HF}}$, we can obtain the energies of the ground and excited states. In order to take the electron–hole correlation into account in the excited state, we define the difference ($\equiv \Delta H$) between the mean field Hamiltonian $H_{\mathrm{HF}}$ and the true one $H$ by

$$\Delta H = H - H_{\mathrm{HF}}\,, \tag{3.25}$$

and diagonalize it within the basis of the one-electron excited states obtained from $H_{\mathrm{HF}}$. This is first-order perturbation theory for the electron–hole correlation. Then, we can determine the new energies of excited states and their wave functions, including the exciton effects. By working this process out for various lattice configurations, we can obtain adiabatic potential energy surfaces.

Our system is highly nonlinear in the sense that the electrons and phonons are strongly correlated with each other. In order to study this nonlinearity from a quantitative point of view, let us now define an index which expresses the degree of nonlinearity of relaxation. We enumerate the number of electrons ($\equiv N_{\mathrm{e}}$) forced to leave the Hilbert space of the CDW ground state by the optical excitation and subsequent lattice distortions. It is written as

$$N_{\mathrm{e}} = \sum_{j=1}^{N/2} \langle \phi_j | \bar{N}_{\mathrm{e}} | \phi_j \rangle\,, \tag{3.26}$$

where the operator $\bar{N}_{\mathrm{e}}$ is defined as

$$\bar{N}_{\mathrm{e}} \equiv \sum_{i=N/2+1}^{N} |\phi_i\rangle\langle\phi_i|\,. \tag{3.27}$$

Here, $N$ is the total number of electrons in the system, $|\phi_i\rangle$ is the $i$th one-electron wavefunction of the CDW ground state, and $|\phi_j\rangle$ is the $j$th wavefunction of the relaxed excited state. The $j$ summation is taken over the occupied states, while $i$ is taken over the unoccupied states. Thus, $N_{\mathrm{e}}$ gives the total number of excited electrons.

Although $N_{\mathrm{e}}$ is almost unity in the Franck–Condon state, it will increase as the lattice relaxation proceeds. This increase is nothing but the nonlinearity of the relaxation. We can also schematically examine the variation of $N_{\mathrm{e}}$

from Fig. 3.7, as follows. $N_e$ is equal to unity in the case of the CT exciton as shown in Fig. 3.7 c. In the case of a soliton pair described in Fig. 3.7 e, however, there are four electrons transferred from the CDW ground state. That is, $N_e$ increases from unity to four as a result of the lattice relaxation. Thus, $N_e$ is the pertinent index that can properly express the degree of nonlinearity of the relaxation.

### 3.3.4 Relaxation in One-Dimensional CDW

Let us now consider the nonlinear lattice relaxation of excitons in one-dimensional CDW. To describe the relaxation path from the free exciton to the soliton pair through the STE, we use the following variational function [425, 426]:

$$Q_l = (-1)^l Q \left\{ 1 + \Delta Q \left[ \tanh \theta \left( |l| - \frac{l_0}{2} \right) - 1 \right] \right\} , \tag{3.28}$$

where $(-1)^l Q$ denotes the Peierls distortion in the CDW ground state, and this $Q$ should be determined beforehand within mean field theory. The curly bracket {...} denotes the local lattice displacement from this ground state. $\Delta Q$ is its amplitude and $\theta$ corresponds to the reciprocal width of a soliton. $l_0$ denotes the inter-soliton distance. When $l_0 = 0$, this pattern just corresponds to the STE-type local lattice distortion. On the other hand, when $l_0 \gg 1$ and $\Delta Q = 1$, the phase of the Peierls distortion is completely inverted in the region $-l_0/2 < l < l_0/2$. This situation corresponds to the soliton–antisoliton pair.

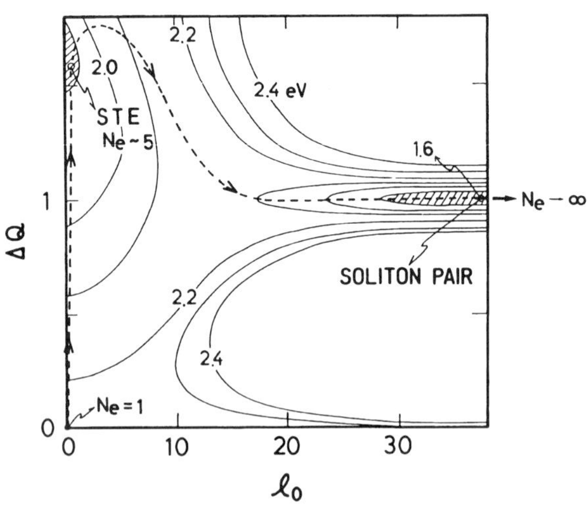


**Fig. 3.12.** The contour map of the adiabatic potential energy surface of an exciton for one-dimensional CDW [426]. $U$ = 1.43 eV, $V$ = 0.79 eV, $S$ = 0.29 eV and $T$ = 1 eV (With permission from The American Physical Society)

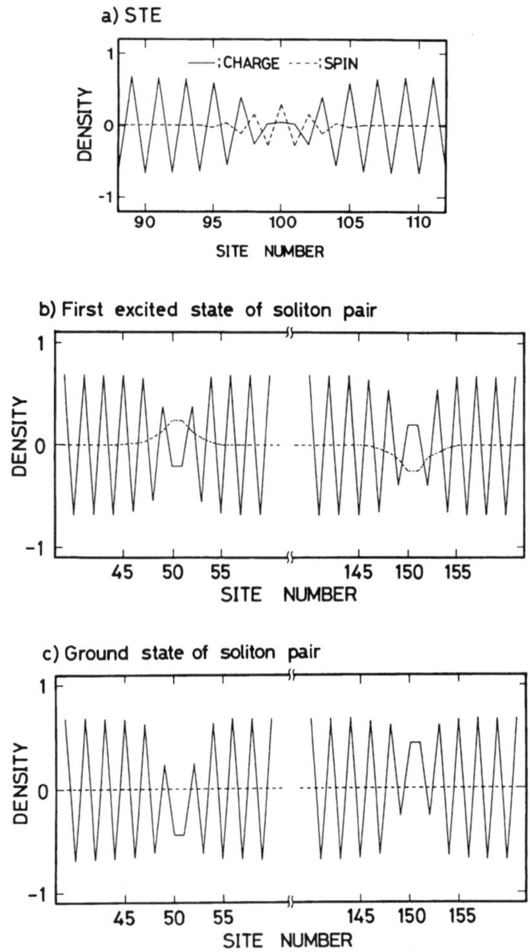


**Fig. 3.13.** Charge- and spin-density profiles of (**a**) STE, (**b**) first excited state of the soliton pair, spin solitons, (**c**) ground state of the soliton pair, charged solitons, in one-dimensional CDW [426]. The *solid line* denotes the charge density $\langle n_{l\alpha}+n_{l\beta}-1\rangle$ and the *dashed line* is the spin density $\langle n_{l\alpha}-n_{l\beta}\rangle$ (With permission from The American Physical Society)

As for the values of parameters, we will investigate the intermediate coupling case, where the inter-site repulsions $V$ and $T$ are relatively large as compared with $S$. We have thus calculated the contour map of the adiabatic potential-energy surface of the exciton, and it is shown in Fig. 3.12 by using the $l_0 - \Delta Q$ plane, wherein $\theta$ is determined to minimize the energy at each point of the surface.

The point $\Delta Q = 0$ and $l_0 = 0$ corresponds to the Franck–Condon state, which is just made by photons from the CDW ground state. In this figure, we can find two energy minima. The minimum at $l_0 = 0$ and $\Delta Q = 1.6$

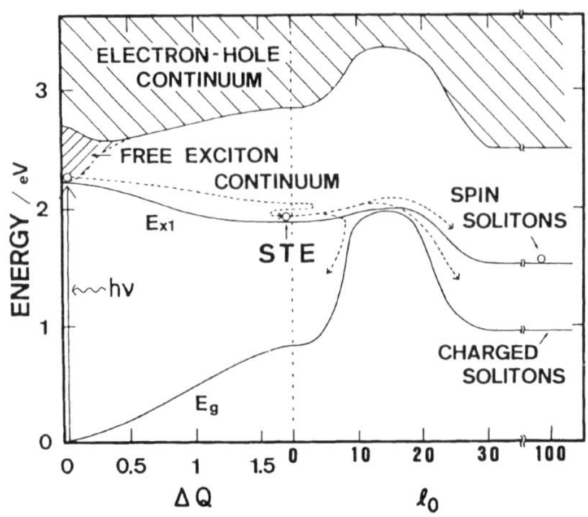


**Fig. 3.14.** Adiabatic potential energy surface along the decay path indicated by the *dashed line* in Fig. 3.12 [426] (With permission from The American Physical Society)

corresponds to the STE, and another minimum at $\Delta Q = 1$ corresponds to the pair of solitons, which is the lowest relaxed excited state of this system. The charge and the spin-density profiles around the STE and the pair of solitons are shown in Fig. 3.13 a and b, respectively. Figure 3.13 a shows that the SDW type spin order has appeared in the STE over about five metallic sites. On the other hand, we can see from Fig. 3.13 b that the soliton pair has spins around the boundaries between two phases. This is a spin soliton. Incidentally, it can be seen from Fig. 3.13 c that the ground state of the soliton has a charge.

Since the energy barrier between the STE and the soliton pair is only about 0.1 eV, the STE is expected to decay into the soliton pair, through the thermal activation process over the barrier, or through the tunneling process under the barrier. When $l_0$ is more than 30, the energy of the soliton pair remains unchanged, being independent of $l_0$. The decay path of the exciton on the adiabatic potential surface is indicated by the dashed line in Fig. 3.12. This figure also shows that $N_e$ increases from unity to about five, as the lattice relaxation proceeds from the Franck–Condon state to the STE. From this STE, relaxation proceeds further with the increase of $N_e$, and finally, the spin soliton is created as the relaxed excited state. In order to visualize the decay path, we have shown in Fig. 3.14 the adiabatic potential energy along the decay path which is indicated by the dashed line in Fig. 3.12. In Fig. 3.14, $E_{X1}$ denotes the energy of the first excited state, namely, the energy of the exciton, and $E_g$ denotes that of the ground state.

The formation energy of the soliton pair depends only on the strain energy to create boundaries between the different phases. In the case of one-dimensional CDW, the size of the boundary region is always no more than microscopic, consisting merely of "points". Hence, as seen from Fig. 3.14, once the soliton has been created, its energy is kept unchanged even if $l_0$ increases. Therefore, the two spin solitons can depart from each other entirely, and $N_e$ will become infinite without any resistance. It means that, in the case of one-dimensional CDW, the other ground state can be created from the starting ground state only by one photon. This is one of the distinctive features of the nonlinear lattice relaxation in one dimension.

Finally, let us briefly discuss these nonlinear lattice relaxations from a somewhat different point of view to the one mentioned before. The collective excited states described in this chapter can never be created by ordinary thermal excitations from the ground state, because a much larger amount of energy is required than the ordinary thermal energy, e.g., at room temperature. It becomes possible only when the energy is supplied by photo-excitation. That is, as a combination of photo-excitation and the subsequent lattice relaxations, we can clarify the multi-stable nature of the ground state, even when thermal excitations can never access it.

## 3.4 Direct and Indirect Excitons in Three-Dimensional CDW State

Let us proceed to the optical properties of the three-dimensional CDW, taking $BaBiO_3$ as a typical example. In the previous section, we have seen the optical properties of the HMMC, and in this case, the lowest optical excitation is almost equal to the excitation across the CDW gap, as shown in Fig. 3.11. In the case of three-dimensional metal oxides, however, the one-body energy $E(\boldsymbol{k})$ of the electron around the Fermi level is usually a complicated function of $\boldsymbol{k}$. It depends on the nature of three-dimensional chemical bonds among the

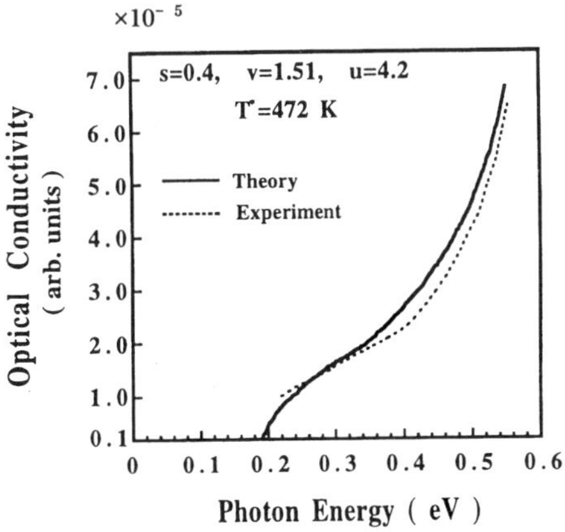


**Fig. 3.15.** The optical conductivity of $BaBiO_3$ in the infrared region. The *solid line* is the calculation [430], and the *dashed line* is the experiment [429]

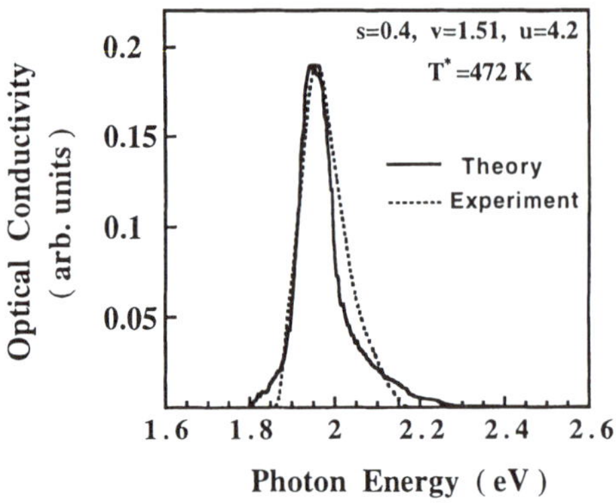


**Fig. 3.16.** The optical conductivity of $BaBiO_3$ in the visible region. The *solid line* is the calculation [430], and the *dashed line* is the experiment [427]

*s*- or *d*-orbitals of the metallic atoms and the *p*-orbitals of Os. As a result, it often occurs that even after the CDW type metal–insulator transition, optical excitations appear below the CDW gap. In this case, we have a "mid-gap absorption", which is not due to the collective excitations in the gap, e.g., solitons, but simply due to the complicated $\boldsymbol{k}$-dependency of the one-electron energy band. Very recently, this type of low energy optical excitation has been found below the CDW gap of $BaBiO_3$.

A wide range of experimental investigations have been carried out on $BaBiO_3$ [427–429]. These results indicate that the frozen breathing mode of O around Bi in the $BiO_6$ octahedron causes the formation of CDW type long-range order. This order leads to a doubling of the unit cell, resulting in the mixed valence state, and opens up an energy gap of about 2 eV at the Fermi surface of this many-electron system. This is nothing but the metal–insulator transition.

Recently, *Uwe* et al. [429] have measured the light absorption spectrum of $BaBiO_3$ in the near-infrared region, and found a long absorption tail as shown in Fig. 3.15. On the other hand, *Tajima* et al. [427] have already measured the optical conductivity in the visible region of this material. In their experiments, a sharp and strong absorption peak is observed near 2 eV, as shown in Fig. 3.16. This sharp and strong absorption is inferred to be the excitation across the CDW gap. However, we also have the absorption tail below this gap.

Thus, the spectral shape of this material is quite different from a typical absorption edge of an ordinary CDW type insulator, such as the HMMC mentioned before. In order to clarify the aforementioned characteristics, *Hasanuzzarman*, *Iwano* and *Nasu* [430] have proposed a model, which can explain both the near-infrared and visible light absorption spectra from a unified point of view, using a three-dimensional extended Peierls–Hubbard model. They have shown that the indirect and direct excitons in this insulator are responsible for the aforementioned two spectral parts. The appearance of these two components is due to the characteristic and complicated nature of the one-electron energy band around the Fermi level, which is brought about through the complicated hybridization between one 6$s$-orbital of Bi and nine 2$p$-orbitals of O in the unit cell.

### 3.4.1 Three-Dimensional Extended Peierls–Hubbard Model for $BaBiO_3$

Let us now examine the three-dimensional extended Peierls–Hubbard model for the electronic state of this material and for these characteristic light absorption spectra [430]. Its Hamiltonian is

$$
\begin{aligned}
H_B = & \sum_{\boldsymbol{k},\sigma} E(\boldsymbol{k}) \eta^{\dagger}_{\boldsymbol{k}\sigma} \eta_{\boldsymbol{k}\sigma} + S \sum_{\boldsymbol{l}} \sum_{i=x,y,z} \frac{Q^2_{\boldsymbol{l}i}}{2} \\
& + S \sum_{\boldsymbol{l},\sigma} \sum_{i=x,y,z} (Q_{\boldsymbol{l}-\boldsymbol{e}_i,i} - Q_{\boldsymbol{l}i}) \, n_{\boldsymbol{l}\sigma} + U \sum_{\boldsymbol{l}} n_{\boldsymbol{l}\alpha} n_{\boldsymbol{l}\beta} \\
& + V \sum_{\boldsymbol{l},\boldsymbol{l}',\sigma} n_{\boldsymbol{l}\sigma} n_{\boldsymbol{l}'\sigma'} \, .
\end{aligned}
\tag{3.29}
$$

The first term is the one-electron energy. $E(\boldsymbol{k})$ is so chosen that it reproduces a single energy band at the Fermi level of the hypothetical metallic state, calculated by *Mattheiss* and *Hamann* [431]. The Wannier function of this energy band is a linear combination between the 6 $s$ orbitals and the 2 $p$ orbitals. It is centered at a certain lattice site $\boldsymbol{l}$ of a simple cubic crystal, but is extended widely to other sites. $Q_{\boldsymbol{l}i}$ denotes the displacement of O toward Bi, and $i(=x, y, z)$ defines the position of each O among three Os in a unit cell, so that it is on each fundamental translation vector $\boldsymbol{e}_i (i = x, y, z)$ of this lattice.

By using this model, the light absorption spectra shown in Figs. 3.15 and 3.16 are calculated within the mean field approximation reinforced by the single configuration interaction (CI) method, which is sufficient to get exciton effects in the excited states. As for $Q_{\boldsymbol{l}i}$, the diplacement with the double period for all three directions $\boldsymbol{e}_i$ $(i = x, y, z)$ is assumed, and the thermal fluctuation around this extremal lattice configuration is also taken into account by the classical Monte Carlo method, which was originally derived by *Iwano* and *Nasu* [424].

### 3.4.2 Direct and Indirect Excitons

Let us now see why we have two spectral parts. The one-electron energy bands and the density of states of the CDW state thus calculated are shown in Fig. 3.17. From this figure, we can see that the direct gap appears at

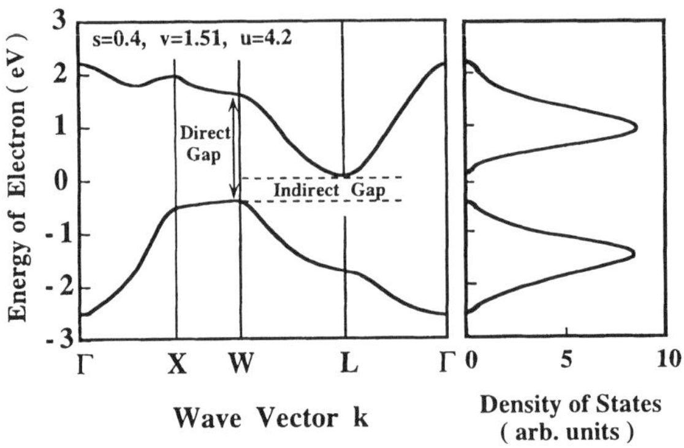


**Fig. 3.17.** The direct and indirect gaps in the one-electron energy band, and the density of states of the CDW state of [430] (With permission from The Physical Society of Japan)

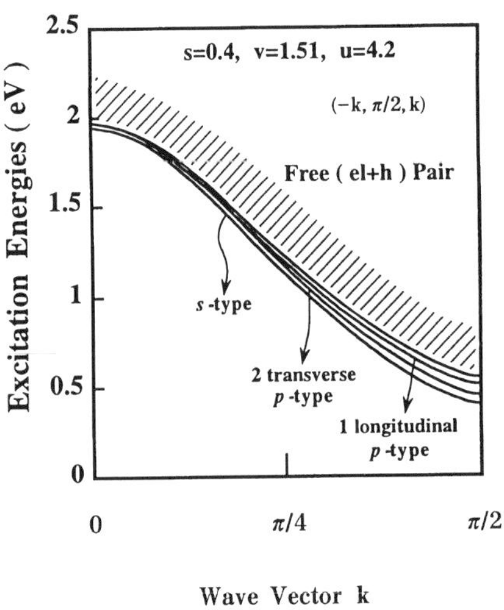


**Fig. 3.18.** The exciton bands along $(-k, \pi/2, k)$ in the CDW state of [430] (With permission from The Physical Society of Japan)

$W(\equiv XM/2)$ point of the first Brillouin zone with energy 1.961 eV, while the indirect gap opens up between the $W$ and $L(\equiv \Gamma R/2)$ points with energy 0.4 eV. The appearance of the direct gap is due to the CDW type long-range order, while the presence of this indirect gap is a consequence of the complicated hybridization between one 6$s$-orbital of Bi and nine 2$p$-orbitals of O in the unit cell.

Let us now turn to the exciton bands obtained by using these new one-electron energy bands. The calculated exciton energy bands are shown in Fig. 3.18, and we can see that various types of exciton bands with a negative mass appear. In our model, the hole can exert its attraction only when the electron is in its six nearest neighboring sites. Therefore, we can obtain six exciton bands at most, whose relative motions have $A_{1g}$ ($s$-type), $T_{1u}$ ($p$-type) and $E_{2g}$ ($d$-type) symmetries in the $O_h$ group, when the total momentum of the exciton ($\equiv K$) is zero. For finite $K$, $T_{1u}$ splits into one longitudinal and two transverse excitons as shown Fig. 3.18.

Therefore, in view of the aforementioned results, we can see that both direct and indirect optical transitions can take place. The direct transition is the dominant part of the spectrum shown in Fig. 3.16, and corresponds to the excitation across the CDW gap, while the indirect part corresponds to the long trail in the infrared region shown in Fig. 3.15. It should be noted that both parts originate from the single Hamiltonian $H_{\mathrm{B}}$ shown in (3.29), although the fomer is due to the frozen part of the Peierls distortion, while the latter is due to the fluctuation therefrom.

## 3.5 Competition Between Superconductivity and CDW State

In the previous sections, we were concerned with the CDW states. However, in the cases of quasi two- and three-dimensional systems, the CDW state is always competing with the SC state, which also originates from the same e–ph coupling. As typical examples of crystals that show this type of competition, we have 2H-$NbSe_2$ [432], the Chevrel phase superconductors such as $PbMo_6S_8$ [433], and $BaPb_{1-x}Bi O_3$ [434]. More recently, we also have the new high-temperature superconductor $Ba_{1-x}K_xBiO_3$ with a $T_c$ of about 30 K, which seems to be just at the boundary between these two phases [435, 436].

As mentioned at the beginning of this chapter, competition between the CDW state and the SC state is nothing more than competition between $T$ and $\omega$ in many-electron systems. When the coupling is weak, $S \ll T$, however, the relative stability between the CDW state and the SC state is mainly determined by the geometric shape of the unperturbed Fermi suface. In the SC state, the energy gap can be opened all over the Fermi surface irrespective of its geometric shape. On the other hand, in the CDW state, it can be opened only in two parts of the Fermi surface, that can be nested with respect to each other by a certain phonon wave vector, and other parts of the Fermi surface are left unchanged. If such a nesting area is very wide, as in the case of a cubic or square Fermi suface shown in Fig. 3.19 a, the CDW always becomes stable. On the other hand, if the nesting area is absent, as in the case of a spherical surface shown in Fig. 3.19 b, the superconducting state is always stable. Moreover, as shown by *Bilbro* et al. [437] in their theory for A15 compounds, these two states can coexist in such a way that the nested area contributes to the CDW and the rest contributes to the SC state. Thus, we can say that in the case of weak coupling, the competition between the adiabatic and inverse adiabatic nature becomes rather obscure.

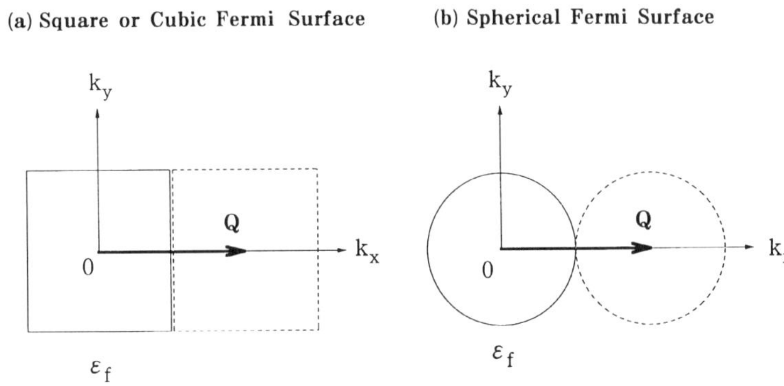


**Fig. 3.19.** The nesting of the Fermi surface. (**a**) The square or cubic case. (**b**) The spherical case. $\boldsymbol{Q}$ is the nesting vector

In the case of strong coupling, $S \gg T$, however, the relative stability between the two phases is rather insensitive to the geometric shape of the unperturbed Fermi suface, since the one-electron energy band itself is greatly altered from the original one. In this case, competition between the two phases is just competition between the frozen part and the moving part of the lattice distortion, and hence the resultant states depend directly on the three basic parameters $T$, $S$ and $\omega$.

In connection with this strong coupling problem, *Chakraverty* has considered the possibility of two phases in a many-polaron system [438]. He has also insisted that any system falls into a crystallized state of bipolarons (or a CDW), provided the e–ph coupling increases sufficiently. On the other hand, *Alexandrov* et al. have insisted that a system with very strong e–ph coupling becomes a superfluid state just like the case of $^4$He, rather than the BCS type superconducting state [439]. These two competing ideas for the strong coupling problem have been a matter of considerable interest. Very recently, however, *Nasu* has clarified the origin of the difference [440]. According to this result, the former conclusion is obtained in the strong coupling and adiabatic case: $\omega \ll T \ll S$, while the latter one is obtained in the strong coupling and inverse adiabatic case: $T \ll S \ll \omega$. We will look at this theory in detail in the next section.

### 3.5.1 The Many-Polaron System

Let us consider a quasi two-dimensional square lattice composed of $N$ sites ($N \gg 1$) and of $N$ electrons interacting with phonons. Its Hamiltonian ($\equiv H_4$) is given as

$$H_4 = -\sum_{l,l',\sigma} T(l-l')\eta^\dagger_{l\sigma}\eta_{l'\sigma} + \sqrt{\frac{S\omega}{2}}\sum_{l\sigma}(\xi^\dagger_l + \xi_l)n_{l\sigma} + \omega\sum_l \xi^\dagger_l\xi_l \,. \tag{3.30}$$

Here $T(\boldsymbol{l}-\boldsymbol{l}')$ denotes the transfer energy of an electron between two lattice sites specified by $\boldsymbol{l}$ and $\boldsymbol{l}'$. To focus our attention mainly on the strong region of e–ph coupling, we will be concerned in this section with the case where the Fermi surface of noninteracting electrons is a complete circle, with no nesting-type instability. Hence, we assume that the transfer energy between nearest-neighbour sites is $T$, that between next-nearest-neighbour sites is $gT$ ($g=0.35$), and zero otherwise. This provides us with an almost completely circular Fermi surface, as shown in Fig. 3.19 b, and its density of states has no singularity near the Fermi level.

Generally speaking, the state of the phonon can be qualitatively divided into two parts: the frozen part that follows only the static charge density, and the moving part that follows the instantaneous motion of each electron. In the adiabatic limit the frozen part will be dominant, while in the inverse adiabatic limit, the moving one will be important. The main idea used by *Nasu* to describe the intermediate region is an interpolation between these two limits [440]. That is, the two parts are assumed to coexist with a ratio determined by a variational method. For this sake, we introduce a displacement operator ($\equiv \bar{M}$) for the phonon:

$$\bar{M} \equiv \exp\left\{-\mathrm{i}s^{1/2}\sum_{\boldsymbol{l}} P_{\boldsymbol{l}}\left[\bar{q}_{\boldsymbol{l}}+\sum_{\boldsymbol{l}',\sigma}\Delta q(\boldsymbol{l}-\boldsymbol{l}')n_{\boldsymbol{l}'\sigma}\right]\right\},$$

$$s \equiv \frac{S}{\omega}, \tag{3.31}$$

$$P_{\boldsymbol{l}} \equiv \frac{\mathrm{i}(\xi_{\boldsymbol{l}}^{\dagger}-\xi_{\boldsymbol{l}})}{2^{1/2}}, \tag{3.32}$$

which can transform $\eta_{\boldsymbol{l}\sigma}^{\dagger}$ and $\xi_{\boldsymbol{l}}^{\dagger}$ as

$$a_{\boldsymbol{l}\sigma}^{\dagger} \equiv \bar{M}^{-1}\eta_{\boldsymbol{l}\sigma}^{\dagger}\bar{M} = \exp\left[\mathrm{i}s^{1/2}\sum_{\boldsymbol{l}'}\Delta q(\boldsymbol{l}-\boldsymbol{l}')P_{\boldsymbol{l}'}\right]\eta_{\boldsymbol{l}\sigma}^{\dagger}, \tag{3.33}$$

$$b_{\boldsymbol{l}\sigma}^{\dagger} \equiv \bar{M}^{-1}\xi_{\boldsymbol{l}\sigma}^{\dagger}\bar{M} = \xi_{\boldsymbol{l}}^{\dagger} + \left(\frac{s}{2}\right)^{1/2}\left[\bar{q}_{\boldsymbol{l}}+\sum_{\boldsymbol{l}',\sigma}\Delta q(\boldsymbol{l}-\boldsymbol{l}')n_{\boldsymbol{l}'\sigma}\right]. \tag{3.34}$$

Here, $\bar{q}_{\boldsymbol{l}}$ denotes the frozen displacement of a phonon at site $\boldsymbol{l}$, since it is not directly related to the occupation $n_{\boldsymbol{l}\sigma}$. While $\Delta q(\boldsymbol{l}-\boldsymbol{l}')$ denotes the moving displacement of phonons at site $\boldsymbol{l}$ created by electrons at site $\boldsymbol{l}'$. It appears and disappears according to the presence or absence of electrons at site $\boldsymbol{l}'$, since it is proportional to $n_{\boldsymbol{l}'\sigma}$. These $\bar{q}_{\boldsymbol{l}}$ and $\Delta q(\boldsymbol{l})$ are variational parameters. $\eta_{\boldsymbol{l}\sigma}^{\dagger}$ is now transformed into a creation operator of a localized polaron $a_{\boldsymbol{l}\sigma}^{\dagger}$ with a phonon cloud whose spatial extent is given by $\Delta q(\boldsymbol{l})$, while $b_{\boldsymbol{l}\sigma}^{\dagger}$ denotes the creation operator of a new phonon whose equilibrium position is already displaced by the aforementioned two parts.

Substituting (3.33) and (3.34) into (3.30), and taking the average with respect to this new phonon vacuum, we get a new Hamiltonian ($\equiv H_{40}$) of our many-polaron system

$$
\begin{aligned}
H_{40} = & -\bar{X} \sum_{l,l',\sigma} T(l-l') a^{\dagger}_{l\sigma} a_{l'\sigma} - S \sum_{l\sigma} \bar{q}_l n_{l\sigma} + S \sum_{l} \frac{\bar{q_l}^2}{2} \\
& - S \sum_{l,l'} \left[ 2\Delta q(l-l') - \Delta q_2(l-l') \right] n_{l\alpha} n_{l'\beta} \\
& - S \sum_{l,l',\sigma} \left\{ \left[ 2\Delta q(l-l') - \Delta q_2(l-l') \right] - \left[ 2\Delta q(0) - \Delta q_2(0) \right] \right\} \frac{n_{l\sigma} n_{l'\sigma}}{2} \\
& - SN \frac{[2\Delta q(0) - \Delta q_2(0)]}{2} ,
\end{aligned} \tag{3.35}
$$

$$
\Delta q_2(l) \equiv \sum_{l'} \Delta q(l-l') \Delta q(l') . \tag{3.36}
$$

Here, $\bar{X}$ is an averaged reduction factor of the transfer energy, coming from the overlap integral between phonon clouds localized at each site, and is given as

$$
\bar{X} = \frac{\sum_l T(l) \exp\{-s[\Delta q_2(0) - \Delta q_2(l)/2]\}}{\sum_l T(l)} . \tag{3.37}
$$

The relation between $x$ defined in (3.6) and this $\bar{X}$ is of course, $x = \bar{X}^{-1}$. We have also assumed the orthogonality between the two parts of the displacement

$$
\sum_{l'} \bar{q_{l'}} \Delta q(l'-l) = 0 , \tag{3.38}
$$

so that their roles do not interfere with each other in the variational method. The first term of (3.35) denotes the transfer of a polaron, the second one is the static potential given by the frozen part $\bar{q}_l$, and the third is its elastic energy. The fourth and fifth terms are the attraction between polarons, minus the increase of elastic energy needed to create the phonon cloud: $\Delta q_2(l-l')$. The sixth is the rest energy. We should note that the frozen displacement $\bar{q}_l$ has no direct effect on $\bar{X}$, since it does not move together with the electron.

### 3.5.2 Phase Diagram

It is not our purpose to study all possible ground states brought about by the model Hamiltonian $H_{40}$, but rather to clarify the nature of the competition between the SC type pairing order and the crystalline order of bipolarons (CDW) in the region where the e–ph coupling is strong. Since our system is a square lattice with $N$ sites and $N$ electrons, the crystalline phase of bipolarons in the strong coupling region, $S \gg T$, is inferred to be such that the frozen displacement occurs with twice the period of the original square lattice, in both directions of the crystal axes. Then

$$\bar{q}_l = \bar{q}\cos(\boldsymbol{Q}\cdot\boldsymbol{l}) + 1\,, \quad \boldsymbol{Q} \equiv (\pi,\pi)\,. \tag{3.39}$$

Here, $\bar{q}(>0)$ denotes the amplitude of this frozen displacement, $\boldsymbol{Q}$ denotes its wave vector with two dimensions, and the constant term denotes the uniform displacement corresponding to the uniform charge density. This gives a bipolaronic crystal with twice the period, shown schematically in Fig. 3.20.

Let us determine the energy of the ground state of a many-polaron system characterized by $H_{40}$, within the mean field theory for interpolaron interactions and by a variational method for $\bar{q}_l$ and $\Delta q(\boldsymbol{l})$. So as to reduce the two-body terms into one-body terms, we approximate $(a^\dagger_{l\alpha}a^\dagger_{l+l',\beta})$ by the partial amplitude of the SC type pairing order ($\equiv f_{l'}$) by

$$\sum_{\boldsymbol{l}} N^{-1}\langle a^\dagger_{l\alpha}a^\dagger_{l+l',\beta}\rangle = f_{l'}\,, \tag{3.40}$$

and according to (3.39), we also approximate the occupation number $n_{l\sigma}$ by

$$n_{l\sigma} \to \langle a^\dagger_{l\sigma}a_{l\sigma}\rangle = \Delta n_\sigma \cos(\boldsymbol{Q}\cdot\boldsymbol{l}) + \frac{1}{2}\,, \quad \Delta n_\alpha = \Delta n_\beta\,. \tag{3.41}$$

Here, $f_{l'}$ and $\Delta n_\sigma$ are parameters and should be determined self-consistently.

The resultant phase diagram is shown in Fig. 3.21. We can see that it is separated into the SC state and the CDW or bipolaronic insulator. We should also note that the SC state, just on the $TS$ and $T\omega$ lines in the triangle, is the ordinary metallic state.

In the adiabatic region, (around the $TS$ line), the boundary between the two phases is given by $S/T = 1.47$ (point C), while as $\omega$ increases, the region of the SC phase expands in the phase diagram, and becomes dominant in the region around the $\omega$ vertex except on the $S$–$\omega$ line. From Figs. 3.22 and 3.23,

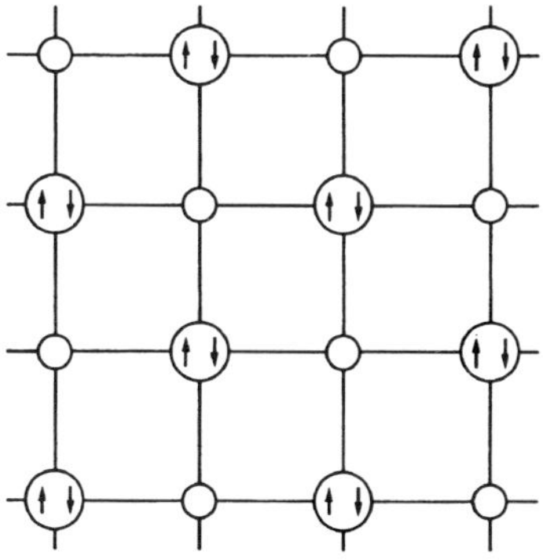

**Fig. 3.20.** The bipolaron lattice (CDW) on the square lattice

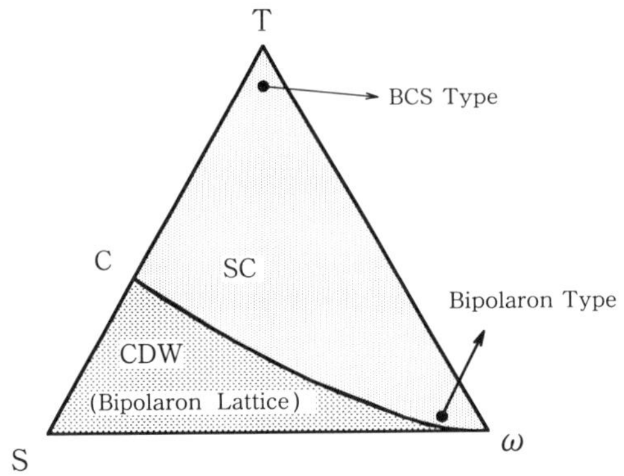


**Fig. 3.21.** The phase diagram on the $T$–$S$–$\omega$ triangle [440]. The ratios of the three parameters are denoted by the lengths of the perpendiculars from a given point to three axes (With permission from The American Physical Society)

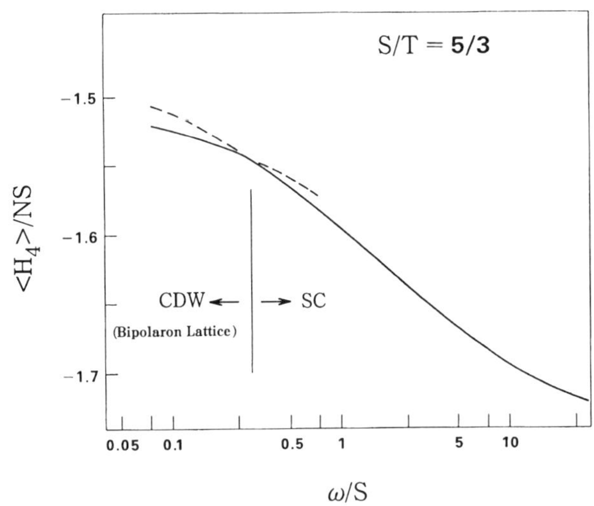


**Fig. 3.22.** The total energy per electron $\langle H_4 \rangle / NS$ as a function of $\omega/S$ [440] (With permission from The American Physical Society)

we can see how the state changes with varying $\omega/S$ (or $\omega/T$) when $S/T$ is fixed at 5/3. As seen in Fig. 3.22, the bipolaronic insulator becomes less stable than the SC state at $\omega/S \sim 0.3$. This is mainly because the moving displacement $\Delta q(0)$ increases with increase in $\omega/S$ (or $\omega/T$) as seen from Fig. 3.23, and hence the bipolaronic crystal is forced to melt. In fact, $\bar{q}$ decreases as $\omega/T$ increases. It is also very interesting that the reduction factor $\bar{X}$, being unity in both limits, is reduced most distinctly in the intermediate region, where the velocity of the electron is of the same order as that of the phonon.

To clarify the SC state near the $\omega$-vertex in the phase diagram, *Nasu* also calculated the energy of the Goldstone mode ($\equiv \Omega(\boldsymbol{k})$) within the random

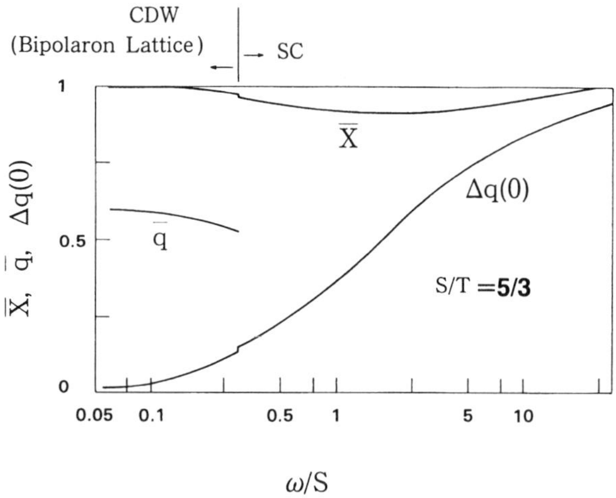


**Fig. 3.23.** $\bar{X}$, $\bar{q}$ and $\Delta q(0)$ as a function of $\omega/S$ [440]

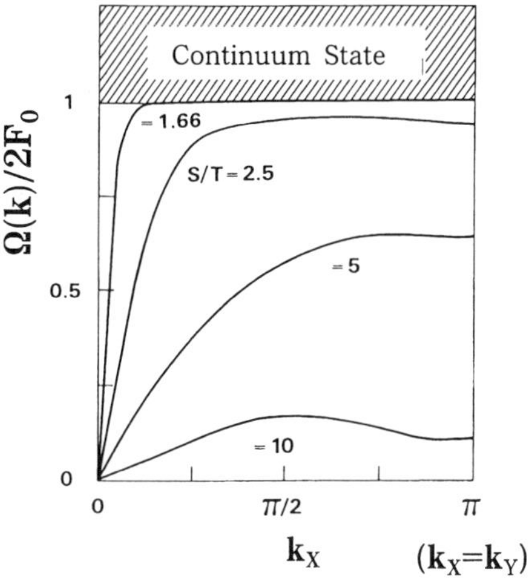


**Fig. 3.24.** The relative energy of the Goldstone mode $\Omega(\boldsymbol{k})/2F_0$ where $F_0$ is the mean field gap [440] (With permission from The American Physical Society)

phase approximation [440]. As shown in Fig. 3.24, it has a sound-wave-like dispersion in the region of small $|\boldsymbol{k}|$. In the case of weak coupling $S/T \ll 1$, as is well known, the sound velocity is almost equal to the Fermi velocity [441]. As $S/T$ increases, however, the velocity gradually decreases, and in the strong coupling case, $S/T \gg 1$, $\Omega(\boldsymbol{k})$ becomes an order of magnitude smaller than the energy gap obtained by mean field theory. Moreover it has a roton-like dispersion, peculiar to the superfluid type state. In the limit of strong coupling, all electrons make singlet pairs (bipolarons), resulting in a Bose condensation. The mean field energy gap $F_0$ in this case corresponds to the pair breaking excitation, while $\Omega(\boldsymbol{k})$ corresponds to the excitation that breaks only the coherence between pairs without changing its internal state. The results of our calculation show that, around the $\omega$-vertex, this change from the weak limit to the strong one is continuous. The possible region of pairing, being limited only around the Fermi level in the case of weak coupling, expands right across the energy band as the coupling increases. That is, the BCS-type state changes to a superfluid-type state. This is consistent with the results by *Pincus* et al. [442], *Leggett* [443], and *Nozieres* et al. [444], which, however, are mainly based on the negative Hubbard model.

Thus, we have seen the interrelation between the BCS-type state and the superfluid-type state, as well as their competition with the bipolaronic insulator (CDW). That is, the strong coupling in the inverse adiabatic limit results in the superfluid-type state, while the strong coupling in the adiabatic limit results in the bipolaronic insulator(CDW), as summarized in the phase diagram.

## 3.6 Superconducting Transition Temperatures of Strongly Coupled Electron–Phonon Systems

### 3.6.1 Expected Behaviour of $T_c$

In the previous section, we have seen the nature of the competition between the SC state and the CDW state, at the absolute zero of temperature. In the present section, we shall look at the nature of the superconducting transition temperatures ($\equiv T_c$s) of systems with very strong e–ph couplings. The problems related to the competition between the SC state and the CDW state at finite temperatures are also very interesting. However, for simplicity, we will exclude the CDW instability completely in this section, and restrict ourselves only to the SC type pairing order at finite temperatures.

Generally speaking, $T_c$ is also determined by the three basic parameters: $T$, $S$ and $\omega$. At present, however, we have no systematic theory for $T_c$ that can cover the whole region of these basic three parameters. The traditional BCS theory [386], being useful only in the weak coupling region $S \ll T$, tells us that $T_c$ increases as $S$ increases. However, in the strong coupling region $S \gg T$, $T_c$ is expected to decrease as $S$ increases [439,442]. In the weak region, only electrons around the Fermi level make singlet pairs called Cooper pairs, while, in the strong limit, almost all electrons in the conduction band make strongly bound pairs, which are nothing but the singlet bipolarons mentioned before. Such bipolarons, once formed, cannot be broken thermally at low temperatures. $T_c$ in this case corresponds to the temperature at which only the superfluid-type coherence between the bipolarons is broken by thermal fluctuation, with no change in each individual bipolaron. According to the general theory of superfluidity, $T_c$ will decrease as the effective mass ($\equiv m^*$) of the bipolaron increases. On the other hand, the standard polaron theory, shown in previous sections, tells us that this effective mass $m^*$ increases as $S$ increases. Thus, we can infer that $T_\mathrm{c}$ becomes a maximum in the transition region $S \sim T$, which lies between the two extreme regions: the BCS region and the bipolaronic region.

In this connection, *Nasu* has recently derived a new theory for $T_\mathrm{c}$ that can cover the whole region of $T$, $S$ and $\omega$ [445]. In this theory, the thermal fluctuation of the phase of the order parameter is taken into account by the single-site coherent potential approximation (CPA) [446], and the resultant $T_\mathrm{c}$ has a maximum at the intermediate region. In the next section, we shall look at this theory in detail.

### 3.6.2 Interpolating Theory for $T_\mathrm{c}$ by CPA

Let us consider the same system as in the previous section, being composed of $N$ lattice sites and $N$ electrons with a Hamiltonian given by (3.30). For simplicity, however, we introduce such a polaron transformation $\bar{M}$ in (3.31) that the frozen lattice distortion is absent, and the phonon cloud is localized only in the same site as the electron,

$$\bar{q}_l = 0\,, \quad \Delta q(\boldsymbol{l}-\boldsymbol{l}') \to \Delta q \delta_{ll'}\,. \tag{3.42}$$

By using $\bar{M}$, thus newly defined, we can rewrite $H_{40}$ into a new Hamiltonian ($\equiv H_{40}'$) as

$$\begin{aligned} H_{40} \to H_{40}' = &-\bar{X} \sum_{\boldsymbol{l},\boldsymbol{l}',\sigma} T(\boldsymbol{l}-\boldsymbol{l}') a_{\boldsymbol{l}\sigma}^{+} a_{\boldsymbol{l}'\sigma} \\ &-\bar{U} \sum_{\boldsymbol{l}} n_{\boldsymbol{l}\alpha} n_{\boldsymbol{l}\beta} - S\left(2\Delta q - (\Delta q)^2\right) \bar{C} N\,, \end{aligned} \tag{3.43}$$

where $\bar{X}$ and $\bar{U}$ are also newly defined as

$$\bar{X} \equiv \exp\left[-\frac{S(\Delta q)^2}{2\omega}\right], \quad \bar{U} \equiv S\left(2\Delta q - (\Delta q)^2\right), \tag{3.44}$$

and $\bar{C}$ in the last term of (3.43) denotes half the average number of electrons per site. It can take not only 0.5, but also any other values. $\Delta q$ is again a variational parameter that denotes the thickness of the phonon cloud. At the present stage it is unknown, but will be determined later. As inferred from Migdal's theorem, it will be very small ($\Delta q \ll 1$) in the weak coupling case, $S \ll T$, or in the adiabatic case $T \gg \omega$, while it will become a maximum ($\Delta q = 1$) in the strong coupling case $S \gg T$ or in the inverse adiabatic case $T \ll \omega$.

At first, we determine this $\Delta q$ by the following variational principle,

$$\frac{\partial \langle H_{40}' \rangle}{\partial \Delta q} = 0\,, \tag{3.45}$$

and, as for the BCS type pairing order, we approximate according to ordinary mean field theory as,

$$-\bar{U}\sum_{l} n_{l\alpha}n_{l\beta} \rightarrow -\bar{U}\bar{f}\sum_{l}(a^{+}_{l\alpha}a^{+}_{l\beta} + a_{l\beta}a_{l\alpha}) , \quad \bar{f} \equiv \langle a^{+}_{l\alpha}a^{+}_{l\beta}\rangle , \tag{3.46}$$

where $\langle \ldots \rangle$ is the average of $\ldots$ . From (3.44) and (3.45), we can easily calculate $\Delta q$, and find that it is an increasing function of $S/T$. We can also easily calculate the superconducting transition temperature ($\equiv T'_{\mathrm{c}}$) self-consistently within this approximation. However, as is already well known, $T'_{\mathrm{c}}$ given by this method increases infinitely as $S$ increases, which is quite unreasonable for the strong-coupling case. The reason behind this unreasonableness is that the thermal fluctuation of the phase of the pairing order is not taken into account.

To overcome this defect, we will now proceed from this mean field theory to the path-integral formalism [388, 390], which was first introduced by *Evenson* et al. to examine $T \leftrightarrow U$ competition [447]. In this path-integral method, the interpolaron attraction term is linearized by the famous Hubbard–Stratonovitch transformation. Cutting short a lengthy mathematical manipulation by this transformation, we finally get the following approximation instead of (3.46):

$$-\bar{U}\sum_{l} n_{l\alpha}n_{l\beta} \rightarrow -\bar{U}\sum_{l} \bar{f}_{l}(a^{+}_{l\alpha}a^{+}_{l\beta} + a_{l\beta}a_{l\alpha}) . \tag{3.47}$$

Here $\bar{f}_l$ is a kind of external field, fluctuating with a statistical weight $\bar{P}_l(\bar{f}_l)$ at each site independently. This $\bar{f}_l$ corresponds to the local order parameter $\langle a^{+}_{l\alpha}a^{+}_{l\beta}\rangle$. Its statistical weight function $\bar{P}_l(\bar{f}_l)$ should be determined self-consistently with the total partition function of our many-polaron system. For this function, we assume the following simple form,

$$\bar{P}_l(\bar{f}_l) = y_{+}\delta(\bar{f}_l - \bar{f}) + y_{-}\delta(\bar{f}_l + \bar{f}) . \tag{3.48}$$

That is, $\bar{f}_l$ becomes $\bar{f}$ or $-\bar{f}$ with the relative weight $y_+$ or $y_-$ ($y_+ + y_- = 1$), respectively. In other words, our many-bipolaron system is regarded to be an Ising spin system, and the local order parameter $\bar{f}_l$ fluctuates only in its sign, keeping its absolute value within the mean field theory. These statistical weights $y_+$ and $y_-$, which are unknown at the present stage, will be determined later, self-consistently. However, once this $y_\pm$ is formally given, we can calculate the total thermodynamic potential and the one-body Green's function of our many-polaron system, by making use of the CPA [446].

To determine $y_\pm$ self-consistently, we artificially stop this static fluctuation at a site $\boldsymbol{l} = \boldsymbol{l}_0$ and fix $\bar{f}_{l_0}$ at $\bar{f}$ or $-\bar{f}$, and still allow fluctuation at all other surrounding sites ($\boldsymbol{l} \neq \boldsymbol{l}_0$) by the distribution $\bar{P}_l(\bar{f}_l)$. Under this fixed condition, we can calculate the total thermodynamic potential ($\equiv \Omega(\bar{f}_{l_0})$) averaged over various configurations of $\bar{f}_l$ in surrounding sites ($\boldsymbol{l} \neq \boldsymbol{l}_0$). This average is the path integral. Since all the sites must be equivalent, we can naively assume that $y_\pm$ in the central site is equal to the Boltzmann distribution

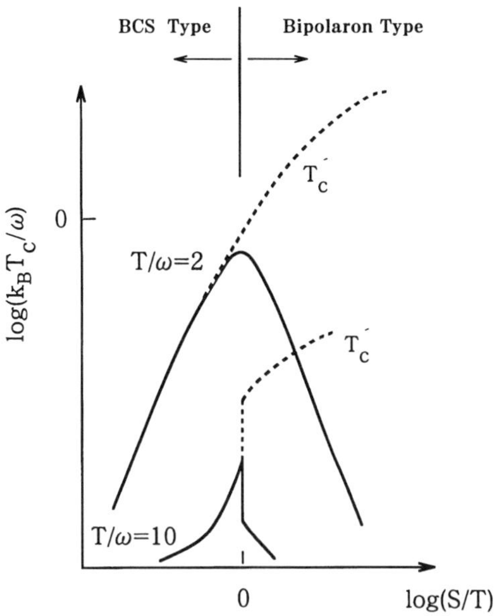


**Fig. 3.25.** The schematic nature of $T_c$ as a function of $S/T$. $T_c'$ is the temperature by the mean field theory [445] (With permission from The American Physical Society)

$$y_{\pm} \propto \exp\left[-\frac{\Omega(\pm\bar{f})}{k_B T_{\rm tem}}\right] . \tag{3.49}$$

From this equation, $T_c$ is given as the maximum temperature to yield a non-trivial solution

$$y_{+} \neq y_{-} . \tag{3.50}$$

The resultant nature of $T_c$ is shown in Fig. 3.25, very schematically. In this calculation the density of states is assumed to be constant with a width $8T$ for simplicity, and the aforementioned conditional thermodynamic potential $\Omega(\pm\bar{f})$ can also be easily calculated by the CPA. We can see that, in the BCS (weak coupling) region, $T_c$ increases as $S/T$ increases, while in the bipolaronic (strong coupling) region, it changes to decrease as $S/T$ increases. Thus, using the path integral and the CPA, we can get an interpolation theory for $T_c$ which can cover all regions of $S/T$. We can conclude that the maximum $T_c$ appears in the intermediate region where the itineracy and the e–ph coupling are well balanced. The transition from the BCS region to the bipolaronic one is continuous when $\omega/T \approx 1$, becoming discontinuous, when $\omega/T \ll 1$. This is nothing but the mass-enhancement effect mentioned before, being common to both the one-electron and the many-electron systems coupling strongly with phonons.

From Fig. 3.25, we can also roughly estimate the maximum of $T_c$ for a given set of $T$ and $\omega$. For example, when $4T = 1\,\mathrm{eV}$ and $\omega = 700\,\mathrm{cm}^{-1}$, $T_c$

can become about 30 K. This situation is expected to be realized in the metal oxides, since the conduction band of this type of material is rather narrow, while the energy of the breathing mode of O is rather high. Thus, this theory is one of the possible mechanisms of high-temperature superconductivity in $Ba_{1-x}K_xBiO_3$ [436] and other ceramics [448].

## 3.7 Many-Electron System Coupling Strongly with Nonlinear Phonons

### 3.7.1 BCS Limit, Nonlinear Phonons and Isotope Effects

In the present section, we will be concerned with quite new aspects of superconductivity due to highly nonlinear or anharmonic phonons. As described by *Ginzburg* [449], the phonon (or BCS) mechanism of superconductivity has been thought not to give higher $T_c$ than 30 K, for the following simple reason. In order to make $T_c$ higher, the e–ph coupling must be strong enough. Nevertheless, when the coupling is too strong, not only the superconducting state, but also the normal metallic state itself becomes unstable, and falls into a CDW type insulator with some structural changes, as shown in Fig. 3.21. This situation is qualitatively described by Fig. 3.26. As the electron–phonon coupling constant ($\equiv S$) increases from zero, $T_c$ increases exponentially, and takes its maximum ($\equiv T_c^*$) at around the critical value of $S$ denoted by $S^*$. As S increases further, the system falls into the CDW type insulator, and hence $T_c$ disappears. In the case of standard inorganic metallic crystals, this maximum ($\equiv T_c^*$) has been thought to be not higher than 30 K, and this maximum value 30 K has been called the "BCS limit".

However, this argument for the presence of the "BCS limit" is based on the tacit assumption that the phonon is perfectly harmonic (Fig. 3.27 a). As

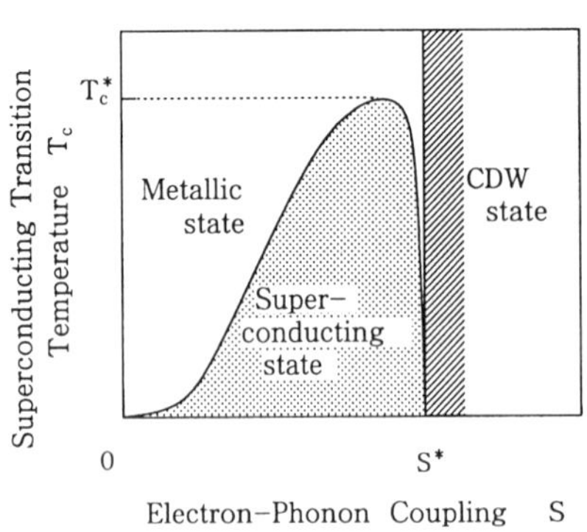


**Fig. 3.26.** Schematic nature of superconductor (metal)–insulator transition and $T_c$, as a function of the electron–phonon coupling constant $S$. $T_c^*$ is the maximum of $T_c$. $S^*$ is the phase boundary

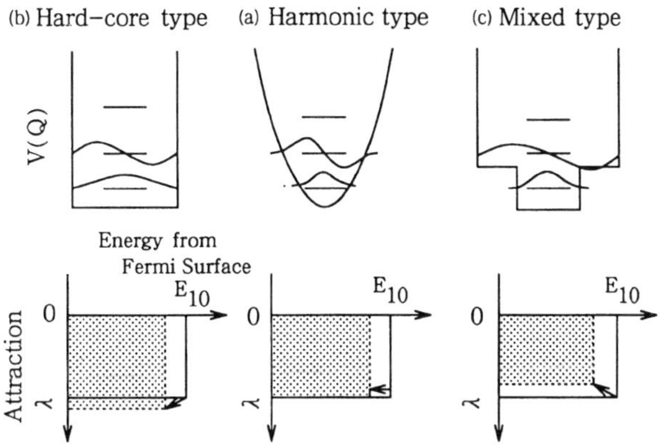


**Fig. 3.27.** Schematic nature of vibronic potentials $V(Q)$. (**a**) Harmonic type. (**b**) Hard–core repulsion type. (**c**) Hard–soft mixed type. Wave functions of ground and first excited states are also illustrated. The lower parts denote the changes in the attraction and the attractive region due to isotopic substitution

far as the atoms in the crystal vibrate only in the vicinity of their equilibrium point, the harmonic approximation may work well. Nevertheless, when they displace largely from their equilibrium point, a hard-core type inter-atomic repulsion (schematically shown in Fig. 3.27 b) begins to act, and hence the elastic energy of the lattice increases much more rapidly than in the harmonic case. Especially, in the case of crystals which contain light atoms such as oxygen, carbon and hydrogen, the amplitude and energy of lattice vibrations become larger than of heavier atoms, and so the aforementioned hard-core repulsion becomes very important.

This hard-core type anharmonicity makes the CDW more unstable, and relatively stabilizes the superconducting (or metallic) state. When the e–ph coupling is very strong, a large lattice distortion must always appear in the CDW state, and this large distortion makes the elastic energy of the lattice increase rapidly. On the other hand, in the case of superconducting (or metallic) states, the lattice distortion is nothing but the quantum zero point fluctuation. This fluctuation does not make the elastic energy increase so rapidly as compared with the CDW, although the amplitude of fluctuation is somewhat enhanced by this anharmonicity.

Consequently, as the hard-core type anharmonicity becomes strong, the critical coupling constant $S^*$ in Fig. 3.26 increases, and accordingly, $T_c^*$ itself also increases. Thus, the occurrence of the CDW is suppressed, even when the e–ph coupling is very strong, and the superconducting (or metallic) region expands in Fig. 3.26. Under this circumstance, the aforementioned “BCS limit” loses its significance.

Moreover, in this new region where the CDW cannot occur in spite of the strong e–ph coupling, the quantum fluctuation of the lattice is enhanced as compared with the harmonic case. Hence, the interelectron attraction mediated by phonons becomes strong. This strong attraction contributes not only to the superconducting state, but also partially cancels the interelectron Coulombic repulsion. Consequently, the SDW type metal–insulator transition is also suppressed, as compared with the harmonic case.

The other new aspect of superconductivity due to highly anharmonic phonons is the isotope effect of $T_c$. According to the standard BCS theory [386], the value of the isotope exponent ($\equiv \alpha$) defined by

$$\alpha = -\frac{\mathrm{d}\ln T_\mathrm{c}}{\mathrm{d}\ln M}\,, \tag{3.51}$$

is equal to 0.5. Here, $M$ denotes the atomic mass. As is well known, the BCS theory does not take the Coulombic repulsion between electrons into account. By adding the effects of repulsion to the BCS theory, *Morel* and *Anderson* (MA) [450] have shown that $\alpha$ becomes smaller than 0.5.

As is well known various new superconducting materials have appeared in recent years, and in the experimental studies of these materials, the isotope effect is one of the most interesting issues. This is because new superconducting materials often show anomalous isotope effects, which seem to be very difficult for us to explain within the framework of the usual BCS theory. That is, new superconducting materials show a variety of isotope effects, sometimes almost non-existent: $\alpha \approx 0$ [451], sometimes negative: $\alpha < 0$ [452], and sometiems even enhanced: $\alpha > 0.5$ [453]. In the case of some old superconducting materials which include hydrogen [454], it is reported that $\alpha$ due to $\mathrm{H} \to \mathrm{D}$ substitution becomes negative. Thus, we have seen that $\alpha$ takes various values in each material, and hence, it seems very difficult to explain this variety only in terms of the above-mentioned BCS-MA theory.

It should be emphasized that the relation $\alpha = 0.5$ does not always come from the phonon mechanism itself, but comes from the harmonic approximation, which is tacitly assumed in the BCS theory. Thus, it becomes possible to change the value of $\alpha$ even within the framework of phonon mechanisms, if we consider anharmonicity.

These new aspects of superconductivity due to highly anharmonic phonons will be seen in detail in the later sections.

### 3.7.2 Anharmonic Peierls–Hubbard Model

Let us consider a many-electron system coupled strongly with anharmonic lattice vibrations in a quasi-two-dimensional square lattice. The Hamiltonian ($\equiv H_\mathrm{A}$) of this system is given by

$$H_\mathrm{A} = -\sum_{l,l',\sigma} T_{l,l'}\eta^+_{l,\sigma}\eta_{l',\sigma} + S\sum_{l,\sigma} Q_l n_{l,\sigma} + U\sum_{l} n_{l\alpha}n_{l\beta} + H_\mathrm{ph}\,, \tag{3.52}$$

$$H_{\text{ph}} \equiv \sum_{l} \left( \frac{P_l^2}{2M} + V(Q_l) \right) ,$$

$$V(Q_l) \equiv \frac{(KQ_l^2 + BQ_l^4 + CQ_l^6)}{2} . \tag{3.53}$$

Here, $P_l$ is the momentum conjugated with $Q_l$, and $M$ denotes the effective mass of the atom corresponding to $Q_l$. $K(> 0)$ indicates the harmonic spring constant, and $B$ and $C$ are the coefficients of quartic and sextic anharmonicities, respectively.

Using this anharmonic model, we will study the metal–insulator transitions and the isotope effects. For this sake, we introduce the harmonic lattice vibration with the spring constant $K_0$ and the atomic mass $M_0$. This vibronic state will act as a reference for various anharmonic vibrations from now on. The phonon energy of this harmonic vibration is given by

$$\omega_0 \equiv \left( \frac{K_0}{M_0} \right)^{1/2} . \tag{3.54}$$

### 3.7.3 Anharmonicity and Metal–Insulator (CDW, SDW) Transitions

As mentioned before, the transition from the metal to the CDW or the SDW is expected to be suppressed in the case of hard-core type anharmonicity. *Nasu* [455] studied this problem, using the the aforementioned anharmonic Peierls–Hubbard model. His method is the mean field approximation for interelectron interactions, and a variational method for phonons, wherein the Holstein type polaron transformation and the squeezing transformation are combined. The Holstein transformation is introduced to take a lattice distortion around an electron into account, while the squeezing transformation compresses or expands the harmonic zero point vibration accoding to the anharmonicity.

The potential functions V($q$) used in this study are shown in Fig. 3.28, where

$$\mathrm{V}(q) \equiv V \left( \frac{q}{\sqrt{M_0 \omega_0}} \right) /T , \quad q \equiv \sqrt{M_0 \omega_0} Q , \tag{3.55}$$

and the site index $l$ of $Q_l$ is omitted for simplicity. We also use the following dimensionless parameters $\gamma, b$ and $c$, hereafter, instead of $K$, $B$ and $C$. They are defined as

$$\gamma \equiv \frac{K}{K_0} , \quad b \equiv \frac{B}{(M_0^2 \omega_0^3)} , \quad c \equiv \frac{C}{(M_0^3 \omega_0^4)} . \tag{3.56}$$

The solid line ($\gamma = 1$, $b = 0$, $c = 0$) in Fig. 3.28 corresponds to the harmonic case (Fig. 3.27 a), while, the broken line ($\gamma = 0.46$, $b = 0$, $c = 0.07$) corresponds to the hard-core case (Fig. 3.27 b).

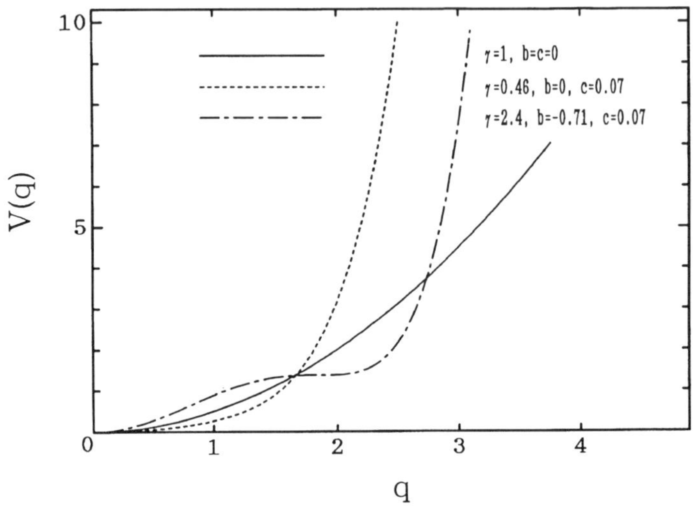


**Fig. 3.28.** Vibronic potentials $\mathrm{V}(q)$ as a function of $q$. The *solid line* denotes the harmonic type ($\gamma = 1, b = c = 0$). The *dashed line* is the hard-core repulsion type ($\gamma = 0.46, b = 0, c = 0.07$). The *dash-dotted line* is the mixed type ($\gamma = 2.4, b = -0.71, c = 0.07$)

There may be various other combinations of $\gamma$, $b$ and $c$, which correspond to the hard-core case. However, we are mainly interested in clarifying how the functional form of $\mathrm{V}(q)$ changes the relative stability between the CDW, the SDW and the metallic states. In order to focus only on this point, we restrict ourselves to the case where the lowest vibrational excitation energy given by $H_{\mathrm{ph}}$ in (3.52) is invariant in spite of the anharmonicity. That is, denoting the eigenvalue and the eigenvector of $H_{\mathrm{ph}}$ by $E_n$ and $|n\rangle$,

$$H_{\mathrm{ph}}|n\rangle = E_n|n\rangle\,, \quad n = 0,\, 1,\, 2,\, \ldots\,, \tag{3.57}$$

we consider only the combinations of $\gamma$, $b$ and $c$, which satisfy the invariant condition

$$\omega_0 = E_{10}\,, \quad E_{nm} = E_n - E_m\,. \tag{3.58}$$

Using such $\mathrm{V}(q)$, *Nasu* has calculated the energies of the CDW, the SDW and the metallic states and completed the phase diagram at absolute zero of temperature, as shown in Fig. 3.29. In this phase diagram, the horizontal and vertical axes denote the dimensionless quantities $u$ and $s$, respectively, which are defined as

$$s \equiv \frac{S}{(T\sqrt{M_0\omega_0})}\,, \quad u \equiv \frac{U}{T}\,. \tag{3.59}$$

In this calculation, the total number of electrons ($\equiv N_{\mathrm{e}}$) and the total number of lattice sites in the crystal ($\equiv N$) are taken to be equal. As for the

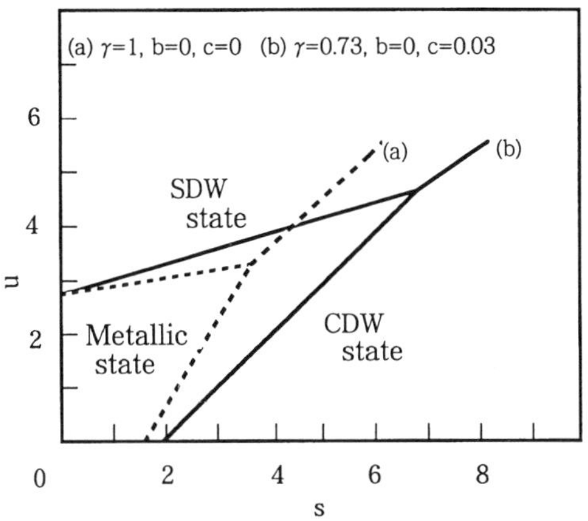


**Fig. 3.29.** The metal–insulator (CDW, SDW) transition and the phase diagram in the $u$–$s$ plane [455]. (**a**) The harmonic case. (**b**) The hard–core case (With permission from The American Physical Society)

unperturbed Fermi surface, it is assumed to be a circle with no other characteristic as in the case of Sect. 3.5. The width of the electron energy band is assumed to be 2 eV, which corresponds to the case of $d$ electrons, and $\omega_0$ is taken as $\omega_0 = 80\,\text{meV}$, which corresponds to the breathing mode of oxygen. We also introduce the dimensionless mass parameter $\mu$ defined as

$$\mu \equiv \frac{M}{M_0}\,, \tag{3.60}$$

and in the calculations shown in Fig. 3.29, $\mu$ is fixed at $\mu = 1$.

The lines (a) and (b) in Fig. 3.29 correspond to the phase boundary of the harmonic and hard-core cases, respectively. Since the Fermi surface is a circle, the nesting type metal–insulator transitions do not occur, and the system remains metallic in the region of small $u$ and $s$. As $u$ and $s$ become large and exceed certain critical values, the CDW or the SDW appears. In this calculation, the charge (or spin) density wave is assumed to undulate with twice the period along both $X$ and $Y$ axes of the crystal, since $N_e = N$.

Let us focus on the horizontal axis ($u = 0$) of Fig. 3.29. In the hard-core case, we can see that the metallic state remains stable in the region with stronger $s$, as compared with the harmonic case. *Nasu* [455] has also calculated $T_c^*$ for this stabilized metallic region, within the BCS theory, and $T_c^*$ is shown to increase exponentially as this new region expands. It easily reaches 100 K. Let us now focus on the borderline between the metallic state and the SDW state. In the hard-core case, we can also see that the metallic region expands into a new region with stronger $u$, as compared with the harmonic case. The reason for this has already been mentioned.

### 3.7.4 Isotope Effects and Anharmonicity by the BCS Theory

Let us now consider the problem of isotope effects. According to the BCS theory, the effective interaction between electrons becomes attractive only in a narrow energy region $\Delta E$ around the Fermi surface. This $\Delta E$ is given by

$$\Delta E = E_{10} , \tag{3.61}$$

while the strength of this attraction $\lambda$ is proportional to

$$\lambda \propto \frac{|\langle 1|q|0\rangle|^2}{E_{10}} . \tag{3.62}$$

The above two quantities $\Delta E$ and $\lambda$ are the two main factors that determine $T_c$.

In (3.62), it is evident that $E_{10}$ simply decreases as the atomic mass increases. However, the nature of the matrix element $|\langle 1|q|0\rangle|$ and its $M$ dependence are very complicated. Let us consider this problem in detail. $|\langle 1|q|0\rangle|$ is given by integration of the following quantity $F(M, q)$ over $q$

$$F(M, q) \equiv \phi_1^*(M, q) q \phi_0(M, q) . \tag{3.63}$$

Here, $\phi_0(M, q)$ and $\phi_1(M, q)$ are the $q$-representations of $|0\rangle$ and $|1\rangle$, respectively. In Fig. 3.30, $F(M, q)$ is illustrated as a function of $q$ together with $|\phi_0(M, q)|^2$. Here the case of the one-dimensional harmonic oscillator is taken as an example ($M$ and the spring constant are taken to be unity). It is clearly seen that the value of $\langle 1|q|0\rangle$ comes mainly from the large $q$ region, with no contribution from the origin, $q \approx 0$. In order to see the dependence of $\langle 1|q|0\rangle$ on $M$, we have also plotted the change in $F$ due to the change in $M$, in the same figure: $F(M + \Delta M, q) - F(M, q)$, ($\Delta M = 0.1$). It is clearly seen that the isotope shift comes from the large $q$ region far from the origin, where $|\phi_0(M, q)|^2$ has only a small value. Generally, in such large $q$ regions, there are some anharmonicities acting, and they become more important than in the neighborhood of the origin. Therefore, a drastic change in the isotope effect is expected due to anharmonicity, even when the vibrational ground state $\phi_0(M, q)$ itself looks like quite ordinary.

In the following, we discuss this effect on $\lambda$, assuming several types of V$(q)$. In the harmonic case, both the denominator and the numerator of $\lambda$ become exactly the same decreasing function of $M$. As a result, the attractive region decreases without change in the strength of attraction $\lambda$, as shown schematically in the lower part of Fig. 3.27 a. Thus, we get the standard isotope shift: $\alpha = 0.5$.

Next, let us consider the hard-core type repulsive potential shown in Fig. 3.27 b, as an example of extremely anharmonic potentials. Generally, the repulsion acts between two neighbouring atoms when they move close enough together. In particular, this anharmonicity is expected to be important in the case of light atoms such as oxygen, carbon and hydrogen, since the

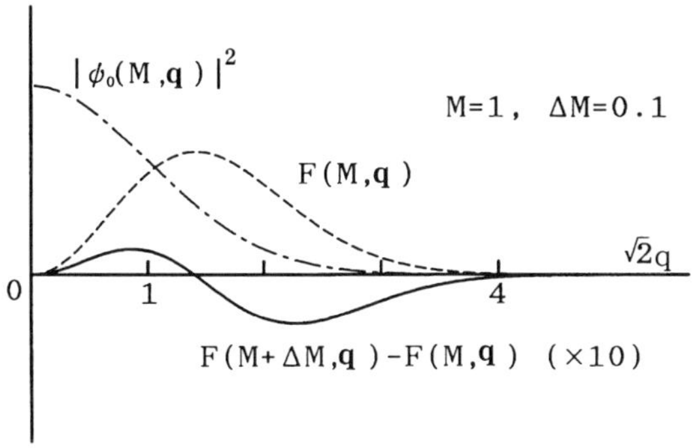


**Fig. 3.30.** Several quantities related to (3.63) as a function of displacement $q$ in the one-dimensional harmonic oscillator; $h \equiv (-\partial^2/\partial q^2 + q^2)/2, M \equiv 1$. The *dash-dotted line* denotes $|\phi_0(M,q)|^2$. The *broken line* denotes the integrand of the matrix element $\langle 1|q|0\rangle$: $F(M,q)$. The *solid line* denotes the difference of $F(M,q)$ after the isotopic substitution: $M \to M + \Delta M, \Delta M = 0.1$. On the *horizontal axis*, scales are marked in the unit of the root-mean-square displacement of the ground state: $(\langle q^2\rangle)^{1/2} = 2^{-1/2}$. From [457] (With permission from The Physical Society of Japan)

amplitudes of these vibrations are expected to become relatively large compared with those of heavier atoms. In this hard-core case, the vibronic wave functions $|n\rangle$ are independent of $M$, being determined only by the boundary condition at the hardcore. As a result, $\lambda$ increases when $M$ increases as shown schematically in the lower part of Fig. 3.27 b. In this case, the decrease in $E_{10}$ is compensated by the increase in $\lambda$. Thus, the value of $\alpha$ becomes smaller than 0.5, and if this compensation is perfect, $\alpha$ becomes zero. In some cases, $\alpha$ can even become negative, owing to over-compensation.

We can also consider of a mixture of a hardcore and a softcore with a step in its potential, as shown schematically in Fig. 3.27 c. This case describes the following situation: when $q$ is large, the hard-core repulsion acts, while, when $q$ is small, a weak repulsion also acts, and makes a shallow well within the deep well due to the hardcore. In this case, the wave function of the ground state $|0\rangle$, being localized strongly in the lower well, shrinks much more rapidly than the first excited state $|1\rangle$ as $M$ increases. This state $|1\rangle$ is almost independent of $M$ for the same reason as mentioned before. Hence, the decrease in $\langle 1|q|0\rangle$ due to the increase in $M$ becomes greater than that of $E_{10}$. Consequently, $\lambda$ as well as $E_{10}$ decreases with the increase in $M$ as shown schematically in the lower part of Fig. 3.27 c. As a result, the isotope effect will be enhanced.

### 3.7.5 Migdal–Eliashberg Theory

We have discussed isotope effects on the basis of three typical types of anharmonic potentials. However, the above discussion is too intuitive. Therefore, a quantitative calculation is required that treats the e–ph interaction within the Migdal approximation [456] together with the anharmonicities. Very recently,

*Inada* and *Nasu* [457] have studied this problem using the model Hamiltonian shown in (3.52).

To calculate $T_c$ by the Migdal–Eliashiberg theory [456], we must determine the electron self-energy and the phonon Green's function. The self-energy is divided into diagonal and off-diagonal parts. The former corresponds to the self-energy of the electron in the normal state just above $T_c$, while the latter corresponds to the superconducting state just below $T_c$. These three quantities must be determined self-consistently. In theoretical papers on superconductivity based on the Eliashiberg equation, we sometimes encounter calculations where the diagonal (normal state) self-energy is not determined self-consistently. Nevertheless, the effects of the diagonal part cannot be ignored so long as we consider isotope shifts of $T_c$, since these shifts are really small quantities. *Inada* and *Nasu* have determined these quantities self-consistently. As for the unperturbed phonon Green's function, it is determined exactly from $E_n$ and $|n\rangle$ in (3.57).

As mentioned before, the isotope effect is also varied by the Coulombic repulsion between electrons even in the harmonic case. However, in order to focus on the isotope shifts due to anharmonicities alone, we will ignore this interaction entirely in the following.

As for $V(q)$, we use three cases shown in Fig. 3.28, where the dash-dotted line ($\gamma = 2.4$, $b = -0.71$, $c = 0.07$) corresponds to the mixed type shown in Fig. 3.27 c. In this case, the value of $\gamma$, $b$ and $c$ are so chosen that the curve $V(q)$ looks like a plateau at around $q \approx 1.8$. This plateau part represents the soft-core. However, this curve is so chosen that it becomes a monotonically increasing (decreasing) function in the range $q > 0$ ($q < 0$), in order to prevent occurrence of the CDW.

Several values of $T_c$ are calculated using these potentials, and relative values $T_c(\mu)/T_c(\mu = 1)$ are plotted as a function of $\mu$ in Fig. 3.31. We have fixed $s$ at 0.24 in these calculations. In the case of the oxide superconductors, $^{18}O$ substitution for $^{16}O$ corresponds to $\mu = 1.125$. In this Fig. 3.31, the case of the harmonic potential ($\gamma = 1$, $b = c = 0$) is shown by the solid line, while the short-dashed line corresponds to the hard-core repulsion ($\gamma = 0.46$, $b = 0$, $c = 0.07$). In this case, $T_c$ is almost unchanged in spite of the increase in $\mu$. It clearly shows the disappearance of the isotope effect ($\alpha \approx 0$). When the hard-core type anharmonicity becomes stronger (the hard-core II: $\gamma = 0.019$, $b = 0$, $c = 0.15$), $T_c$ increases with the increase in $\mu$ as shown by the long dashed line, contrary to the usual isotope shifts. This is the inverse isotope effect ($\alpha < 0$). On the other hand, the dash-dotted line corresponds to the mixed case ($\gamma = 2.4$, $b = -0.71$, $c = 0.07$). It is observed that $T_c$ decreases more rapidly than in the harmonic case as $\mu$ increases. This is the enhancement of the isotope effect ($\alpha > 1/2$).

We have thus obtained a wider variety of isotope effects than expected from the BCS–MA theory. We have also seen the new aspects of a many-electron system coupling strongly with nonlinear phonons. Hard-core type

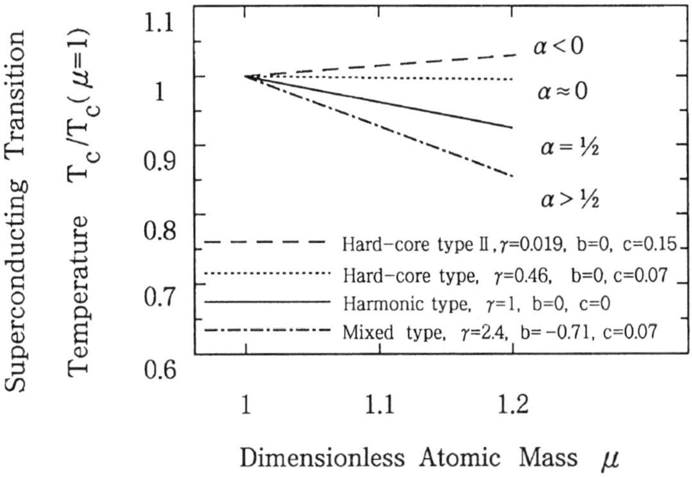


**Fig. 3.31.** Relative superconducting transition temperature $T_c/T_c(\mu = 1)$ as a function of $\mu$ for several vibronic potentials. $s = 0.24, \omega/t = 0.32$. The *solid line* is the case of the harmonic type ($\gamma = 1, b = c = 0$). The *short dashed line* is the hard–core type ($\gamma = 0.46, b = 0, c = 0.07$). The *long dashed line* is the hard–core type II ($\gamma = 0.019, b = 0, c = 0.15$). The *dash-dotted line* is the mixed type ($\gamma = 2.4, b = -0.71, c = 0.07$). From [457] (With permission from The Physical Society of Japan)

anharmonicity is shown to suppress the metal–insulator transitions, enhance the supeconductivity and reduce the isotope effect.

Finally, it should be noted that the inter-atomic potential in each crystal is always anharmonic, in the sense that it is not a simple function of interatomic distance. Its functional form is determined by the nature of the cohesive mechanism of each crystal. The harmonic approximation, on which we conventionally rely, should not be overrated, since it is but one possible approximation for such interatomic interactions.

## 3.8 Non-Grassmann Path Integral Theory for Long-Range Coulomb Repulsion

According to QED due to *Feynman*, the Coulombic repulsion is mediated by photons as mentioned earlier. This means that even the ordinary long-range Coulombic repulsion can also be represented in the same context as that of the e–ph coupling, if we introduce some hypothetical bosonic field which linearly couples to the electrons. The aim of this section is to present general and mathematical aspects of this theory. In this theory, repulsions with arbitrary spatial extension can be reduced to a simple time-dependent one-body potential through the path-integral over the hypothetical bosonic field [458–460], resulting in quantum fluctuations of spins and charges. To evaluate multibody Green's functions in this fluctuating system, the conventional Bloch–De

Dominicis theorem [461, 462] is extended so that it can be used even in cases where the original one-body Hamiltonian is time-dependent [390], as well as in the well-known cases with no time dependence [463].

Using this theorem, the optical response function is formulated, and its numerical results are represented and compared with experiment.

### 3.8.1 Quadratic Form for Long-Range Coulomb Interaction

Let us start from a Hamiltonian ($\equiv H_{\mathrm{L}}$) that includes the long-range Coulomb repulsion, whose spatial extension is arbitrary. It is given as

$$H_{\mathrm{L}} = H_0 + H_{\mathrm{EE}} , \tag{3.64}$$

where $H_0$ is an original one-body term, and is given by

$$H_0 = -\sum_{\langle \boldsymbol{l},\boldsymbol{l}'\rangle}\sum_{\sigma} T(\boldsymbol{l}-\boldsymbol{l}')\eta^{\dagger}_{\boldsymbol{l}\sigma}\eta_{\boldsymbol{l}'\sigma} . \tag{3.65}$$

$H_{\mathrm{EE}}$ denotes the long-range Coulomb repulsion

$$H_{\mathrm{EE}} = \frac{1}{2}\sum_{\boldsymbol{l},\boldsymbol{l}'} \mathcal{U}(\boldsymbol{l}-\boldsymbol{l}')n_{\boldsymbol{l}}n_{\boldsymbol{l}'} , \tag{3.66}$$

where $\mathcal{U}(\boldsymbol{l}-\boldsymbol{l}')$ is the strength of the repulsion between two sites $\boldsymbol{l}$ and $\boldsymbol{l}'$. It can be arbitrary in its force range. The Fourier transform ($\equiv \nu(\boldsymbol{q})$) of $\mathcal{U}(\boldsymbol{l}-\boldsymbol{l}')$ with a wave vector $\boldsymbol{q}$ is given as

$$\nu(\boldsymbol{q}) = \sum_{\boldsymbol{l}} e^{-i\boldsymbol{q}\cdot\boldsymbol{l}}\mathcal{U}(\boldsymbol{l}) , \tag{3.67}$$

while its inverse transformation becomes

$$\mathcal{U}(\boldsymbol{l}) = N^{-1}\sum_{\boldsymbol{q}} e^{i\boldsymbol{q}\cdot\boldsymbol{l}}\nu(\boldsymbol{q}) , \tag{3.68}$$

where $N$ is the total number of sites, assumed equal to the total number of electrons. Using these relations, we can rewrite $H_{\mathrm{EE}}$ in the following form

$$H_{\mathrm{EE}} = \frac{1}{2}\sum_{\boldsymbol{l},\boldsymbol{l}'}\left\{\nu(0)\delta_{\boldsymbol{l},\boldsymbol{l}'} - N^{-1}\sum_{\boldsymbol{q}}\left[\nu(0)-\nu(\boldsymbol{q})\right]e^{i\boldsymbol{q}\cdot(\boldsymbol{l}-\boldsymbol{l}')}\right\} n_{\boldsymbol{l}}n_{\boldsymbol{l}'} . \tag{3.69}$$

In order to proceed to the path integral form, we have to cast this repulsion $H_{\mathrm{EE}}$ into a negative quadratic form. For this reason, we define a new one-electron operator ($\equiv C_{\boldsymbol{l}}$) as

$$C_{\boldsymbol{l}} = \sum_{\boldsymbol{l}'} f(\boldsymbol{l}-\boldsymbol{l}')n_{\boldsymbol{l}'} ,$$

$$f(\boldsymbol{l}-\boldsymbol{l}') \equiv N^{-1}\sum_{\boldsymbol{q}}\sqrt{\frac{\nu(0)-\nu(\boldsymbol{q})}{\nu(0)}}\cos\left[\boldsymbol{q}\cdot(\boldsymbol{l}-\boldsymbol{l}')\right] , \tag{3.70}$$

where we have tacitly assumed that $\mathcal{U}(\boldsymbol{l}-\boldsymbol{l}')$ is not an ill-natured force, but an ordinary good-natured one in the following sense

$$\nu(0) > 0\,, \quad \nu(0) \geq \nu(\boldsymbol{q})\,, \quad \nu(\boldsymbol{q}) = \nu(-\boldsymbol{q})\,. \tag{3.71}$$

Using $C_l$ thus defined, we can rewrite $H_{\mathrm{EE}}$ into a new negative quadratic form as

$$H_{\mathrm{EE}} = -\frac{\nu(0)}{2}\sum_l (C_l^2 + M_l^2) + \nu(0)N\,, \quad M_l \equiv n_{l\alpha} - n_{l\beta}\,. \tag{3.72}$$

One can easily see that $M_l$, thus defined, corresponds to the short-range Coulombic correlation, while $C_l$ corresponds to the long-range one.

### 3.8.2 Path-Integral for Both Short- and Long-Range Parts

We are now ready to proceed to the path integral form of the Boltzmann operator $e^{-\theta H_{\mathrm{L}}}, \theta \equiv 1/k_{\mathrm{B}}T_{\mathrm{temp}}$. By Trotter's formula, $\theta$ is divided into infinitesimal parts as

$$e^{-\theta H_{\mathrm{L}}} = \lim_{W\to\infty}(e^{-\Delta H_{\mathrm{L}}}\dots e^{-\Delta H_{\mathrm{L}}})\,, \quad \Delta \equiv \theta/W\,. \tag{3.73}$$

In each time interval $\Delta$, $C_l^2$ and $M_l^2$ of $H_{\mathrm{L}}$ can be linearized by the Gaussian integral formula [459] as

$$e^{-\Delta H_{\mathrm{EE}}} \to \prod_l \left[\frac{\Delta\nu(0)}{2\pi}\int\!\!\int \mathrm{d}\mathrm{x}_{Ml}\mathrm{d}\mathrm{x}_{Cl}\right]$$
$$\times \exp\left\{-\Delta\nu(0)\sum_l\left[-(\mathrm{x}_{Cl}C_l + \mathrm{x}_{Ml}M_l) + \frac{\mathrm{x}_{Cl}^2 + \mathrm{x}_{Ml}^2}{2}\right]\right\}\,, \tag{3.74}$$

wherein $\mathrm{x}_{Cl}$ and $\mathrm{x}_{Ml}$ are ordinary $c$-numbers. Using (3.73) and (3.74), we get a path integral form of the Boltzmann operator as

$$e^{-\theta H_{\mathrm{L}}} \to \int \mathcal{D}\mathrm{x}\left\{T_+ \exp\left[-\int_0^\theta \mathrm{d}\tau\left(H_0(\tau) + \nu(0)\right.\right.\right. \tag{3.75}$$
$$\left.\left.\left.\times\sum_l\left(-[\mathrm{x}_{Ml}(\tau)M_l(\tau) + \mathrm{x}_{Cl}(\tau)C_l(\tau)] + \frac{\mathrm{x}_{Ml}^2(\tau) + \mathrm{x}_{Cl}^2(\tau)}{2}\right)\right)\right]\right\}\,.$$

Here $\int \mathcal{D}\mathrm{x}$ means the path integrals over $\mathrm{x}_{Ml}(\tau)$ and $\mathrm{x}_{Cl}(\tau)$, in the sense written in (3.74). In this equation, $T_+$ is the time-ordering operator, and the time-argument $\tau(\theta \geq \tau \geq 0)$ of all the operators $H_0(\tau)$, $C_l(\tau)$ and $M_l(\tau)$ is only for this time ordering. Hence, these operators have no real time-dependence, in contrast to the usual Heisenberg representation, while the $c$-numbers $\mathrm{x}_{Ml}(\tau)$ and $\mathrm{x}_{Cl}(\tau)$ are really time-dependent. Thus, both the short- and long-range parts are treated on an equal footing, and reduced into the time-dependent one-body Hamiltonian, finally resulting in time-dependent charge- and spin-fluctuations. We should note that only two degrees of freedom per site, $\mathrm{x}_{Ml}(\tau)$ and $\mathrm{x}_{Cl}(\tau)$, are enough to linearize arbitrary $\mathcal{U}(\boldsymbol{l}-\boldsymbol{l}')$.

### 3.8.3 One-Body Green's Function Free from Grassmann Algebra

Let us now formulate the one-body Green's function. For this sake, we here define several mathematical notations. At first, we define a time dependent one-body Hamiltonian ($\equiv h(\tau,\mathrm{x})$) along a path $\mathrm{x}(\tau)$ as

$$h(\tau,\mathrm{x}) = H_0(\tau) \tag{3.76}$$

$$+\nu(0)\sum_{\boldsymbol{l}}\left\{-\left[\mathrm{x}_{M\boldsymbol{l}}(\tau)M_{\boldsymbol{l}}(\tau)+\mathrm{x}_{C\boldsymbol{l}}(\tau)C_{\boldsymbol{l}}(\tau)\right]+\frac{1}{2}\left[\mathrm{x}_{M\boldsymbol{l}}^2(\tau)+\mathrm{x}_{C\boldsymbol{l}}^2(\tau)\right]\right\},$$

where x symbolically denotes $\mathrm{x}_{M\boldsymbol{l}}$ and $\mathrm{x}_{C\boldsymbol{l}}$. We now define a $2N\times 2N$ matrix ($\equiv \boldsymbol{H}(\tau,\mathrm{x})$) which is the one-body version of $h(\tau,\mathrm{x})$. Its element is given as

$$\left[\boldsymbol{H}(\tau,\mathrm{x})\right]_{j,j'} \equiv \langle 0|\eta_j(\tau)h(\tau,\mathrm{x})\eta_{j'}^\dagger(\tau)|0\rangle\,,$$

$$|0\rangle \equiv \text{true electron vacuum}\,, \tag{3.77}$$

where index $j$ symbolically denotes $\boldsymbol{l}$ and $\sigma$, and hence the total number of $j$ is $2N$.

Using this $\boldsymbol{H}(\tau,\mathrm{x})$, we can also define the time evolution operator ($\equiv \mathcal{R}(\tau,\mathrm{x})$) from 0 to $\tau$ along the path. It is given as

$$\mathcal{R}(\tau,\mathrm{x}) = T_+\exp\left\{-\int_0^\tau \mathrm{d}\tau''\boldsymbol{A}^\dagger(\tau'')\boldsymbol{H}\left[\tau'',\mathrm{x}(\tau'')\right]\boldsymbol{A}(\tau'')\right\},$$

$$\theta \geq \tau \geq 0\,, \tag{3.78}$$

where $\boldsymbol{A}^\dagger$ is a $2N$-dimensional vector operator whose component is $\eta_j^\dagger$, defined by

$$\boldsymbol{A}^\dagger \equiv (\eta_1^\dagger,\ldots,\eta_j^\dagger,\ldots\eta_{2N}^\dagger)\,. \tag{3.79}$$

The one-body time-evolution matrix ($\equiv \boldsymbol{R}(\tau,\mathrm{x})$) corresponding to this operator $\mathcal{R}(\tau,\mathrm{x})$ can also be given as

$$\boldsymbol{R}(\tau,\mathrm{x}) = T_+\exp\left\{-\int_0^\tau \mathrm{d}\tau''\boldsymbol{H}\left[\tau'',\mathrm{x}(\tau'')\right]\right\}, \tag{3.80}$$

and it is also a $2N\times 2N$ matrix.

In terms of this operator $\mathcal{R}(\tau,\mathrm{x})$, we can formally define the free energy $[\equiv \Omega(\mathrm{x})]$ along a given path x as

$$e^{-\theta\Omega(\mathrm{x})} = \mathrm{Tr}\left[\mathcal{R}(\theta,\mathrm{x})\right], \tag{3.81}$$

and the total free energy ($\equiv \Omega$) is given by its path integral

$$e^{-\theta\Omega} = \int \mathcal{D}\mathrm{x}e^{-\theta\Omega(\mathrm{x})}\,. \tag{3.82}$$

Using the operator $\mathcal{R}(\tau, \mathrm{x})$, we can also obtain the average ($\equiv \langle ... \rangle_{\mathrm{x}}$) of an ordinary operator ... along the path, as

$$\langle ... \rangle_{\mathrm{x}} = \frac{\mathrm{Tr}\left[\mathcal{R}(\theta, \mathrm{x}) ...\right]}{\mathrm{Tr}\left[\mathcal{R}(\theta, \mathrm{x})\right]} . \tag{3.83}$$

From these notations, we can now define the one-body Green's function ($\equiv G(j\tau, j'\tau', \mathrm{x})$) along a path x as

$$G(j\tau, j'\tau', \mathrm{x}) = -\mathrm{sign}(\tau - \tau') \langle T_{+} \overrightarrow{\eta}_{j} (\tau) \overrightarrow{\eta}^{\dagger}_{j'} (\tau') \rangle_{\mathrm{x}} , \tag{3.84}$$

where an operator $\overrightarrow{O}(\tau)$ with an arrow on its head denotes the Heisenberg representation of an operator $O$, and is defined by

$$\overrightarrow{O}(\tau) \equiv \mathcal{R}^{-1}(\tau, \mathrm{x}) O \mathcal{R}(\tau, \mathrm{x}) . \tag{3.85}$$

This notation will be used very frequently in the following. The ordinary one-body Green's function ($\equiv G(j\tau, j'\tau')$) is given after the path integral of $G(j\tau, j'\tau', \mathrm{x})$ as

$$G(j\tau, j'\tau') = \int \mathcal{D}\mathrm{x} e^{-\theta[\Omega(\mathrm{x}) - \Omega]} G(j\tau, j'\tau', \mathrm{x}) . \tag{3.86}$$

From (3.78), (3.80) and (3.85), we can easily derive the differential equation for $\overrightarrow{\eta}_{j}(\tau)$ as

$$\frac{\partial \overrightarrow{\eta}_{j}(\tau)}{\partial \tau} = -\sum_{j_2} \left[\boldsymbol{H}(\tau, \mathrm{x})\right]_{jj_2} \overrightarrow{\eta}_{j_2}(\tau) , \tag{3.87}$$

and the solution of this equation is given by

$$\overrightarrow{\eta}_{j}(\tau) = \sum_{j_2} \left[\boldsymbol{R}(\tau, \mathrm{x})\right]_{jj_2} \eta_{j_2} . \tag{3.88}$$

In the case of the creation operator, the procedure is the same, and we also get

$$\overrightarrow{\eta}^{\dagger}_{j'}(\tau') = \sum_{j_2'} \eta^{\dagger}_{j_2'} \left[\boldsymbol{R}^{-1}(\tau', \mathrm{x})\right]_{j_2'j'} . \tag{3.89}$$

These results can be substituted into (3.84), and $G(j\tau, j'\tau', \mathrm{x})$ becomes

$$G(j\tau, j'\tau', \mathrm{x}) = \sum_{j_2, j_2'} [\boldsymbol{R}(\tau, \mathrm{x})]_{jj_2} [\boldsymbol{R}^{-1}(\tau', \mathrm{x})]_{j_2'j'} \begin{cases} -\langle \eta_{j_2} \eta^{\dagger}_{j_2'} \rangle_{\mathrm{x}}, \tau > \tau' \\ \langle \eta^{\dagger}_{j_2'} \eta_{j_2} \rangle_{\mathrm{x}}, \tau < \tau' \end{cases} . \tag{3.90}$$

Let us now focus on the last factor of (3.90). When $\tau > \tau'$, the last factor $-\langle\eta_{j_2}\eta^\dagger_{j'_2}\rangle_{\rm x}$ can be rewritten in terms of $\langle\eta^\dagger_{j'_2}\eta_{j_2}\rangle_{\rm x}$ using the commutation relation:

$$-\langle\eta_{j_2}\eta^\dagger_{j'_2}\rangle_{\rm x} = -\delta_{j_2j'_2} + \langle\eta^\dagger_{j'_2}\eta_{j_2}\rangle_{\rm x}\,. \tag{3.91}$$

From the definition of the trace, and also from (3.83), (3.85) and (3.89), $\langle\eta^\dagger_{j'_2}\eta_{j_2}\rangle_{\rm x}$ is rewritten as

$$\begin{aligned}\langle\eta^\dagger_{j'_2}\eta_{j_2}\rangle_{\rm x} &= \frac{{\rm Tr}[\eta_{j_2}\mathcal{R}(\theta,{\rm x})\eta^\dagger_{j'_2}]}{{\rm Tr}[\mathcal{R}(\theta,{\rm x})]} = \langle\vec{\eta}_{j_2}(\theta)\eta^\dagger_{j'_2}\rangle_{\rm x}\\ &= \sum_{j''_2}[\boldsymbol{R}(\theta,{\rm x})]_{j_2j''_2}\langle\eta_{j''_2}\eta^\dagger_{j'_2}\rangle_{\rm x}\,.\end{aligned} \tag{3.92}$$

Combining these two results, we can obtain the following equation for $\langle\eta_{j''_2}\eta^\dagger_{j'_2}\rangle_{\rm x}$:

$$\delta_{j_2j'_2} = \sum_{j''_2}[1+\boldsymbol{R}(\theta,{\rm x})]_{j_2j''_2}\langle\eta_{j''_2}\eta^\dagger_{j'_2}\rangle_{\rm x}\,. \tag{3.93}$$

Regarding this equation as a kind of matrix equation, we can obtain its solution as

$$\langle\eta_{j''_2}\eta^\dagger_{j'_2}\rangle_{\rm x} = \left(\frac{1}{1+\boldsymbol{R}(\theta,{\rm x})}\right)_{j_2j'_2}\,. \tag{3.94}$$

When $\tau < \tau'$, the procedure in (3.90) is identical, and we finally get

$$G(j\tau,j'\tau',{\rm x}) = \left(\boldsymbol{R}(\tau,{\rm x})\left\{\begin{array}{l} -\frac{1}{1+\boldsymbol{R}(\theta,{\rm x})},\,\tau>\tau',\\ \frac{1}{1+\boldsymbol{R}^{-1}(\theta,{\rm x})},\,\tau<\tau'\end{array}\right\}\boldsymbol{R}^{-1}(\tau',{\rm x})\right)_{jj'}\,. \tag{3.95}$$

At a glance, we can understand that it is a straightforward extension of a time-independent case, since this formula is quite analogous to the ordinary Green's function.

Incidentally, the unequal-time commutation relation becomes

$$\vec{\eta}_j(\tau)\,\vec{\eta}^\dagger_{j'}(\tau') + \vec{\eta}^\dagger_{j'}(\tau')\,\vec{\eta}_j(\tau) = [\boldsymbol{R}(\tau,{\rm x})\boldsymbol{R}^{-1}(\tau',{\rm x})]_{jj'}\,, \tag{3.96}$$

and this equation will be used later.

As for the free energy, using the conventional mathematical technique [463], we can cast $\Omega({\rm x})$ into the following form:

$$\begin{aligned}&\Omega({\rm x}) = \Omega_1({\rm x}) = \Omega_0({\rm x}) + \int_0^1 {\rm d}\lambda\frac{\partial\Omega_\lambda({\rm x})}{\partial\lambda}\,,\\ &\Omega_\lambda({\rm x}) \equiv -\theta^{-1}\ln\left\{{\rm Tr}\left[\mathcal{R}_\lambda(\theta,{\rm x})\right]\right\}\,,\end{aligned} \tag{3.97}$$

$$\mathcal{R}_\lambda(\theta, \mathrm{x}) \equiv T_+ \exp\left\{ -\lambda \int_0^\theta \mathrm{d}\tau'' \boldsymbol{A}^\dagger(\tau'') \boldsymbol{H}(\tau'', \mathrm{x}(\tau'')) \boldsymbol{A}(\tau'') \right\} , \tag{3.98}$$

and from (3.95),

$$\Omega(\mathrm{x}) = \Omega_0(\mathrm{x}) + \theta^{-1} \int_0^\theta \mathrm{d}\tau \int_0^1 \mathrm{d}\lambda$$

$$\times \mathrm{Tr}\left[ \boldsymbol{R}_\lambda^{-1}(\tau, \mathrm{x}) \boldsymbol{H}(\tau, \mathrm{x}) \boldsymbol{R}_\lambda(\tau, \mathrm{x}) \frac{1}{1 + \boldsymbol{R}_\lambda^{-1}(\theta, \mathrm{x})} \right] , \tag{3.99}$$

$$\boldsymbol{R}_\lambda(\tau, \mathrm{x}) \equiv T_+ \exp\left\{ -\lambda \int_0^\tau d\tau'' \boldsymbol{H}[\tau'', \mathrm{x}(\tau'')] \right\} . \tag{3.100}$$

Making use of the partial integration method with respect to $\lambda$, we finally get the following simple form of $\Omega(\mathrm{x})$:

$$\Omega(\mathrm{x}) = -\theta^{-1} \mathrm{Tr}\left\{ \ln[1 + \boldsymbol{R}(\theta, \mathrm{x})] \right\} . \tag{3.101}$$

and its practical calculations can be performed by the following mathematical identity

$$\mathrm{Tr}\left\{ \ln[1 + \boldsymbol{R}(\theta, \mathrm{x})] \right\} \equiv \ln\left[ \det |1 + \boldsymbol{R}(\theta, \mathrm{x})| \right] . \tag{3.102}$$

It is also wellknown that $\Omega(\mathrm{x})$ can be calculated by the path integral method based on the Grassmann algebra [460, 464, 465]. However, the theory, developed in this section, depends only on the ordinary equation of motion method, being completely free from the complicated path integral based on the Grassmann algebra [465]. We will see, in the next section, that the present equation of motion method is very simple and convenient in the case of multi-body Green's functions for more than one body.

### 3.8.4 Time-Dependent Bloch–De Dominicis Theorem

Whenever we have to calculate multi-body Green's functions associated with a certain perturbation theory, we usually collect Feynman diagrams corresponding to higher order terms of this perturbation expansion. The rule for collecting various diagrams is already well established, and it is based on the Bloch–De Dominicis theorem, which proves that the thermal average of higher order terms can be decoupled into multiple products of the unperturbed one-body Green's functions. However, this theorem tacitly assumes that the unperturbed one-body Hamiltonian is time-independent [463].

Let us now extend this theorem to time-dependent cases. We first consider the calculation of the following average of $2m$ ($m \geq 2$) time-ordered field operators

$$\langle \overrightarrow{O}_{j_1}(\tau_1) \overrightarrow{O}_{j_2}(\tau_2) \ldots \overrightarrow{O}_{j_{2m-1}}(\tau_{2m-1}) \overrightarrow{O}_{j_{2m}}(\tau_{2m}) \rangle_{\mathrm{x}} ,$$

$$\tau_1 > \tau_2 > \ldots \tau_{2m-1} > \tau_{2m} , \tag{3.103}$$

where $O$ denotes only the creation or the annihilation operator, and the total numbers of these two kinds of operators in this average are equal. The commutator ($\equiv (j\tau, j'\tau')$) between $\vec{O}_j(\tau)$ and $\vec{O}_{j'}(\tau')$ can be formally given as

$$(j\tau, j'\tau') = \vec{O}_j(\tau)\,\vec{O}_{j'}(\tau') + \vec{O}_{j'}(\tau')\,\vec{O}_j(\tau)\,, \tag{3.104}$$

and it always becomes a $c$-number, being not zero only when one of these two operators is a creation operator and the other is an annihilation operator, as shown in (3.96). By this commutation relation, the operator at the left end of (3.103) can be moved one step toward the right end:

$$\begin{aligned}
&\langle \vec{O}_{j_1}(\tau_1)\,\vec{O}_{j_2}(\tau_2)\ldots\vec{O}_{j_{2m-1}}(\tau_{2m-1})\,\vec{O}_{j_{2m}}(\tau_{2m})\rangle_{\mathrm{x}} \\
&= -\langle \vec{O}_{j_2}(\tau_2)\,\vec{O}_{j_1}(\tau_1)\,\vec{O}_{j_3}(\tau_3)\ldots\vec{O}_{j_{2m}}(\tau_{2m})\rangle_{\mathrm{x}} \\
&\quad + (j_1\tau_1, j_2\tau_2)\langle \vec{O}_{j_3}(\tau_3)\ldots\vec{O}_{j_{2m}}(\tau_{2m})\rangle_{\mathrm{x}}\,.
\end{aligned} \tag{3.105}$$

Repeating this procedure, we can finally move $\vec{O}_{j_1}(\tau_1)$ to the right end to give

$$\begin{aligned}
&\langle \vec{O}_{j_1}(\tau_1)\,\vec{O}_{j_2}(\tau_2)\ldots\vec{O}_{j_{2m}}(\tau_{2m})\rangle_{\mathrm{x}} \\
&= -\langle \vec{O}_{j_2}(\tau_2)\ldots\vec{O}_{j_{2m}}(\tau_{2m})\,\vec{O}_{j_1}(\tau_1)\rangle_{\mathrm{x}} + F_{j_1}(\tau_1)\,,
\end{aligned} \tag{3.106}$$

$$F_{j_1}(\tau_1) \equiv \sum_{i=2}^{2m} \operatorname{sgn}\begin{pmatrix} 1,2,3,\ldots,i,i+1,\ldots,2m \\ 1,i,2,\ldots,i-1,i+1,\ldots,2m \end{pmatrix} \tag{3.107}$$

$$\times (j_1\tau_1, j_i\tau_i)\langle \vec{O}_{j_2}(\tau_2)\ldots\vec{O}_{j_{i-1}}(\tau_{i-1})\,\vec{O}_{j_{i+1}}(\tau_{i+1})\ldots\vec{O}_{j_{2m}}(\tau_{2m})\rangle_{\mathrm{x}}\,.$$

Here $\operatorname{sgn}(c)$ denotes the sign of the permutation $c$. We should note that $F_{j_1}(\tau_1)$, thus defined, includes only the average of $2(m-1)$ field operators, in which $\vec{O}_{j_1}(\tau_1)$ and $\vec{O}_{j_i}(\tau_i)$ are already missing, as seen in the last factor of (3.107). On the other hand, from the definition of the trace, $\vec{O}_{j_1}(\tau_1)$ in the right end can return to the left again as

$$\begin{aligned}
&\langle \vec{O}_{j_2}(\tau_2)\ldots\vec{O}_{j_{2m}}(\tau_{2m})\,\vec{O}_{j_1}(\tau_1)\rangle_{\mathrm{x}} \\
&= \langle \mathcal{R}^{-1}(\theta,\mathrm{x})\,\vec{O}_{j_1}(\tau_1)\mathcal{R}(\theta,\mathrm{x})\,\vec{O}_{j_2}(\tau_2)\ldots\vec{O}_{j_{2m}}(\tau_{2m})\rangle_{\mathrm{x}}\,.
\end{aligned} \tag{3.108}$$

Let us now restrict the rest of our derivation to the case where $\vec{O}_{j_1}(\tau_1)$ is a creation operator. In this case, $\mathcal{R}^{-1}(\theta,\mathrm{x})\,\vec{O}_{j_1}(\tau_1)\mathcal{R}(\theta,\mathrm{x})$ in (3.108) can be rewritten using (3.89) twice:

$$\begin{aligned}
\mathcal{R}^{-1}(\theta,\mathrm{x})\,\vec{\eta}^{\,\dagger}_{j_1}(\tau_1)\mathcal{R}(\theta,\mathrm{x}) &= \sum_{j_1'} \vec{\eta}^{\,\dagger}_{j_1'}(\theta)[\boldsymbol{R}^{-1}(\tau_1,\mathrm{x})]_{j_1'j_1} \\
&= \sum_{j_1'} \eta^{\dagger}_{j_1'}[\boldsymbol{R}^{-1}(\theta,\mathrm{x})\boldsymbol{R}^{-1}(\tau_1,\mathrm{x})]_{j_1'j_1}\,,
\end{aligned} \tag{3.109}$$

and substituting this result into (3.108), we get

$$\langle \vec{O}_{j_2}(\tau_2) \ldots \vec{O}_{j_{2m}}(\tau_{2m})\, \vec{\eta}^{\dagger}_{j_1}(\tau_1)\rangle_{\mathrm{x}}$$

$$= \sum_{j_1'} \langle \eta^{\dagger}_{j_1'}\, \vec{O}_{j_2}(\tau_2) \ldots \vec{O}_{j_{2m}}(\tau_{2m})\rangle_{\mathrm{x}} \left[\boldsymbol{R}^{-1}(\theta,\mathrm{x})\boldsymbol{R}^{-1}(\tau_1,\mathrm{x})\right]_{j_1' j_1} . \tag{3.110}$$

On the other hand, the original average given in (3.103) can also be rewritten using (3.89):

$$\langle \vec{\eta}^{\dagger}_{j_1}(\tau_1)\, \vec{O}_{j_2}(\tau_2) \ldots \vec{O}_{j_{2m}}(\tau_{2m})\rangle_{\mathrm{x}}$$

$$= \sum_{j_1'} \langle \eta^{\dagger}_{j_1'}\, \vec{O}_{j_2}(\tau_2) \ldots \vec{O}_{j_{2m}}(\tau_{2m})\rangle_{\mathrm{x}} [\boldsymbol{R}^{-1}(\tau_1,\mathrm{x})]_{j_1' j_1} . \tag{3.111}$$

Substituting (3.110) and (3.111) into (3.106), we get

$$\sum_{j_1'} \langle \eta^{\dagger}_{j_1'}\, \vec{O}_{j_2}(\tau_2) \ldots \vec{O}_{j_{2m}}(\tau_{2m})\rangle_{\mathrm{x}} \left\{\left[1+\boldsymbol{R}^{-1}(\theta,\mathrm{x})\right] \boldsymbol{R}^{-1}(\tau_1,\mathrm{x})\right\}_{j_1' j_1}$$

$$= F_{j_1}(\tau_1) . \tag{3.112}$$

Regarding this equation as a set of simultaneous equations to determine $\langle \eta^{\dagger}_{j_1'}\, \vec{O}_{j_2}(\tau_2) \ldots \vec{O}_{j_{2m}}(\tau_{2m})\rangle_{\mathrm{x}}$ with various values of $j_1'$, we can get its solution

$$\langle \eta^{\dagger}_{j_1'}\, \vec{O}_{j_2}(\tau_2) \ldots \vec{O}_{j_{2m}}(\tau_{2m})\rangle_{\mathrm{x}}$$

$$= \sum_{j_1''} F_{j_1''}(\tau_1) \left[\boldsymbol{R}(\tau_1,\mathrm{x}) \frac{1}{1+\boldsymbol{R}^{-1}(\theta,\mathrm{x})}\right]_{j_1'' j_1'} . \tag{3.113}$$

Combining this equation with (3.89), (3.96) and (3.104), we finally get

$$\langle \vec{\eta}^{\dagger}_{j_1}(\tau_1)\, \vec{O}_{j_2}(\tau_2) \ldots \vec{O}_{j_{2m}}(\tau_{2m})\rangle_{\mathrm{x}}$$

$$= \sum_{i=2}^{2m}{}' \operatorname{sgn} \begin{pmatrix} 1,2,3,\ldots,i,i+1,\ldots,2m \\ 1,i,2,\ldots,i-1,i+1,..,2m \end{pmatrix}$$

$$\times \left[\boldsymbol{R}(\tau_i,\mathrm{x}) \frac{1}{1+\boldsymbol{R}^{-1}(\theta,\mathrm{x})} \boldsymbol{R}^{-1}(\tau_1,\mathrm{x})\right]_{j_i j_1}$$

$$\times \langle \vec{O}_{j_2}(\tau_2) \ldots \vec{O}_{j_{i-1}}(\tau_{i-1})\, \vec{O}_{j_{i+1}}(\tau_{i+1}) \ldots \vec{O}_{j_{2m}}(\tau_{2m})\rangle_{\mathrm{x}} , \tag{3.114}$$

where $\sum'$ means this summation should be taken only when $\vec{O}_{j_i}$ is an annihilation operator. Thus the average of $2m$ field operators is proved to be decoupled into the one-body Green's function and the average of the residual $2(m-1)$ field operators. In the case when $\overrightarrow{O}_{j_1}(\tau_1)$ is an annihilation operator, we can also prove that this decoupling is possible by the same procedure as mentioned above. Consequently, repeating this decoupling, we can finally describe the original average given in (3.103) as multiple products of the Green's functions.

Moreover, we can easily see that the conclusion shown in (3.114) accords precisely with the time-independent Bloch–De Dominicis theorem, except that the Green's function should be replaced by the one given by (3.95). For example, the problem of the sign of the permutation in (3.114) is common to both the time-dependent and the time-independent cases. It comes only from the exchange between fermion operators, regardless of whether the unperturbed one-body Hamiltonian is time-independent or not.

Thus we can conclude that the rule for collecting Feynman diagrams is the same as that for the time-independent case as long as we use the new one-body Green's function shown in (3.95).

### 3.8.5 Light Absorption Spectrum of the SDW State

As a primitive application of this new diagram technique, let us now focus on the light absorption spectrum of the one-dimensional SDW state given by (3.23). In this case, we have to calculate the dipole correlation function, which is a kind of two-body Green's function, whose Fourier transform gives the light absorption spectrum. The dipole operator ($\equiv D$) is given by

$$D = i \sum_{\langle \boldsymbol{l},\boldsymbol{l}'\rangle} \sum_{\sigma} D(\boldsymbol{l}-\boldsymbol{l}')\eta^{\dagger}_{\boldsymbol{l}\sigma}\eta_{\boldsymbol{l}'\sigma'} , \quad D(\boldsymbol{l}-\boldsymbol{l}') \equiv \boldsymbol{p}\cdot\boldsymbol{e}(\boldsymbol{l}-\boldsymbol{l}')T' , \tag{3.115}$$

where $\boldsymbol{p}$ is the light polarization vector, $\boldsymbol{e}(\boldsymbol{l}-\boldsymbol{l}')$ is the direction vector from $\boldsymbol{l}'$ to $\boldsymbol{l}$, and $T'$ is the dipole matrix element between two neighboring orbitals. In terms of this $D$, the dipole correlation function ($\equiv K(\tau-\tau')$) is written as

$$K(\tau-\tau') = \int \mathcal{D}\mathrm{x}e^{-\theta[\Omega(\mathrm{x})-\Omega]}\langle T_{+}[\overrightarrow{D}(\tau)\,\overrightarrow{D}^{\dagger}(\tau')]\rangle_{\mathrm{x}} . \tag{3.116}$$

This path integral form is decomposed into a two-body Green's function ($\equiv K^{\sigma\sigma'}(\boldsymbol{l}_1\boldsymbol{l}'_1\tau,\boldsymbol{l}_2\boldsymbol{l}'_2\tau',\mathrm{x})$) along each path x, as

$$K(\tau-\tau') = -\int \mathcal{D}\mathrm{x}e^{-\theta[\Omega(\mathrm{x})-\Omega]} \sum_{\langle \boldsymbol{l}_1,\boldsymbol{l}'_1\rangle} \sum_{\langle \boldsymbol{l}_2,\boldsymbol{l}'_2\rangle} D(\boldsymbol{l}_1-\boldsymbol{l}'_1)D(\boldsymbol{l}_2-\boldsymbol{l}'_2)$$

$$\times \sum_{\sigma\sigma'} K^{\sigma\sigma'}(\boldsymbol{l}_1\boldsymbol{l}'_1\tau,\boldsymbol{l}_2\boldsymbol{l}'_2\tau',\mathrm{x}) , \tag{3.117}$$

where this two-body Green's function is given by

$$K^{\sigma\sigma'}(\boldsymbol{l}_1\boldsymbol{l}'_1\tau,\boldsymbol{l}_2\boldsymbol{l}'_2\tau',\mathrm{x}) = -\langle T_{+}[\overrightarrow{\eta}_{\boldsymbol{l}_1\sigma}(\tau)\,\overrightarrow{\eta}^{\dagger}_{\boldsymbol{l}'_1\sigma}(\tau)\,\overrightarrow{\eta}_{\boldsymbol{l}_2\sigma'}(\tau')\,\overrightarrow{\eta}^{\dagger}_{\boldsymbol{l}'_2\sigma'}(\tau')]\rangle_{\mathrm{x}} . \tag{3.118}$$

From (3.118), we get two type diagrams. However, the contractions between $l_1$ and $l'_1$, and also between $l_2$ and $l'_2$ are always irrelevant to the optical response. Hence, $K^{\sigma\sigma'}(l_1 l'_1 \tau, l_2 l'_2 \tau', \mathrm{x})$ is given by

$$K^{\sigma\sigma'}(l_1 l'_1 \tau, l_2 l'_2 \tau', \mathrm{x}) = \delta_{\sigma\sigma'} G(l'_1 \sigma\tau, l_2 \sigma\tau', \mathrm{x}) G(l'_2 \sigma\tau', l_1 \sigma\tau, \mathrm{x}) . \tag{3.119}$$

In practical calculations, we can use the quantum Monte Carlo method [460] to obtain various numerical results. To perform this calculation numerically, the following linearization by a discrete random number is sometimes more tractable than (3.74):

$$\begin{aligned} e^{-\Delta H_{EE}} &\to \prod_l \left\{ \left( \frac{1}{2} \sum_{\mathrm{x}_{Ml}=1,-1} \right) \left( \frac{1}{2} \sum_{\mathrm{x}_{Cl}=1,-1} \right) \right\} \\ &\times \exp \left[ \Delta \sum_l \sqrt{\frac{\nu(0)}{\Delta}} \left( \mathrm{x}_{Cl} C_l + \mathrm{x}_{Ml} M_l \right) \right] + O(\Delta^2) , \end{aligned} \tag{3.120}$$

wherein $\mathrm{x}_{Cl}$ and $\mathrm{x}_{Ml}$ are discrete numbers, which take only the values $\pm 1$ randomly. In this case, the path integral is reduced to the summation over these random numbers at each time interval $\Delta$.

On the other hand, we can also rewrite (3.74) into the mean field approximation and a discrete random fluctuation therefrom:

$$\begin{aligned} e^{-\Delta H_{EE}} &\to \exp \left\{ -\Delta \nu(0) \sum_l \left[ -(\bar{\mathrm{x}}_{Cl} C_l + \bar{\mathrm{x}}_{Ml} M_l) + \frac{\bar{\mathrm{x}}_{Cl}^2 + \bar{\mathrm{x}}_{Ml}^2}{2} \right] \right\} \\ &\times \prod_l \left[ \left( \frac{1}{3} \sum_{\mathrm{x}_{Ml}=1,0,-1} \right) \left( \frac{1}{3} \sum_{\mathrm{x}_{Cl}=1,0,-1} \right) \right] \\ &\times \exp \left\{ \Delta \sum_l \sqrt{\frac{3\nu(0)}{2\Delta}} \left[ \mathrm{x}_{Cl} (C_l - \bar{\mathrm{x}}_{Cl}) + \mathrm{x}_{Ml} (M_l - \bar{\mathrm{x}}_{Ml}) \right] \right\} + O(\Delta^2) . \end{aligned} \tag{3.121}$$

Here, $\bar{\mathrm{x}}_{Cl}$ and $\bar{\mathrm{x}}_{Ml}$ are averages of $C_l$ and $M_l$, respectively, which should be determined beforehand by some approximation such as the Hartree type mean field theory. $\mathrm{x}_{Cl}$ and $\mathrm{x}_{Ml}$ are still discrete numbers, but take the value 0 as well as $\pm 1$, randomly, so that we can refer to this mean field theory. Depending on the problem which interests us, we can choose either (3.74), (3.120) or (3.121).

In Fig. 3.32 a, we have shown one of the calculated results for a one-dimensional SDW state by *Tomita* and *Nasu* [466]. In this figure, the lowest energy peak is the CT exciton in the SDW state. It has a sharp infrared-divergence type spectral shape coming from the multi-magnon excitations induced by this charge transfer exciton. The highest energy peak comes from the two-exciton excitation. In Fig. 3.32 b, we have also shown the light absorption spectrum of the HMMC ($M$ = Ni) mentioned before [418], and we

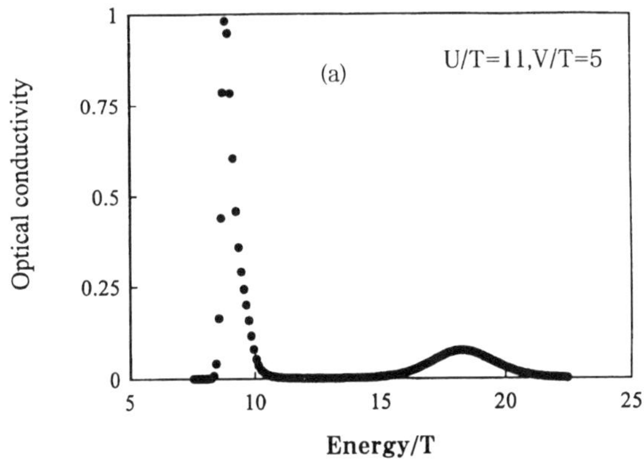


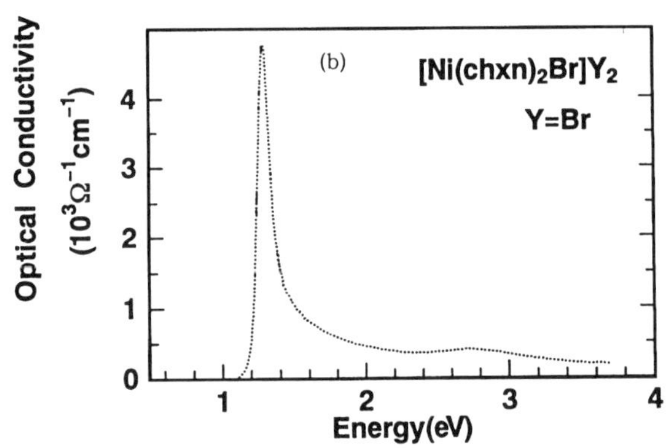


**Fig. 3.32.** The light absorption spectra of the one-dimensional SDW state. (**a**) The spectral shape calculated using the extended Hubbard model and the quantum Monte Carlo method (3.120) [466]. $T = 0.16\,\text{eV}$. (**b**) The light absorption spectral shape of HMMC ($M$ = Ni, $X$ = Br) in the quasi-one-dimensional SDW state [418] (With permission from The American Physical Society)

can also see the infrared-divergence type spectral shape of the lowest energy peak. Thus theory and experiment clearly show the strong nonlinear coupling between the charge- and spin-excitations in the system with strong Coulombic correlation. In other words, charge and spin in this system are inseparable from one another.

Thus, we have seen that almost all phenomena in solid state physics can be described in the context of the linear coupling between electrons and phonons (or bosons), even including the strong interelectron Coulombic repulsion.

# 4. Electron–Electron Interaction and Electron Correlation

The problem of electronic states in a solid is a many-body problem of a macroscopic number of particles since the electrons interact with each other through Coulomb repulsion. Since it is obviously impossible to solve such a formidable problem exactly, one has to resort to approximate methods. Band theory is one such method based on a mean field approximation, namely, on the one-electron approximation, in which each electron is assumed to move in the crystal potential made up of all the electrons in the solid. The present success of condensed-matter physics has been brought about by band theory because the one-electron approximation has greatly simplified the problem. In this book, however, we deal with oxides with $d$ electrons in which the conventional band theory breaks down due to strong Coulomb interaction. The electronic properties of such compounds can be understood only if one considers more explicitly the electron–electron interaction and resulting electron–electron correlation. Since the electron–electron interaction is the origin of many interesting phenomena in these materials, such as metal–insulator transitions, magnetic ordering and superconductivity, the successful establishment of a new, efficient tool for treating the interaction leads to a new paradigm of condensed-matter physics. In this chapter, we describe various methods to treat the electron–electron interaction and electron–electron correlation in solids, including the one-electron approximation.

## 4.1 Introduction

In band theory, one considers the energy levels of a single electron placed in the crystal potential which is determined by the charge distribution of the remaining $\sim 10^{23}$ electrons as well as the ion cores. The energy levels are given as a function of crystal momentum $\boldsymbol{k}$, and therefore form energy bands $E_{\boldsymbol{k}}$. The $\sim 10^{23}$ electrons occupy the energy bands up to the electron chemical potential $\mu$ or the Fermi energy $E_{\mathrm{F}}$ in accordance with the Fermi–Dirac statistics. On this level of approximation, the motion of each electron is independent of the motion of the others, and therefore the theory is called the mean field approximation, independent-particle approximation or one-electron approximation. By the one-electron approximation, the many-body problem of the $\sim 10^{23}$ electrons is greatly simplified to the problem of a single electron in the self-consistent mean field potential. In most first-principles band-structure calculations, the electron–electron interaction term in the mean field potential is further simplified by using the so-called local-density approximation (LDA) (see Sect. 4.3.2).

One-electron band theory has been a powerful tool in understanding the physical properties of many solids and has been the most useful basic concept in the development of solid-state physics. On the other hand, some limitations of the band theory have been pointed out in systems with strong electron–electron interactions, particularly in metal oxides with $d$ electrons, whose wavefunctions are rather localized at each atomic site and which interact strongly with each other. (Among them, $3d$ electrons are most strongly localized and interacting.) In order to correctly predict the electronic properties of these systems, the electron–electron interaction has to be treated more rigorously than in conventional band theory.

In real systems with strong electron-electron interaction, the electron–phonon interaction cannot generally be neglected. This is because electron–electron interaction will enhance interaction between the electron and lattice distortion and electron–phonon interaction, because electron–electron interaction tends to localize electrons. In this chapter, however, we restrict ourselves to electron–electron interactions and refer the reader to Chap. 3 for the electron–phonon interaction problem.

## 4.2 Microscopic Models of Interacting Electrons

The most general form of the Hamiltonian for outer-shell electrons in a solid is given by

$$\mathcal{H} = \sum_{i}^{N} \left[ \frac{\boldsymbol{p}_i^2}{2m} + v(\boldsymbol{r}_i) \right] + \sum_{i,j}^{N'} \frac{e^2}{8\pi\epsilon_0 r_{ij}} \equiv \sum_{i}^{N} H_i + \sum_{i,j}^{N'} \frac{e^2}{8\pi\epsilon_0 r_{ij}} , \tag{4.1}$$

where the first term on the right-hand side is the non-interacting (one-electron) part and the second term is the interaction energy among the outer electrons. Here, $\boldsymbol{r}_i$ and $\boldsymbol{p}_i$ are the spatial coordinate and the momentum of the $i$-th electron, $r_{ij} \equiv |\boldsymbol{r}_i - \boldsymbol{r}_j|$ is the distance between the $i$-th and $j$-th electrons, $V(\boldsymbol{r})$ is the periodic potential induced by ion cores (= nuclei plus inner-shell electrons) and $N$ ($\sim 10^{23}$) is the total number of outer-shell electrons in the solid.

In the practical application of (4.1) to actual problems, one has to choose an appropriate basis set of one-electron wavefunctions. Atomic orbitals (such as oxygen $2p$ and transition-metal $3d$ orbitals) are usually taken as the basis functions in metal oxides since electrons are relatively localized around the ion cores. The atomic orbitals are denoted as $\phi_{nl\mu\sigma}(\tau) = R_{nl}(r)Y_{l\mu}(\theta,\phi)\chi_\sigma(s)$, where $Y_{l\mu}(\theta,\phi)$ is a linear combination of spherical harmonics $Y_{lm}(\theta,\phi)$ and, in the cubic crystal field, $\mu = u(d_{3z^2-r^2}),\ v(d_{x^2-y^2}),\ \xi(d_{yz}),\ \eta(d_{zx}),\ \zeta(d_{xy})$ (see (2.7), (2.8)). $\chi_\sigma(s)$ is an eigenfunction of spin, $\sigma$ takes eigenvalues of $s_z$: $\sigma = \frac{1}{2}(\uparrow), -\frac{1}{2}(\downarrow)$, $s$ is the spin coordinate, and $\tau \equiv (\mathbf{r}, s)$. Using the above basis set, the Hamiltonian (4.1) is rewritten in second quantization form as

$$\mathcal{H} = \sum_I E_I n_I + \sum_{I\neq J} t_{IJ} a_I^\dagger a_J + \frac{1}{2} \sum_{I,J,I',J'} \langle IJ||I'J'\rangle a_I^\dagger a_J^\dagger a_{J'} a_{I'} \,, \tag{4.2}$$

where $a_I^\dagger$ and $a_I$ are the creation and annihilation operators of atomic spin orbital $I$ (identified by site $i$ and quantum numbers $nl\mu\sigma$) and $n_I \equiv a_I^\dagger a_I$ (= 0 or 1) is the occupation number of atomic spin orbital $I$. Here, $E_I$ is the "bare" energy of atomic spin orbital $I$ and

$$t_{IJ} \equiv \langle I|H|J\rangle = \int \mathrm{d}\tau \phi_I(\tau)^* H \phi_J(\tau) \tag{4.3}$$

is the transfer integral which represents the transfer of an electron from atomic spin orbitals $I$ to $J$. It follows from (4.3) that $t_{IJ} \neq 0$ only if $I$ and $J$ have the same spin. The third term of (4.2) has the matrix elements of the electron–electron interaction $\sum'_{ij} e^2/8\pi\varepsilon r_{ij}$, namely, of two-electron integrals,

$$\langle IJ||I'J'\rangle \equiv \int\int \mathrm{d}\tau_1 \mathrm{d}\tau_2 \phi_I(\tau_1)^* \phi_J(\tau_2)^* \frac{1}{4\pi\epsilon_0 r_{12}} \phi_{I'}(\tau_1)\phi_{J'}(\tau_2) \,. \tag{4.4}$$

The first and second terms of (4.2) can be diagonalized by using Bloch orbitals

$$\psi_{\boldsymbol{k}\alpha\sigma} = \left(\frac{1}{N}\right)^{1/2}$$

$$\sum_{jznl\mu} c_{\boldsymbol{k}\alpha:znl\mu\sigma} \exp[i\boldsymbol{k}\cdot(\boldsymbol{R}_j + \boldsymbol{d}_z)]\phi_{nl\mu}(\boldsymbol{r} - \boldsymbol{R}_j - \boldsymbol{d}_z)\chi_\sigma(s) \,, \tag{4.5}$$

where $\boldsymbol{R}_j$ is the position of the $j$-th unit cell, $\boldsymbol{d}_z$ is the relative position of the $z$-th atom in the unit cell and $\alpha$ is the band index. One thus obtains

$$\mathcal{H} = \sum_{\alpha\boldsymbol{k}} E_{\boldsymbol{k}\alpha} a_{\boldsymbol{k}\alpha\sigma}^\dagger a_{\boldsymbol{k}\alpha\sigma} + \frac{1}{2} \sum_{I,J,I',J'} \langle IJ||I'J'\rangle a_I^\dagger a_J^\dagger a_{J'} a_{I'} \,. \tag{4.6}$$

The two-electron integrals (4.4), which are of the central importance in the present chapter, are classified as follows:

1. $\langle IJ||IJ\rangle \equiv U_{IJ}$ (> 0) represents the Coulomb repulsion between electrons at $I$ and $J$ and is called the Coulomb integral.
2. $\langle IJ||JI\rangle \equiv J_{IJ}$ ($I \neq J$) (> 0) is called the exchange integral and is nonzero only if $I$ and $J$ have parallel spins. Because the exchange interaction contributes $-\langle IJ||JI\rangle \equiv J_{IJ}$ to the total energy, electrons at spin-orbitals $I$ and $J$ tend to have parallel spins (Hund's rule).

3. $\langle IJ||I'J'\rangle$ $(I \neq J \neq I' \neq J')$ causes simultaneous transitions of two electrons at $I$ and $J$ into $I'$ and $J'$. It is non-zero only if $I$ and $I'$ have parallel spins and $J$ and $J'$ have parallel spins.

The above two-electron integrals are generally the largest when all the orbitals are on the same atom. Therefore, interatomic two-electron integrals are usually ignored except for the Coulomb integrals, which fall off rather showly with atomic distance and sometimes remain significant at a long distance (Sect. 4.7).

Now we apply the Hamiltonian (4.2) or (4.6) to transition-metal oxides. Only oxygen $p$ orbitals and transition-metal $d$ orbitals are considered as basis atomic orbitals. If we retain two-electron integrals only for $d$ electrons on the same metal atom, the Hamiltonian consists of the $d$-electron term $\mathcal{H}_d$, the $p$-electron term $\mathcal{H}_p$ and the $d$-$p$ hybridization term $\mathcal{H}_{pd}$:

$$\mathcal{H} = \mathcal{H}_{\mathrm{d}} + \mathcal{H}_{\mathrm{p}} + \mathcal{H}_{\mathrm{dp}}\,, \tag{4.7}$$

$$\begin{aligned}\mathcal{H}_{\mathrm{d}} &= \sum_{i,\mu,\sigma} E_{\mathrm{d}}^{\mu} n_{\mathrm{d}i\mu\sigma} \\ &+\frac{1}{2} \sum_{i,\mu,\mu',\nu,\nu'\sigma,\sigma',\sigma'',\sigma'''} \langle \mu_\sigma \nu_{\sigma'} || \mu'_{\sigma''} \nu'_{\sigma'''} \rangle d^\dagger_{i\mu\sigma} d^\dagger_{i\nu\sigma'} d_{i\nu'\sigma'''} d_{i\mu'\sigma''}\,,\end{aligned} \tag{4.8}$$

$$\begin{aligned}\mathcal{H}_{\mathrm{p}} &= \sum_{i,\mu,\sigma} E_{\mathrm{p}}^{\mu} n_{\mathrm{p}i\mu\sigma} + \sum_{i\neq j,\mu,\nu,\sigma} t_{i\mu j\nu} p^\dagger_{i\mu\sigma} p_{j\nu\sigma} \\ &= \sum_{\boldsymbol{k},\alpha,\sigma} E_{\boldsymbol{k}\alpha} p^\dagger_{\boldsymbol{k}\alpha\sigma} p_{\boldsymbol{k}\alpha\sigma}\,,\end{aligned} \tag{4.9}$$

$$\mathcal{H}_{\mathrm{dp}} = \sum_{\boldsymbol{k},\alpha,i,\mu,\sigma} T_{\boldsymbol{k}\alpha i\mu} (d^\dagger_{i\mu\sigma} p_{\boldsymbol{k}\alpha\sigma} + p^\dagger_{\boldsymbol{k}\alpha\sigma} d_{i\mu\sigma})\,, \tag{4.10}$$

where $d$ and $d^\dagger$ are the annihilation and creation operators of the $d$ electron, $p$ and $p^\dagger$ are the annihilation and creation operators of the $p$ electron, and $T$s in (4.10) are transfer integrals between neighboring $d$ and $p$ orbitals. In many transition-metal oxides, the interatomic distance between transition-metal ions are large and transfer integrals between transition-metal ions are neglected. The model (4.7) is called the Anderson-lattice model or multi-band Hubbard ($d$–$p$) model.

If one projects out the almost-filled $p$ orbitals, the model is called the (degenerate) Hubbard model:

$$\begin{aligned}\mathcal{H} &= \sum_{i,\mu,\sigma} E_{\mathrm{d}}^{\mu} n_{\mathrm{d}i\mu\sigma} + \sum_{i\neq j,\mu,\nu,\sigma} t_{i\mu j\nu} d^\dagger_{i\mu\sigma} d_{j\nu\sigma} \\ &+\frac{1}{2} \sum_{i,\mu,\mu',\nu,\nu'\sigma,\sigma',\sigma'',\sigma'''} \langle \mu_\sigma \nu_{\sigma'} || \mu'_{\sigma''} \nu'_{\sigma'''} \rangle d^\dagger_{i\mu\sigma} d^\dagger_{i\nu\sigma'} d_{i\nu'\sigma'''} di\mu'\sigma''\,.\end{aligned} \tag{4.11}$$

Here, the $d$ orbitals are not pure $d$ orbitals as in (4.7) but have a significant admixture of oxygen $p$ orbitals. Therefore, the $d$-$d$ hybridization term in (4.11) stems largely from the $d$-$p$ hybridization.

Equation (4.11) is still complicated, although the $p$ orbitals have been projected out, because of the degeneracy of the $d$ orbitals. If the degeneracy of the $d$ orbitals has been neglected, one obtains the single-band Hubbard model:

$$\mathcal{H} = \sum_{i,j,\sigma} t_{ij} d^{\dagger}_{i\sigma} d_{j\sigma} + \frac{1}{2} \sum_{i\sigma} U n^{\dagger}_{i\sigma} n_{i-\sigma} \,. \tag{4.12}$$

Most theoretical studies on the Hubbard model so far have been made on this single-band model. Here, $U$ ($\equiv U_{\sigma,-\sigma}$) is the on-site Coulomb repulsion energy.

## 4.3 One-Electron Theories and Electron Correlation

The effects of electron–electron interaction can be divided into mean field effects and correlation effects. The mean field effects are those effects which can be explained within the one-electron approximation, where each electron is assumed to move in a mean field potential resulting from all the other electrons in the system. Correlation effects are those effects which go beyond the mean field effects. That is, correlation effects arise from the correlated motion of electrons. Hence, the potential which each electron feels is fluctuating in space and time according to the motion of the other electrons. The correlated motion occurs in real space, of course, since electrons repel each other through Coulomb repulsion, but it also occurs in spin space. Thus electron correlation is nothing but quantum mechanical fluctuations of electron motion in real and spin space.

It has often been stated that strong electron–electron interaction (Coulomb repulsion) leads to the breakdown of the one-electron approximation. As typical examples, Mott insulators such as NiO and the parent compounds of high-$T_{\mathrm{c}}$ cuprates have been referred to as strongly correlated systems (Sect. 5.6). However, in these Mott insulators, the strong interaction causes strong spin and orbital polarization, which makes the one-electron approximation for the description of ground-state properties as valid as in the limit of weak electron–electron interaction, as we shall see below. The reason for the success of the one-electron approximation in Mott insulators is that the almost fully spin- and orbital-polarized states have very little quantum fluctuation. The one-electron approximation breaks down when the interaction strength is intermediate or the spin and orbital are not fully polarized as in the low-spin states, because in those cases quantum fluctuations become strong. The one-electron approximation also breaks down in the treatment of excited-state properties and finite-temperature properties. Before going into electron correlation problems, we describe the one-electron approximation, namely, the Hartree–Fock approximation.

### 4.3.1 Hartree–Fock Approximation

In the one-electron (Hartree–Fock) approximation, electrons interact with each other through the mean field potential. The wave function of the ground state of a many-electron system is written as a single Slater determinant as in a non-interacting electron system:

$$\Psi_{N,\mathrm{g}}^{\mathrm{HF}}(\tau_1,\tau_2,...,\tau_N) = \frac{1}{\sqrt{N!}} \begin{vmatrix} \psi_{k1}(\tau_1) & \dots & \psi_{k1}(\tau_N) \\ \psi_{k2}(\tau_1) & \dots & \psi_{k2}(\tau_N) \\ \dots & & \dots\dots \\ \psi_{kN}(\tau_1) & \dots & \psi_{kN}(\tau_N) \end{vmatrix} \equiv |\psi_{k_1}\psi_{k_2}\dots\psi_{k_N}|\,, \tag{4.13}$$

where $\psi_k(\tau_i)$ means that the $i$-th electron occupies one-electron state $k$. The Slater determinant (4.13) defines the electron configuration $\{k_1, k_2, ...., k_N\}$. In band theory, one uses Bloch orbitals $\psi_{\alpha k\sigma}(\tau)$ (4.5) for the one-electron states $\psi_k(\tau)$ in (4.13). In the following, we will not impose the condition that the orbital part of the one-electron states is the same for the spin-up and spin-down states. This is called the unrestricted Hartree–Fock approximation as opposed to the restricted Hartree–Fock approximation, in which the orbital part of $\psi_k(\tau)$ is assumed to be the same for both spins.

The energy $\mathcal{E}_{N,\mathrm{g}}^{\mathrm{HF}}$ of the $N$-electron Hartree–Fock ground state $\Psi_{N,\mathrm{g}}^{\mathrm{HF}}$ is given by

$$\mathcal{E}_{N,\mathrm{g}}^{\mathrm{HF}} \equiv \langle\Psi_{N,\mathrm{g}}^{\mathrm{HF}}|\mathcal{H}|\Psi_{N,\mathrm{g}}\rangle = \sum_k^N H_k + \sum_{k>l}^N (U_{kl} - J_{kl})\,, \tag{4.14}$$

where $H_k \equiv \langle k|H|k\rangle$, and $U_{kl} \equiv \langle kl||kl\rangle$ and $J_{kl} \equiv \langle kl||lk\rangle$ are the Coulomb and exchange integrals, respectively, between one-electron states $k$ and $l$. By applying a variational principle to (4.13), one obtains an equation (Hartree–Fock equation) which the one-electron wave function $\psi_k(\tau)$ should satisfy:

$$\left[H + \sum_l^N (U_l - J_l)\right]\psi_k(\tau) = E_k\psi_k(\tau)\,. \tag{4.15}$$

Here $U_l$ and $J_l$ are defined by

$$U_l\psi_k(\tau_1) \equiv \int \mathrm{d}\tau_2\psi_l^*(\tau_2)\psi_l(\tau_2)\frac{1}{r_{12}}\psi_k(\tau_1)\,, \tag{4.16}$$

$$J_l\psi_k(\tau_1) \equiv \int \mathrm{d}\tau_2\psi_l^*(\tau_2)\psi_k(\tau_2)\frac{1}{r_{12}}\psi_l(\tau_1)\,, \tag{4.17}$$

and yield the Coulomb potential $V_\mathrm{c}(\boldsymbol{r})$ and the exchange potential $\int \mathrm{d}\tau' V_\mathrm{x}(\tau,\tau')$ in the form

$$V_{\rm c}(\boldsymbol{r})\psi_k(\tau) \equiv \sum_{l\neq k}^{N} U_l \psi_k(\tau)\,, \tag{4.18}$$

$$\int {\rm d}\tau' V_{\rm x}(\tau,\tau')\psi_k(\tau') \equiv -\sum_{l\neq k}^{N} J_l \psi_k(\tau)\,. \tag{4.19}$$

Thus $U_l$ and $J_l$ are Coulomb and exchange potentials induced by the electron in state $l$. $J_l$ is non-zero only for electrons whose spins are parallel to the electron in $l$. The Coulomb potential is a function of only the spacial coordinate of the electron $\boldsymbol{r}$ while the exchange potential is nonlocal and spin-dependent (4.19). These potentials are determined by the one-electron wave functions $\psi_k(\tau)$, which are in turn determined by the potentials through (4.15). Therefore, the potentials and the wave functions have to be determined self-consistently. It should be noted that in (4.15) the $l = k$ term is automatically eliminated from the potential $\sum_l (U_l - J_l)$ because $U_k = J_k (= \langle kk||kk\rangle)$. Therefore, in the Hartree–Fock approximation, it is guaranteed that each electron does not interact with itself.

Multiplying (4.15) by the complex conjugate of $\psi_k(\tau)$ and integrating over $\tau$, one obtains

$$E_k = H_k + \sum_{l}^{N} (U_{kl} - J_{kl})\,. \tag{4.20}$$

Therefore, one can interpret $E_k$ as the sum of the bare electron energy $H_k$ and the Coulomb and exchange energies coming from the rest of the electrons and therefore as the energy of the electron in state $k$. Indeed, if the electron is removed from the one-electron state $k'$ of the ground state $\Psi^{\rm HF}_{N,{\rm g}}$, the energy $\mathcal{E}^{\rm HF}_{N-1,{\rm f}}$ of the $(N-1)$-electron system is given by

$$\mathcal{E}^{\rm HF}_{N-1,{\rm f}} = \sum_{k\neq k'}^{N} H_k + \sum_{k>l(\neq k')}^{N} (U_{kl} - J_{kl}) = \mathcal{E}^{\rm HF}_{N,{\rm g}} - E_{k'}\,, \tag{4.21}$$

(if the remaining orbitals do not relax with the electron removal) and $-E_{k'}$ can be regarded as the ionization potential of the electron in state $k'$. Equation (4.21) is called Koopmans' theorem. In the same way, the energy $E_{k''}$ can be regarded as the affinity level of state $k''$.

Equation (4.20) again guarantees that each electron does not interact with itself because $U_{kk} \equiv J_{kk}$ and hence $k = l$ terms are zero. The total energy $\mathcal{E}^{\rm HF}_{N,{\rm g}}$ of the $N$-electron system is given by the sum of $E_k$ with the double-counted interaction energy being subtracted.

$$\mathcal{E}^{\rm HF}_{N,{\rm g}} = \sum_{k}^{N} E_k - \sum_{k>l}^{N} (U_{kl} - J_{kl})\,. \tag{4.22}$$

### 4.3.2 Local Density Approximation

The nonlocal nature of the exchange potential in the Hartree–Fock equation complicates calculations and makes it impractical to apply the Hartree–Fock approximation to actual band-structure calculations. In most first-principles band-structure calculations, the local-density approximation (LDA) has been used. In the LDA, the exchange energy at $\boldsymbol{r}$ is given in terms of the electron density $n(\boldsymbol{r})$ at that point. The spin-independent part of the exchange potential is derived from $V_{\mathrm{x}}(\boldsymbol{r}) = \partial\mathcal{E}_{\mathrm{x}}(\boldsymbol{r})/\partial n(\boldsymbol{r})$, where $\mathcal{E}_{\mathrm{x}}(\boldsymbol{r})$ is the density of exchange energy calculated for the free-electron gas. *Kohn*, *Sham* and *Gaspar* have given

$$V_{\mathrm{x}}(\boldsymbol{r}) = -\frac{1}{\pi}\left[3\pi^2 n(\boldsymbol{r})\right]^{1/3} . \tag{4.23}$$

Correlation energy is added to the exchange energy in

$$V_{\mathrm{xc}}(\boldsymbol{r}) = \beta[n(\boldsymbol{r})]V_x(\boldsymbol{r}) , \tag{4.24}$$

where $\beta$ is a monotonically decreasing function of $n(\boldsymbol{r})$ and $\beta \to 1$ in the high density limit ($n(\boldsymbol{r}) \to \infty$). When the electrons are spin-polarized, $V_{\mathrm{xc}}(\boldsymbol{r})$ becomes a function of both $n_\uparrow(\boldsymbol{r})$ and $n_\downarrow(\boldsymbol{r})$.

The most serious drawback of LDA is that each electron necessarily interacts with itself because the potential is give by the electron (and spin) density, which includes that electron itself. Because of this spurious self-interaction, the LDA potential is not able to describe the different potential energy between occupied and unoccupied states, which are produced by $N-1$ electrons and $N$ electrons, respectively. This leads to a serious problem especially when the electrons are strongly interacting. This is why band gaps are generally underestimated in LDA band-structure calculations.

### 4.3.3 Electron Correlation Effects

Correlation effects are defined as those effects which cannot be treated within a single Slater determinant, that is, by a single configuration of electrons. Electron correlation can generally be described by a linear combination of Slater determinants, that is, by a superposition of many electronic configurations which interact with each other through configuration interaction (CI). Because of the CI, two-electron integrals of the form $\langle IJ||I'J'\rangle$ ($I \neq I', J \neq J', I \neq J', J \neq I'$) (Sect. 4.2) enter the matrix elements of the Hamiltonian and mix Slater determinants of different configurations.

One of the problems which cannot be treated within the Hartree–Fock approximation is quantum spin fluctuations. For example, two-dimensional antiferromagnets with $S = 1/2$ local spins at each site, as in the parent insulators of high-$T_{\mathrm{c}}$ cuprates, have strong quantum spin fluctuations. As an illustrative example, let us consider two antiferromagnetically coupled

$S = 1/2$ sites. The ground state of this system is a spin singlet ($S = 0$) and has the wave function $\Psi = (1/\sqrt{2})(|1_\uparrow 2_\downarrow| - |1_\downarrow 2_\uparrow|)$. Since this state cannot be reduced to a single Slater determinant, it is certainly beyond the Hartree–Fock approximation. Indeed, a Hartree–Fock state $\Psi^{\mathrm{HF}} = |1_\uparrow 2_\downarrow|$ is not an eigenstate but is a 1:1 mixture of the $S = 0$ and $S = 1$ states.

In a metallic state with strong electron–electron interaction, conduction electrons repel each other through the Coulomb repulsion (in addition to the Pauli principle) and the ground state cannot then be described by a single Slater determinant. Let us consider a diatomic system with two valence electrons. If there were no electron–electron interaction, two electrons would occupy the bonding orbital $b \equiv (1/\sqrt{2})(|1\rangle + |2\rangle)$ in the ground state: $\Psi^{\mathrm{HF}} = |b_\uparrow b_\downarrow|$. This ground state can be expanded to

$$\Psi^{\mathrm{HF}} = \frac{1}{2}\left(|1_\uparrow 1_\downarrow| + |1_\uparrow 2_\downarrow| + |2_\uparrow 1_\downarrow| + |2_\uparrow 2_\downarrow|\right) . \tag{4.25}$$

With increasing Coulomb repulsion $U$, the first and fourth terms in (4.25) increase their energies and therefore lose their weight. This situation can be described by adding another Slater determinant $|a_\uparrow a_\downarrow|$ to (4.25):

$$\Psi^{\mathrm{CI}} = c_b |b_\uparrow b_\downarrow| - c_a |a_\uparrow a_\downarrow| , \tag{4.26}$$

where $a \equiv (1/\sqrt{2})(|1\rangle - |2\rangle)$ is the antibonding orbital. Equation (4.26) thus describes the ground state in which electrons repel each other through Coulomb repulsion, and a single Slater determinant is not able to describe such a correlated ground state.

## 4.4 Electronic Structure of Transition-Metal Ions

### 4.4.1 Hartree–Fock Scheme

Let us treat the interaction term for the $d$ electrons [(4.8) or (4.11)] in the Hartree–Fock approximation [467]. We assume that the transition metal ion is coordinated to six oxygen ions forming an octahedron and that the $d$ electrons are localized at that transition-metal site. The crystal field splits the $d$ level into the $e_g$ and $t_{2g}$ levels (Sect. 2.1), whose bare energies are $E_d^\mu = E_d^0 + 6Dq$ and $E_d^0 - 4Dq$, respectively. The Coulomb and exchange integrals for the $d$ electrons are given in terms of Kanamori parameters:

$$U \equiv \langle \mu\mu || \mu\mu \rangle, \quad U' \equiv \langle \mu\nu || \mu\nu \rangle, \quad J \equiv \langle \mu\nu || \nu\mu \rangle \quad (\mu \neq \nu) . \tag{4.27}$$

These parameters have to satisfy the relationship $U - U' = 2J$ in order to ensure the rotational invariance of the electron–electron interaction term in (4.8) and (4.11). $U$ and $U'$ are several eV (e.g., 2–4 eV for Ti and V, and 7–9 eV for Ni and Cu) and $J$ is $\approx$ eV or smaller.

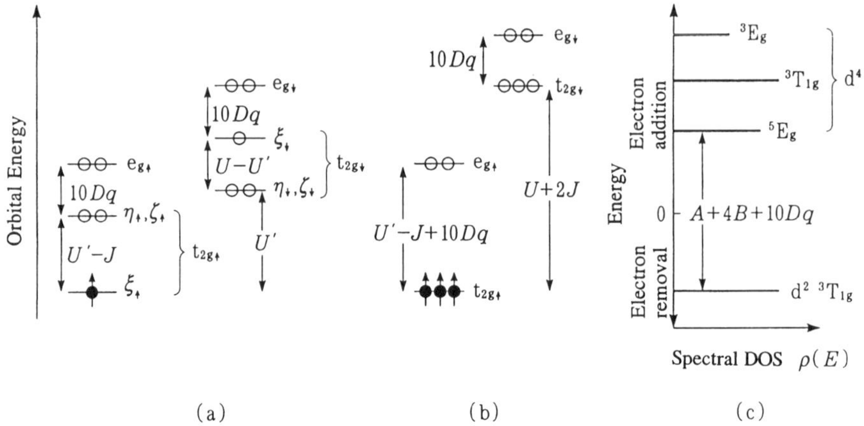


**Fig. 4.1.** Energy-level diagrams of the $d^1$ configuration (**a**) and the $d^3$ configuration (**b**) in the cubic crystal field according to the Hartree-Fock approximation. *Filled* and *open* circles denote occupied and unoccupied one-electron states, respectively. (**c**): Density of states of the $d^3$ configuration according to ligand-field theory

Let us consider a simplest case of the $d^1$ ground state, in which an electron occupies one (say $\xi_\uparrow$) of the $t_{2g}$ orbitals. According to (4.20), the energy of each one-electron state is given by

$$\begin{aligned} &E_{\xi\uparrow} = E_d^0 - 4Dq\,, && E_{\xi\downarrow} = E_d^0 - 4Dq + U \\ &E_{\eta\uparrow} = E_{\zeta\uparrow} = E_d^0 - 4Dq + U' - J\,, && E_{\eta\downarrow} = E_{\zeta\downarrow} = E_d^0 - 4Dq + U'\,, \\ &E_{u\uparrow} = E_{v\uparrow} = E_d^0 + 6Dq + U' - J\,, && E_{u\downarrow} = E_{v\downarrow} = E_d^0 + 6Dq + U'\,. \end{aligned} \tag{4.28}$$

The energy-level diagram of these states is shown in Fig. 4.1 a. The diagram shows the splitting between the $e_g$ and $t_{2g}$ orbitals ($= 10Dq$), $\uparrow$ and $\downarrow$ states (proportional to $J$) and the splitting between the $\xi$ and other $t_{2g}$ orbitals ($= U - U'$). It should be noted that the energy spectrum has a gap of the order of the Coulomb integral between the occupied and unoccupied states, $\mathcal{E}_{\text{gap}} = U' - J$. A gap of the same magnitude is opened for the $d^2$ ground state. For the $d^3$ ground state ($Cr^{3+}$, $Mn^{4+}$), the $t_{2g\uparrow}$ orbitals are fully occupied, resulting in a simpler energy diagram as shown in Fig. 4.1 b. In this case, the lowest unoccupied level is either the $e_{g\uparrow}$ or the $t_{2g\downarrow}$ state, depending on the magnitude of the crystal-field splitting, i.e., depending on whether $10Dq < U - U' + 3J \sim 5J$ or $10Dq > 5J$. Accordingly, the $d^4$ ground state takes either the high-spin ($S = 2$) state or low-spin ($S = 1$) state, respectively, depending on $10Dq$.

### 4.4.2 Ligand-Field Theory

Two-electron integrals of the form $\langle \mu_\sigma \nu_{\sigma'} || \mu'_{\sigma''} \nu'_{\sigma'''} \rangle$ ($\mu_{\sigma'} \neq \mu'_{\sigma''}$, $\nu_\sigma \neq \nu'_{\sigma'''}$, $\mu_\sigma \neq \nu_{\sigma'}$, $\mu'_{\sigma''} \neq \nu'_{\sigma'''}$) which do not appear in the above Hartree–Fock treatment, induce mixing between two Slater integrals (i.e., between two electronic configurations) through the two-electron excitation process $\mu_\sigma \nu_{\sigma''} \leftrightarrow \mu'_{\sigma''} \nu'_{\sigma'''}$. We have mentioned above the importance of this type of mixing in the context of low-spin states, in which two-electron integrals of the form $\uparrow\downarrow\leftrightarrow\downarrow\uparrow$ play an important role. Even in high-spin states, two-electron processes such as $t_{2g\uparrow}^2 \leftrightarrow e_{g\uparrow}^2$ and $t_{2g\uparrow}^2 \leftrightarrow t_{2g\uparrow}e_{g\uparrow}$ may exist. In order to take into account those electron correlation effects within the transition-metal ion, the electronic structure of the transition-metal ion has to be treated using the ligand-field theory or multiplet theory.

In ligand-field theory in general, a large number of Slater determinants are hybridized with each other, leading to a large dimension of the Hamiltonian matrix to be diagonalized. In order to reduce the size of the matrix, one often makes use of group theory. Each irreducible matrix is denoted by an irreducible representation $\Gamma$ of the point group ($O_h$ group in the case of cubic symmetry) and the spin degeneracy $2S+1$ as ${}^{2S+1}\Gamma$. For details, see [468].

When there is electron correlation, the energy of each one-electron state loses its meaning. The total energies of eigenstates of the many-electron system and their differences (such as excitation energies, ionization energies and electron affinity) are the only relevant, observable physical quantities. Nevertheless, if one plots ionization energies downwards and affinity levels upwards, the density of states (DOS) $\rho(E)$ of the many-electron system, a many-electron analog of the one-electron energy-level scheme, is obtained:

$$\rho(E) \propto \sum_{f,\mu,\sigma} |\langle \Psi_{N-1,f}|d_{\mu\sigma}|\Psi_{N,g}\rangle|^2 \delta(\mathcal{E}_{N-1,f} + E_\mathrm{F} - \mathcal{E}_{N,\mathrm{g}} + E)$$

$$+ \sum_{f,\mu,\sigma} |\langle \Psi_{N+1,f}|d^+_{\mu\sigma}|\Psi_{N,\mathrm{g}}\rangle|^2 \delta(\mathcal{E}_{N+1,f} - E_\mathrm{F} - \mathcal{E}_{N,\mathrm{g}} - E) \quad (4.29)$$

Here, the first term corresponds to electron removal (ionization level, $E < 0$) and the second term to electron addition (affinity level, $E > 0$), where the energy $E$ is measured from the electron chemical potential or the Fermi level ($E_\mathrm{F}$). That is, $-E$ in the first term is the energy of the $(N-1)$-electron state plus an electron at $E_\mathrm{F}$ ($= \mathcal{E}_{N-1,f} + E_\mathrm{F} - \mathcal{E}_{N,\mathrm{g}}$) and $E$ in the second term is the energy of the $(N+1)$-electron state minus an electron at $E_\mathrm{F}$ ($= \mathcal{E}_{N+1,f} - E_\mathrm{F} - \mathcal{E}_{N,\mathrm{g}}$) measured from the $N$-electron ground state. The $E < 0$ and $E > 0$ parts of the DOS $\rho(E)$ are the photoemission and inverse-photoemission spectra, respectively. The matrix elements of the creation and annihilation operators in (4.29) yield the spectral intensities. In the Hartree–Fock approximation, (4.29) is a set of delta functions whose intensity ratio is given by the orbital degeneracy ratio. Final states $\Psi_{N\pm1,f}$ which can be reached from the ground state $\Psi_{N,\mathrm{g}}$ by a two-electron excitation or a spin-flip transition have zero intensity in the DOS (4.29). Figure 4.1 c shows the DOS of the $d^3$ configuration in the cubic field according to ligand-field theory. The different energy positions and spectral intensities between the Hartree–Fock energy-level scheme (Fig. 4.1 b) and the ligand-field theoretical

DOS (Fig. 4.1 c) signal the effect of electron correlation. Most remarkably, the intensity ratio of the two peaks arising from the addition of an $e_g$ electron (denoted $E_g$) differs from the orbital degeneracy ratio 1 : 1.

The energy gap $\mathcal{E}_{\mathrm{gap}}$ of a many-electron system is defined by the sum of the minimum energy required to remove an electron from the $N$-electron ground state and the minimum energy required to add an electron to it:

$$\begin{aligned}\mathcal{E}_{\mathrm{gap}} &= (\mathcal{E}_{N+1,\mathrm{g}} - \mathcal{E}_{N,\mathrm{g}}) + (\mathcal{E}_{N-1,\mathrm{g}} - \mathcal{E}_{N,\mathrm{g}}) \\ &= \mathcal{E}_{N+1,\mathrm{g}} + \mathcal{E}_{N-1,\mathrm{g}} - 2\mathcal{E}_{N,\mathrm{g}} \, . \end{aligned} \tag{4.30}$$

Therefore, the band gap of an insulating transition-metal compound is the energy required to cause the charge transfer $d_i^n + d_j^n \to d_i^{n+1} + d_j^{n-1}$, if the $d$ electrons are localized and the $p$ electrons can be neglected. This energy is referred to as the effective Coulomb energy $U_{\mathrm{eff}}$. In the Hartree–Fock approximation, i.e., if $\mathcal{E}_{N,\mathrm{g}}^{\mathrm{HF}}$ is used for $\mathcal{E}_{N,\mathrm{g}}$, $U_{\mathrm{eff}}$ is nothing but the energy difference between the lowest occupied and highest unoccupied one-electron states as described in Sect. 4.4.1.

$U_{\mathrm{eff}}$ is related to the average $\bar{U}$ $[= (8/9)U' + (1/9)U - (4/9)J]$ of the Coulomb-exchange energy between two electrons $\langle \mu_\sigma \nu_{\sigma'} || \mu_\sigma \nu_{\sigma'} \rangle - \langle \mu_\sigma \nu_{\sigma'} || \nu_{\sigma'} \mu_\sigma \rangle$ through

$$U_{\mathrm{eff}} = \bar{U} + 2\Delta\mathcal{E}_n - \Delta\mathcal{E}_{n+1} - \Delta\mathcal{E}_{n-1} \tag{4.31}$$

where $\Delta\mathcal{E}_n (> 0)$ is the energy difference between the lowest multiplet component (Hund rule ground state) and the center of gravity of the $d^n$ multiplet, that is, $\Delta\mathcal{E}_n$ is the Hund rule stabilization energy. We refer to $U_{\mathrm{eff}} - U$ as a multiplet correction. The multiplet correction is maximized for the half-filled ($n = 5$) case because the total spin takes the maximum value, and therefore $\Delta\mathcal{E}_n$ shows a peak at $n = 5$. Therefore, $U_{\mathrm{eff}} - \bar{U}$ shows a sharp peak at $n = 5$ as shown in Fig. 4.2 a while it is slightly negative for $n \neq 5$. (Fig. 4.2 shows the case where $10\,Dq = 0$.) Figure 4.2 shows multiplet corrections calculated both using the Hartree–Fock approximation and the multiplet theory (i.e., ligand-field theory with $10\,Dq = 0$). The two methods have given similar results although there are subtle differences. For example, for the $d^2$ ground state, $U_{\mathrm{eff}} - \bar{U}$ is larger by $\sim 0.5$ eV in the multiplet theory than in the Hartree–Fock theory. This difference represents an effect of electron correlation within the $d^2$ configuration and is absent in the $d^1$ configuration. Indeed, $d^2$ compounds tend to be insulating (e.g., $LaVO_3$) while $d^1$ compounds tend to be metallic (e.g., $SrVO_3$, $CaVO_3$) or narrow-gap semiconductors (e.g., $LaTiO_3$).

The Kanamori parameters $U'$ and $J$ have been defined as the averages of the Coulomb and exchange integrals ($\langle \mu\nu || \mu\nu \rangle$ and $\langle \mu\nu || \nu\mu \rangle$), respectively, while in ligand-field theory one uses Coulomb and exchange integrals without averaging. One uses Slater integrals $F_0$, $F_2$ and $F_4$ defined by

$$\left.\begin{aligned} F^k &= \int r_1^2 dr_1 \int r_2^2 dr_2 R_{23}^2(r_1) R_{23}^2(r_2) \frac{r_<^k}{r_>^{k+1}} \\ F_0 &= F^0 \, , \quad F_2 = \frac{1}{49} F^2 \, , \quad F_4 = \frac{1}{144} F^4 \end{aligned}\right\} \tag{4.32}$$

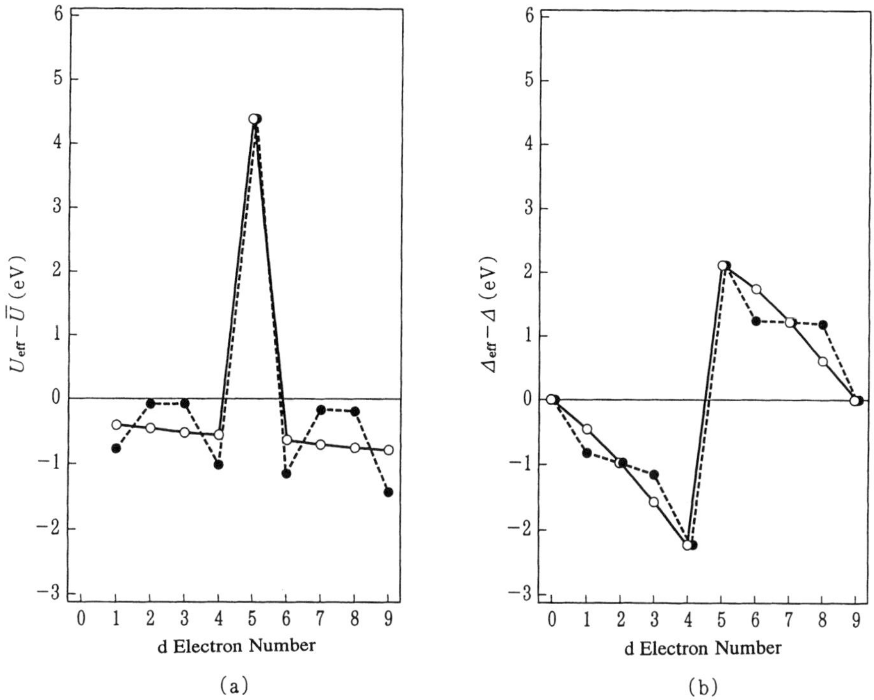


**Fig. 4.2. a, b.** Multiplet corrections $U_{\rm eff} - \bar{U}$ and $\Delta_{\rm eff} - \Delta$ for the Coulomb repulsion and the charge transfer energy, respectively [471]. *Open circles:* Hartree-Fock approximation, *filled circles:* multiplet theory. (With permission from The American Physical Society)

or Racah parameters $A$, $B$ and $C$ defined by

$$A = F_0 - 49F_4\,, \quad B = F_2 - 5F_4\,, \quad C = F_2 - 5F_4\,. \tag{4.33}$$

Kanamori parameters are related to Racah parameters through

$$U = A + 4B + 3C\,, \quad U' = A - B + C\,, \quad J = \frac{5}{2}B + C\,. \tag{4.34}$$

### 4.4.3 *d* Bands and Carrier Doping in Mott Insulators

We shall see below that the energy gap of $\sim U' - J$ ($\sim U + 4J$ in the case of the $d^5$ configuration) for a transition-metal ion obtained using the Hartree–Fock approximation survives in Hartree–Fock band theory. Let us consider a simple case in which one $d$ electron occupies each transition-metal site. For simplicity, we assume that in the ground state the electrons occupy the same atomic orbitals and their spins are aligned ferromagnetically. The Slater determinant of the Bloch orbitals $\psi_{k\xi\uparrow}(\tau) = (\frac{1}{N})^{1/2} \sum_i {\rm e}^{{\rm i}k\cdot R_i} \phi_{i\xi\uparrow}(\tau)$,

$$\Psi_{N,\mathrm{g}}^{\mathrm{HF}} = |\psi_{k1\xi\uparrow}...\psi_{kN\xi\uparrow}| \,, \tag{4.35}$$

is equivalent to that of localized atomic orbitals $\phi_{i\xi\uparrow}(\tau)$,

$$\Psi_{N,\mathrm{g}}^{\mathrm{HF}} = |\phi_{1\xi\uparrow}...\phi_{N\xi\uparrow}| \,. \tag{4.36}$$

Therefore, the energy gap on the single ion remains essentially unchanged in the band state. The energy difference between the highest occupied Bloch orbital $\psi_{k\xi\uparrow}$ and the lowest unoccupied Bloch orbital $\psi_{k'\eta\uparrow}$ or $\psi_{k'\zeta\uparrow}$ is given by $\sim U' - J - W_d$, where $W_d$ is the $d$-band width, or $\sim U' - J$ if the band is sufficiently narrow. The same argument holds for cases where two or more of the ten $d$ bands are completely occupied and the remaining bands are empty. The antiferromagnetic insulating state and its band gap can be described in an analogous way. In order for an energy gap of $\sim U' - J$ to be opened, each ion should have fully spin- and orbital-polarized electrons. If one moves from the Hartree–Fock approximation to the LDA, the magnitude of the gap becomes of order $J$ or $10\,Dq$, making it more difficult to maintain the finite gap.

We have seen that the Hartree–Fock approximation properly describes the ground states and the band gaps of Mott insulators, except for effects of intraatomic correlation (which change the band gaps by about 0.5 eV or so). However, we note that carriers doped into the Mott insulator do not occupy the lowest unoccupied Bloch states or the highest unoccupied Bloch states as one would expected from Koopmans' theorem. The reason is that the presence of doped carriers itself modifies the Bloch orbital through strong Coulomb interaction, invalidating Koopmans' theorem. For example, when a hole is introduced into an antiferromagnet described by the half-filled single-band Hubbard model on a square lattice with only nearest-neighbor transfer

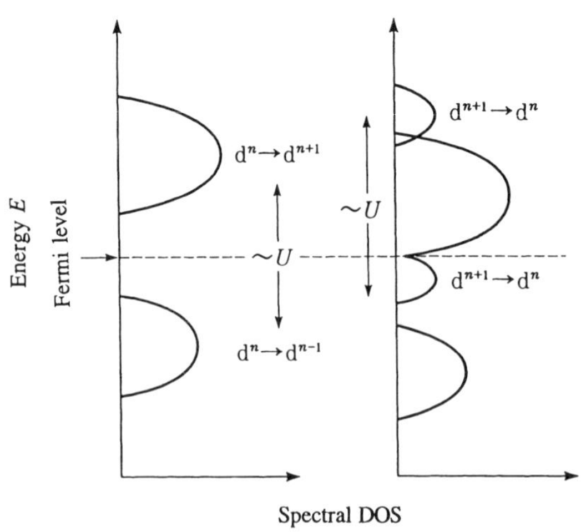


**Fig. 4.3.** Changes in the electronic structure induced by electron doping into a Mott-Hubbard-type insulator

integrals, the hole moves around through the lattice by flipping spins in the system. When a certain number of electrons or holes are doped into a Mott insulator, the same number of mobile $d^{n+1}$ or $d^{n-1}$ sites are created in the lattice of $d^n$ ions. Crudely speaking, the DOS of such a system then becomes a superposition of the DOS of $d^{n+1}$ or $d^{n-1}$ and that of $d^n$ weighted by the corresponding weight in the ground state. Figure 4.3 shows the case of electron doping, in which the lowest unoccupied states of the $d^n$ ground state ($d^n \to d^{n+1}$ spectral weight) exist just above $E_\mathrm{F}$ and the highest occupied states of the $d^{n+1}$ ground state ($d^{n+1} \to d^n$ spectral weight) exist just below $E_\mathrm{F}$. It should be noted that the doping of a small number of electrons changes the DOS at energies $\sim U$ away from $E_\mathrm{F}$. Also, electron doping creates as many as $n+1$ states per electron just below $E_\mathrm{F}$. These features cannot be understood within the rigid band model of carrier doping into an uncorrelated band insulator and are characteristics of strongly correlated systems.

## 4.5 Hybridization Between *d* and *p* Electrons

### 4.5.1 Mott–Hubbard Type and Charge-Transfer Type

In order to consider the effects of hybridization between the transition-metal $d$ and oxygen $p$ orbitals, we consider an octahedral $MO_6$ cluster consisting of the central transition-metal ion $M$ and neighboring six oxygen ions, and treat it within the Hartree–Fock approximation. We assume that the cluster has an electric charge determined by the formal ionic charges of the constituent ions. As one-electron states, we consider molecular orbitals (MOs) consisting of linear combinations of the $d$ and $p$ atomic orbitals. Out of the $6 \times 6 = 36$ (including the spin degrees of freedom) MOs derived from the $p$ orbitals, ten MOs ($\phi^L_{\mu\sigma}(\tau)$: $\mu = \xi, \eta, \zeta$; $\sigma = \uparrow, \downarrow$, see Fig. 2.3) have the same symmetry as the $d$ orbitals ($\phi^d_{\mu\sigma}(\tau)$) and are hybridized with them while the remaining 26 MOs do not hybridize with the $d$ orbitals. If we denote the $d$-$p$ bonding and antibonding MOs ($\psi^\mathrm{B}_{\mu\sigma}$ and $\psi^\mathrm{A}_{\mu\sigma}$) by

$$\left.\begin{aligned} \psi^\mathrm{B}_{\mu\sigma}(\tau) &= \cos\theta_{\mu\sigma}\phi_{\mathrm{L}\mu\sigma}(\tau) + \sin\theta_{\mu\sigma}\phi_{\mathrm{d}\mu\sigma}(\tau) \\ \psi^\mathrm{A}_{\mu\sigma}(\tau) &= \sin\theta_{\mu\sigma}\phi_{\mathrm{L}\mu\sigma}(\tau) - \cos\theta_{\mu\sigma}\phi_{\mathrm{d}\mu\sigma}(\tau) \end{aligned}\right\} \quad \left(0 \le \theta \le \frac{\pi}{2}\right), \tag{4.37}$$

the Hartree–Fock ground state of the cluster is written as

$$\Psi^\mathrm{HF}_{N,\mathrm{g}} = \left| \prod_{l,\sigma}^{26} \phi_{\mathrm{L}l\sigma} \prod_{\mu,\sigma}^{10} \psi^\mathrm{B}_{\mu\sigma} \prod_{\mu,\sigma}^{n} \psi^\mathrm{A}_{\mu\sigma} \right| . \tag{4.38}$$

Here $n$ denotes the formal $d$-electron number. The first product in (4.38) denotes 26 electrons in the 26 non-bonding oxygen $p$ MOs. The coefficients $\cos\theta_{\mu\sigma}$ and $\sin\theta_{\mu\sigma}$ in (4.38) and the energies of the bonding and antibonding

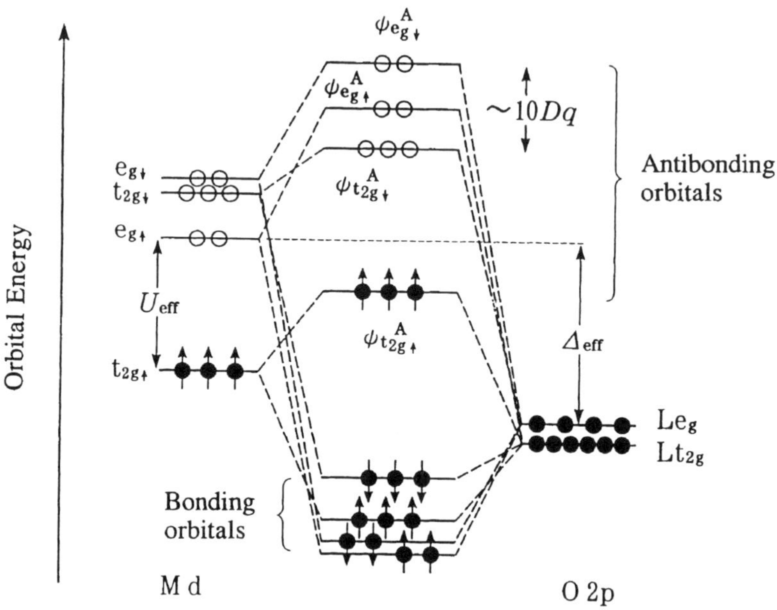


**Fig. 4.4.** One-electron energy levels of the octahedral $MO_6$ cluster in the Hartree-Fock approximation. The $M$ ion has the $d^3$ configuration

orbitals, $E^{\mathrm{B}}_{\mu\sigma}$ and $E^{\mathrm{A}}_{\mu\sigma}$, are to be determined self-consistently by solving a set of Hartree–Fock equations (4.15). Transfer integrals between a $d$ orbital and a $p$ MO are finite if the MO and the $d$ orbital have the same symmetry:

$$\langle \mu_\sigma | H | L_{\mu\sigma} \rangle \equiv T_e(\mu = u, v), T_{t2}(\mu = \xi, \eta, \zeta) . \tag{4.39}$$

If the $d$ orbitals lie above the $p$ orbitals, the bonding and antibonding orbitals are derived predominantly from the $p$ and $d$ orbitals, respectively, as shown in Fig. 4.4. The $\underline{d}$ orbitals used in ligand-field theory were actually those antibonding orbitals. Because the $e_g$ orbitals have larger overlaps with the oxygen orbitals than the $t_{2g}$ orbitals ($|T_{e_g}/T_{t_{2g}}| \sim 2$), the $e_g$ orbitals are more strongly hybridized with the $p$ orbitals than the $t_{2g}$ orbitals and hence the antibonding states are shifted above the antibonding states of $t_{2g}$-symmetry. This indicates that $d-p$ hybridization contributes to the observed crystal-field splitting $10\,Dq$ between the $t_{2g}$ and $e_g$ orbitals. Other contributions to $10\,Dq$ such as electrostatic fields and $d-p$ orbital non-orthogonality are relatively small.

If we denote the position of the lowest unoccupied $d$ orbital relative to the occupied $p$ orbitals (with the hybridization being switched off) by the effective charge-transfer energy $\Delta_{\mathrm{eff}}$ (Fig. 4.4), the position of the highest occupied $d$ orbital relative to the $p$ orbitals is given by $\Delta_{\mathrm{eff}} - U_{\mathrm{eff}}$. With increasing atomic number, the $d$-level position tends to be lowered. This reduces $\Delta_{\mathrm{eff}}$, thereby increasing the $d$–$p$ covalency. Increase in the valence of the transition-metal

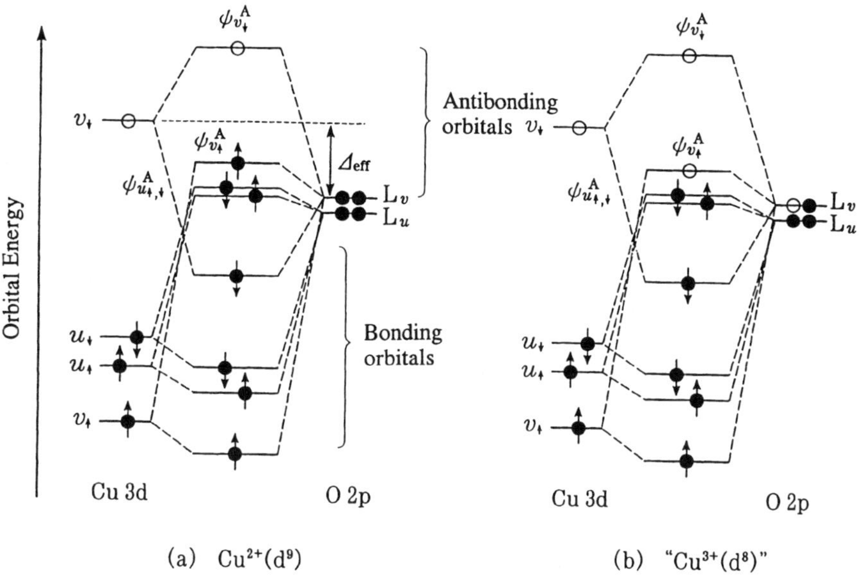


**Fig. 4.5.** One-electron energy levels of the square planer $CuO_4$ cluster in the Hartree-Fock approximation. Only the $u$ ($d_{x^2-y^2}$) and $v$ ($d_{3z^2-r^2}$) orbitals of Cu and those $p$ MO's which hybridize with them are shown. (**a**) $M = Cu^{2+}(d^9)$, (**b**) $M$ = '$Cu^{3+}(d^8)$'

ion also reduces $\Delta_{eff}$ and increases the covalency. If $\Delta_{eff}$ is further lowered, $\Delta_{eff} - U_{eff}$ becomes negative, the energies of the $p$ and highest occupied $d$ orbitals are interchanged and the $p$ orbitals become the highest unoccupied states. As a typical example of the latter case, the one-electron energy levels of the square planar $CuO_4^{6-}$ cluster, a relevant model for $Cu^{2+}$ oxides, are shown in Fig. 4.5 a. The energy gap is not $\sim U_{eff}$ as discussed in Sect. 4.4 but is $\sim \Delta_{eff}$. Such an insulator with a $p$-to-$d$ gap is called a charge-transfer type insulator [469], as opposed to the Mott-Hubbard type insulator with a $d$-to-$d$ gap which has been discussed in Sect. 4.4. Oxides of Co, Ni and Cu as well as those of Cr, Mn and Fe with high valences ($Cr^{4+}$, $Mn^{3+}$, $Mn^{4+}$, $Fe^{3+}$) belong to the charge-transfer type and oxides of Ti and V belong to the Mott–Hubbard type. $Mn^{2+}$ oxides are intermediate between the two types.

If both the atomic number and the valence of the transition-metal ion are large, $\Delta_{eff}$ may become vanishingly small or even negative. Then one might expect the band gap to be closed and electrons to be transferred from the $p$ orbitals to the empty $d$ orbitals, making the system metallic. However, strong $d$–$p$ hybridization prevents the gap from closing for a certain range of negative $\Delta_{eff}$. The character of the band gap for the negative charge-transfer-energy insulator is not of the $p$-to-$d$ nor $d$-to-$d$ type but of the $p$–$p$ type [470]. Oxides of $Cu^{3+}$ and $Fe^{4+}$ belong to this class of insulators and $Ni^{3+}$ oxides are intermediate between the charge-transfer type and the negative charge-transfer-energy insulators. Figure 4.5 b shows the energy levels of the $CuO_4^{5-}$ cluster, which represents the non-magnetic insulator $NaCuO_2$ with the formal valence of $Cu^{3+}(d^8)$. The figure shows that both the highest occupied and lowest unoccupied states are predominantly of the oxygen $p$ character.

### 4.5.2 Configuration-Interaction Theory

If one removes an electron from or adds an electron to the metal-ligand $MO_6$ cluster, considerable orbital relaxation (i.e., changes in $\cos\theta_{\mu\sigma}$ and $\sin\theta_{\mu\sigma}$ of (4.37)) may be expected because of the strong $d-d$ Coulomb interaction. This makes Koopmans' theorem useless. Under this circumstance, it is convenient to expand the wavefunction $\Psi_{N,\mathrm{g}}$ of the many-electron ground state of the cluster using a set of configurations starting from the purely ionic configuration and transferring electrons from the ligand to $d$ orbitals:

$$\Psi_{N,\mathrm{g}}^{\mathrm{CI}} = c_{\mathrm{g}0}\Psi(d^n) + c_{\mathrm{g}1}\Psi(d^{n+1}\underline{L}) + c_{\mathrm{g}2}\Psi(d^{n+2}\underline{L}^2) + \dots \tag{4.40}$$

Here, $\Psi(d^n)$ is the ionic state where the transition metal has the $d^n$ configuration and the ligand orbitals are completely filled ($O^{2-}$), and $\Psi(d^{n+1}\underline{L})$, $\Psi(d^{n+2}\underline{L}^2)$, etc, are successive ligand-to-metal charge-transfer states, where $\underline{L}$ denotes a hole in the ligand orbitals. Between two adjacent configurations (e.g., between $d^n$ and $d^{n+1}\underline{L}$), the one-electron part (the $\sum H_i$ term in (4.1)) of the Hamiltonian has a finite matrix element, i.e., a transfer integral,

$$\langle d^n|\mathcal{H}|d^{n+1}\underline{L}\rangle \sim \begin{cases} \sqrt{(4-n_e)}T_e & (\text{if } L \text{ has } E_{\mathrm{g}} \text{ symmetry}), \\ \sqrt{(6-n_{t2})}T_{t2} & (\text{if } L \text{ has } T_{2\mathrm{g}} \text{ symmetry}). \end{cases} \tag{4.41}$$

This configuration-interaction (CI) theoretical treatment can be successfully applied to the cluster model (i.e., the metal-ligand clusters considered above) as well as to the Anderson-impurity model, in which the $p$ band is treated as a continuum.

Not only the ground state but also the excited states of the cluster can be expanded in the CI form of (4.40). The excited states include those dominated by the $d^n$ configuration and those dominated by the charge-transfer ($d^{n+1}\underline{L}$, etc) configurations. The former excited states correspond to the intraatomic $d$–$d$ excitations which have been treated by ligand-field theory while the latter excited states correspond to $p$-to-$d$ charge-transfer excitations. The diagram of many-electron energy levels in the CI picture is depicted in Fig. 4.6. (Note that the vertical scale of Fig. 4.6 shows total energies of the many-electron system and not one-electron energies.)

We define the effective charge-transfer energy $\Delta_{\mathrm{eff}}$ as the energy difference from the lowest multiplet term of the $d^n$ configuration to that of the $d^{n+1}\underline{L}$ configuration. This $\Delta_{\mathrm{eff}}$ is the same as that defined for the Hartree–Fock levels in Sect. 4.5.1 and in Fig. 4.5. One can also define the (average) charge-transfer energy $\Delta$ as the energy difference from the center of gravity of the multiplet

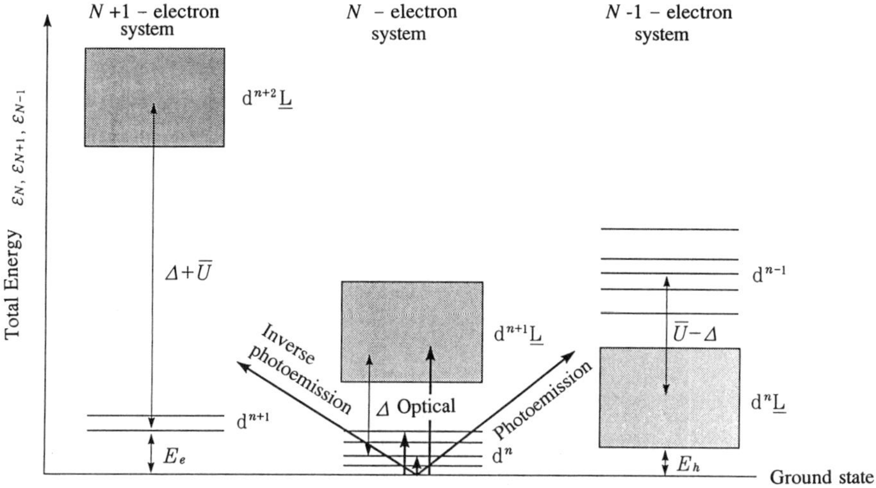


**Fig. 4.6.** Energy-level diagram of the transition-metal oxide in the configuration-interaction picture [469]. The compound is assumed to be of the charge-transfer type ($\Delta < \bar{U}$). For simplicity, energy shifts due to $d$-$p$ hybridization is not shown. $E_h + E_e$ is equal to the band gap

of the $d^n$ configuration to that of the $d^{n+1}\underline{L}$ configuration. In the definition of $\Delta$, the anisotropic part of the Coulomb repulsion and the exchange energy has been averaged out, and therefore $\Delta$ shows a smooth variation as a function of the atomic number and the valence of the transition-metal ion [471]. Empirically, it has been found that $\Delta$ decreases by about 0.6 eV for a unit increase of the atomic number and by about 2.5 eV for a unit increase of the valence. $\Delta_{\text{eff}}$ contains multiplet effects (i.e., contributions from the exchange interaction and the anisotropic part of the Coulomb repulsion). $\Delta_{\text{eff}}$ and $\Delta$ are related to each other through a relationship analogous to that between $U_{\text{eff}}$ and $U$ (4.31):

$$\Delta_{\text{eff}} = \Delta + \Delta\mathcal{E}_n - \Delta\mathcal{E}_{n+1} \,. \qquad (4.42)$$

Since $\Delta\mathcal{E}_n$ takes the maximum at $n = 5$, the multiplet correction $\Delta_{\text{eff}} - \Delta$ takes a minimum at $n = 4$ and a maximum at $n = 5$ (see Fig. 4.2). In charge-transfer-type insulators, therefore, the multiplet effects reduce the band gaps of $d^4$ compounds and increase the band gaps of $d^5$ compounds. This explains the experimental fact that $d^4$ compounds have narrow band gaps (e.g., $LaMnO_3$) or are metallic (e.g., $SrFeO_3$).

In the presence of electron correlation, because one-electron energy levels cannot be measured by experiment and only the total energies of the many-electron system are measurable, the formula (4.29) for the DOS is used. The DOS of the non-bonding $p$ orbitals, namely, the DOS of $p$ orbitals that are not hybridized with $d$, is given by the DOS of the one-electron ligand molecular

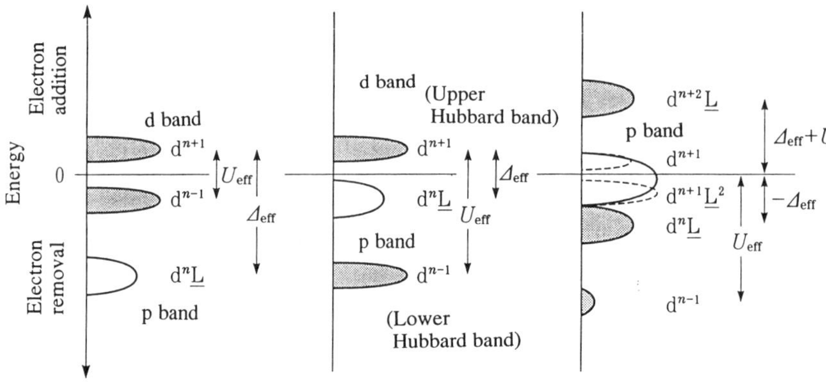


**Fig. 4.7.** Schematic density of states of transition-metal oxides in the configuration-interaction picture. (**a**) Mott-Hubbard type, (**b**) charge-transfer type, (**c**) negative charge-transfer-energy type

orbitals or of the $p$-band states. The schematic DOS are shown in Fig. 4.7 for various cases depending on the relative magnitudes of $\Delta_{\rm eff}$ and $U_{\rm eff}$. If $U_{\rm eff} < \Delta_{\rm eff}$, the DOS both just above and below $E_{\rm F}$ are transition-metal $d$-like and the magnitude of the band gap is given by $\sim U_{\rm eff}$ (Mott–Hubbard-type insulator). If $U_{\rm eff} > \Delta_{\rm eff}$, the DOS just above $E_{\rm F}$ is $d$-like while the DOS just below $E_{\rm F}$ is oxygen $p$-like, and the magnitude of the band gap is given by $\sim \Delta_{\rm eff}$ (charge-transfer-type insulator). To be more precise, energy shifts due to $d$–$p$ hybridization should be included in the magnitude of the band gap:

$$\mathcal{E}_{\rm gap} = U_{\rm eff} - W_d + 2\delta_N - \delta_{N-1} - \delta_{N+1} \tag{4.43}$$

for a Mott-Hubbard-type insulator or

$$\mathcal{E}_{\rm gap} = \Delta_{\rm eff} - \frac{1}{2}(W_p + W_d) + 2\delta_N - \delta_{N-1} - \delta_{N+1} \tag{4.44}$$

for a charge-transfer-type insulator. Here, $W_d$ and $W_p$ are the widths of the $d$ and $p$ bands, respectively, and $\delta_N$ is the energy lowering of the ground state of the $N$-electron system caused by $d$–$p$ hybridization.

In Fig. 4.8, various transition-metal oxides are plotted against $U_{\rm eff}$ and $\Delta_{\rm eff}$. This plot is called the Zaanen–Sawatzky–Allen diagram [469]. In this diagram, the diagonal line $U_{\rm eff} = \Delta_{\rm eff}$ separates the Mott–Hubbard regime from the charge-transfer regime, but because of the $d$–$p$ hybridization the boundary is not a sharp phase boundary but rather a crossover line. In the Mott–Hubbard regime, the band gap becomes small or zero near the $\Delta_{\rm eff}$-axis, realizing a $d$-band metal.

In the charge-transfer regime, it was originally thought that the band gap $\Delta_{\rm eff}$ becomes small or zero near the $U_{\rm eff}$-axis. However, a finite band gap remains in a certain $\Delta_{\rm eff} < 0$ range as shown in Fig. 4.8 [470]. The

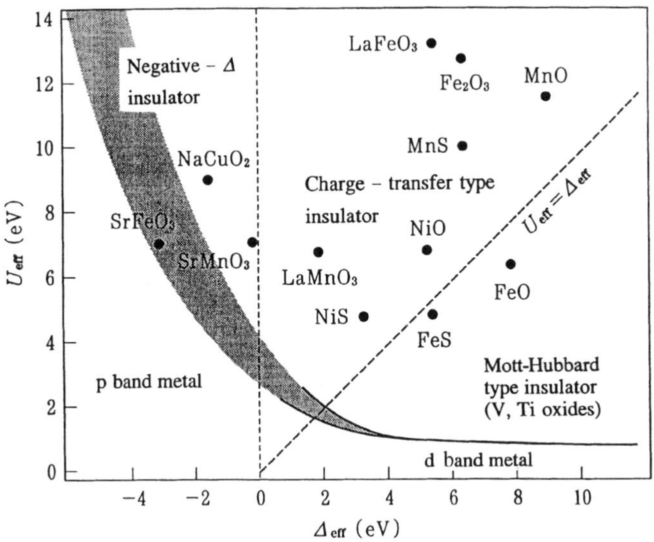


**Fig. 4.8.** Plot of various transition-metal compounds in the $\Delta_{\rm eff}$-$U_{\rm eff}$ plane ('Zaanen-Sawatzky-Allen' diagram)

opening of the band gap in the negative $\Delta_{\rm eff}$ regime is made possible by strong $d$–$p$ hybridization. If we apply the Hartree–Fock approximation to the $CuO_4^{5-}$ cluster (Fig. 4.5 b), which represents the nonmagnetic insulator $NaCuO_2$ (with $Cu^{3+}$), the spins of the $d$ and $p$ electrons become antiparallel:

$$\Psi_{N,{\rm g}}^{\rm HF} = |\underline{\psi}_{v\uparrow}^{\rm A}\underline{\psi}_{v\downarrow}^{\rm A}| ,$$

$$\left.\begin{aligned} \psi_{v\uparrow}^{\rm A} &= \sin\theta_\uparrow\phi_{{\rm L}v\uparrow} - \cos\theta_\uparrow\phi_{v\uparrow} \sim -\phi_{v\uparrow} \\ \psi_{v\downarrow}^{\rm A} &= \sin\theta_\downarrow\phi_{{\rm L}v\downarrow} - \cos\theta_\downarrow\phi_{v\downarrow} \sim \phi_{{\rm L}v\downarrow} \end{aligned}\right\} , \tag{4.45}$$

($\underline{\psi}_{v\uparrow}^{\rm A}$ denotes a hole in the $\psi_{v\uparrow}^{\rm A}$ orbital) which are not nonmagnetic. In CI theory, the $d(v)$ hole and the ligand ($L_v$) hole in the dominant configuration $d^9\underline{L}$ can form a Heitler–London type spin singlet:

$$\Psi_{N,{\rm g}} \sim \Psi(d^9\underline{L}) \equiv \frac{1}{\sqrt{2}}(|\underline{v}_\uparrow\underline{L}_{v\downarrow}| - |\underline{v}_\downarrow\underline{L}_{v\uparrow}|) . \tag{4.46}$$

Indeed, the energy of this nonmagnetic state is given by

$$\mathcal{E}_{N,{\rm g}} \cong -E_d' - E_p - \frac{2T_{\rm e}^2}{|\Delta_{\rm eff}|} - \frac{2T_{\rm e}^2}{|\Delta_{\rm eff} + U_{\rm eff}|} , \tag{4.47}$$

where $-E_d'$ is the energy of the bare $d$ hole, and is lower than the energy of the Hartree–Fock ground state:

$$\mathcal{E}_{N,{\rm g}}^{\rm HF} \cong -E_d' - E_p - \frac{T_{\rm e}^2}{|\Delta_{\rm eff}|} - \frac{T_{\rm e}^2}{|\Delta_{\rm eff} + U_{\rm eff}|} . \tag{4.48}$$

## 4.6 Magnetic Interactions

### 4.6.1 Superexchange Interaction

Since the ground state and low-energy excited states of a Mott insulator do not conduct electricity, one often employs the effective spin Hamiltonian (Heisenberg Hamiltonian), which retains only the spin degrees of freedom instead of the Anderson model (4.7) and the Hubbard model (4.11) and (4.12). The Heisenberg Hamiltonian

$$\mathcal{H} = \sum_{i,j} J_{ij} \boldsymbol{S}_i \cdot \boldsymbol{S}_j \tag{4.49}$$

represents superexchange interaction with a coupling constant $J_{ij}$ between the local spins $\boldsymbol{S}_i$ of transition-metal ions. We shall consider below $J_{ij}$ between transition-metal ions mediated by an oxygen through a 180° metal–oxygen–metal bond, which exists in NaCl-type and perovskite-type oxides [472].

Let us consider a linear $Cu^{2+}$-$O^{2-}$-$Cu^{2+}$ cluster and take the $z$-axis in the bond direction. If there is a hole in the $u$ ($d_{3z^2-r^2}$) orbital of each Cu ion, then one has to consider hybridization of this orbital only with the O $2p_z$ orbital. For the $Ni^{2+}$, $Co^{2+}$, $Fe^{2+}$ and $Mn^{2+}$ ions, there are also holes in the $v$ ($d_{x^2-y^2}$) and $t_{2g}$ orbitals, but these orbitals have small or zero overlap with the oxygen $p$ orbitals and may be ignored. We assume the antiferromagnetic state and take the CI picture. Starting from the ionic configuration,

$$\Psi_1^{\mathrm{AF}} \equiv |\underline{\psi}_{1d\uparrow}\underline{\psi}_{2d\downarrow}| \quad (\mathcal{E} = -2E_d'), \tag{4.50}$$

we consider states in which one hole is transferred from the Cu to oxygen atoms

$$\Psi_2^{\mathrm{AF}} \equiv |\underline{\psi}_{\mathrm{L}\uparrow}\underline{\psi}_{2d\downarrow}|, \Psi_3^{\mathrm{AF}} \equiv |\underline{\psi}_{1d\uparrow}\underline{\psi}_{\mathrm{L}\downarrow}| \quad (\mathcal{E} = -E_d' - E_p), \tag{4.51}$$

a state in which the two holes are transferred to the oxygen

$$\Psi_4^{\mathrm{AF}} \equiv |\underline{\psi}_{\mathrm{L}\uparrow}\underline{\psi}_{\mathrm{L}\downarrow}|, \quad (\mathcal{E} = -2E_p) \tag{4.52}$$

and states in which one Cu has two holes

$$\Psi_5^{\mathrm{AF}} \equiv |\underline{\psi}_{1d\uparrow}\underline{\psi}_{1d\downarrow}|, \quad \Psi_6^{\mathrm{AF}} \equiv |\underline{\psi}_{2d\uparrow}\underline{\psi}_{2d\downarrow}| \quad (\mathcal{E} = -2E_d' + U). \tag{4.53}$$

These states form a linear combination to constitute the ground state. Those two states which are connected through the transfer of one hole (e.g., $\Psi_1^{\mathrm{AF}}$ and $\Psi_2^{\mathrm{AF}}$) have a transfer integral $T(= T_{\mathrm{e}}/\sqrt{3})$ between them. Taking $\Psi_1^{\mathrm{AF}}$ as the unperturbed state, fourth order perturbation with respect to $T$ gives the ground-state energy of

$$\mathcal{E}_{\mathrm{g}}(\mathrm{AF}) \approx -2E_d' - \frac{2T^2}{\Delta}\left[1 + \frac{T^2}{\Delta^2} + \frac{T^2}{U\Delta}\right] . \tag{4.54}$$

(Here, note $\Delta = E_d' - E_p$.) On the other hand, the ferromagnetic state is a linear combination only of

$$\Psi_1^{\mathrm{F}} \equiv |\underline{\psi}_{1d\uparrow}\underline{\psi}_{2d\uparrow}| \quad (\mathcal{E} = -2E_d') ,$$

$$\Psi_2^{\mathrm{F}} \equiv |\underline{\psi}_{\mathrm{L}\uparrow}\underline{\psi}_{2d\uparrow}|, \quad \Psi_3^{\mathrm{F}} \equiv |\underline{\psi}_{1d\uparrow}\underline{\psi}_{\mathrm{L}\uparrow}| \quad (\mathcal{E} = -E_d' - E_p) , \tag{4.55}$$

and therefore has a higher energy than the antiferromagnetic state:

$$\mathcal{E}_{\mathrm{g}}(\mathrm{F}) \approx -2E_d' - \frac{2T^2}{\Delta} . \tag{4.56}$$

Because the expectation values of Hamiltonian (4.49) in the antiferromagnetic and ferromagnetic states are $-J_{ij}/4$ and $J_{ij}/4$, respectively, (4.54) and (4.56) yield

$$J_{ij} \approx \frac{4T^4}{\Delta}\left[\frac{1}{\Delta^2} + \frac{1}{U\Delta}\right] , \tag{4.57}$$

which is antiferromagnetic ($J_{ij} > 0$). The Néel temperature $T_{\mathrm{N}}$ is given in the molecular-field approximation by

$$T_{\mathrm{N}} = \frac{2J_{\mathrm{eff}}ZS(S+1)}{3k_{\mathrm{B}}} , \tag{4.58}$$

where $k_{\mathrm{B}}$ is the Boltzmann constant, $Z$ is the number of nearest neighbor metal ions, and $J_{\mathrm{eff}} \equiv J_{ij}/(2S)^2$.

In the charge-transfer-type insulator ($\Delta < U$), $J_{ij}$ rapidly increases with decreasing $\Delta$ as indicated by (4.57). In the Mott–Hubbard-type insulator ($\Delta > U$), the second term in (4.57) becomes dominant and

$$J_{ij} \sim \frac{4t^2}{U} , \tag{4.59}$$

where $t \equiv T^2/\Delta$ can be considered as the transfer integral between the $d$ orbitals appearing in (4.11) and (4.12). Indeed, if we ignore the intervening oxygen orbital and consider the direct transfer integral $t$ between the $d$ orbitals, we obtain

$$\mathcal{E}_{\mathrm{g}}(\mathrm{AF}) \cong -2E_d' - \frac{2t^2}{U} , \tag{4.60}$$

$$\mathcal{E}_{\mathrm{g}}(\mathrm{F}) \cong -2E_d' , \tag{4.61}$$

which gives (4.59) directly.

Next, we consider cases where the degeneracy of the $d$ orbitals is important. Let us consider one of the simplest examples in which $d^1$ ions such as

$Ti^{3+}$ and $V^{4+}$ and oxygen form 180° $M$–O–$M$ bonds. Here one of the $t_{2g}$ orbitals is occupied at each site [473]. Electrons can hop between the $\eta$ ($3d_{zx}$) or $\xi$ ($3d_{yz}$) orbitals through the $p_x$ or $p_y$ orbitals of the intervening anion, respectively. Since the $Ti^{3+}$ and $V^{4+}$ oxides are Mott–Hubbard type, we employ the degenerate Hubbard model (4.11) and use the transfer integral $t$ between neighboring $d$ orbitals. There are orbital degrees of freedom $(\eta, \xi)$ in addition to the spin degrees of freedom. According to whether the orbital is the same or different between the nearest-neighbor metal atoms, the system is called ferro-orbital or antiferro-orbital. Within the second-order perturbation with respect to $t$, the energy of the antiferromagnetic and antiferro-orbital pair is given by

$$\mathcal{E}_{\mathrm{g}}(\mathrm{AF}, \mathrm{AF}) \approx 2E_d - \frac{2t^2}{U'} . \tag{4.62}$$

Similarly,

$$\text{antiferromagnetic ferro-orbital}: \quad \mathcal{E}_{\mathrm{g}}(\mathrm{AF}, \mathrm{F}) \approx 2E_d - \frac{2t^2}{U} , \tag{4.63}$$

$$\text{ferromagnetic antiferro-orbital}: \quad \mathcal{E}_{\mathrm{g}}(\mathrm{F}, \mathrm{AF}) \approx 2E_d - \frac{2t^2}{U' - J} , \tag{4.64}$$

$$\text{ferromagnetic ferro-orbital}: \quad \mathcal{E}_{\mathrm{g}}(\mathrm{F}, \mathrm{F}) \approx 2E_d , \tag{4.65}$$

and from the relationship $U > U' > U' - J$, it follows that $\mathcal{E}_{\mathrm{g}}(\mathrm{AF}, \mathrm{AF}) < \mathcal{E}_{\mathrm{g}}(\mathrm{F}, \mathrm{AF}) < \mathcal{E}_{\mathrm{g}}(\mathrm{AF}, \mathrm{F})$. Therefore, if one decreases the temperature from a high temperature, orbitals would first be ordered antiferromagnetically and then spins would be ordered ferromagnetically. In general, orbital ordering induces a distortion of neighboring oxygen atoms, namely, Jahn–Teller distortion. In fact, in $LaMnO_3$ (which has the $d^4$ configuration with one $e_g$ electron), spins are ferromagnetically ordered within the (001) plane accompanied by a Jahn–Teller distortion. $YTiO_3$ ($d^1$) shows ferromagnetism with orbital ordering [474].

Two transition-metal ions forming a 90° bond with a non-metal anion are coupled ferromagnetically ($J_{ij} < 0$). This is due to Hund rule coupling at the anion site between the spins of electrons partially transferred from the transition-metal ions.

### 4.6.2 Local Moment in Metals

When the system undergoes a transition from an insulator to a metal, in addition to the superexchange interaction mediated by the filled $p$ band, there appears interaction between the local moments mediated by the Fermi sea. However, the situation is not so simple in many transition-metal compounds because $d$ electrons are responsible for both magnetic and transport properties.

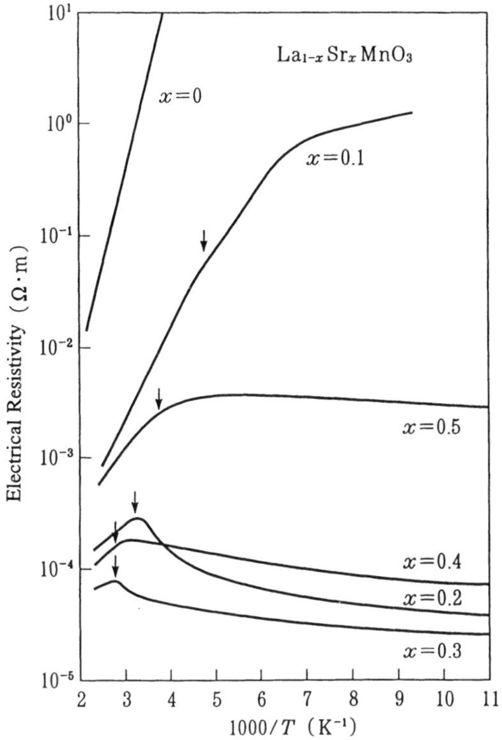


**Fig. 4.9.** Electrical resistivity of $La_{1-x}Sr_xMnO_3$ [475]. The arrows indicate Curie temperatures. (With permission from North-Holland, Elsevier)

When the local moment of the $d$ electrons is large and also part of the $d$ electrons somehow become itinerant metallic electrons, the spins of the itinerant electrons are ferromagnetically coupled to the local moments through Hund rule coupling. Since the spin direction is conserved when an electron is transferred from one atom to another, the kinetic energy of the electrons is increased when the local spins are aligned parallel to each other. This ferromagnetic coupling mechanism between local moments is called double exchange interaction. A typical example of the double exchange systems is $La_{1-x}Sr_xMnO_3$. The end members $LaMnO_3$ and $SrMnO_3$ are both antiferromagnetic insulators. The La→Sr (Sr→La) substitution introduces holes (electrons) into $LaMnO_3$ ($SrMnO_3$) and these holes (electrons) induce double exchange interaction, making the system a ferromagnetic metal. As shown in Fig. 4.9, the electrical resistivity is semiconductive above the Curie temperature while it is metallic below it [475]. The saturation magnetization of the metallic sample is consistent with the mixed valence of $Mn^{3+}$ ($d^4, S = 2$) and $Mn^{4+}$ ($d^3, S = 3/2$) in $La_{1-x}Sr_xMnO_3$, where the atomic ratio is $Mn^{3+}$:$Mn^{4+}$ $= 1 - x : x$ (for details, see Sect. 5.7).

For well-localized magnetic moments like those of the rare-earth ions, the Kondo effect, which arises from coupling between the local moments and conduction electrons, and RKKY (Ruderman–Kittel–Kasuya–Yosida) interaction, which is interaction between local moments mediated by conduction electrons, both become important. If the charge-transfer degrees of freedom between the conduction band and the localized level are projected out from the Anderson model (4.7), one obtains the interaction Hamiltonian

$$\mathcal{H}_{\mathrm{K}} = \sum_{j} J_{\mathrm{K}} \boldsymbol{S} \cdot \boldsymbol{s}_j \,, \tag{4.66}$$

where $\boldsymbol{S}$ is the spin of the localized electrons and $\boldsymbol{s}_j$ is the spin of the $j$-th conduction electron. Here, $J_{\mathrm{K}} > 0$, meaning that the magnetic coupling is antiferromagnetic. Therefore, the ground state of the Hamiltonian (4.66) is a spin singlet (Kondo singlet) formed by the local moment and the conduction electrons; the local moment thus disappears (Kondo effect).

RKKY interaction between local moments $\boldsymbol{S}_i$ and $\boldsymbol{S}_j$ is given by

$$\mathcal{H}_{\mathrm{RKKY}} = \sum_{ij} J_{ij} \boldsymbol{S}_i \cdot \boldsymbol{S}_j \,,$$

$$J_{ij} \propto N_{\mathrm{F}}(0) \frac{2k_{\mathrm{F}}|\boldsymbol{r}_i - \boldsymbol{r}_j| \cos 2k_{\mathrm{F}}|\boldsymbol{r}_j| - \sin 2k_{\mathrm{F}}|\boldsymbol{r}_i - \boldsymbol{r}_j|}{(2k_{\mathrm{F}}|\boldsymbol{r}_i - \boldsymbol{r}_j|)^4} \,, \tag{4.67}$$

where $k_{\mathrm{F}}$ is the Fermi wave number and $N_{\mathrm{F}}(0)$ is the DOS at $E_{\mathrm{F}}$. $J_{ij}$ in (4.67) is oscillating in real space and becomes positive (antiferromagnetic) or negative (ferromagnetic) depending on the atomic distance $|\boldsymbol{r}_i - \boldsymbol{r}_j|$.

The typical Kondo effect and RKKY interaction are usually not seen in transition-metal oxides. However, interaction between the $Cu^{2+}$ local spins and $p$ holes in the high-$T_{\mathrm{c}}$ cuprates has been discussed in the context of the Kondo effect. Here, interaction between the $d$ spin and $p$ hole is extremely strong: $J_{\mathrm{K}} \sim 1\,\mathrm{eV}$. In the same system, the Cu–Cu superexchange interaction is $J_{ij} \sim 0.1\,\mathrm{eV}$.

## 4.7 Correlated Metals

### 4.7.1 Metal–Insulator Transition

There are two scenarios for insulator-to-metal transitions:

1. the band gap is closed (bandwidth-control metal–insulator transition),
2. carriers are doped (band-filling-control metal–insulator transition).

Since the magnitude of the band gap is given by (4.43) or (4.44), a band width-control insulator-to-metal transition can be induced by decreasing $\bar{U}(U_{\mathrm{eff}})$ or $\Delta$ ($\Delta_{\mathrm{eff}}$) for a Mott–Hubbard-type insulator and a charge-transfer insulator, respectively. In the insulators of either type, an increase in the band width ($W_d$ or $W_p$) reduces the band gap.

There are two types of filling control. One is $B$-site substitution, in which transition-metal atoms themselves are replaced (e.g., $Li_xNi_{1-x}O$, $BaPb_{1-x}Bi_xO_3$), and the other is $A$-site substitution, in which the other atoms are replaced and supply the transition-metal atoms with carriers (e.g., high-$T_c$ cuprates, $La_{1-x}Sr_xMnO_3$). The conduction path is strongly disturbed in the $B$-site substitution while it is not in $A$-site substitution; therefore, it is easier to cause an insulator-to-metal transition by $A$-site substitution.

### 4.7.2 Hubbard Model

The Hubbard model without orbital degeneracy (4.12) is the simplest model by which one can study the effect of electron correlation in the metallic state. Electron correlation in metals makes electrons repel each other. In order to represent this situation, *Gutzwiller* proposed a wavefunction in which the weight of double occupancy is reduced relative to the Hartree–Fock wavefunction:

$$\Psi^{\mathrm{G}} = \prod_i (1 - \eta n_{i\uparrow} n_{i\downarrow})\, \Psi^{\mathrm{HF}} \quad (0 < \eta < 1)\,, \tag{4.68}$$

where $\Psi^{\mathrm{HF}}$ is the nonmagnetic Hartree–Fock ground state with one electron per site, $1 - \eta n_{i\uparrow} n_{i\downarrow}$ is an operator which reduces the double occupancy of site $i$ by factor $1 - \eta$: double occupancy is totally prohibited for $\eta = 1$. *Brinkman* and *Rice* [476] have discussed the metal–insulator transition using Gutzwiller's wavefunction (4.68). They have shown that when the system approaches the metal–insulator transition from the metallic side, i.e., when $U$ approaches a critical value $U_0$ from below it, the effective mass of conduction electrons diverges as

$$\frac{m^*}{m} = \left[1 - \left(\frac{U}{U_0}\right)^2\right]^{-1} . \tag{4.69}$$

Both the electronic specific heat coefficient $\gamma$ and the Pauli paramagnetic susceptibility $\chi$ diverge in proportion to $[1-(U/U_0)]^{-1}$, and hence the Wilson ratio $\chi/\gamma$ retains a constant value.

Electron correlation also plays an important role in magnetism at finite temperatures. If the ground state of the single-band Hubbard model with one electron per atom is almost fully spin polarized, Hartree–Fock calculations yield the splitting of $\sim U$ between the $\uparrow$ and $\downarrow$ bands. If we consider the magnetic properties at finite temperatures in terms of one-electron excitations between the two bands, the magnetic-to-nonmagnetic transition should occur at $T_c \sim U/k_B$, at which temperature a sufficient number of electrons would be excited from the $\uparrow$ band to the $\downarrow$ band and destroy the magnetic ordering

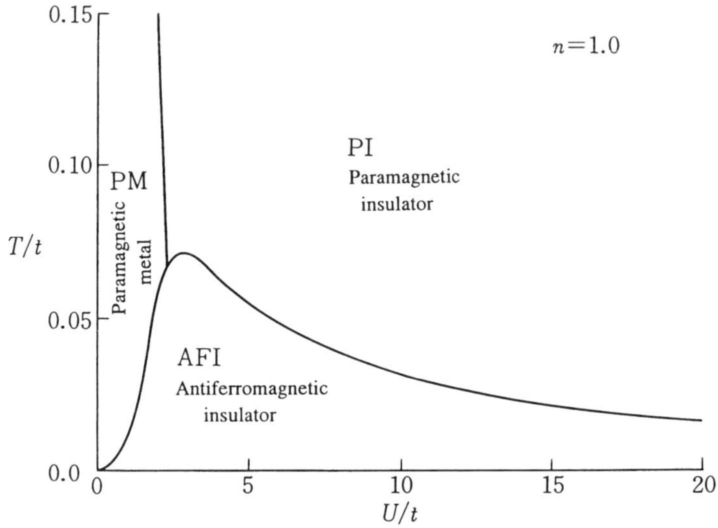


**Fig. 4.10.** Phase diagram of the single-band Hubbard model with occupation 1 calculated using spin-fluctuation theory [477]. (With permission from The Physical Society of Japan)

(Stoner model). That $T_c$, however, easily exceeds 10 000 K for transition metals and their oxides, and does not explain the experimentally observed Curie and Néel temperatures of a few 100 K or lower. The Stoner model fails there because it neglects attractive interaction of order $U$ between the excited electron and hole. This effect has been considered in Moriya's spin fluctuation theory, in which effects of spin fluctuations are self-consistently taken into account with $U$ as perturbation. The theory has successfully explained local-moment magnetism and itinerant-electron magnetism in a unified way by considering spin fluctuations on general grounds. Figure 4.10 shows a phase diagram of the single-band Hubbard model calculated for one electron per site [477]. For small $U/t$, since double occupancy is suppressed with increasing $U$, spin polarization increases with $U$ and hence $T_N$ increases. As $U$ further increases and the spin polarization saturates, the magnetic interaction becomes that between local spins (4.59). Therefore $J_{ij} \propto 1/U$ decreases and hence $T_N$ decreases with $U$.

### 4.7.3 Fermi-Liquid Properties

If one gradually increased the electron–electron interaction strength starting from a nonmagnetic metal in the weak interaction limit, the system would exhibit qualitatively the same behavior as noninteracting electrons unless the system underwent a phase transition such as a magnetic phase transition or a metal–insulator transition. This forms the basis of Landau's Fermi-liquid theory [478]. Wide band metals with weak electron correlation such as

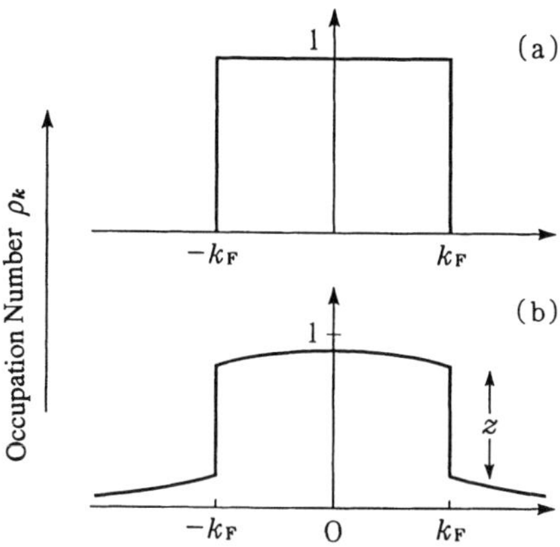


**Fig. 4.11.** Momentum distribution function for non-interacting (**a**) and interacting (**b**) electrons

$ReO_3$ are obviously Fermi liquids. In addition, overdoped high-$T_c$ cuprates, perovskite-type $SrVO_3$ and $La_{1-x}Sr_xTiO_3$ are known to exhibit Fermi-liquid properties [479].

In Fermi-liquid theory, one considers quasi-particles as elementary excitations. The quasi-particle is not a bare electron or hole but is an electron or hole dressed by an interaction cloud. When the number of quasi-particles with momentum $\boldsymbol{k}$ and spin $\sigma$ is changed by $\Delta n_{\boldsymbol{k}\sigma}$, the change in the total energy of the system will be given by

$$\Delta \mathcal{E} = \sum_{\boldsymbol{k},\sigma} E_{\boldsymbol{k},\sigma} \Delta n_{\boldsymbol{k},\sigma} + \frac{1}{2} \sum_{\boldsymbol{k},\boldsymbol{k}',\sigma,\sigma'} f_{\boldsymbol{k}\sigma\boldsymbol{k}'\sigma'} \Delta n_{\boldsymbol{k},\sigma} \Delta n_{\boldsymbol{k}',\sigma'} \tag{4.70}$$

up to the second order in $\Delta n_{\boldsymbol{k}\sigma}$. Here, $E_{\boldsymbol{k}\sigma}$ is the energy of a quasiparticle with momentum $\boldsymbol{k}$ and spin $\sigma$, and $f_{\boldsymbol{k}\sigma\boldsymbol{k}'\sigma'}$ is the interaction between two quasi-particles $\boldsymbol{k}_\sigma$ and $\boldsymbol{k}'_{\sigma'}$. Since the Fermi liquid considered here is non-magnetic, the index $\sigma$ will be omitted hereafter. Without electron–electron interaction, each electron occupies a Bloch orbital with a fixed momentum $\boldsymbol{k}$, and therefore the momentum distribution function $\rho_{\boldsymbol{k}}$ becomes $\rho_{\boldsymbol{k}} = 1$ inside the Fermi surface and $\rho_{\boldsymbol{k}} = 0$ outside it, as shown in Fig. 4.11 a. Electron–electron interaction scatters electrons between different momenta, making $\rho_{\boldsymbol{k}} \leq 1$ inside the Fermi surface and $\rho_{\boldsymbol{k}} \geq 0$ outside it as in Fig. 4.11 b. Note, however, that the finite jump in $\rho_{\boldsymbol{k}}$ at the Fermi surface persists and the volume inside the Fermi surface remains unchanged. The latter important statement is called Luttinger's theorem. Most direct experimental techniques to study Fermi surfaces are de Haas–van Alphen effect and angle-resolved photoemission spectroscopy. Those measurements have given evidence for the existence of Fermi surfaces in interacting electron systems.

The quasi-particle has a finite lifetime due to collision with other quasi-particles. Each quasi-particle decays into a quasi-particle with a lower en-

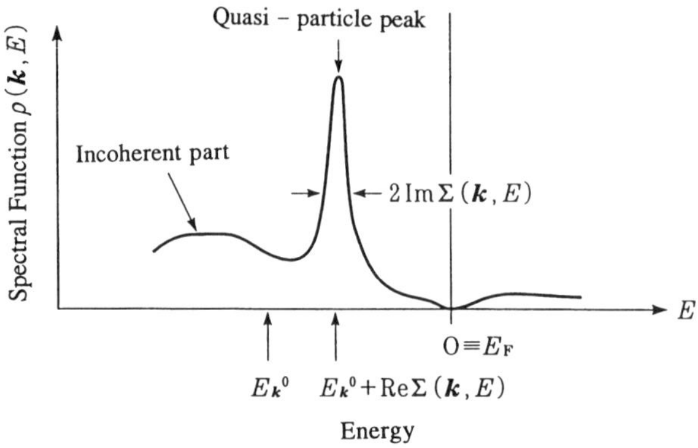


**Fig. 4.12.** Single particle spectral function $\rho(\boldsymbol{k}, E) \equiv (1/\pi)\mathrm{Im}G(\boldsymbol{k}, E)$

ergy by exciting low-energy electron–hole pairs. The decay probability of a quasi-particle is proportional to the number of possible electron–hole pair excitations, i.e., $\propto E^2$ for small $E$, where $E$ is the energy of the quasi-particle measured from the Fermi level. That is, the quasi-particle energy $E_{\boldsymbol{k}}$ is a complex number with an imaginary part $\hbar/\tau_{\boldsymbol{k}} \propto E^2$. One can thus define the self-energy $\Sigma(\boldsymbol{k}, \omega)$ of a quasi-particle by $E_{\boldsymbol{k}} = E_{\boldsymbol{k}}^0 + \Sigma(\boldsymbol{k}, \omega)$, where $E_{\boldsymbol{k}}^0$ is the electron energy derived from the one-electron approximation. Because the real and imaginary parts of the self-energy should satisfy the causality or Kramers–Kronig relation, it follows that $\mathrm{Re}\Sigma(\boldsymbol{k}, \omega)$ is proportional to $E$ for small $E$. The self-energy is energy dependent because the potential energy felt by an electron depends on the distribution of other electrons which is dynamically fluctuating, influenced by the motion of the electron considered. (In the one-electron approximation, the distribution of electrons is static and therefore each electron feels an energy independent potential.)

The spectral function $\rho(\boldsymbol{k}, E)$ can be measured by angle-resolved photoemission spectroscopy and is given by the imaginary part of the single-particle Green's function

$$G(\boldsymbol{k}, E) \equiv \frac{1}{E - E_{\boldsymbol{k}}^0 - \Sigma(\boldsymbol{k}, E)} \tag{4.71}$$

as

$$\begin{aligned} \rho(\boldsymbol{k}, E) &\equiv -\frac{1}{\pi}\mathrm{Im}G(\boldsymbol{k}, E) \\ &= -\frac{\mathrm{Im}\Sigma(\boldsymbol{k}, E)}{\pi\{[E - E_{\boldsymbol{k}}^0 - \mathrm{Re}\Sigma(\boldsymbol{k}, E)]^2 + [\mathrm{Im}\Sigma(\boldsymbol{k}, E)]^2\}} \,. \end{aligned} \tag{4.72}$$

As shown in Fig. 4.12, (4.72) gives a peak at $E = E_{\boldsymbol{k}}^0 + \mathrm{Re}\Sigma(\boldsymbol{k}, E)$.

Let us consider a simple case in which the self-energy is local, i.e., independent of $\boldsymbol{k}$ : $\Sigma(\boldsymbol{k}, E) \equiv \Sigma(E)$. This is supposed to be a relatively good approximation for the Hubbard model, where the interaction is shortrange. $\Sigma(E)$ can be expanded near the Fermi level $E = 0$ as

$$\Sigma \cong -aE - \mathrm{i}bE^2 \quad (a, b > 0)\,. \tag{4.73}$$

From (4.72) and (4.73), one obtains

$$\rho(\boldsymbol{k}, E) \cong -\frac{bE^2}{\pi[(E - E^0_{\boldsymbol{k}} + aE)^2 + b^2E^4]} = \frac{z^2bE^2}{\pi[(E - zE^0_{\boldsymbol{k}})^2 + z^2b^2E^4]}\,, \tag{4.74}$$

where $z \equiv 1/(1+a) < 1$. In going from the non-interacting case

$$\rho_0(\boldsymbol{k}, E) = \lim_{\eta \to 0} \frac{\eta}{\pi[(E - E^0_{\boldsymbol{k}})^2 + \eta^2]} \equiv \delta(E - E^0_{\boldsymbol{k}}) \tag{4.75}$$

to the interacting case [(4.74)], the peak at $E = E^0_{\boldsymbol{k}}$ is shifted to $E = zE^0_{\boldsymbol{k}}$, reducing the dispersional band width by the factor $z$ ($< 1$). Because of this reduction of the band width, the density of quasi-particles at the Fermi level ($\Sigma_{\boldsymbol{k}}\delta(E - \mathrm{Re}E_{\boldsymbol{k}})$) is enhanced by the factor $1/z$ ($> 1$) and hence the electronic specific $\gamma T$ is enhanced by $1/z$. This can be interpreted as an enhancement of the conduction-electron effective mass $m^*$ by a factor $1/z$ as a result of electron correlation.

The spectral weight of each quasi-particle in (4.74) has been reduced to $z$ compared with the spectral weight 1 of the non-interacting case (4.75). This reduction is canceled with the enhancement of the quasi-particle density,

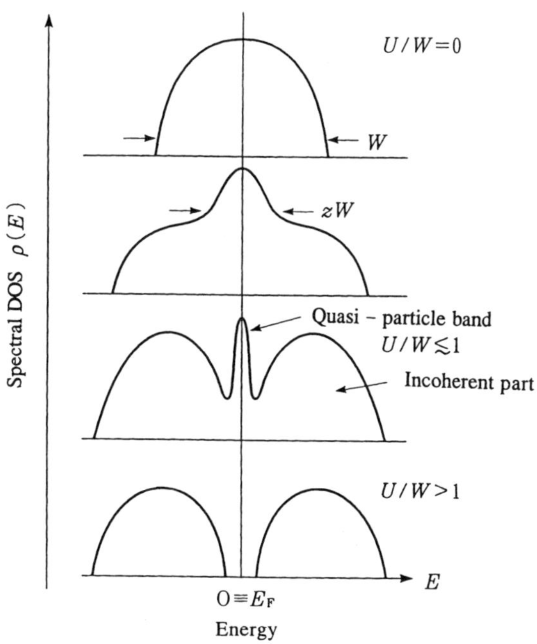


**Fig. 4.13.** Schematic picture of changes in the density of states $\rho(\varepsilon)$ caused by electron correlation [480]. Here, the self-energy is assumed to be local

giving the same DOS at the Fermi level $\rho(E_{\rm F}) \equiv \int \rho(\boldsymbol{k}, E_{\rm F})\mathrm{d}\boldsymbol{k}$ as that of the non-interacting system. Thus, as shown in Fig. 4.13, when one approaches a bandwidth-control metal–insulator transition from the metallic side, the DOS at the Fermi level remains essentially the same up to the transition point and then suddenly disappears at the transition. Since the total width of the quasi-particle band is reduced by factor $z$, the integrated intensity of the quasi-particle band is reduced by $z$. At the same time, spectral weight is transferred from the quasi-particle band to the satellite or the incoherent part of the spectral function as shown in Fig. 4.13. These incoherent features become the lower and upper Hubbard bands in the insulating limit. Recently, this spectroscopic and thermodynamic behavior of the correlated metals has been corroborated by dynamical mean field treatment of the Hubbard model [480].

The enhancement of the quasi-particle density near the Fermi level enhances not only the electronic specific heat coefficient $\gamma$ but also the Pauli-paramagnetic susceptibility $\chi$. The enhancement factor for $\chi$ is $(m^*/m)(1 + F_a^0)$ rather than $(m^*/m)(\equiv 1/z)$, where $F_a^0$ is the spin-dependent isotropic part of $f_{\boldsymbol{k}\sigma\boldsymbol{k}'\sigma'}$ in (4.70). The ratio between the Pauli-paramagnetic susceptibility and the electronic specific heat coefficient $R \equiv (3\pi^2 k_{\rm B}^2/\mu_0\mu_{\rm B})(\chi/\gamma)$ (Wilson ratio) is a number which characterizes the Fermi liquid. $R = 1$ for the non-interacting case. In a correlated Fermi liquid, $R$ takes a value between 1 and 2 although the mass enhancement factor $m^*/m$ varies over a very wide range: from $\sim$1 in weakly correlated systems to $\sim$ 1000 in some heavy Fermion rare-earth and uranium compounds. If the system is close to a ferromagnetic instability, only $\chi$ is strongly enhanced (Stoner enhancement) and hence $R \gg 1$. If the electron-phonon interaction is a dominant factor that enhances the conduction-electron mass, then $R < 1$. Also, it has been found that the coefficient $A$ in the electrical resistivity $\rho(T) = \rho_0 + AT^2$ is enhanced as the mass is enhanced; $A$ is approximately proportional to $\gamma^2$ (Kadowaki–Woods relation).

### 4.7.4 Long-Range Coulomb Interaction

The original idea of metal–insulator transitions by *Mott* was based on the effect of long-range Coulomb interaction [481]. The attractive potential for an electron from the positive charge of an ion core or a hole in the electron gas is given by

$$V(r) = -\frac{e^2}{4\pi\epsilon r}\exp(-\kappa_{\rm s} r)\,, \tag{4.76}$$

according to Thomas–Fermi theory. Here, $\varepsilon$ is the dielectric constant of the medium, and $\kappa_{\rm s}^{-1} \equiv (m^* e^2 n^{1/3}/\pi\varepsilon\hbar)^{-1/2}$ is the Thomas–Fermi screening length. The potential (4.76) has a bound state if

$$\kappa_{\rm s} < \frac{m^* e^2}{4\pi\epsilon\hbar}\,, \tag{4.77}$$

which can be rewritten using the Bohr radius $a_B \equiv 4\pi\hbar^2\varepsilon/m^*e^2$ in the medium as

$$a_B n^{1/3} < 0.25\,. \tag{4.78}$$

When electron localization starts, the density $n$ of free electrons decreases and $\kappa_s^{-1}$ increases, accelerating the electron localization. Thus the metal–insulator transition occurs as a first-order phase transition. Most recent theoretical studies of electron correlation have been made using the Hubbard model and the Anderson model, which ignore the long-range Coulomb interaction. Accordingly, the metal–insulator transition occurs rather continuously as a function of temperature in many theoretical studies.

The long-range part of the Coulomb interaction is almost perfectly screened in ordinary metals and alloys. In transition-metal oxides, interatomic Coulomb energy is significantly screened, e.g., to ~1 eV between adjacent transition-metal and oxygen atoms. This is small compared to the intra-atomic Coulomb interaction of several eV, but is not always negligible. The effect of long-range Coulomb interaction has not been well understood so far, but we shall describe below a few cases in which the long-range Coulomb interaction becomes essential.

### 4.7.5 Mixed Valence States

Carriers doped into an insulator interact with each other not only through the on-site Coulomb repulsion $U$ but also through the intersite Coulomb repulsion $U_1$, $U_2$, ..., where suffixes denote the first, second, ..., nearest neighbors, respectively. If $U$ is sufficiently large compared to the band width, double occupancy by the doped carriers is prohibited and the system can be described as a mixture of two valence states. For example, the transition-metal ion $M$ in $La_{1-x}Sr_xMO_3$ is in the mixed-valent state of 3+ and 4+ with the ratio $M^{3+} : M^{4+} = 1 - x : x$. If double occupancy is prohibited, $U_1$, $U_2$, ..., instead of $U$ play the central roles in the thermodynamic and transport properties. Magnetite $Fe_3O_4$ is a classical example of mixed-valent systems. The Fe ions in the $B$-site (octahedral site) of $Fe_3O_4$ are in the mixed valent state of $Fe^{2+} : Fe^{4+} = 1 : 1$. $Fe_3O_4$ shows a conductivity transition (Verwey transition) at $T_V = 124\,K$, above which the system becomes highly conductive. In filling-control systems such as $La_{1-x}Sr_xMO_3$, the substitution at the $A$-site induces randomness, though small, and affects their electronic properties. Effects of interatomic Coulomb interaction in $Fe_3O_4$ are described in detail in Sect. 5.7.

Charge ordering is a phenomenon in which the long-range Coulomb interaction may play an important role. In general, itinerant carriers gain kinetic energy at the cost of potential energy as they approach each other. Therefore, for carriers in a narrow band the kinetic energy gain is small and they may gain potential energy by forming a charge-ordered state. The interest in charge ordering goes back to the prediction by *Wigner* that the free-electron gas forms an ordered lattice called a Wigner crystal when the electron density is very low. The energy gap of the Wigner crystal is of order $U_1$ or $U_2$. The phase below the Verwey transition in $Fe_3O_4$ (Sect. 5.8) can be thought of as a Wigner crystal. However, the interatomic Coulomb interaction is not the only driving force for the Verwey transition: charge ordering induces lattice distortion and is stabilized through electron–lattice interaction. Also, interatomic superexchange interaction (Sect. 4.6.1) stabilizes particular forms of charge-ordered state in some transition-metal oxides including the perovskite-type Mn oxides (Sect. 5.7).

### 4.7.6 Anderson Localization

In order to introduce extra holes or electrons in Mott insulators as well as in band insulators (e.g., $WO_3$, $TiO_2$, $SrTiO_3$), one has to substitute atoms or introduce extra atoms or vacancies in the crystal, thereby creating impurity potentials in the crystal. In the limit of low carrier density, therefore, carriers are bound to the impurity centers which have electrical charges of opposite sign. As the impurity concentration increases, overlap between the wavefunctions of the impurity states increases. Then the carriers tend to itinerate

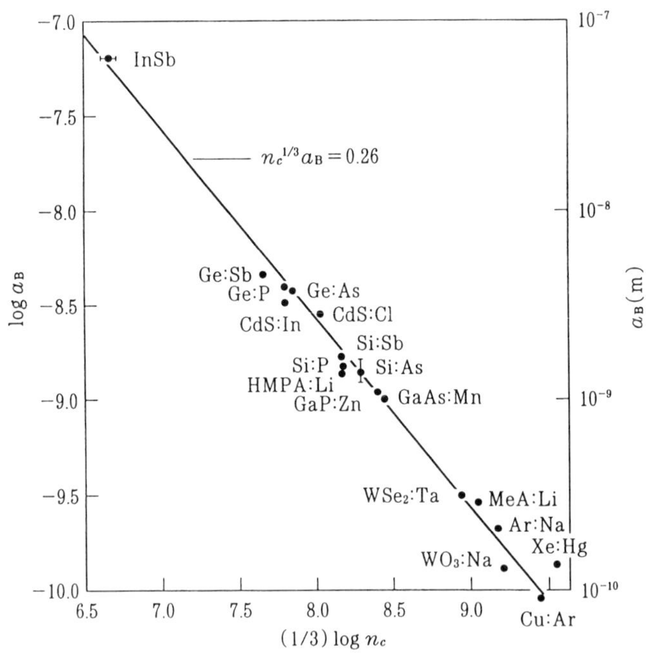


**Fig. 4.14.** Relationship between the impurity-orbital radius $a_B$ and the critical carrier concentration $n_c$ for metal-insulator transition [482]. The solid line represents $a_B n_c^{1/3} = 0.26$. (With permission from The American Physical Society)

between the impurities and the system makes a transition from insulator (semiconductor) to metal. In analogy with (4.78), there is a relationship between the radius $a_B$ of the bound-state orbital (of the donor or acceptor) and the critical concentration $n_c$ for the metal–insulator transition:

$$a_B n_c^{1/3} \cong 0.25 \,. \tag{4.79}$$

Figure 4.14 shows that the relationship (4.79) holds for a wide range of filling-control systems [482]. Electron correlation is certainly important near the metal–insulator boundary since carriers are barely localized or itinerant in this region.

If one views the metal-insulator transition from the metallic side, a random potential causes electron localization as pointed out for the first time by *Anderson* [481]. This phenomenon is called Anderson localization. With increasing strength of random potential, the localization starts from the band edge and then proceeds to the middle of the band as shown in Fig. 4.15. According to Mott, localized and delocalized states do not coexist at the same energy: there exists a boundary which separates the localized and delocalized states called the mobility edge (see Fig. 4.15). As the Fermi level crosses the mobility edge with filling-control, a metal–insulator transition called Anderson transition occurs.

Considering the finite Coulomb repulsion between charge carriers, a donor (acceptor) can accommodate a limited number of (one to a few) electrons (holes) depending on the valence of the impurity. Therefore, a gap is opened between the occupied donor level (empty acceptor level) and the bottom of the empty conduction band (top of the occupied valence band), and the Fermi level lies within this gap. In real systems, the donors (acceptors) are partially compensated by a small number of acceptors (donors) which are created simultaneously. For instance, when Sr ions are substituted for La in

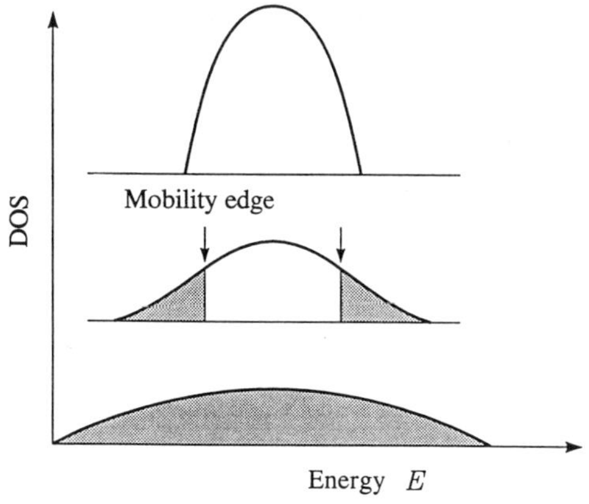


**Fig. 4.15.** Electron localization caused by random potential. The degree of disorder increases from the top to the bottom. The shaded areas represent localized states

a compound of the type La$M$O$_3$ in order to dope the system with extra holes, then a small number of oxygen vacancies are sometimes created to produce minority electron carriers and some of the holes are compensated by these electrons. Then the Fermi level is shifted to the impurity band, which is derived from the overlapping donor or acceptor levels. If the Fermi level lies on the localization side of the mobility edge, hopping conduction (impurity conduction) occurs within the impurity band [481].

Now we consider the effect of Coulomb interaction in the case where the Fermi level lies on the localized side of the mobility edge. If there were no Coulomb interaction, the DOS at the Fermi level would be finite as shown in Fig. 4.15. If one switches on the interatomic Coulomb interaction, electrons tend to be distributed so that they are separated from each other as much as possible to minimize the total potential energy of the system. In a periodic

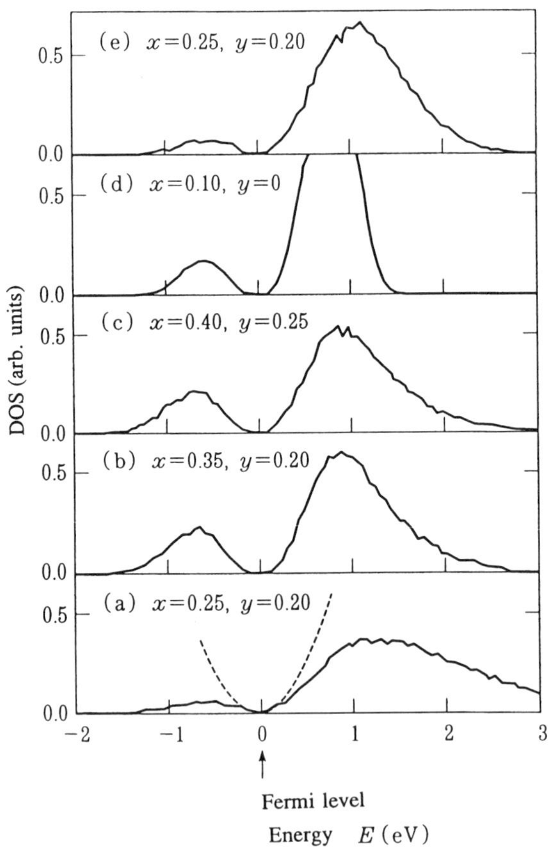


**Fig. 4.16.** Computer simulations for the density of states of Na$_x$W$_{1-y}$Ta$_y$O$_3$ showing a Coulomb gap at the Fermi level [484] (With permission from The American Physical Society)

system such interatomic Coulomb interaction leads to a Wigner crystallization with a finite gap, whereas in a disordered system the gap becomes a pseudogap (a dip in the DOS at the Fermi level). *Efros* and *Shklovskii* [483] pointed out that for the long-range Coulomb interaction which decays as $1/r$, the density of states becomes zero at the Fermi level. This type of gap is called the Coulomb gap. The Coulomb gap results from the instability of a finite DOS at the Fermi level in the presence of long-range interaction. If the DOS at the Fermi level were finite and given by $N(E_{\rm F})$, then the mean distance between electrons within a small energy window $E$ around the Fermi level would be given by $r \simeq (N(E_{\rm F})E)^{1/3}$ since $N(E_{\rm F})Er^3 \simeq 1$. On the other hand, an electron and a hole created within $E$ of the Fermi level would feel an attractive Coulomb potential of $-e^2/4\pi\varepsilon r \simeq -e^2[N(E_{\rm F})E]^{1/3}/4\pi\varepsilon$. This attractive potential would exceed the bare electron–hole pair energy $E$ for sufficiently small $E$, resulting in an instability of the system towards spontaneous electron–hole pair formation. For such an instability not to occur, $\nu$ in $N(E) \propto |E|^\nu$ must be larger than 2, meaning that the DOS at the Fermi level should vanish as a soft gap. However, the electronic specific heat $\gamma T$ of the same system remains finite as in the non-interacting case because the density of states for electron–hole pair excitations in the low-energy limit remains finite owing to the electron–hole attraction.

Figure 4.16 shows the result of computer simulations for Anderson-localized states in the $d$ band of $Na_xW_{1-y}Ta_yO_3$ [484]. In $Na_xW_{1-y}Ta_yO_3$, Na is ionized to $Na^+$ and donates an electron to the W $5d$ band whereas Ta has one less electron than W and acts as a deep acceptor compensating the electrons supplied by Na ions. Therefore $x-y$ electrons are distributed at the $B$ sites which feel random potential from the randomly distributed Na and Ta ions. As predicted by *Efros* and *Shklovskii*, a Coulomb gap is found to be formed at the Fermi level, which also explains photoemission experiments on the same system.

# 5. Representative Conducting Oxides

In this chapter, various representative oxides will be discussed in detail to present useful ideas concerning their electronic transport phenomena.

They are representative by virtue of the following characteristic features:

- $ReO_3$ (Sect. 5.1): The structure is simple cubic and it shows the highest conductivity in the normal oxides. The conduction band is a simple $d\epsilon$-$O_{2p}$ type.
- $SnO_2$ and $TiO_2$ (Sect. 5.2): $SnO_2$ is sometimes called a transparent metal, and it is a broad $s$–$p$ band semiconductor for which the electron mobility is large. $TiO_2$ has the same lattice structure but the mobility is small, and it has been disputed whether the electrons form large polarons or localized small polarons.
- $LiTi_2O_4$ and $LiV_2O_4$ (Sect. 5.3): $LiTi_2O_4$ is superconducting below 13.7 K, and until the discovery of Cu oxides, it was highest among the oxides. The magnetic susceptibility of metallic $LiV_2O_4$ depends on temperature while it is temperature-independent for $LiTi_2O_4$.
- $WO_3$ and $M_xWO_3$ (Sect. 5.4): The carriers may be large polarons in $WO_3$. They are heavily doped in $M_xWO_3$ where $M$ ions distribute randomly and there a metal–insulator transition occurs at certain carrier concentrations. Percolation theory will be useful here.
- $M_xV_2O_5$ and $M_xMoO_3$ (Sect. 5.5): These are low-dimensional substances. The former is quasi-one-dimensional and the carriers may be small polarons. A bipolaron state has been reported. Mo bronzes form various low-dimensional lattices whose metallic conductivity is suppressed by the charge density wave at lower temperatures.
- NiO (Sect. 5.6): NiO is an insulator while the local density approximation mean field theory predicts that it should be metallic, and the origin of the gap has been discussed. The importance of the electron–electron interaction is noted.
- Perovskite-type Mn oxides (Sect. 5.7): These are metallic when ferromagnetic. The recently discovered very large magnetoresistance renewed interest in these substances.
- $Fe_3O_4$ (Sect. 5.8): This is ferrimagnetic below 860 K and shows a transition at 123 K with a jump in electrical conductivity, which was ascribed to the order–disorder transition of $Fe^{2+}$ and $Fe^{3+}$. Many results have been

accumulated on the nature of correlated polarons in this fluctuating valence material.

- High-$T_c$ Cu oxides (Sect. 5.9): These have a layered structure, and the superconducting transition temperature increases with the layer number at least up to three and is higher than 150 K. The mechanism of superconductivity has been disputed over ten years but is not yet clear.

## 5.1 $ReO_3$: The Most Conductive $d\epsilon$ Conductor

$ReO_3$ has the simplest structure among $d$ electron conductors and its conductivity is comparable with that of metallic Ag. The color is purple red and it is an ionic crystal. Electrons are scattered by both acoustic and optical phonons. The Fermi level lies in the $d\epsilon$ band which can be approximately represented analytically and is two-dimensional though the lattice is three-dimensional. The electron–phonon interaction is weak and $ReO_3$ is not superconducting down to 20 mK.

### 5.1.1 Crystal Structure

In Fig. 5.1, the crystal structure of $ReO_3$ is shown. The lattice parameter is 0.374 nm at 300 K. If the vacant cube corner is occupied by a cation, it becomes the perovskite structure, and it is possible to dope Na and K into this site. Re is in the cubic symmetry position but O is in an axial symmetry position and the environment of O perpendicular to the axis is vacant.

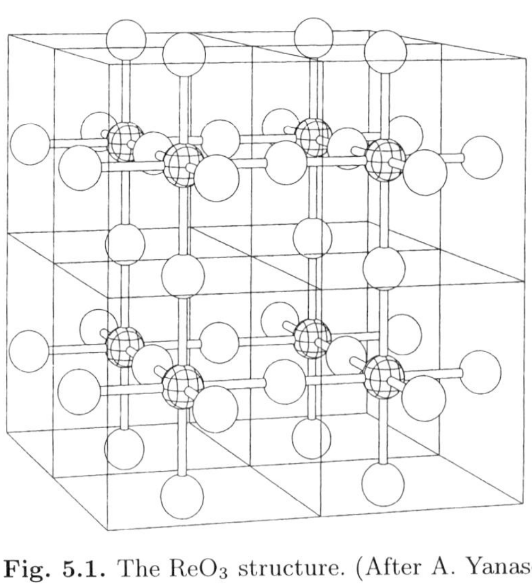

**Fig. 5.1.** The $ReO_3$ structure. (After A. Yanase)

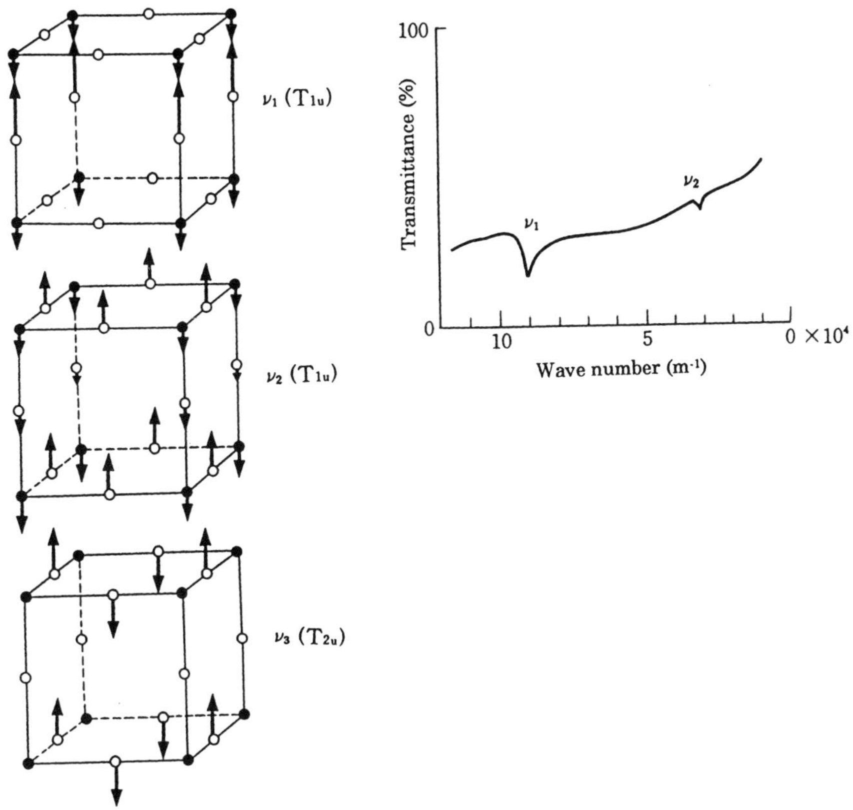


**Fig. 5.2.** Infrared absorption in $ReO_3$ and the normal modes of a $ReO_3$ molecule [486]. The $\nu_3$ mode is infrared inactive. The length of an arrow is proportional to $\sqrt{\text{mass}}\times$ amplitude and the Re ions vibrate very little compared with the O ions. (With permission from The Physical Society of Japan)

Due to this vacant structure, oxygen ions can oscillate with large amplitudes perpendicular to the axis. This oscillation attracts Re ions on both sides and makes $ReO_3$ one of the least thermally expansive oxides. The coefficient of linear thermally expansion is $(1.7 \pm 1) \times 10^{-6}$ $K^{-1}$ at 300 K [485].

$ReO_3$ has an ionicity and there are infrared active vibrations as shown in Fig. 5.2. The normal modes at $\Gamma$ are shown in Fig. 5.2 [486]. Of these $T_{2u}$ is not infrared active. $\nu_1$ is the highest frequency and its Einstein temperature is as high as 1303 K. The elastic constants were determined from the free vibration of a tiny single crystal [485], ultrasonically [487], and from Brillouin light scattering [488]. $C_{11} = 5.7$, $C_{12} = 0.07$, and $C_{44} = 0.68$ in units of $10^{11}\,\mathrm{Pa} = 10^{12}$ dyne/cm$^2$ at 303 K [485]. Then the velocity of the longitudinal wave is 8780 ms$^{-1}$ along the $\langle 100 \rangle$ axis. *Zumsteg* et al. determined the Debye temperature from the low temperature specific data and it is 460 K [489]. From the elastic constant it is 544 K at 78 K [485].

### 5.1.2 Electronic Properties

In Fig. 5.3, the temperature dependence of the resistivity is shown [490]. When the scattering of electrons occurs independently, the Matthiessen rule holds as

$$\tau^{-1} = \tau_{\rm imp}^{-1} + \tau_{\rm ac}^{-1} + \tau_{\rm op}^{-1} , \tag{5.1}$$

where $\tau_i$ is the relaxation time of electron momentum by impurities, acoustic phonons and optical phonons. The resistivity by the acoustic phonons is expressed by (2.17) and is proportional to temperature above the Debye temperature [491]. *Howarth* and *Sondheimer* have calculated the resistivity of a polar semiconductor by optical phonons using the Frölich Hamiltonian [492, 493]. In a degenerate semiconductor it is written briefly as

$$\rho_{\rm op} = C\left[\frac{T}{\theta_{\rm E}}\sinh^2\left(\frac{\theta_{\rm E}}{2T}\right)\right]^{-1} . \tag{5.2}$$

The solid line in Fig. 5.3 is calculated from $\rho = \rho_{\rm imp} + B(T/\Theta_{\rm D})^5 J_5(\Theta_{\rm D}/T) + \rho_{\rm op}$. In that $\rho_{\rm imp} = 4.3\times10^{-11}\Omega m$, $B = 2.72\times10^{-7}\Omega m$, $C = 8.90\times10^{-8}\Omega m$, $\Theta_{\rm D} = 400$ K, and $\Theta_{\rm E} = 910$ K [490].

The Hall coefficient is $-(3.28\pm0.15)\times10^{-10}$ m$^3$ C$^{-1}$ = $-(3.65\pm0.16)\times10^{-24}$ cgs esu [494]. For an isotropic single band it is $1/ne$. From this relation, $n$ becomes $1.90\times10^{28}$ m$^{-3}$ whereas $n$ is calculated to be $1.91\times10^{28}$ m$^{-3}$ from one electron per $ReO_3$. However, this coincidence of $n$ will be accidental since there are three Fermi surfaces in $ReO_3$ as shown in Fig. 5.4. Two of them are

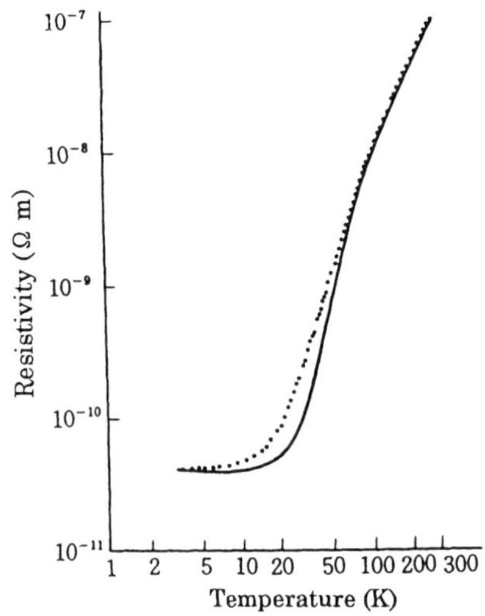


**Fig. 5.3.** Resistivity of $ReO_3$ [490]. The *solid line* is the calculated resistivity. (With permission from The Institute of Physics Publishing)

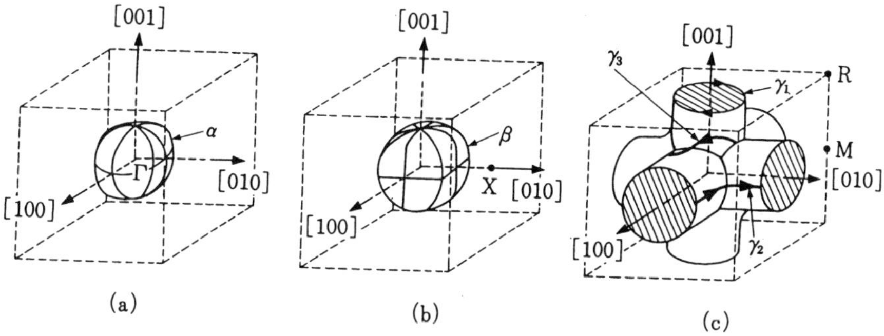


**Fig. 5.4.** Three sheets of the Fermi surface [499]. (With permission from The American Physical Society)

nearly spherical and 0.093 and 0.171 electron/Re are contained inside each sphere, respectively. These two have electron type conductivity. The third is called a junglegym because of its appearance when represented in the repeated zone scheme. It contains 0.736 electron per Re. The $\gamma_2$ orbital is hole-type.

From $R_H$ and a $\sigma$ of $1\times10^7$ $Sm^{-1}$ at 300 K, the Hall mobility $\mu_H = R_H\sigma$ is given as $3\times10^{-3}$ $m^2V^{-1}s^{-1} = 30$ $cm^2V^{-1}s^{-1}$. This is a rather large value for oxides and shows that conduction electrons can be treated as band electrons. Compared to this, the drift mobility of Cu is $3\times10^{-3}$ $m^2V^{-1}s^{-1}$. The drift mobility $\mu_D$ is defined by

$$v = \mu_D E \,, \tag{5.3}$$

where $v$ is the drift velocity in the electric field $E$. The Hall and the drift mobilities are not the same, especially for hopping conduction but in an isotropic band and for the energy-independent relaxation time, they are the same.

The electron–phonon interaction enhances the band density of states as explained in Sect. 2.4.3. The enhanced density of states at the Fermi level is determined from the temperature-proportional specific heat $C_e = \gamma T$ where $\gamma$ is

$$\gamma = \frac{2\pi^2}{3} k_B^2 N^*(E_F) \,. \tag{5.4}$$

$N^*(E_F)$ is the enhanced density of states of (2.22). $\gamma$ is $2.83\times10^4$ mJ $mol^{-1}K^{-2}$ [495]. Then $N^*(E_F)$ is $2.26\times10^{42}$ $J^{-1}mol^{-1}$(one spin)$^{-1}$ = 8.15 $Ryd^{-1}$(unit cell)$^{-1}$(one spin)$^{-1}$. $N(E_F)$ is the bare density of states at the Fermi level and $\lambda$ is the electron–phonon interaction parameter of (2.23) [496].

$N(E_F)$ for one spin deirection is experimentally determined from the Pauli paramagnetic susceptibility $\chi_s$ by

$$\chi_s = 2\mu_0\mu_B^2 N(E_F) \,. \tag{5.5}$$

$\chi_s$ is obtained from the observed susceptibility, $\chi_{obs}$, by

$$\chi_{obs} = \chi_s^* + \chi_{VV} + \chi_{orb} , \tag{5.6}$$

$$\chi_s = \frac{3\chi_s^*}{2} , \tag{5.7}$$

where $\chi_{VV}$ represents the Van Vleck paramagnetism of $Re^{7+}$ and $\chi_{orb}$ is the orbital diamagnetism. $\chi_{orb} = -12 \times 10^{-6}$ emu mol$^{-1}$ ($O^{2-}$) and $-12 \times 10^{-6}$ emu mol$^{-1}$ ($Re^{7+}$) [497, 498]. $\chi_s^*$ is a contribution from the 5$d$ conduction electrons. If we assume that the Landau orbital diamagnetism of the conduction electron is 1/3 of the spin susceptibility, (5.7) is valid.

The observed paramagnetic susceptibility is $2 \times 10^{-5}$ emu mol$^{-1}$ [489]. Assuming that the Van Vleck term is the same as that of $W^{6+}$ ($35 \times 10^{-6}$ emu mol$^{-1}$) [489], $\chi_s$ becomes $5 \times 10^{-5}$ emu mol$^{-1}$. Then $N(E_F)$ is 10 Ryd$^{-1}$ (unit cell)$^{-1}$(one spin)$^{-1}$ = $2.8 \times 10^{42}$ J$^{-1}$mol$^{-1}$(one spin)$^{-1}$. This $N(E_F)$ is larger than $N(E_F)^*$ and it is improbable. When the effective mass is large, the orbital diamagnetism of the conduction electron will be small and if we neglect it, $N(E_F)$ will be 6.7 Ryd$^{-1}$(unit cell)$^{-1}$(one spin)$^{-1}$. Then $\lambda$ becomes 0.22. The band calculation of *Mattheiss* gives $N(E_F)$ as 7 Ryd$^{-1}$(unit cell)$^{-1}$(one

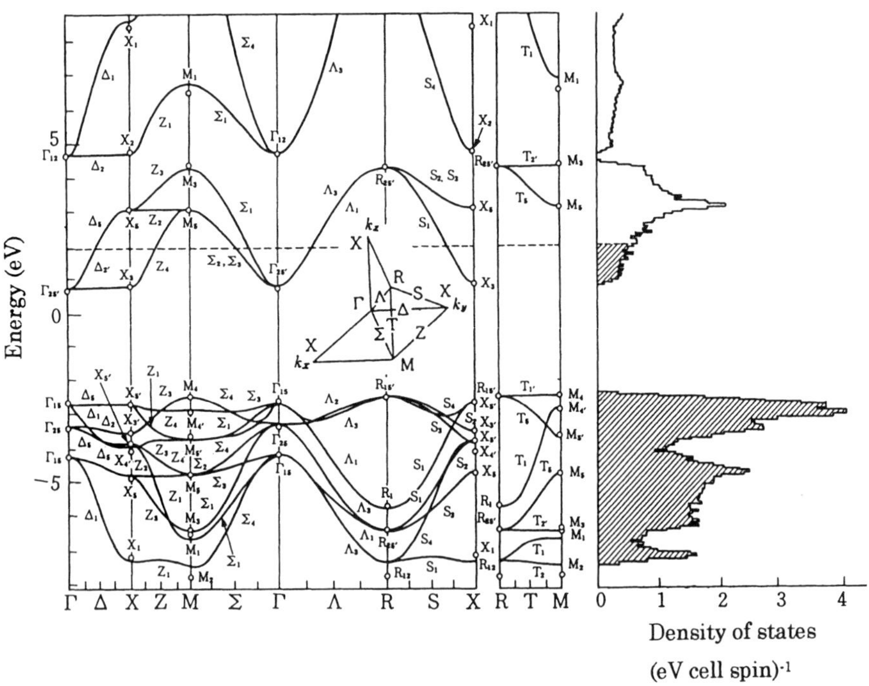


**Fig. 5.5.** Band structure of $ReO_3$ calculated by *Mattheiss* [499]. The *broken line* is the Fermi level. (With permission from The American Physical Society)

spin)$^{-1}$ as shown in Fig. 5.5 [499, 500]. Then $\lambda$ is 0.16. $\lambda$ may also be estimated from the superconducting transition temperature, $T_c$, which is lower than 20 mK. Using the McMillan formula of (5.38), $\lambda$ is estimated to be less than 0.29 where $\Theta_D$ and $\mu^*$ were put as 460 K and 0.13, respectively. Thus $\lambda$ may be smaller than 0.3, which is smaller than that in a simple metal. For instance, $\lambda$ is 0.38 (Al), 0.60 (Sn) and 1.12 (Pb) [496]. The small $\lambda$ is consistent with the high conductivity and the low $T_c$.

The effective mass of the conduction electron is determined from cyclotron resonance or the de Haas–van Alven effect. *Razari* et al. have observed the dH–vA oscillations and obtained the cyclotron mass as shown in Fig. 5.6 [501].

Figure 5.5 shows the energy band structure calculated by *Mattheiss* [499, 500]. The symbols $\Gamma$, $X$, $R$, $M$, etc., represent particular values of the crystalline momentum as shown in the inset. Table 5.1 shows the Bloch functions at each symmetry point [502].

*Wolfram* et al. pointed out that the eigenvalues and eigenfunctions of electrons are rather well expressed analytically in the $ReO_3$ structure [503]. In the tight binding approximation, the orbital is expressed by a linear combination of atomic orbitals. In this case, fourteen orbitals are enough to describe the electronic states: five from Re5$d$ and $3 \times 3$ from O2$p$. The number of orbitals is further reduced when $T_{2g}$ symmetry orbitals are considered. For instance the following combination has the $T_{2g}$ symmetry at the $\Gamma$ point:

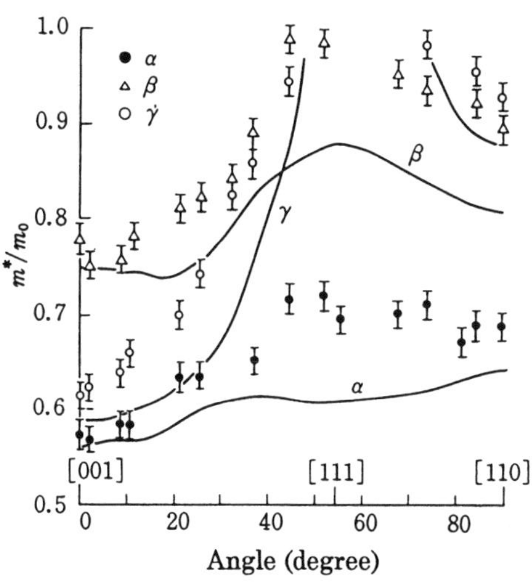


**Fig. 5.6.** The cyclotron mass of $ReO_3$ [501]. The magnetic field is in the $(1\bar{1}0)$ plane and its angle from the [001] axis is shown. The *solid lines* are those calculated by *Mattheiss*. (With permission from The National Research Council of Canada)

**Table 5.1.** Bloch functions composed of $Ti_{3d}$ and $O_{2p}$ orbitals at each special symmetry point in the band calculation of $SrTiO_3$ after *Kahn* and *Leyendecker* [502]. Ti is at the origin of the cell and three O1, O2, and O3 ions are at $(a/2, 0, 0)$, $(0, a/2, 0)$ and $(0, 0, a/2)$, respectively, where $a$ is the lattice period. For instance, $(y, -x, 0)$ stands for the Bloch function composed of the $p_y$ orbital of O1 and the $-p_x$ orbital of O2 and $\frac{1}{\sqrt{2}}(y, -x, 0) = \frac{1}{\sqrt{2N}} \sum_j \left[ e^{\mathrm{i}\boldsymbol{k}\cdot\boldsymbol{R}_{j1}} \phi_y(\boldsymbol{r} - \boldsymbol{R}_{j1}) - e^{\mathrm{i}\boldsymbol{k}\cdot\boldsymbol{R}_{j2}} \phi_x(\boldsymbol{r} - \boldsymbol{R}_{j2}) \right]$

| Representation | *d*-orbitals | *p*-orbitals |
|---|---|---|
| $2\Gamma_{12}$ | $(3z^2 - r^2)$, $(x^2 - y^2)$ | |
| $3\Gamma_{25'}$ | $(xy)$, $(yz)$, $(zx)$ | |
| $3\Gamma_{15}$ | | $(x, 0, 0)$, $(0, y, 0)$, $(0, 0, z)$ |
| $3\Gamma_{15}$ | | $(0, x, x)$, $(y, 0, y)$, $(z, z, 0)$ |
| $3\Gamma_{25}$ | | $(z, -z, 0)$, $(0, x, -x)$, $(-y, 0, y)$ |
| $1X_1$ | $(3y^2 - r^2)$ | $(0, y, 0)$ |
| $1X_2$ | $(z^2 - x^2)$ | |
| $1X_3$ | $(xz)$ | |
| $2X_5$ | $(xy)$, $(yz)$ | $(0, x, 0)$, $(0, z, 0)$ |
| $1X_{3'}$ | | $(y, 0, -y)$ |
| $1X_{4'}$ | | $(y, 0, y)$ |
| $2X_{5'}$ | | $(x, 0, 0)$, $(0, 0, z)$ |
| $2X_{5'}$ | | $(0, 0, x)$, $(z, 0, 0)$ |
| $1M_1$ | $(3z^2 - r^2)$ | $(x, y, 0)$ |
| $1M_2$ | $(x^2 - y^2)$ | $(x, -y, 0)$ |
| $1M_3$ | $(xy)$ | $(y, x, 0)$ |
| $1M_4$ | | $(y, -x, 0)$ |
| $2M_5$ | $(yz)$, $(zx)$ | $(0, z, 0)$, $(z, 0, 0)$ |
| $1M_{4'}$ | | $(0, 0, z)$ |
| $2M_{5'}$ | | $(0, 0, x)$, $(0, 0, y)$ |
| $1R_1$ | | $(x, y, z)$ |
| $2R_{12}$ | $(x^2 - y^2)$, $(3z^2 - r^2)$ | $(x, -y, 0)$, $(-x, -y, 2z)$ |
| $3R_{25'}$ | $(yz)$, $(zx)$, $(xy)$ | $(0, z, y)$, $(z, 0, x)$, $(y, x, 0)$ |
| $3R_{15'}$ | | $(0, z, -y)$, $(-z, 0, x)$, $(-y, x, 0)$ |
| $1\Delta_1$ | $(3y^2 - r^2)$ | $(0, y, 0)$ |
| $1\Delta_1$ | | $(y, 0, y)$ |
| $1\Delta_2$ | $(z^2 - x^2)$ | $(y, 0, -y)$ |
| $1\Delta_{2'}$ | $(xz)$ | |
| $2\Delta_5$ | $(xy)$, $(yz)$ | $(x, 0, 0)$, $(0, 0, z)$ |
| $2\Delta_5$ | | $(0, x, 0)$, $(0, z, 0)$ |
| $2\Delta_5$ | | $(0, 0, x)$, $(z, 0, 0)$ |

$$\psi_{xy} = \sum_j \mathrm{e}^{\mathrm{i}\boldsymbol{k}\cdot\boldsymbol{R}_j}[a_{xy}\phi_{xy}(\boldsymbol{r}-\boldsymbol{R}_j)$$

$$+a_y\phi_y(\boldsymbol{r}-\boldsymbol{R}_j-\boldsymbol{d}_1)\mathrm{e}^{\mathrm{i}\boldsymbol{k}\cdot\boldsymbol{d}_1}+a_x\phi_x(\boldsymbol{r}-\boldsymbol{R}_j-\boldsymbol{d}_2)\mathrm{e}^{\mathrm{i}\boldsymbol{k}\cdot\boldsymbol{d}_2}], \qquad (5.8)$$

where $\phi_{xy}(\boldsymbol{r}-\boldsymbol{R}_j)$ is $d_{xy}$ centered at $\boldsymbol{R}_j$ and the associated $p$ orbitals are $p_y(\boldsymbol{r}-\boldsymbol{r}_j-\boldsymbol{d}_1)$ and $p_x(\boldsymbol{r}-\boldsymbol{R}_j-\boldsymbol{d}_2)$, as shown schematically in Fig. 2.3. $|\boldsymbol{d}_i| = a$. The $p_z$ orbital is absent for symmetry reasons. The unknown amplitudes $a_{xy}$, $a_x$, $a_y$ and the eigenenergy are obtained from

$$\begin{pmatrix} E_t - E & 2\mathrm{i}(pd\pi)\sin k_x a & 2\mathrm{i}(pd\pi)\sin k_y a \\ -2\mathrm{i}(pd\pi)\sin k_x a & E_\perp - E & 0 \\ -2\mathrm{i}(pd\pi)\sin k_y a & 0 & E_\perp - E \end{pmatrix} \begin{pmatrix} a_{xy} \\ a_y \\ a_x \end{pmatrix} = 0\,. \qquad (5.9)$$

where $(pd\pi)$ is the overlap integral [504]

$$(pd\pi) = \int \phi^*_{xy}(\boldsymbol{r}-\boldsymbol{R}_j)H\phi_y(\boldsymbol{r}-\boldsymbol{R}_j-\boldsymbol{d}_1)\mathrm{d}\tau\,. \qquad (5.10)$$

$\sigma$, $\pi$, $\delta$, etc., represent the angular momentum around the axis combining the two ions, that is, $m = 0, 1, 2$, etc. in $\exp(\mathrm{i}m\phi)$, respectively, where $\phi$ is the angle around the axis. $E_t$ and $E_\perp$ are the self energies of $\phi_{xy}$ and $p_y$ or $p_x$, respectively. In the above matrix, overlaps between $p$ functions were neglected. This neglected $p$–$p$ overlap is believed to play an important

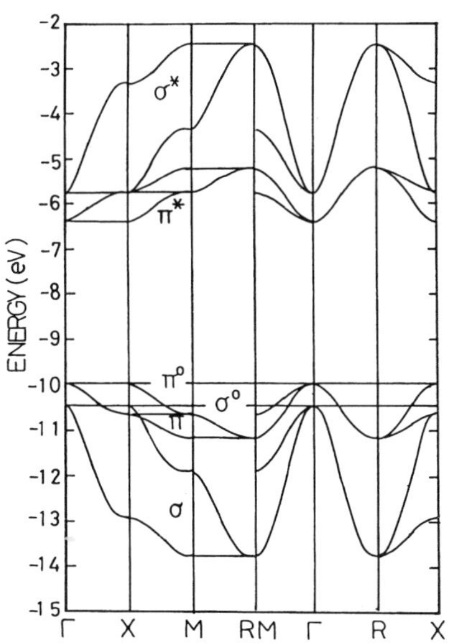


**Fig. 5.7.** Band structure of $SrTiO_3$ in the absence of the ($p$–$p$) overlap [505]. $\pi, \pi^0$, and $\pi^*$ represent the bonding, nonbonding, and antibonding character of the $d\epsilon$ band. (With permission from The American Physical Society)

role for high-$T_c$ cuprates giving a $p$ band. Then the problem can be solved analytically. Thus determined eigenfunctions and eigenvalues are

$$E_1 = E_\perp \,, \tag{5.11}$$

$$a_{xy} = 0, a_y = \sin k_y a, a_x = -\sin k_x a \,, \tag{5.12}$$

$$E_2 = \frac{(E_t + E_\perp)}{2} + \sqrt{\frac{(E_t - E_\perp)^2}{4} + 4(pd\pi)^2(\sin^2 k_x a + \sin^2 k_y a)} \,, \tag{5.13}$$

$$a_{xy} = E_\perp - E_2(<0),\ \ a_y = 2i(pd\pi)\sin k_x a,\ \ a_x = 2i(pd\pi)\sin k_y a \,, \tag{5.14}$$

$$E_3 = \frac{(E_t + E_\perp)}{2} - \sqrt{\frac{(E_t - E_\perp)^2}{4} + 4(pd\pi)^2(\sin^2 k_x a + \sin^2 k_y a)} \,, \tag{5.15}$$

$$a_{xy} = E_\perp - E_3(>0),\ \ a_y = 2i(pd\pi)\sin k_x a,\ \ a_x = 2i(pd\pi)\sin k_y a \,, \tag{5.16}$$

where $E_2$ is for the conduction band orbital, $E_1$ is for the nonbonding orbital and $E_3$ is for the valence band orbital. $E_2$ and $E_3$ orbitals are the antibonding and bonding orbitals, respectively, in terms of the molecular orbitals. Similar orbitals are obtained for the $d_{yz}$ and $d_{zx}$ orbitals. For the $d\gamma$-orbitals $d_{3z^2-r^2}$, $d_{x^2-y^2}$, $p_x(1)$, $p_y(2)$, and $p_z(3)$ orbitals are involved, and the matrix is then $5 \times 5$. The $d\gamma$ band is broader than the $d\epsilon$ band because $(pd\sigma)$ is larger than $(pd\pi)$. In Fig. 5.7, a band structure thus obtained is shown after *Ellialtioglu* and *Wolfram* [505]. It is a little different from that of *Mattheiss* due to the above approximations.

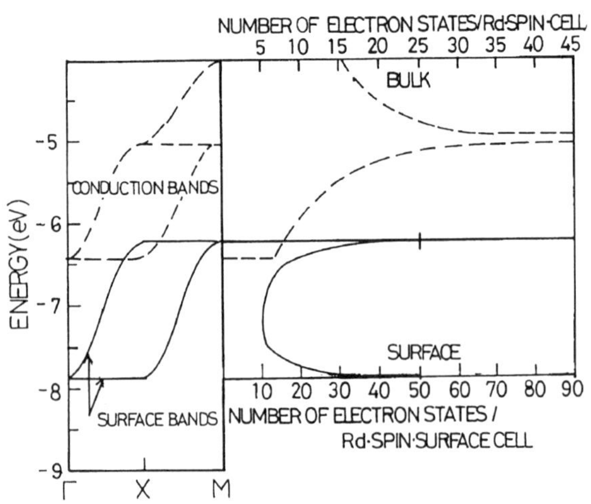


**Fig. 5.8.** Density of surface states of $SrTiO_3$ [506]. (With permission from The American Physical Society)

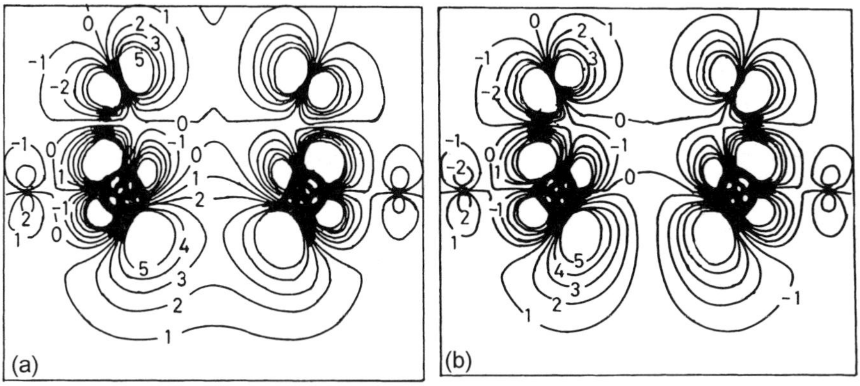

**Fig. 5.9. a, b.** Surface states around an oxygen vacancy on the surface of ReO$_3$ [507]. The vacancy is at the center and the lower half is a vacuum in each figure. The bonding and the antibonding nature is clearly seen from the equiamplitude map of the wave function. (With permission from The Physical Society of Japan)

As seen from (5.13, 5.15), $E(k)_{xy}$ depends on $k_x$ and $k_y$ and not on $k_z$. Thus the $E_{xy}$ band has a two-dimensional character and there is a logarithmic peak in the density of states of the conduction band [505]. *Morin* and *Wolfram* have pointed out that on the surface of a two-dimensional lattice in the $xy$ plane, $d_{yz}$ and $d_{zx}$ bands become one-dimensional, that is, they depend only on one $k_i$, and the density of states shows an $E^{-1/2}$ divergence at the edge (Fig. 5.8) [506]. Such a surface state has almost pure $d$ character and they suggested that such a state would be active for catalysis [506]. In many catalytic reactions, however, oxygen vacancies become the active points. *Tsukada* has made a cluster calculation for the vacant structure and Fig. 5.9 shows how the orbitals are modulated at such a vacant site [507].

*Chiba* et al. have studied the momentum distribution of electrons by the positron annihilation method [508]. In that method, the $z$ component of the momentum of electron–positron pairs is given by

$$I(p_z) = A \int \left| \int \psi_e \phi_p e^{i\boldsymbol{p}\cdot\boldsymbol{r}} d^3r \right| dp_x dp_y \,, \tag{5.17}$$

where $\psi_e$ and $\phi_p$ are the electron and the positron wavefunctions, respectively. The range of the $p$ integration depends on the experimental method. In principle, it reveals the momentum distribution not only of the Fermi surface electrons but also of the electrons deep in the bands. Fig. 5.10 shows the momentum distribution along the ⟨100⟩ direction, integrated perpendicular to the axis. The difference curves are compared with the calculated results in which the positron wavefunction was assumed to be constant and the electron wavefunctions were analytically represented, following *Wolfram* [503]. It is a very rough approximation, but the characteristic structures of the wavefunctions are reproduced.

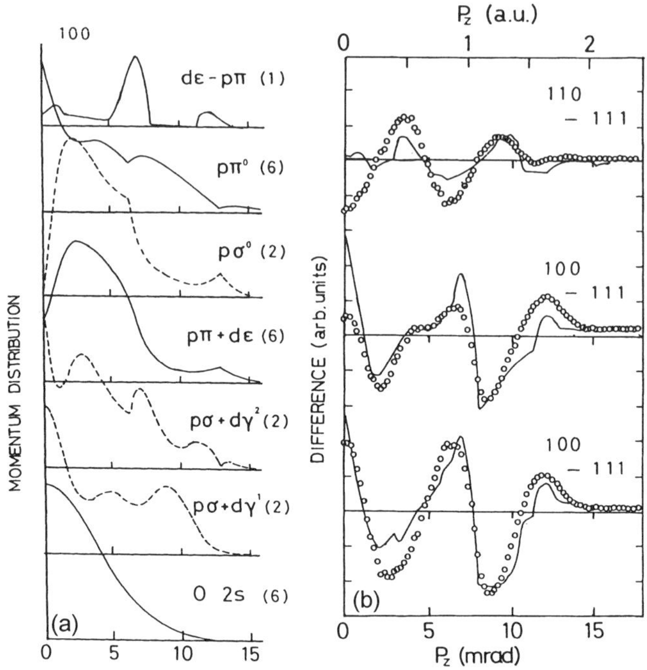


**Fig. 5.10.** Density of the $p_z$ component of electron–positron pairs in $ReO_3$ and the differences [508]. $p_z$ is the momentum component along a direction shown in the figure. *Lines* are calculated results and *circles* are observed ones. (**a**) Calculated momentum distribution of one electron along $\langle 100 \rangle$ in each band. $p_z = mc\theta$ where $\theta$ is the angle between two emitted photons. (**b**) Differences in the density of the $p_z$ component. (With permission from The Japan Institute of Metals)

## 5.2 $SnO_2$ and $TiO_2$: Oxide Semiconductors

$SnO_2$ is sometimes called a transparent metal and has been used as a transparent electrode and as a gas sensor. $SnO_2$ is one of the most representative broad band semiconductors in which $s$ and $p$ electrons propagate with a large mobility. Nowadays $In_2O_3$ doped with $SnO_2$ is utilized for a transparent electrode and is known as ITO.

On the other hand, the isostructural $TiO_2$ is known as a poor-mobility semiconductor and it has been disputed whether the electrons are small polarons or large polarons. The low mobility may be due to the $d$ character of the conduction electrons.

In this chapter, the conduction mechanism of the two extreme oxides which have the same lattice structure will be discussed.

### 5.2.1 Electronic Energy Band Structure of $SnO_2$

$SnO_2$ has the rutile structure shown in Fig. 5.11. The $MO_6$ octahedra share edges along the $c$ axis. *Robertson* calculated the energy band structure of this material using an LCAO calculation of $s$ and $p$ orbitals (Fig. 5.12) [509]. As shown in Fig. 5.13, the valence band is mainly composed of $O_{2p}$ orbitals and the conduction band consists of $Sn_{5s}$ and $Sn_{5p}$ orbitals. The bottom of the conduction band is more than 90% $s$ type. The bottom of the conduction band and the top of the valence band are located at the $\Gamma$ point $k = 0$. Thus it is a direct gap semiconductor [510] in which the electrons can be excited to the conduction band without the help of phonons to conserve the crystal momentum. The intrinsic energy gap is as large as 3.6 eV and stoichiometric $SnO_2$ is an insulator. So the high conductivity is due to deviation from stoichiometry.

From the band structure, the effective mass of the conduction electron is expected to be small. Experimentally, from the cyclotron resonance, the mass tensor components were obtained as $m^*_{\mathrm{p}\perp} = 0.299m_0$ and $m^*_{p\|} = 0.234m_0$ where $\perp$ and $\|$ are with respect to the $c$ axis [511], $m_0$ is the free electron mass and $m^*_{\mathrm{p}}$ is the phonon enhanced band mass

$$m^*_{\mathrm{p}} = m^* \left(1 + \frac{\alpha}{6}\right) . \tag{5.18}$$

Here $m^*$ is the band mass and $\alpha$ is the electron–phonon interaction parameter as in (2.19).

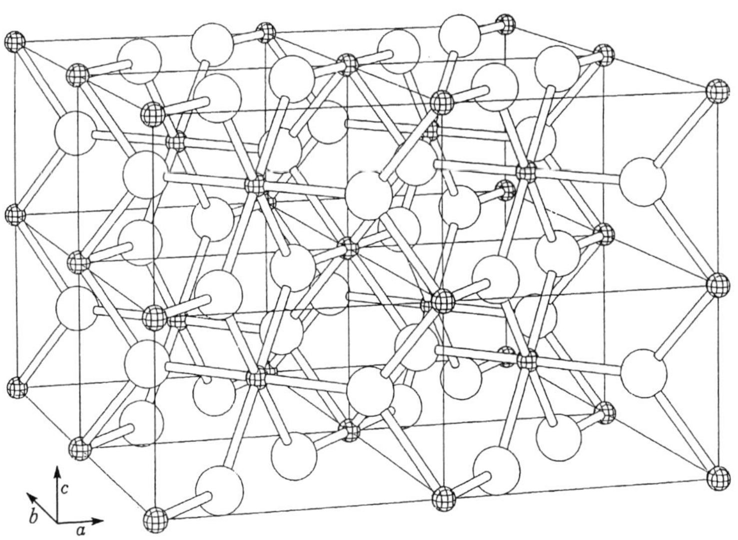


**Fig. 5.11.** Rutile structure. Eight unit cells are shown. $a$ and $c$ are 0.459 nm and 0.296 nm for $TiO_2$ and 0.474 nm and 0.319 nm for $SnO_2$, respectively. A chain of edge sharing octahedrons extends along the $c$-axis. (After A. Yanase)

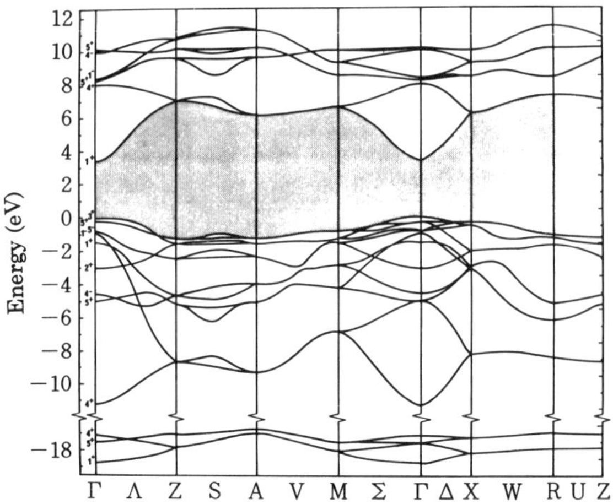


**Fig. 5.12.** Band structure of $SnO_2$ after *Robertson* [509]. (With permission from The Institute of Physics Publishing)

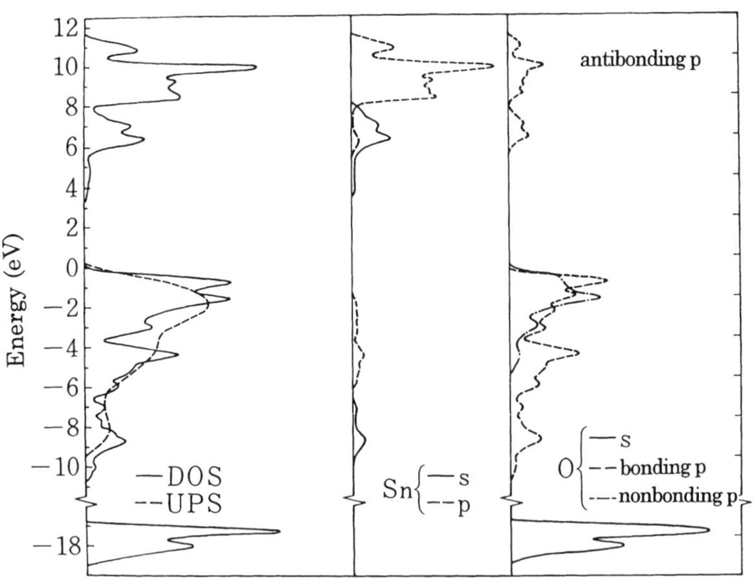


**Fig. 5.13.** Density of states and distribution of *s* and *p* orbitals in the $SnO_2$ band [509]. DOS: calculated density of states, UPS: intensity of the ultraviolet photoelectron emission. (With permission from The Institute of Physics Publishing)

### 5.2.2 Electrical Conductivity of $SnO_2$

The determination of the intrinsic conductivity is difficult because of the difficulty in preparaing stoichiometric $SnO_2$ [512–514]. *Nagasawa*, *Shionoya*, and *Fonstad* prepared single crystals and studied the optical and electrical

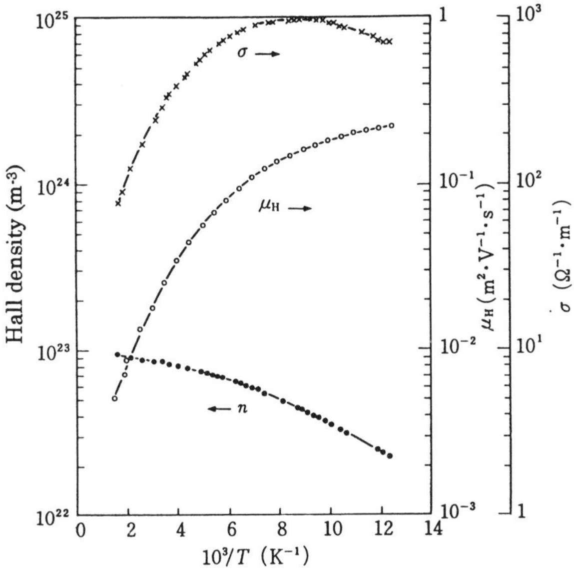


**Fig. 5.14.** Conductivity, carrier density and Hall mobility [518]. (With permission from The American Institute of Physics)

properties [511, 515–518]. Figure 5.14 shows the conductivity, the Hall mobility, and the carrier density [518]. These values depend on the purity and the stoichiometry of the sample. The carrier density and the Hall mobility are deduced from the conductivity and the Hall coefficient [491]

$$\mu_{\mathrm{H}} = R_{\mathrm{H}}\sigma \,, \tag{5.19}$$

$$R_{\mathrm{H}} = \begin{cases} 3\pi/8ne & \text{nondegenerate} \\ 1/ne & \text{degenerate} \end{cases} . \tag{5.20}$$

This is valid when the conduction band is an isotropic and single band and $\langle\tau^2\rangle$ is $\langle\tau\rangle^2$ where $\tau$ is the relaxation time of the momentum distribution. In Fig. 5.14, $n$ decreases at lower temperatures but by slightly changing the preparation conditions, the degenerate sample is easily prepared. *Fonstad* and *Rediker* determined the effective mass $m_{\mathrm{p}}^*$ from the relation

$$\frac{n(n+N_{\mathrm{A}})}{N_{\mathrm{D}}-N_{\mathrm{A}}-n} = \left(\frac{2\pi m_{\mathrm{p}}^* k_{\mathrm{B}} T}{h^2}\right)^{3/2} \exp\left(-\frac{E_{\mathrm{D}}}{k_{\mathrm{B}}T}\right) , \tag{5.21}$$

$$m_{\mathrm{p}}^* = \left(m_{\mathrm{p}\perp}^* m_{\mathrm{p}\parallel}^{*2}\right)^{1/3} , \tag{5.22}$$

where $N_{\mathrm{A}}$ and $N_{\mathrm{D}}$ are the acceptor and the donor concentrations, respectively, $E_{\mathrm{D}}$ is the activation energy of the donor equal to 27.5 meV, and $m_{\mathrm{p}}^*$

is $0.39\,m_0$ [518]. The effective mass can be compared with that in Si where $m^*_{\mathrm{p}\|} = 0.97\,m_0$ and $m^*_{\mathrm{p}\perp} = 0.19\,m_0$. The subscripts refer to the long axis of the constant energy ellipsoid. Then $m^*_{\mathrm{p}} = 0.33m_0$. Thus the conduction electrons in $SnO_2$ are as light as those in Si. The mobility is $2.4 \times 10^{-2}\,\mathrm{m^2V^{-1}s^{-1}}$ at 300 K (Fig. 5.14). Though it differs from sample to sample, it is not so large as in Si ($1.6 \times 10^{-1}$), InSb (7.7) and PbS ($6.0 \times 10^{-2}$) at 300 K. The units are $\mathrm{m^2V^{-1}s^{-1}}$. The smaller mobility may be due to the ionicity of $SnO_2$. Following *Robertson*, the polarity is 0.8 [509].

The mobility depends on various scattering mechanisms. The mobility due to impurity scattering is expressed for non-degenerate semiconductors as

$$\mu_{\mathrm{imp}} = \frac{2^{7/2}(4\pi\epsilon)^2(k_{\mathrm{B}}T)^{3/2}}{\pi^{3/2}m^{*1/2}e^3N}\left[\ln(1+b) - \frac{b}{1+b}\right]^{-1}, \tag{5.23}$$

$$b \equiv \frac{6}{\pi}\frac{4\pi\epsilon m^* k_{\mathrm{B}}^2T^2}{n\hbar^2e^2},$$

where $N$ and $n$ are the concentration of the ionized impurity and the carrier density, respectively, and $\epsilon$ is the dielectric constant [519]. The scattering rate is constant for a metal. The longitudinal acoustic phonon has been treated by *Bardeen* and *Schokley* in their deformation potential approximation [520]. The mobility is

$$\mu_{\mathrm{ac}} = \frac{\sqrt{8\pi}e\hbar^4\rho v_{\mathrm{l}}^2}{3E_{\mathrm{l}}^2m^{*5/2}(k_{\mathrm{B}}T)^{3/2}}, \tag{5.24}$$

where $\rho$ is the density of the matrix and $v_{\mathrm{l}}$ is the velocity of the longitudinal acoustic phonon. $E_{\mathrm{l}}$ is the constant of (2.16).

Scattering by optical phonons has been treated by *Howarth* and *Sondheimer* [492]. When the carriers are not degenerate the Fermi energy $E_{\mathrm{F}}$ is given by

$$n = \frac{2(2\pi m^* k_{\mathrm{B}}T)^{3/2}}{h^3}\exp\left(\frac{E_{\mathrm{F}}}{k_{\mathrm{B}}T}\right), \tag{5.25}$$

and when $\hbar\omega_0 \gg k_{\mathrm{B}}T$, the conductivity is given by

$$\sigma = (4\pi\epsilon_0)^2\frac{2\sqrt{\pi}a^3M}{e^2}\left(\frac{\nu_0k_{\mathrm{B}}T}{h}\right)^{3/2} \left[\exp\left(\frac{\hbar\omega_0}{k_{\mathrm{B}}T}\right) - 1\right]\exp\left(\frac{E_{\mathrm{F}}}{k_{\mathrm{B}}T}\right). \tag{5.26}$$

When $\hbar\omega_0 \ll k_{\mathrm{B}}T$, it is

$$\sigma = (4\pi\epsilon_0)^2 5.3a^3M\nu_0\left(\frac{k_{\mathrm{B}}T}{eh}\right)^2\left[\exp\left(\frac{\hbar\omega_0}{k_{\mathrm{B}}T}\right) - 1\right]\exp\left(\frac{E_{\mathrm{F}}}{k_{\mathrm{B}}T}\right). \tag{5.27}$$

When the carriers are degenerate

$$E_F \approx \frac{\hbar}{2m^*}(3\pi^2 n)^{2/3}\,, \tag{5.28}$$

and the conductivity is

$$\sigma = (4\pi\epsilon_0)^2 \frac{2.67a^3 M E_F^2}{e^2 h^3} k_B T \left[\exp\left(\frac{\hbar\omega_0}{k_B T}\right) + \exp\left(-\frac{\hbar\omega_0}{k_B T}\right) - 2\right]\,, \tag{5.29}$$

where $M^{-1} = M_1^{-1} + M_2^{-1}$ and $M_1$ and $M_2$ are the ionic masses separated by a distance $a$ and the electric charge is $\pm e$, which vibrates optically with a frequency $\nu_0$. It should be noted that these expressions were obtained under the assumption that the electron–phonon interaction is weak enough to be treated as a perturbation.

*Lee* et al. have calculated the drift mobility of the electron for $1 < \alpha < 5$, where $\alpha$ is the electron–phonon interaction parameter [521–523]. The drift mobility is

$$\mu_D = \frac{1}{2\alpha\omega_0}\frac{e}{m^*}\left(\frac{m^*}{m_p}\right)^3 f(\alpha)\left[\exp\left(\frac{\hbar\omega_0}{k_B T}\right) - 1\right]\,, \tag{5.30}$$

where $k_B T$ was assumed to be much less than $\hbar\omega_0$, and $f(\alpha)$ is nearly one. The energy of the electron interacting with phonons is given by

$$E(k) \approx -\alpha\hbar\omega_0 + \frac{\hbar^2 k^2}{2m^*(1+\frac{\alpha}{6})} \tag{5.31}$$

and $m_p^* = m^*(1+\alpha/6)$. Such a quasi-particle is called a large polaron.

*Fonstad* and *Rediker* represented the observed $\mu_{op}$ by

$$\begin{aligned}\mu_{op} &= 1.30\times10^{-1}\left[\exp\left(\frac{390}{T}\right) - 1\right] + 4.95\times10^{-2}\left[\exp\left(\frac{515}{T}\right) - 1\right] \\ &\quad +1.58\times10^{-4}\left[\exp\left(\frac{1080}{T}\right) - 1\right]\end{aligned} \tag{5.32}$$

in units of $m^2V^{-1}s^{-1}$ [518]. For $\mu_{ac}$, they assumed that $E_1$ is 8 eV. Their results are shown in Fig. 5.15. As seen from the figure, the mobility is mainly determined by the optical phonon scattering at 300 K. The discrepancy between the calculated and the observed mobility may be due to impurity scattering for it becomes larger as the impurity concentration increases.

The next problem is to determine the dopant. Usually in semiconducting oxides, the most effective impurity is not the foreign atom but defects due to the deviation from stoichiometric composition. In the case of $SnO_2$, the oxygen vacancy or the interstitial Sn can be the donor. Their concentration obeys the mass-action law, and this is utilized to determine the dopant. When oxygen ions leave the solid there are two electrons/ion left behind, and if there is no oxygen vacancy, which traps one electron, then the carrier density will be

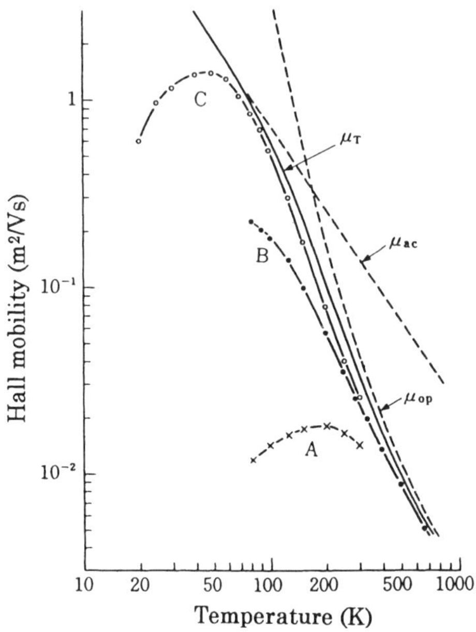


**Fig. 5.15.** Hall mobility of $SnO_2$ [518]. Calculated mobility $\mu_{op}$: due to optical phonon scattering, $\mu_{ac}$: due to acoustic phonon scattering. $\mu_{tot}$: total calculated mobility in which impurity scattering has been neglected. The carrier density is $2.2 \times 10^{24}$(A), $8.6 \times 10^{23}$(B), and $8.5 \times 10^{21}$(C) per $m^3$ at 300 K. (With permission from The American Institute of Physics)

$$n \propto p_{O_2}^{-1/6}, \tag{5.33}$$

where $p_{O_2}$ is the oxygen partial pressure in the atmosphere in which $SnO_2$ is heat treated. On the other hand, if the interstitial Sn is the donor as Sn $\rightarrow$ $Sn^4 + 4e$, then

$$n \propto p_{O_2}^{-1/5}. \tag{5.34}$$

*Samson* and *Fonstad* consider that the oxygen vacancy is the donor because the conductivity is proportional to $p_{O_2}^{-1/6.5}$ [524].

$SnO_2$ is usually used as a film. Figure 5.16 shows the conductivity of the film and can be seen from the figure, it depends strongly on preparation conditions.

$SnO_2$ is also used as a gas sensor. An electron transfer occurs between any adsorbed molecule and the solid because the chemical potentials are not the same. Moreover, a chemical reaction occurs between the two at higher temperatures. For instance, the reduction of $SnO_2$ by $H_2$ gas. Then the carriers are injected. The change in the carrier density is detected by a change in the conductivity. For practical applications, $SnO_2$ is used in air and the surface is already covered by various molecules. Molecules such as $H_2$, CO, or $C_3H_8$

are adsorbed and detected by substitution for the already adsorbed species or by chemical reactions with them. The addition of Pt or Pd enhances the sensitivity greatly due to their catalytic action. When the exothermic reaction occurs, $SnO_2$ is heated locally and this also changes the conductivity. *Yamazoe* and *Seiyama* found that adsorption of $O_2$ decreases the conductivity of $SnO_2$ above 423 K [526]. When the reducing gas is in contact with the oxygen layer, it removes the adsorbed oxygen which leaves electrons behind and the conductivity increases. At higher temperatures, the reduction of $SnO_2$ itself occurs and the conductivity again increases. In Fig. 5.17, an example of the conductivity change measured using a constant voltage source is shown [527].

### 5.2.3 Optical Properties of $SnO_2$

$SnO_2$ is transparent in the visible region from 300–700 nm (Fig. 5.18) [528]. The absorption edge of highly conductive films is 3.82 eV and that of poorly conductive films is 3.71 eV. The electromagnetic wave whose frequency is larger than the plasma frequency $(ne^2/\epsilon_0 m^*)^{1/2}$ is not attenuated in the

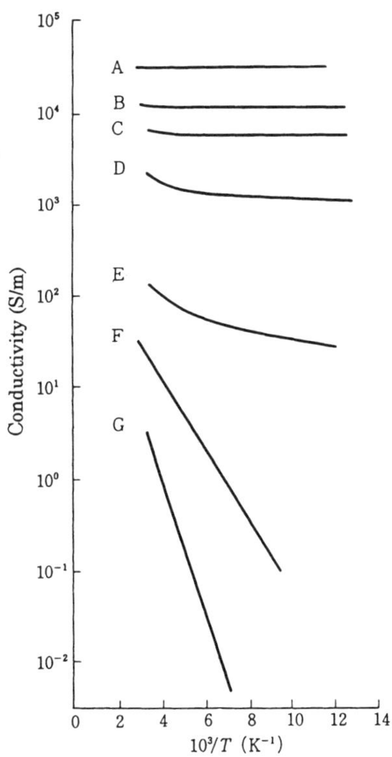


**Fig. 5.16.** Conductivity of $SnO_2$ films [525]. The carrier density is $2 \times 10^{26}$ (A), $8 \times 10^{25}$ (B), $5 \times 10^{25}$ (C), $2 \times 10^{25}$ (D), $2 \times 10^{24}$ (E), $5 \times 10^{23}$ (F), and $9 \times 10^{22}$ (G) per 1 $m^3$. (With permission from Akademie-Verlag, Wiley–VCH)

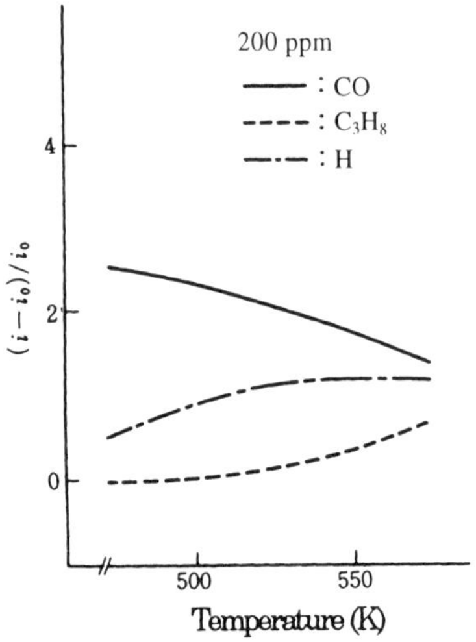


**Fig. 5.17.** Change in the resistivity of $SnO_2$ under various atmospheres [527]. $i_0$ is the current under dry air. $SnO_2$ was doped. (With permission from The Minerals, Metals and Materials Society)

electron gas. The plasma oscillation which corresponds to a wavelength of 650 nm is attained by a free electron density of $2.6 \times 10^{27}\ \mathrm{m}^{-3}$ and as it is much larger than that in $SnO_2$ ($10^{26}\ \mathrm{m}^{-3}$), $SnO_2$ is transparent to visible light. *Nagasawa* and *Shionoya* expressed the attenuation of the polarized light whose electric field is perpendicular to the $c$ axis as

$$K = K_0(\hbar\omega - E_g)^{3/2} \tag{5.35}$$

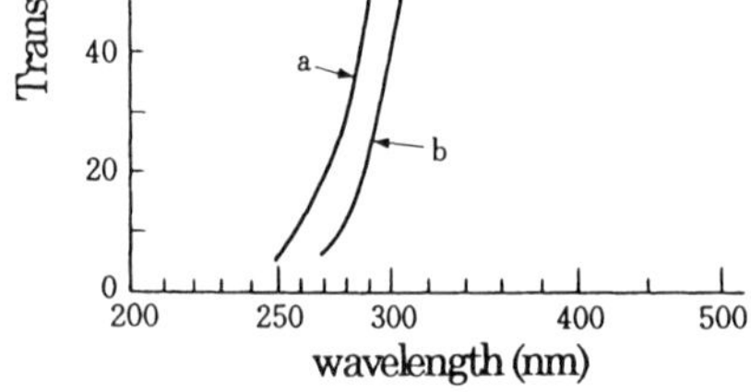


**Fig. 5.18.** Transmission of light through an $SnO_2$ film at 303 K [528]. $a$ is a conductive and $b$ a resistive film. (With permission from The Physical Society of Japan)

and determined $E_g$ to be 3.59 eV [529]. This is the direct and forbidden transition. The result does not conflict with the band calculation of *Robertson* [509]. *Fröhlich* et al. showed that the top of the valence band is correctly given by this calculation, using two-photon absorption [530].

### 5.2.4 $TiO_2$

There are three structures in $TiO_2$ minerals: rutile, brookite and anatase. In Fig. 5.11, the structure of rutile is shown. It is an insulator in which the mobility is much smaller than in $SnO_2$. Some researchers consider that the band picture is not applicable and carriers may be in localized states [531] though this is not conclusive. Thus $TiO_2$ and $SnO_2$ are two extreme substances whose structures are the same. In the former, the $3d$ electrons participate in conduction in contrast to the $5s$ electrons in the latter. $TiO_2$ is also expected to be a useful photocatalyst [532].

*Breckenridge* and *Hosler* [533] and *Cronemeyer* [534] have investigated the resistivity of a single crystal of rutile as a function of temperature (Figs. 5.19, 5.20). The resistivity becomes smaller as $TiO_2$ is reduced as in $SnO_2$ and is larger in the direction perpendicular to the $c$ axis than along the $c$ axis. The Hall constants are shown in Fig. 5.21. From these quantities, the Hall mobilities are calculated and are shown in Fig. 5.22. They differ from sample to sample but it can be seen that they are smaller than that in $SnO_2$ by a factor of more than 1/100. This small mobility is the most characteristic feature of $TiO_2$. The mobility becomes even smaller as temperature is raised. This is attributed to optical-phonon scattering (Fig. 5.22) which suggests that the carriers are large polarons and not localized small polarons because the

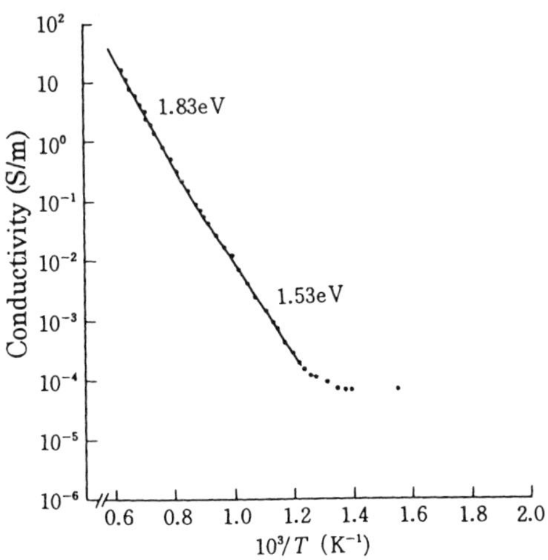


**Fig. 5.19.** Conductivity of $TiO_2$ along the $c$ axis [534]. The activation energy is also shown. (With permission from The American Physical Society)

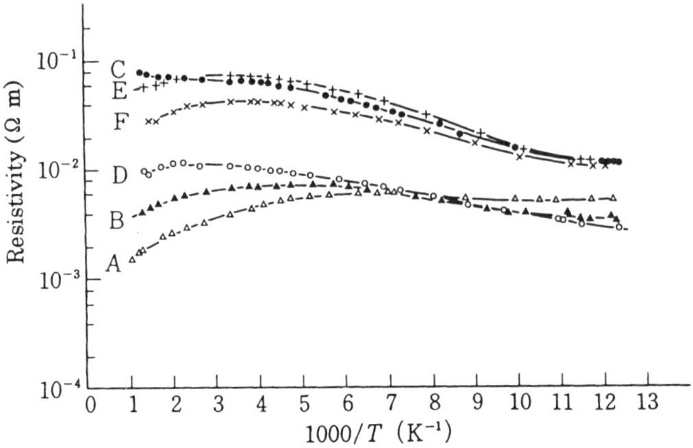


**Fig. 5.20.** Resistivity of reduced $TiO_2$ [533]. Five minutes' reduction in $H_2$ at 1173 K (A), 1123 K (B), 1073 K (C), 973 (K) (D), 1023 K (E), and 923 K (F). The resistivity was measured along the *c* aixs for A–D and perpendicular to the *c* axis for E and F. (With permission from The American Physical Society)

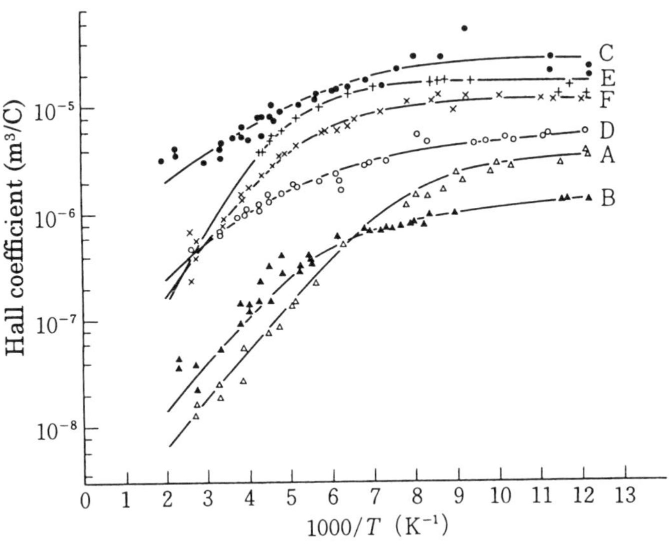


**Fig. 5.21.** Hall coefficient of reduced $TiO_2$ [533]. A–F are defined in Fig. 5.20. (With permission from The American Physical Society)

latter become more mobile at higher temperatures in the hopping process. However, if the small polarons diffuse by tunneling, the mobility may decrease as the temperature is raised.

Mobility can be considered as a measure of localization, though this is not a strict criterion. The critical mobility is given by $2a^2e/\hbar$ from the condition as to whether the mean free path $l$ is longer or shorter than the nearest

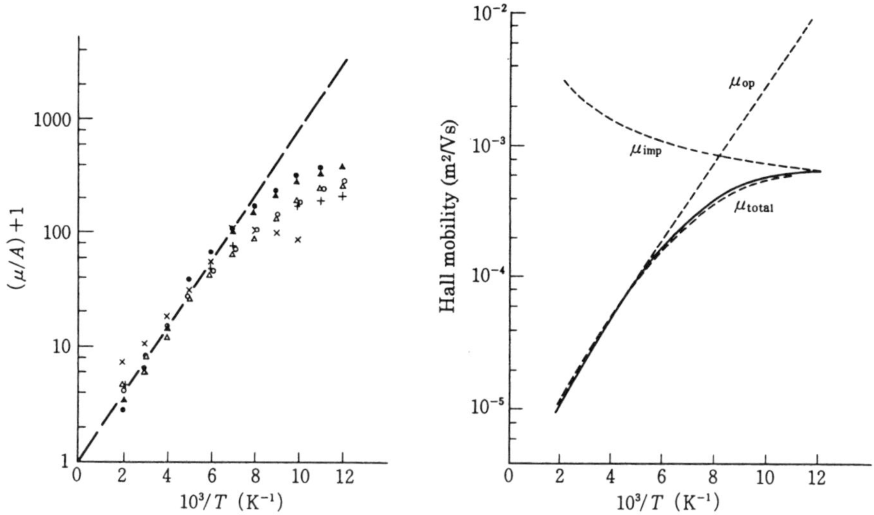


(a) (b)

**Fig. 5.22.** (**a**) Hall mobility of reduced $TiO_2$ [533]. ○: D of Fig. 5.20, • : F of Fig. 5.20, △: reduced at 973 K for 5 min and measured along the $c$ aixs, ▲: 973 K for 15 min and along the $a$ axis, +: 1173 K for 10 min, ceramics, ×: 873 K for 10 min, ceramics. $A$ is 6.3(○), 3.3(•), 9.4(△), 4.4(▲), 2.1(+), and 2.3 × in unit of $10^{-6}\,m^2V^{-1}\,s^{-1}$. The *broken line* represents the relation $\mu = A[\exp(670/T) - 1]$. (**b**) Hall mobility of ceramics [533]. The *solid line* represents the experimental results. The *broken lines* are calculated ones: $\mu_{op}$ due to optical phonon scatterings, $\mu_{imp}$ due to impurities, $\mu_{tot}$: total calculated mobility. (With permission from The American Physical Society)

neighbor distance $a$:

$$\mu = \frac{e\tau}{m}, \quad l = v\tau \leq a, \quad mv = \hbar k \leq \frac{h}{2a}\,.$$

If $a = 0.3$ nm, $\mu$ is $4 \times 10^{-5}\ m^2V^{-1}s^{-1}$. It should be noted that the condition has no firm foundation but is only an intuitive measure. It can be shown by calculation that even for an itinerant electron, the mobility can be infinitely small as a result of multiple scattering and there is no critical value which divides itinerant and localized states. So the above numerical value only really tells us that when the mobility is as low as this, we must consider the possibility that the electrons are in localized states. In $TiO_2$ it is in this range around 300 K.

The localized polaron, which is called a small polaron because the wavefunction localizes within the unit cell, is formed when the energy gain due to the interaction with phonons is larger than the band width. According to *Fröhlich* [535], the interaction strength is expressed by (2.19). In the following, we will compare the physical quantities which are concerned with $\alpha\hbar\omega_0$ in both materials.

*Jourdan* et al. [536] and *Daude* et al. [537] calculated the band structure of $TiO_2$ and the effective mass may be as large as ten times the mass in $SnO_2$ if we estimate it from the figures. *Frederikse* estimated the effective mass of the carriers as $m_p^* = 12 - 32m_0$ from the conductivity, the Hall coefficient, and the thermopower [538].

The frequencies of the longitudinal optical phonons are $SnO_2 : 2.73(\perp)$, $3.68(\perp)$, $7.57(\perp)$ and $7.04(\|)$ [539]; $TiO_2 : 3.73(\perp)$, $4.58(\perp)$, $8.06(\perp)$ and $8.11(\|)$ [540, 541]; in units of $10^4\,\mathrm{m}^{-1}(= 12.4\,\mathrm{meV})$. It can be seen that they are very similar.

The dielectric constants are [542]

$$\begin{aligned}
&SnO_2 : \epsilon_{rs}(\perp) = 14.0,\, \epsilon_{r\infty}(\perp) = 3.8,\, \epsilon_{rs}(\|) = 9.9,\, \epsilon_{r\infty}(\|) = 4.2\,;\\
&TiO_2 : \epsilon_{rs}(\perp) = 89,\, \epsilon_{r\infty}(\perp) = 6.0,\, \epsilon_{rs}(\|) = 173,\, \epsilon_{r\infty}(\|) = 7.2\,.
\end{aligned}$$

The subscripts are: s: static, $\infty$: high frequency limit, $\perp$: perpendicular to the $c$ axis, $\|$: along the $c$ axis. They do not give very different values of $\alpha$.

*Toyozawa* has postulated that acoustic phonons produce the small polarons and found that the interaction parameter is proportional to $E_1^2 m^*/C$ where $E_1$ is the coefficient of the deformation potential as in (2.16) and (5.24) and $C$ is the elastic constant [543]. The interaction is larger for larger effective mass as for the optical phonons. *Chang* and *Graham* found that the elastic constants are nearly the same for $SnO_2$ and $TiO_2$ [544].

Thus if the polaron is large in $SnO_2$ and small in $TiO_2$, the difference will be in the band width or the effective mass of electrons and not in the above quantities.

*Eagles* has calculated the electron–phonon interaction in $TiO_2$ and has found that the largest interaction is with the highest-frequency longitudinal optical mode for which $\alpha_i(m_0/m^*)^{1/2}$ is 1.61 where $m^*/m_0$ is the relative band mass [541]. The polaron mass is then $m^*(1 + (\sum_i \alpha_i)/6)$. Neglecting much smaller $\alpha_i$ of other modes, $m_p$ becomes $18m_0$ for $m^*/m_0 = 10$ or 44 for 20, which is very large.

*Bogomolov* et al. argued that the polaron in $TiO_2$ would be a small polaron [531]. In that case, the observed mobility, which becomes smaller as temperature is raised, has to be explained by tunneling motion. In a real sample, various defects may exist which lead to polaron trapping and the analysis of the observed results will not be simple. The small polaron is really a self-trapped localized state [545, 546]. Its radius will be $\alpha^{-1}u^{-1}$ where $u$ is the wavenumber of an electron with kinetic energy $\hbar\omega_0$, that is, $u = (2m\omega_0/\hbar)^{1/2}$. The radius is 0.03 nm when $\epsilon_{r\infty} = 6$ and $m^*$ is assumed to be 10 $m_0$. This radius is shorter than the distance between Ti ions, viz., 0.296 nm.

Following *Holstein* [547, 548], and *Yamashita* and *Kurosawa* [549] the electron–phonon interaction of a small polaron in a simplified model is

$$\gamma = \frac{1.25}{\pi} \frac{c^2}{\hbar\omega_0 a} \frac{1}{4\pi\epsilon_0} \left( \frac{1}{\epsilon_{r\infty}} - \frac{1}{\epsilon_{rs}} \right), \tag{5.36}$$

where $a$ is the lattice parameter and $\hbar\omega_0$ is the phonon energy.

For $SnO_2$ the lattice parameters are: $a = b = 0.4737\,\mathrm{nm}$, $c = 0.3185\,\mathrm{nm}$, and for $TiO_2$ they are: $a = b = 0.459\,\mathrm{nm}$, $c = 0.296\,\mathrm{nm}$. Thus the $\gamma$ of $TiO_2$ can only be as large as twice that of $SnO_2$ along the $c$ axis. So if the polaron is small in $TiO_2$, it will not be due to a larger $\gamma$ but may be due to the smaller band width, that is, the larger effective mass.

The motion of the small polaron will be described by thermally activated hopping motion when the temperature is higher than half the Debye temperature and by tunneling motion below that temperature [548]. In the tunnelling range, the mobility becomes smaller as temperature is raised. The mobility has been calculated by *Holstein* [547, 548], *Yamashita* and *Kurosawa* [549], and *Böttger* and *Bryksin* [550, 551] in a defect-free lattice.

Following *Böttger* and *Bryksin* [550,551] the drift mobility of the hopping motion is given as follows when $\hbar\omega_0 \ll k_\mathrm{B}T$, $J \ll E_\mathrm{a}$, and $J^2 \ll E_\mathrm{a}k_\mathrm{B}T$:

$$\mu_\mathrm{D} = \frac{z}{4\pi}\frac{ea^2}{\hbar}\frac{\hbar\omega_\mathrm{D}}{k_\mathrm{B}T}\exp\left(-\frac{E_\mathrm{a}}{k_\mathrm{B}T}\right) \times \begin{cases} 1 & b > 1 \\ \pi^{3/2}b & b \ll 1 \end{cases}, \tag{5.37}$$

$$b = \frac{J^2}{\hbar\omega_0\sqrt{E_\mathrm{a}k_\mathrm{B}T}},$$

where $E_\mathrm{a}$ is the activation energy and $z = 2$ or 3 in a cubic or hexagonal lattice. In the real crystal, there are many optical modes, thus many values of $\hbar\omega_0$. $J$ is the transfer integral before the electron–phonon interaction is introduced. The drift mobility is in general not the same as the Hall mobility [550,551]. Especially in hopping conduction, not only the magnitude but also the sign may differ. In a real crystal, defects will trap polarons and mobility will be reduced.

In conclusion, the mobility is smaller in $TiO_2$ than in $SnO_2$ and this may be due to the larger band mass. The larger mass makes the electron–phonon interaction larger and the polaron may be in the boundary region between the localized small-polaron and the itinerant large-polaron state. Thus the difference between the two extremal substances $SnO_2$ and $TiO_2$ may come from the difference in the contribution of the $5s$ and $3d$ electrons to the conduction band.

## 5.3 $LiTi_2O_4$ and $LiV_2O_4$: Weak-Coupling Superconductor and Temperature-Dependent Magnetism

$LiTi_2O_4$ is a superconductor below 13.7 K which is the highest $T_\mathrm{c}$ among $d\epsilon$ conductors. This substance may be considered as $TiO_2$ doped with a high density of Li ions. As the electron–phonon interaction is rather large in $TiO_2$, the high $T_\mathrm{c}$ may be a natural consequence of that interaction. There occurs a metal–insulator transition at $x = 0.16$ for $Li_{1+x}Ti_{2-x}O_4$, and metal-like properties are found for the insulating $Li_{1+x}Ti_{2-x}O_4$.

$LiV_2O_4$ is a normal metal down to 1.8 K. It shows a Curie–Weiss type magnetic susceptibility but no long-range magnetic ordering. There is a metal–insulator transition in $Li_xZn_{1-x}V_2O_4$ and the conduction electrons have a large effect on the magnetic properties.

### 5.3.1 Crystal Structure

$LiTi_2O_4$ is a superconductor with a $T_c$ of 13.7 K [552, 553] and one of a few metallic spinels together with $LiV_2O_4$ and $Fe_3O_4$. In the rhombohedral primitive cell, there are two molecules. The conventional cell is cubic with eight molecules as shown in Fig. 5.83. The lattice parameter of the cubic cell is 0.8405 nm at 300 K. Li is in a tetrahedral $A$ site and Ti is in an octahedral $B$ site. Some of the cation sites in this structure are vacant. If they were completely filled, the spinel structure would turn to the NaCl structure as $LiTi_2O_4$+Li$\rightarrow$ 2($LiTiO_2$) [554].

The octahedra share edges as they do in the rutile structure along the $c$ axis as shown in Fig. 5.11. In the rutile structure, the cation arrays run along the $c$ axis without crossing but in the spinel structure, three arrays cross at each $B$ site as shown in Fig. 5.83. Thus each Ti has six neighbouring Ti ions with a separation of 0.297 nm, whereas in the rutile structure the separation is 0.296 nm along the $c$ axis. Thus as far as the arrangement of the octahedra is concerned, $LiTi_2O_4$ and $TiO_2$ are similar oxides except that the cation array does not cross in the latter. As for the six neighbouring Ti ions, they are again their nearest nearbors as shown in Fig. 5.83 and there occurs a frustration when spins are to be arranged antiparallel.

In this respect, $LiTi_2O_4$ can be regarded as a heavily doped $TiO_2$. In Si, a P doping of $4\times10^{24}$ m$^{-3}$ does not destroy its diamond structure but in $TiO_2$, Li doping as much as $1.4\times10^{28}$ m$^{-3}$ destroys the rutile structure and changes it to the spinel structure. As explained in Sect. 5.2, the electrons are in the intermediate state between large and small polarons in $TiO_2$. Thus $LiTi_2O_4$ may be thought of as a metallic oxide in which such polarons are supplied by as much as $1.4\times10^{28}$ m$^{-3}$. The oxide thus obtained is superconducting below 13.7 K. Despite the electron–phonon interaction, the isotope effect of Li on $T_c$ has been reported to be absent [555].

When Li is doped further in $Li_{1+x}Ti_{2-x}O_4$, it enters the $B$ sites and all the Ti ions become $Ti^{4+}$ at $x = 1/3$. It is then an insulator. *Harrison* et al. have suggested that the metal–insulator transition occurs at $x = 0.16$ [552,556,557]. The oxide is unstable in air. Figure 5.23 shows that the lattice parameter of the oxides prepared under dynamic vacuum decreases linearly with an increase in $x$. Such a linear relationship is called Vegard's law. On the other hand, oxides prepared in a sealed vacuum show a nonlinear dependence of the lattice parameter of $x$. *Harrison* et al. attribute these phenomena to the formation of a stable layer at the surface [556]. They suggest also

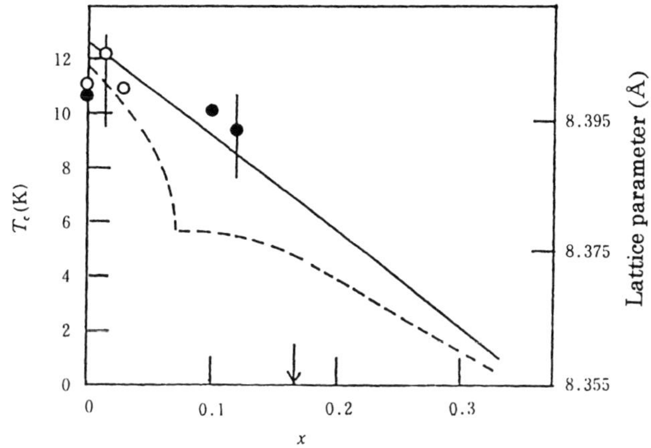


**Fig. 5.23.** The lattice parameter (*lines*) and superconducting transition temperature $T_c$ (*circles*) in relation to $x$ in $Li_{1+x}Ti_{2-x}O_4$ [552, 556]. The *arrow* shows a metal–insulator boundary. The *bar* shows the transition width between $0.9\rho_n$ and $0.1\rho_n$. The *solid line* shows a lattice parameter of a "stable" oxide and the *broken line* an "unstable" oxide. (With permission from Taylor & Francis and Plenum)

that a spinodal segregation of $LiTi_2O_4$ occurs near the critical composition. However, due to a lack of single crystals, the nature of the transition is not clear. $LiTi_2O_4$ disproportionates above about 1230 K.

### 5.3.2 Electronic Properties

*Johnston* and *McCallum* et al. have measured the specific heat and the magnetic susceptibility [552, 553, 558]. From the temperature-proportional specific heat the enhanced density of states $N(E_F)^*$ was determined to be 9.1 states/(eV molecule). This includes both spin states. From the $T^3$ term, the Debye temperature was determined to be 685 K at about 10 K. From the magnetic susceptibility, $N(E_F)$ was found to be 6.8 states/(eV molecule). This gives an electron–phonon interaction parameter $\lambda$ of 0.34 which is smaller than that of a simple metal. For instance it is 0.38 in Al, 0.60 in Sn, and 1.12 in Pb [559] (1.5 in [560]). $\lambda$ is also estimated from $T_c$. *McMillan* expresses $T_c$ of Nb as [559]

$$T_c = \frac{\theta_D}{1.45} \exp\left[-\frac{1.04(1+\lambda)}{\lambda(1-0.62\mu^*)-\mu^*}\right], \tag{5.38}$$

where $\theta_D$ is the Debye temperature and $\mu^*$ is the Coulomb repulsion in the band. It was found that the expression is applicable for many materials when $\lambda$ is less than 1.5. For larger $\lambda$, $T_c$ is proportional to $\sqrt{\lambda\langle\omega^2\rangle}$ where $\omega$ is the phonon frequency [560]. When (5.38) is applied to $LiTi_2O_4$, $\lambda$ is found to be 0.68. Where $T_c$, $\theta_D$, and $\mu^*$ were taken to be 13.7 K, 685 K and 0.13, respectively. The last value is usually found for the $d$ band electrons. It must be noted that $T_c$ is as high as 13 K even for such a moderate $\lambda$. This may be due to the high frequency of optical phonons in oxides. The softening of phonons has been observed and it will be due to the interaction with electrons [561].

*Satpathy* and *Martin* have calculated the band structure of $LiTi_2O_4$ as shown in Fig. 5.24 [562]. Their $N(E_F)$ is 3.3 states/(eV molecule) which is smaller than that deduced from the magnetic susceptibility. According to *Brinkman* and *Rice* the electron correlation enhances both the spin susceptibility and the effective mass by $(1-(U/U_0)^2)^{-1}$ where $U$ is the intra-atomic Coulomb repulsion [563, 564]. If the discrepancy between the observed and the calculated density of states were due to electron correlation, the conduction electrons in $LiTi_2O_4$ would be highly correlated, and $\lambda$ would be large, inconsistent with the small experimental value.

A much smaller density of states of 0.25 states/(eV molecule) appeared in the UPS (ultraviolet photoelectron spectroscopy) spectrum and this was attributed to a strong electron–phonon interaction [565]. The reduction in the density of states may also occur for the electron–electron interaction. *Efros* et al. have claimed that the long-range Coulomb interaction between

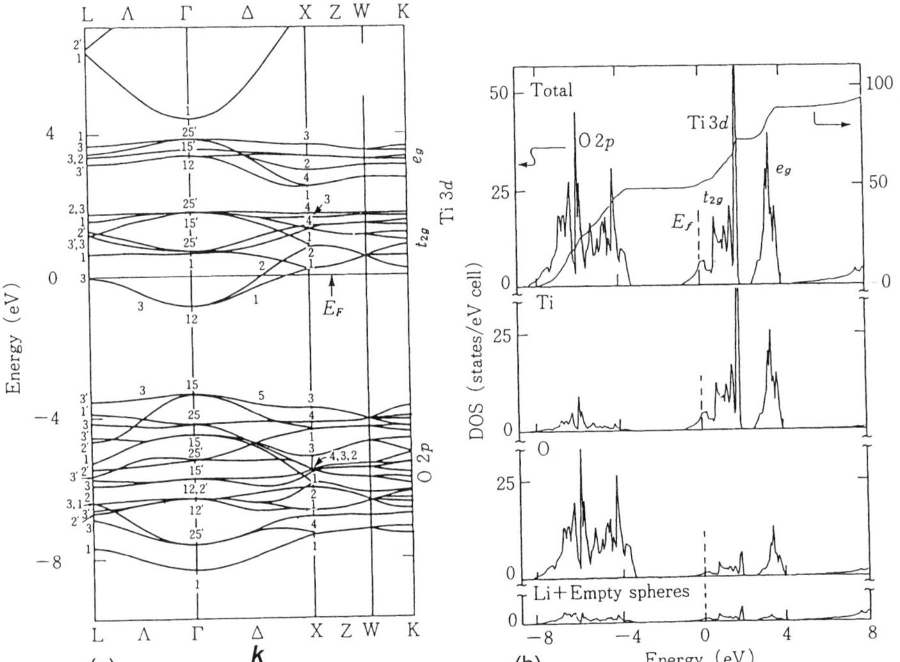


**Fig. 5.24. a, b** Band structure and density of states of $LiTi_2O_4$ [562]. (With permission from The American Physical Society)

electrons in a random potential depletes the states at the Fermi level, and the Coulomb gap appears in the single particle density of states, which is obtained by adding an extra particle to the ground state without relaxing the localized electrons in the random potential [566–569]. This density of states will correspond to the UPS density of states. Such reduction in the density of states was also observed for $Li_xZn_{1-x}V_2O_4$ [570] and for $Na_xTa_yW_{1-y}O_3$ [569]. However, the small density of states of the unstable Li oxides may possibly be due to the poor quality of the samples at the surface region.

### 5.3.3 Superconducting Properties

$LiTi_2O_4$ is a superconductor of the second kind. When a magnetic field is applied, it begins to penetrate into the bulk at $H_{c1}$ and the sample becomes normal at $H_{c2}$ except for a surface region parallel to the field in which superconductivity persists up to $H_{c3}$. These critical fields depend on the mean free path of a normal electron through [571,572]

$$H_{c1} = \frac{\ln\kappa}{\sqrt{2}\kappa} H_c \,, \tag{5.39}$$

$$H_{c2} = \sqrt{2}\kappa H_c \,, \tag{5.40}$$

$$H_{c3} \approx 2.4\kappa H_c \,, \tag{5.41}$$

$$\kappa = \frac{\lambda}{\xi} \begin{cases} = 0.96\frac{\lambda_L(0)}{\xi_0} \ (l \gg \xi_0) \\ = 0.75\frac{\lambda_L(0)}{l} \ (l \ll \xi_0), \end{cases} \tag{5.42}$$

where $l$ is the mean free path, $\xi_0$ is the intrinsic coherence length, and $\lambda_L$ is the London penetration depth. The thermodynamical critical field $H_c$ is a material constant and is given by

$$H_c^2 - -2\int_0^{H_{c2}} M(H)dH \,, \tag{5.43}$$

where $M$ is the magnetization.

In a weak coupling BCS superconductor where $\lambda$ is much smaller than 1, as in Al, there are general relations among the various thermodynamical quantities, such as

$$\frac{H_c}{H_c(0)} = 1 - 1.07\left(\frac{T}{T_c}\right)^2 \,, \tag{5.44}$$

$$\Delta C = 1.43\gamma T_c \,, \tag{5.45}$$

where $\Delta C$ is the jump in the specific heat at $T_c$, and $\gamma$ is the coefficient of the electronic specific heat. When $\lambda$ is as large as 1 or larger than 1 such as for

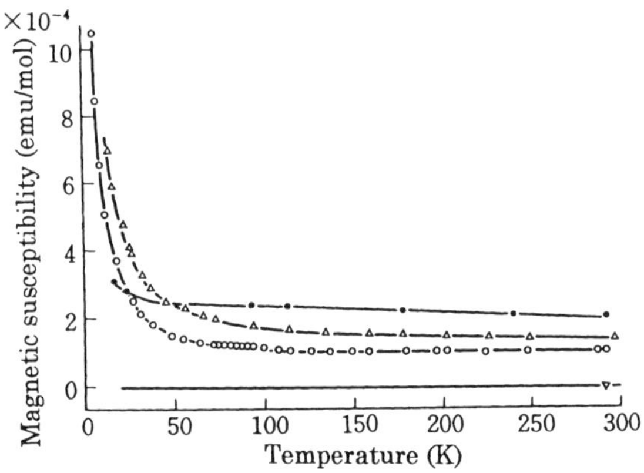


**Fig. 5.25.** Susceptibility of $Li_{1+x}Ti_{2-x}O_4$ [552]. $\bullet : x = 0, \triangle : x = 0.1, \circ : x = 0.2, \nabla : x = 1/3$. A sharp rise at the lowest temperature may be due to localized $Ti^{3+}$ ions. Their concentration would be 0.35 mol% ($x = 0$) and 1.63 mol% ($x = 0.1$). (With permission from Plenum)

Nb ($\lambda = 0.82$) and $V_3Si$ ($\lambda = 1.10$), the superconductor is called a strong-coupling superconductor and some deviations appear in the above relations. In Sect. 5.9.3 a, various strong-coupling limits will be discussed, based on the work of *Carbotte* et al.

In $LiTi_2O_4$, *Johnston* has found that [552]

$$H_c(T) = 0.197 \left[1 - 1.07 \left(\frac{T}{T_c}\right)^2\right] \quad \text{(tesla)}, \tag{5.46}$$

and

$$\Delta C = 1.6\gamma T_c\,. \tag{5.47}$$

From these results, $LiTi_2O_4$ seems to be a weak-coupling superconductor. In other words, in the oxides, even a weak-coupling superconductor can have a $T_c$ as high as 13 K.

### 5.3.4 Insulating Properties: Nonzero Density of States

The properties of the insulating $Li_{1+x}Ti_{2-x}O_4$ are not simple and show a variety of interesting behaviors. Below about 50 K, the paramagnetic susceptibility increases rapidly as the temperature decreases, as shown in Fig. 5.25. After subtracting this as being due to localized $Ti^{3+}$ ions, the nearly temperature-independent paramagnetic susceptibility is obtained (Fig. 5.26) [552, 558]. *Harrison* et al. noted that the small temperature coefficient changes its sign at the transition [558]. Such a nearly temperature-independent paramagnetism in an insulator is usually attributed to Van Vleck paramagnetism.

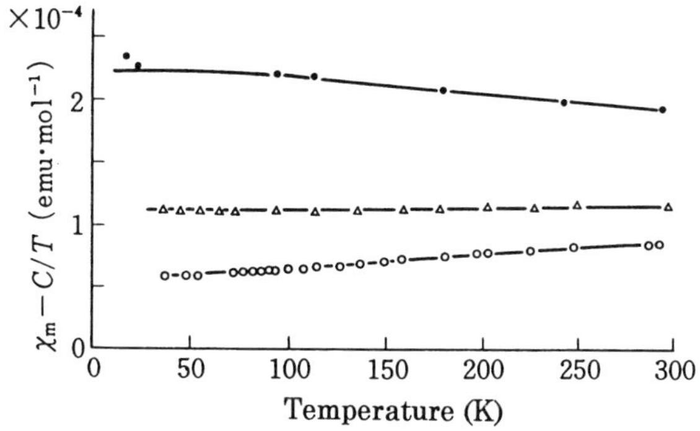


**Fig. 5.26.** The Curie–Weiss term in Fig. 5.25 has been subtracted [552]. (With permission from Plenum)

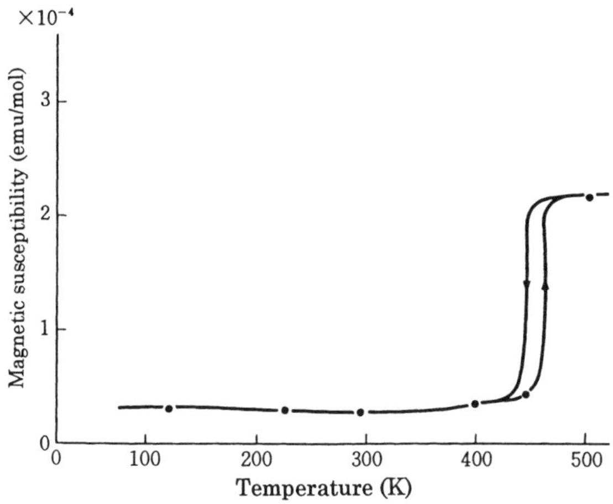


**Fig. 5.27.** Susceptibility of $Ti_3O_5$ [576]. It changes drastically at the M-I transition temperature. In the higher temperature range, the oxide is metallic. (With permission from The American Institute of Physics)

However, it is not the case here for it has been already subtracted. As shown in Fig. 5.25, the susceptibility of the $x = 1/3$ sample is nearly zero and is composed of the Van Vleck term and the orbital diamagnetism. The numbers are $-0.6$ (Li+), $-5$ ($Ti^{4+}$) and $-12$ ($O^{2-}$) for the orbital diamgnetism in units of $10^{-6}$ emu/mol, and $33 \times 10^{-6}$ emu/mol for the Van Vleck paramagnetism of $Ti^{4+}$. Thus the susceptibility shown in Fig. 5.25 is approximately due to the $3d$ electrons on the $Ti^{3+}$ ions. The core term is smaller than 10% of this.

This temperature-independent paramagnetism is the most characteristic feature of $Li_{1+x}Ti_{2-x}O_4$. The electron spins must be antiparallel in the zero field to elucidate such magnetism. For Pauli paramagnetism, up and down spins occupy the same orbital which extends over the whole lattice. For Van Vleck paramagnetism, up and down spins occupy the same orbital which

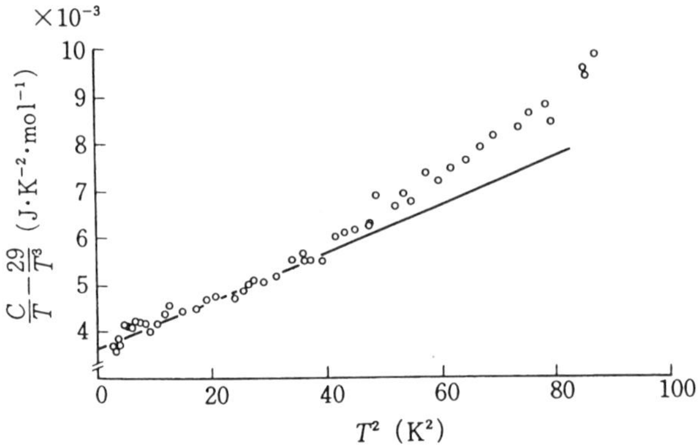


**Fig. 5.28.** Specific heat of $Li_{1.2}Ti_{1.8}O_4$ [553]. The Schottky term has been subtracted. The coefficient of the temperature-proportional term is 3.6 mJ/mol K$^2$. (With permission from Plenum)

localizes in one atom. These two orbitals are the extreme examples. There may be other states where both spins occupy the same orbital which may extend over a certain range larger than an atomic size and smaller than the entire sample. An example of this is an intersite bipolaron in which electrons locate on neighbouring ions in a spin singlet state. This is sometimes called the Chakraverty intersite bipolaron [573]. When the attractive interaction between electrons is strong enough, two electrons can reside on the same ion. This is called the Anderson on-site bipolaron [574]. Experimentally the intersite bipolaron has been identified by *Nagasawa* in $Na_xV_2O_5$ which is a rare example of a small-polaron oxide [575] together with $Ti_4O_7$ (Sect. 3.2.1).

Similar temperature-independent susceptibilities are found in other oxides. In Fig. 5.27, that of $Ti_3O_5$ is shown [576]. This is an insulator with an energy gap of 0.087 eV below about 450 K and a metal above that temperature. Below the transition temperature, the susceptibility is constant as if in a metal. *Mulay* and *Danley* explain this by antiparallelly coupling spins [576].

In the insulating range of $Li_{1+x}Ti_{2-x}O_4$, another characteristic feature appears in the specific heat. *McCallum* et al. found that there is a metal-like $\gamma T$ term for the $x = 0.2$ sample which does not show superconductivity down to 1 K (Fig. 5.28) [553]. The $\gamma T$ term means that low-lying excited states must exist continuously around the Fermi level.

The Seebeck coefficient also suggests such low-lying excitations exist in the insulating region. Figure 5.29 shows the Seebeck coefficient of $Li_{1+x}Ti_{2-x}O_4$ prepared under a dynamic vacuum [577]. Other than $LiTi_2O_4$, the oxides in the figure may be insulators. Their Seebeck coefficients are small and seem to be proportional to temperature. Usually these are the characteristic features of metals and originate in densely distributed excited states around the Fermi level. *Brenig* et al. have calculated the Seebeck coefficient of a random system in which variable-range hopping (VRH) conduction occurs. (VRH will be

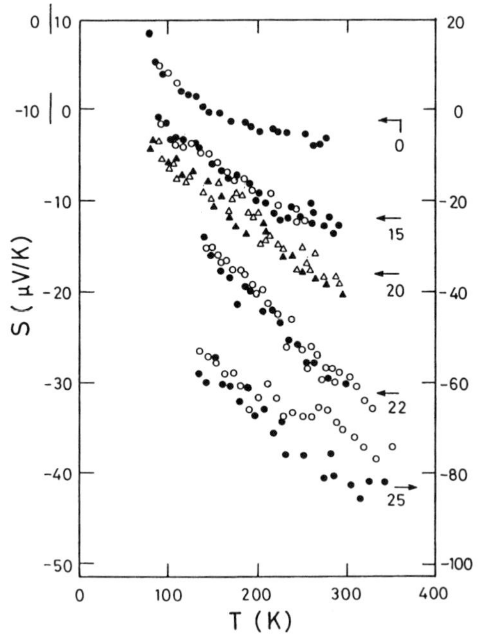


**Fig. 5.29.** The Seebeck coefficient of $Li_{1+x}Ti_{2-x}O_4$ [577]. Samples are insulators except for the $x = 0$ sample. The number beside each line is $100x$. (With permission from The Physical Society of Japan)

explained in Sect. 5.5.2). Following them, $S$ is generally given as [578]

$$S \sim \frac{W^2}{Te}\frac{2}{W}\frac{\sigma(E_F+W)-\sigma(E_F-W)}{\sigma(E_F+W)+\sigma(E_F-W)} \sim \frac{W^2}{Te}\left(\frac{\partial}{\partial E}\ln\sigma\right)_{E_F} . \tag{5.48}$$

In the three-dimensional lattice,

$$\sigma \sim \frac{e^2}{k_B T} NWR^2\gamma_0 \exp\left[-\left(\frac{T_0}{T}\right)^{1/4}\right] , \tag{5.49}$$

where

$$W = \frac{k_B}{4}T_0^{1/4}T^{3/4} , \quad k_B T_0 = \delta\frac{\alpha^3}{N(E_F)} , \quad R = \frac{3}{4\alpha}\left(\frac{T_0}{T}\right)^{1/4} .$$

If we assume that only the density of states is energy-dependent, $S$ will be

$$S \sim \frac{k_B^2}{64e}T_0^{1/2}T^{1/2}\left(\frac{N'}{N}\right)_{E_F}\left[1+\left(\frac{T_0}{T}\right)^{1/4}\right] . \tag{5.50}$$

In the temperature range of Fig. 5.29, no clear difference appears among $T$, $T^{1/2}$ or $T^{1/4}$ dependences [577]. However, at least $S$ is not proportional to $1/T$ which is the characteristic feature of a broad-band semiconductor.

Thus the magnetic susceptibility, the specific heat and the Seebeck coefficient behave as if $Li_{1+x}Ti_{2-x}O_4$ were a metal even in the insulating region. As was explained in Sect. 2.5.1, a random potential produces densely distributed localized states near the band edge. If the above properties are due to such states, the localized orbital must contain two electrons with up and down spins to explain the susceptibility. Then such orbitals would have to be large enough to maintain the charge neutrality naturally. For instance, such a localized orbital would have to extend more widely than the unit cell when $x$ is larger than 0.25.

The unknown inhomogeneity, which could be spinodal segregation of $LiTi_2O_4$ according to *Harrison* et al. [556], might be another cause of these phenomena. It is difficult to distinguish the random states from microscopic inhomogeneity and it remains a cumbersome problem. Similar problems occur in the high-$T_c$ oxides as will be explained in Sect. 5.9.

### 5.3.5 $LiV_2O_4$-$ZnV_2O_4$

$LiV_2O_4$ is metallic. *Rogers* et al. succeeded in preparing a single crystal by the hydrothermal method [579]. The resistivity is shown in Fig. 5.30 in relation to temperature. On the other hand, $ZnV_2O_4$ is an insulator. The activation energy of the conduction of the sintered material is reported as 0.13 [580], 0.16 [581], 0.19 [582] and 0.2 eV [583]. There are no states at the Fermi level in the UPS spectrum [570], so there must be a metal–insulator transition in the system $Li_xZn_{1-x}V_2O_4$. *Reuter* and *Müller* have estimated the critical composition as $x_c = 0.55$ from thermoelectric and infrared absorption measurements [583] whereas *Kawakami* et al. estimated it as 0.35 from the Seebeck coefficient [582].

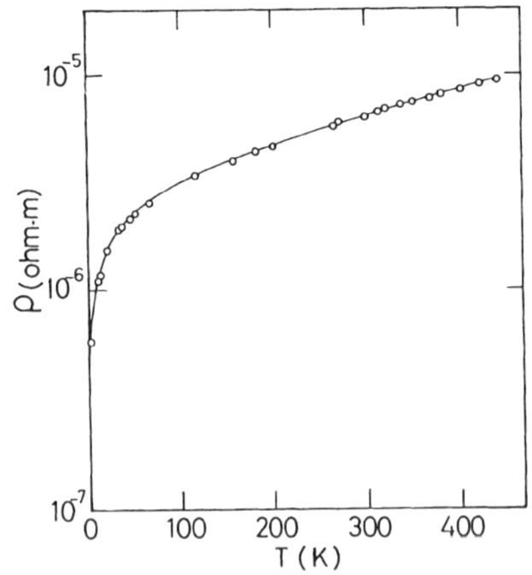


**Fig. 5.30.** Resistivity of single crystalline $LiV_2O_4$ [579]. (With permission from Pergamon, Elsevier)

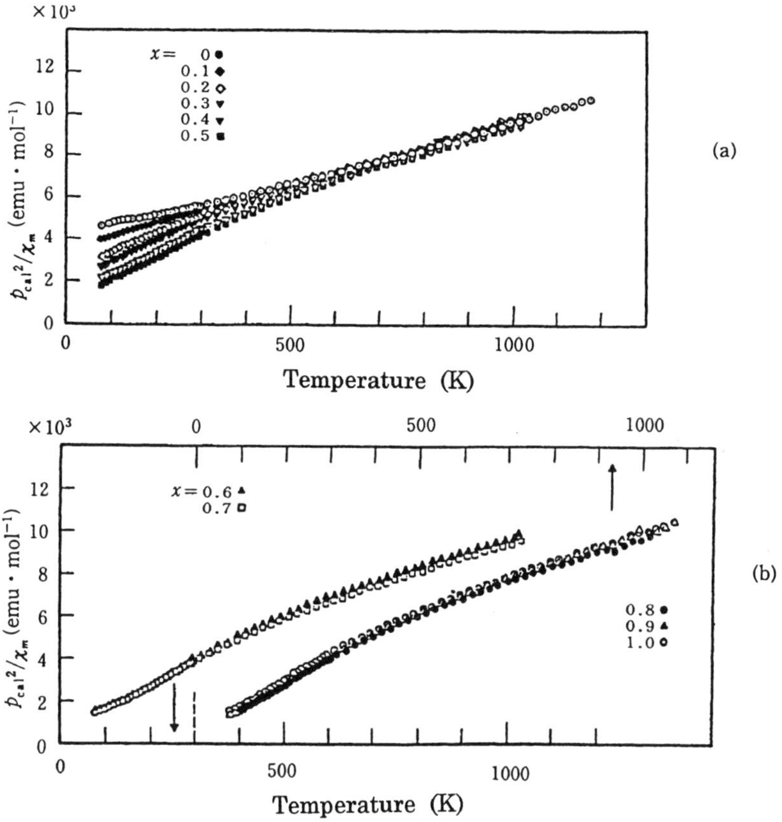


**Fig. 5.31. a, b.** Inverse molar susceptibility of a metal–insulator system $Li_xZn_{1-x}V_2O_4$ [584]. $\chi_m$ has been normalized by $p_{eff}^2$ using (2.33). The oxide is metallic for $x > 0.35$. (With permission from The Physical Society of Japan)

The largest difference between $LiTi_2O_4$ and $LiV_2O_4$ appears in the magnetic behavior. The susceptibility of the latter is temperature dependent as shown in Fig. 5.31 [584] and the localization nature of 3*d* electrons becomes stronger in V oxides than in Ti oxides. As shown in Fig. 5.31, the MIT seems to affect the magnetic property significantly. However, the interrelation of magnetic and conduction electrons is not yet clear in these V oxides, and the nature of conduction electrons is widely disputed [584]. Even a heavy fermionic property has been pointed out [585].

## 5.4 $WO_3$ and $M_xWO_3$: Large Polarons

In $WO_3$ the conduction electrons may form large polarons. With Na doping, $Na_xWO_3$ becomes metallic at $x = 0.25$. As Na ions distribute randomly electrons diffuse under the disturbed potential and localization is expected. Phenomenologically a percolation model is applicable.

The poor screening of the Coulomb interaction will be the cause of the depletion of the UPS density of states at the Fermi level of $Na_xTa_yW_{1-y}O_3$. This is called the Coulomb gap.

$Na_xWO_3$ is superconducting and the critical temperature depends on the carrier concentration. It is explained by the screening effect on the electron–phonon interaction.

$M_xWO_3$ shows various colors from purple red to blue and these are called the tungsten bronzes. H and Li tungsten bronzes are electrochromic materials.

### 5.4.1 Structure

Figure 5.32 is the phase diagram of $Na_xWO_3$ [586, 587]. According to *Clarke* [588], $Na_{0.81}WO_3$ is cubic above 423 K and the transition to the tetragonal phase occurs at 423 K and at about 343 K as the temperature decreases. It is again cubic below 293 K but it contains eight molecules in the unit cell. *Sato* et al. have studied the successive transitions by neutron scattering and have confirmed that the transitions are due to the softening of the $M_3$ mode [589]. This mode involves a rotation of the $WO_6$ octahedra around the $z$ axis, for instance, and the neighbouring octahedra on the $xy$-plane rotate in the reverse direction. As for $WO_3$, it is reported to be triclinic between 290 K and 233 K and monoclinic below 233 K [590]. At room temperature $a = 0.730$ nm, $b = 0.753$ nm, $c = 0.768$ nm, and $\beta = 90°54'$. Thus it is almost the $ReO_3$ structure.

Generally the symmetry is higher in the conductive materials than in the insulators as seen from the comparison of $ReO_3$ and $WO_3$. The lower symmetry is usually caused by softening of the optical phonon. So the above phenomenon may be due to suppression of the softening of the optical phonons by the conduction electrons. *Fujimori* explained the transformation of $WO_3$

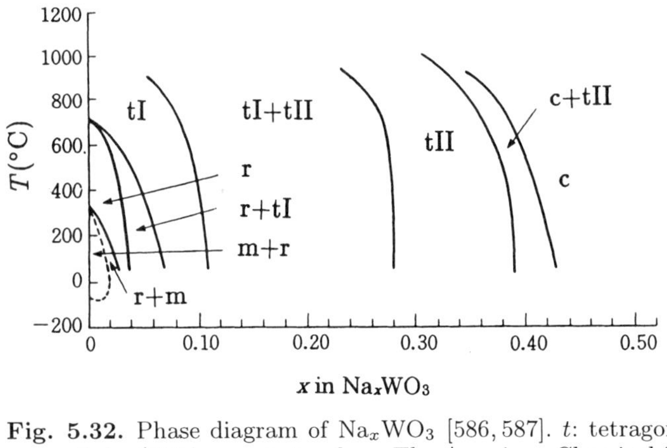


**Fig. 5.32.** Phase diagram of $Na_xWO_3$ [586, 587]. $t$: tetragonal, $c$: cubic, and $m$: monoclinic. (With permission from The American Chemical Society)

by calculating the energy gain by the electron–phonon interaction $H_{ep}$ using the second order process [591]

$$\Delta E_{\boldsymbol{q}} = \sum_{\boldsymbol{k}} \sum_{\substack{n \\ \text{occ}}} \sum_{\substack{m \\ \text{unocc}}} \frac{|\langle \boldsymbol{k}+\boldsymbol{q}, n|H_{ep}|\boldsymbol{k}, m\rangle|^2}{E_{\boldsymbol{k}+\boldsymbol{q},n} - E_{\boldsymbol{k},m}} , \qquad (5.51)$$

where $E_{\boldsymbol{k}}$ is the one-electron energy and $\boldsymbol{q}$ is the phonon wavevector. If the energy gain is larger than the loss in elastic energy, the lattice may distort. Equation (5.51) indicates that if the top of the valence band and the bottom of the vacant conduction band are connected by a wavevector $\boldsymbol{q}$, there is a possibility that the softening of such a phonon with that wavevector may occur. The displacement of a W ion is expressed as $u\cos[(\pi/a)x + (\pi/a)y + 0]$ at the cubic–tetragonal transition. At that point, $\boldsymbol{q} = (\pi/a, \pi/a, 0)$. Then the following pairs of occupied and unoccupied states will be responsible for the distortion (Fig. 5.5): $\Gamma'_{25} - M'_5$, $X_3 - X'_3$, $X_3 - X'_4$, $\Delta'_2 - Z_3$. *Fujimori* explained the various transformations in $WO_3$ using the band structure of $ReO_3$ calculated by *Mattheiss* [499].

$WO_3$ is antiferroelectric. *Nagasawa* found that the dielectric constant is very large [592] but due to the low resistivity, measurement of the dielectric constant is difficult. It is about $1000\epsilon_0$ at room temperature and drops to a few hundred $\epsilon_0$ at 233 K.

### 5.4.2 Electronic Properties in the Insulating Range and the Metal–Insulator Transition

*Berak* and *Sienko* studied the electrical conductivity of $WO_{3-x}$ [593]. Figure 5.33 shows the resistivity of the nearly stoichiometric ($x = 0$) sample in relation to temperature. The jumps in resistivity are due to lattice transformations. At room temperature, the Hall mobility, $\mu_H = R_H\sigma$ is as large as $1 \times 10^{-3}\mathrm{m^2V^{-1}s^{-1}}$ which is comparable with those of $ReO_3$ and Cu and is larger than that of $TiO_2$. This indicates that the conduction electrons in stoichiometric $WO_3$ may form large polarons. *Langreth* calculated the drift mobility of a large polaron as

$$\mu_D = \frac{e\hbar}{2\alpha k_B \theta_E m_p^*}\left(1 + 1.53\frac{T}{\theta_E}\right)\left[\exp\left(\frac{\theta_E}{T}\right) - 1\right] , \qquad (5.52)$$

where $\alpha$ is the electron–phonon coupling constant of (2.19), $m_p^*$ is the polaron mass

$$m_p^* = \frac{1 - 0.0008\alpha^2}{1 - \alpha/6 + 0.0034\alpha^2} m^* , \qquad (5.53)$$

and $m^*$ is the band mass [594]. $\theta_E$ is the Einstein temperature of the optical phonon. *Langreth* suggests that the formula is applicable for $\alpha < 3$

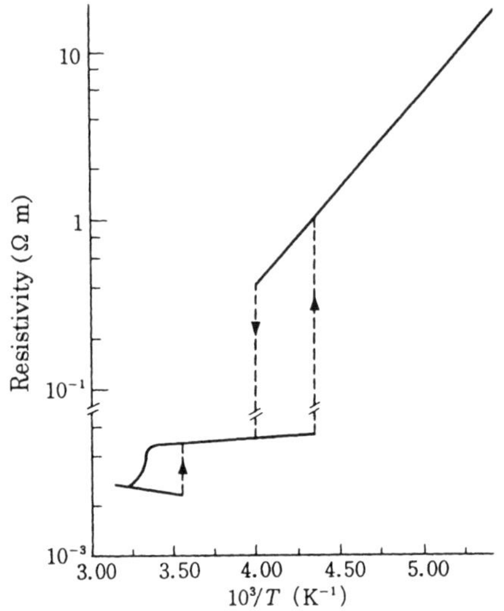


**Fig. 5.33.** Resistivity of $WO_{3.0}$ [593]. (With permission from Academic Press)

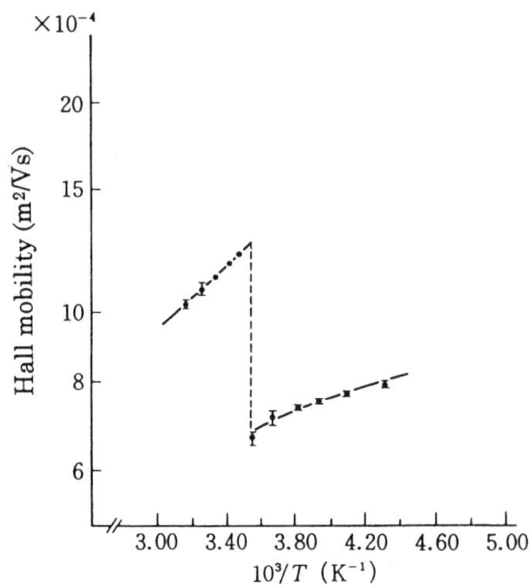


**Fig. 5.34.** Mobility of $WO_{3.0}$ [593]. Lower temperature (*solid line*) represents $\mu = 1.92(1 + 1.53T/225)[\exp(225/T) - 1] \times 10^{-4} \mathrm{m^2V^{-1}s^{-1}}$. (With permission from Academic Press)

and $\theta_E/T > 2$. *Berak* et al. analyzed the experimental results with this formula below about 290 K as shown in Fig. 5.34. The parameters are: $\alpha = 3.0$, $\theta_E = 561$ K and $m^* = 1.75\, m_0$. When oxygen vacancies increase slightly, $\mu_H$ decreases considerably and reaches $1 \times 10^{-4}\, \mathrm{m^2V^{-1}s^{-1}}$ at 318 K [593]. Moreover it increases as temperature increases according to $T^{-1.5} \exp(-E/k_BT)$,

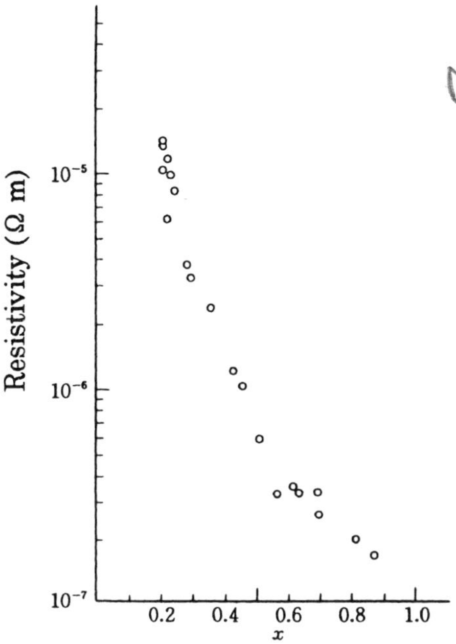


**Fig. 5.35.** Resistivity of $Na_xWO_3$ in the metallic range [598,599]. (With permission from The American Physical Society)

with an activation energy $E$ of 0.035 eV. The large polarons may have been trapped at the oxygen vacancies.

Electrons are supplied by doping other elements whose ionicity is larger than W or by reducing $WO_3$. $Na_xWO_3$ turns metallic at about $x_c = 0.25$ [595, 596]. Practically the critical value depends on how the sample is prepared. The materials are all cubic in the metallic range as explained previously. *Lightsey* quenched the single crystalline $Na_xWO_3$ from 1198 K to room temperature to extend the cubic phase range to as low an $x$ as possible, and found that it was cubic even at $x = 0.2$ though the $x$ value fluctuated by about 20% from point to point. The MIT occurs at about $x = 0.2$. The resistivity and the Hall concentration ($n = 1/R_{\mathrm{H}}e$) in the metallic range are shown in Figs. 5.35 and 5.36 [597–599]. When $x$ is less than 0.25, the carrier concentration decreases as temperature decreases suggesting that there are localized states just below the Fermi level.

A similar MIT is also observed in amorphous $H_xWO_3$ which has attracted attention as an electrochromic material [600]. The $WO_3$ powders look light green or nearly white whereas a single crystal is dark green or nearly black. When the powders are wetted in acid and negatively biased, protons diffuse in and they turn blue. This occurs reversibly by altering the bias potential. According to *Crandall* and *Faughnan*, the MIT occurs at $x = 0.32$ [601]. The logarithm of the conductivity is proportional to $T^{-1/4}$ in the insulating

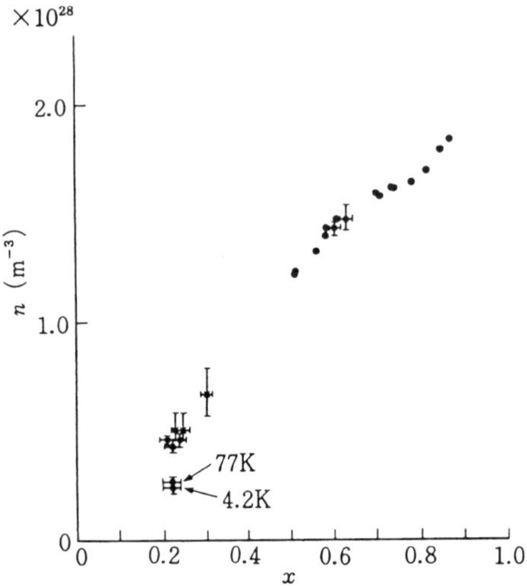


**Fig. 5.36.** Hall carrier density of cubic $Na_xWO_3$. At the lowest Na concentration, $n$ depends on temperature suggesting a localization of part of the carriers [598,599]. (With permission from The American Physical Society)

range (Fig. 5.37), suggesting variable-range hopping conduction in the three-dimensional lattice. As will be explained in Sect. 5.5.2 this type of conduction is characteristic of a random system.

*Lightsey* also studied the concentration dependence of the conductivity and found that it is expressed as

$$\sigma = A(x - x_c)^b \tag{5.54}$$

in the metallic range, where $x_c = 0.16 \pm 0.03$ and $b = 1.8 \pm 0.2$. *Kirkpatrick* showed that the conductivity is expressed this way near the critical concentration and $b$ = 1.6–1.5 from a computer simulation of the percolation model [602].

In the percolation model tiny conductive regions are distributed randomly in the insulating background. When the conductive regions make a thread which percolates from one end to the other of the specimen, it is considered to be a metal.

As for the randomness of Na ions, the specific heat, the magnetic susceptibility, NMR, and XPS suggest that their distribution is random [603,604] except on the surface [605]. Then if the large polarons are, for instance, trapped by $Na^+$ ions, the percolation model is applicable. *Fuchs* [606] and *Lightsey* [597,598] tried to calculate the critical concentration. *Lightsey* assumes a localized state which extends to the six nearest neighbors, the twelve second neighbors and eight third neighbors with a weighting of 1, 1/2, and 1/4, respectively, to obtain the observed $x_c$ of 0.16. This localized state is rather large, and the carrier will be a large polaron.

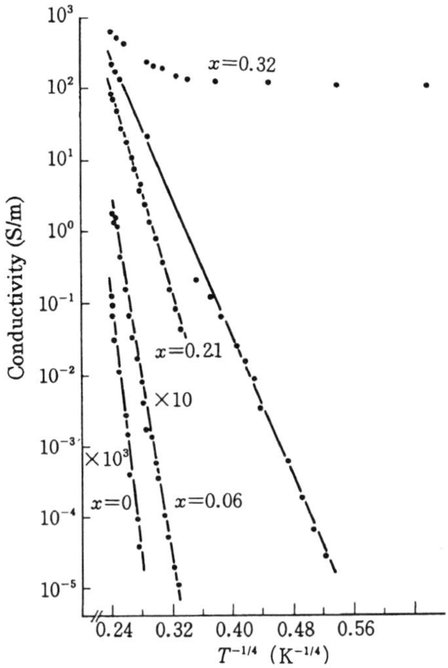


**Fig. 5.37.** Conductivity of amorphous $H_xWO_3$ [601]. (With permission from The American Physical Society)

The percolation model does not identify the localization mechanism. It merely assumes a random distribution of localized states. *Mott*, *Hollinger* et al., and *Hill* and *Egdell* have attributed the localization to the random potential, and claim that the metal–insulator transition is the Anderson transition [607–610].

Randomness is more evident in the $Na_xTa_yW_{1-y}O_3$ system where the compositional metal–insulator transition occurs at $x - y = 0.18$ [611, 612]. Tantalum is believed to be an acceptor, and in the insulating range, a depletion of the density of states has been noticed near the Fermi level in the UPS spectra [569, 609]. *Efros* and *Shklovski* claimed that in such a random system, the long-range Coulomb interaction depletes the single-particle density of states at the Fermi level and a Coulomb gap appears. It is not a finite gap but only a depletion of states at the Fermi level caused by the long-range interaction which tends to order the random electric charges [566–569, 613]. The single-particle density of states means the density of states for one particle which is to be added or removed from the ground state, while the background remains unchanged. Thus it may correspond to the UPS density of states in which electrons are taken out in $10^{-15}$ s or so. *Davies* and *Franz* explain the small density of states of $Na_xTa_yW_{1-y}O_3$ with this model

(Sect. 4.7.6) [569,609]. This effect is important in such a system in which the screening of the Coulomb interaction is insufficient.

### 5.4.3 Superconductivity and Screening of the Electron–Phonon Interaction

Figure 5.38 shows the resistivity of metallic $Na_xWO_3$ [614]. It is not proportional to temperature due to scattering by optical phonons, as explained in Sect. 2.4.2. *Zumsteg* studied the specific heat and the magnetic susceptibility and found that they are characteristic of band electrons [603]. NMR results show that the electronic states in metallic $Na_xWO_3$ are similar to those in $ReO_3$ [615]. Therefore (2.26) is applicable. *Tralshawala* et al. applied (2.26) to the resistivity of NCCO by using the spectral function of the electron–phonon interaction which was determined from the tunneling conductance (Sect. 5.9.3) [616]. *Shimada* et al. also did it for $Bi_2Sr_2CaCu_2O_8$ (Sect. 5.9.4) [617]. The temperature dependence of the resistivity was well fitted by choosing a plasma frequency as a parameter. The metallic oxide is superconducting (Fig. 5.39) [618] and its characteristic feature is that $T_c$ increases as the material approaches the MIT boundary in contrast to the high-$T_c$ cuprates. *Salchow* et al. [619, 620] have studied the screening effect of the conduction electrons on the electron–phonon interaction using the two-dimensional Lindhard dielectric function whose derivative is singular at $q = 2k_F$, where $\hbar k_F$ is the Fermi momentum. As explained in Sect. 5.1.2, the conduction band is two-dimensional though the lattice is three-dimensional. Thus the strong concentration dependence appears in the electron–phonon interaction. As the carrier density decreases, the electron–phonon interaction becomes larger due to the poor screening and $T_c$ increases. This is shown by the solid line in Fig. 5.39.

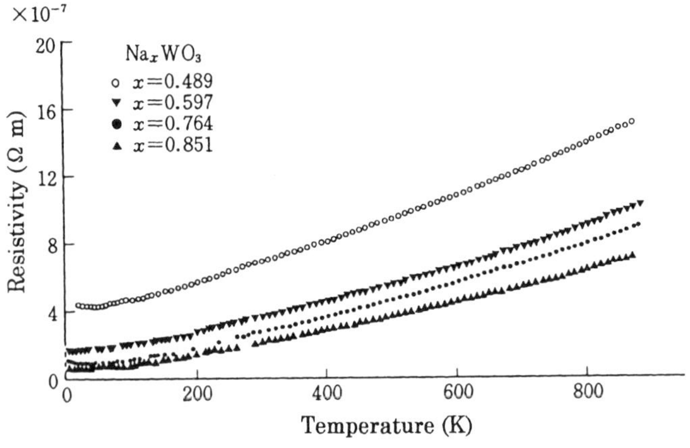


**Fig. 5.38.** Resistivity of metallic $Na_xWO_3$ [614]. (With permission from The American Institute of Physics)

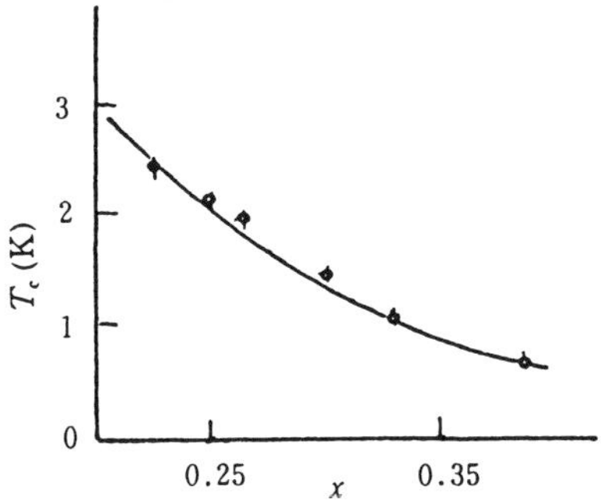


**Fig. 5.39.** Superconducting critical temperature of $Na_xWO_3$ in relation to $x$ [618,619]. The *solid line* is the calculated one [620]. (With permission from North-Holland, Elsevier)

## 5.5 $M_xV_2O_5$ and $M_xMoO_3$: Low-Dimensional Oxides

Low-dimensional substances have attracted attention in the search for a non-electron–phonon superconductor. For example, the incommensurate charge density wave was expected to show Fröhlich superconductivity, and low-dimensionality is favorable for the appearance of such a CDW. Another example is the Little-type superconductor in which electronic polarization was expected to couple electrons in a one-dimensional conduction path as in certain organo-metallic compounds. One-dimensional oxides have also attracted attention stimulated by such organic compounds.

$\beta$-$M_xV_2O_5$ are oxides in which both the small polaron and Chakraverty bipolaron have been reported. Also, in the more extended 4$d$ oxides of Mo, CDWs have been observed.

### 5.5.1 Crystal Structure of $\beta$-$Na_xV_2O_5$

Figure 5.40 shows the structure of $\beta$-$Na_{1/3}V_2O_5$ at room temperature [621]. The cell constants are $a = 1.008\,\mathrm{nm}$, $b = 0.361\,\mathrm{nm}$, $c = 1.544\,\mathrm{nm}$, and $\beta = 109.6°$. There are three kinds of vanadium sites and they extend along the $b$ axis as $V_1$–$O_2$–$V_1$–$O_2$, $V_2$–$O_3$–$V_2$–$O_3$, and $V_3$–$O_7$–$V_3$–$O_7$. $V_1$ and $V_2$ are at the body centers of the distorted octahedra which share edges as shown in Fig. 5.40. The $V_3$ site is five coordinated as one oxygen ion is further away, and the V–V separation is longer than Goodenough's critical separation of 0.294 nm for vanadium oxide [622]. Na is in the tunnel which lies along the $b$ axis. The unit cell contains six molecules. As there are four Na sites in the cell, two of them are occupied with Na when $x = 1/3$. Na may be substituted by Li, K, Ca, Cu, Ag, Cd or Pb [623–625], of which the Li bronze is the fast ionic conductor. The cation sites in the tunnel are not of one kind and modifications of the structure can occur at certain temperatures. For instance the period $b$ is doubled below about 200 K in $Na_{0.4}V_2O_5$ [626,627].

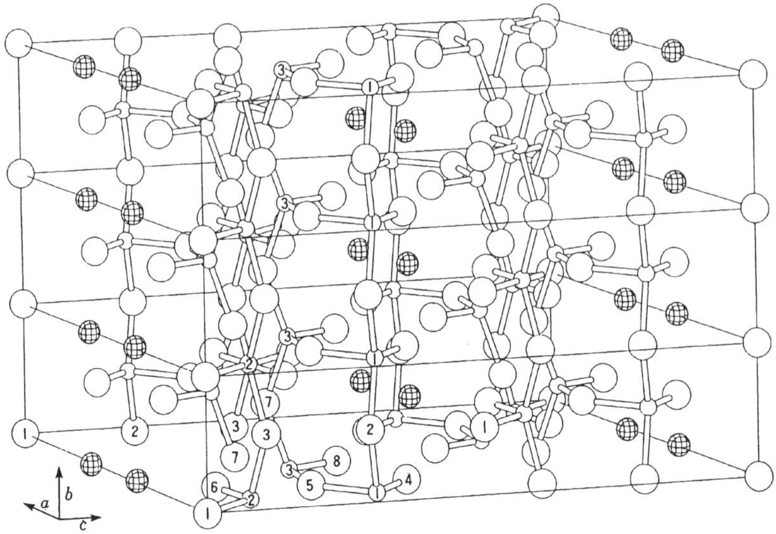

**Fig. 5.40.** Crystal structure of $\beta$-$Na_{0.33}V_2O_5$ [621]. Three units are shown along the $b$ axis. (After A. Yanase)

### 5.5.2 Electronic Properties of Na-Vanadium Bronze

Na is the donor and 1/6 of the V ions are $V^{4+}$ when $x = 1/3$. Thus even if the electrons accumulate on one kind of the three V sites, $V_1$, $V_2$, and $V_3$, they are half occupied and good conductivity would be expected if the electrons were itinerant. $\beta$-$Na_{0.33}V_2O_5$ is, however, not metallic and the conductivity at 300 K is about $10^4$ $Sm^{-1}$ [628, 629]. An ESR study by *Takahashi* and *Nagasawa* shows that most electrons occupy the $V_1$ sites and the remaing few the $V_3$ sites [575, 630]. This was expected by *Goodenough* by the following simple logic [631]. When the conduction electron enters an antibonding orbital, it will search for the one with lowest energy. The lowest antibonding energy means the shallowest bonding energy and thus the longest bond distance. Thus the electrons will enter those vanadium sites which are far from the oxygen. These are the $d_{xy}$ orbitals of $V_1$–$O_5$ (Fig. 5.41) [632]. Later it was found from an analysis of the $g$-factor that $d_{xy}$ is not the lowest orbital, perhaps due to pair formation [575].

The most characteristic feature of this material is its one-dimensional conductivity (Fig. 5.42), which is larger by 100 times along the $b$ axis than perpendicular to it [629]. Figure 5.43 shows that there is a kink in the conductivity at about 140 K which is expressed as $\sigma \propto T^{-1}\exp(-E/k_BT)$. $E$ is 0.049 eV above 140 K and 0.067 eV below that [628].

Another characteristic feature is the low mobility. *Perlstein* and *Sienko* determined the Hall mobility of $\beta$-$Na_{0.33}V_2O_5$ along the $b$ axis to be $1.8 \times 10^{-5} m^2V^{-1}s^{-1}$ at 298 K and $9 \times 10^{-6}$ at 106 K in the same units [628]. This

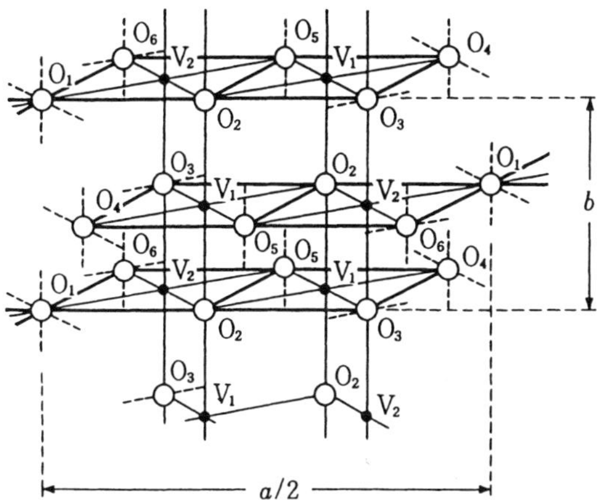


**Fig. 5.41.** Part of the V–O lattices in $Na_xV_2O_5$ [632]. *Dotted lines* through O ions represent $p$ orbitals which interact with the $t_{2g}$ orbitals of V ions. (With permission from Academic Press)

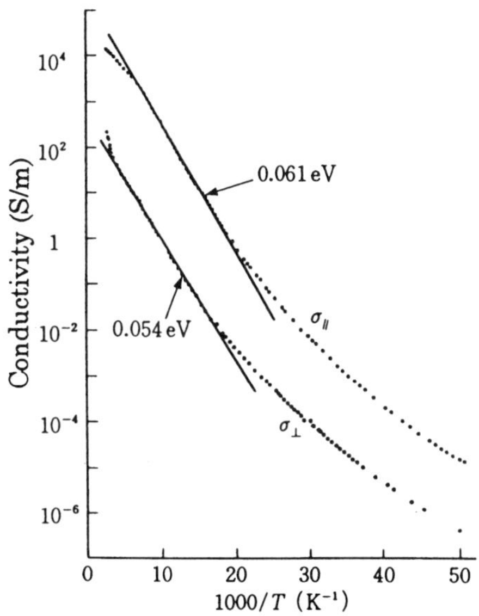


**Fig. 5.42.** Conductivity of single crystalline $Na_xV_2O_5$ [629]. ∥ and ⊥ stand for parallel and perpendicular to the $b$ axis. (With permission from Pergamon)

mobility is so low that the itinerant nature of the electrons must be doubted as was stated in Sect. 5.2.4. The carrier density is $2.6-2.2\times10^{27}\ \mathrm{m}^{-3}$ between 189 K and 298 K and is nearly equal to the Na concentration of $3.8\times10^{27}\ \mathrm{m}^{-3}$, implying that Na behaves as the donor. This is also indicated by the NMR of $^{23}$Na in which the shift is zero [633]. The shift is relative shift of the resonance

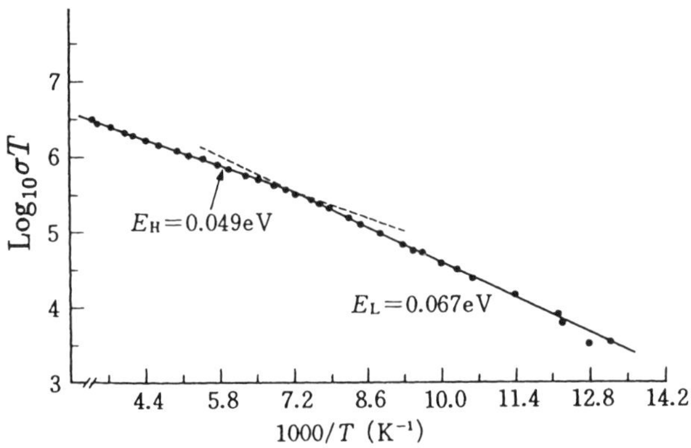


**Fig. 5.43.** Log $\sigma T$ of $\beta$-$Na_{0.33}V_2O_5$ along the $b$ aixs. The activation energy is also shown [628]. (With permission from The American Institute of Physics)

position from the diamagnetic substance. In normal metals it is positive and is known as the Knight shift. Positive means a shift to higher frequency if the applied field is the same. It is expressed as

$$K = -\frac{2}{3}\chi_e|\psi(0)|^2 , \tag{5.55}$$

where $\chi_e$ is the spin susceptibility of the conduction electrons per unit volume and $\psi(0)$ is the wavefunction of the conduction electron normalized in the unit volume at the nucleus. Thus a zero shift means that there are no $s$ electrons on the Na nucleus. The zero shift is also observed for Li [634].

The low mobility for activated-type conduction suggests that the electrons are in a localized state. As will be explained later this may be an inter-site bipolaron. As the V–V separation is large, the mean field band width may be smaller than the electron–phonon or electron–electron interaction. It is not clear how much the electron correlation contributes to localization. *Gunning* et al. suggest the importance of the intersite correlation [635]. In $VO_2$, the importance of electron correlation for pair formation has been stressed, as was explained in Sect. 2.8.

A localized state due to the electron–phonon interaction is called a small polaron. *Holstein*, *Yamashita* and *Kurosawa*, and *Böttger* and *Bryksin* have calculated the mobility of the small polaron as (5.37) [547–551]. Figure 5.43 shows that $\sigma \propto T^{-1}\exp(-E/k_BT)$ and as the carrier concentration is constant, (5.37) seems to hold. However, since the temperature range is 80 K to 250 K, it is not certain whether the necessary condition $\hbar\omega_0 \ll k_BT$ is fulfilled or not (Sect. 5.2.4). This relation is nothing other than the Einstein relation

$$\mu = \frac{eD}{k_BT} . \tag{5.56}$$

Usually the diffusion constant $D$ is expressed using the correlation time $\tau$. In one-dimensional diffusion with a jumping distance $a$

$$D = \frac{a^2}{\tau} = \frac{a^2}{\tau_0} \exp\left(-\frac{E}{k_B T}\right) \tag{5.57}$$

and

$$\sigma = ne\mu = \frac{ne^2a^2}{k_B T \tau_0} \exp\left(-\frac{E}{k_B T}\right), \tag{5.58}$$

where $n$ is the carrier density, $e$ the electronic charge, $\tau = \tau_0 \exp(E/k_B T)$ is the average staying period at one site, $\tau_0^{-1}$ is the attempting frequency, and $E$ is the activation energy. *Nagasawa* and *Onoda* have applied (5.58) to the conductivity of $\beta$-$Na_{0.4}V_2O_5$ with an activation energy $E = 886$ K and $E = 574$ K for $\perp b$ and $\parallel b$, respectively [625,630]. Then the anisotropy in the conductivity is

$$\frac{\sigma_\parallel}{\sigma_\perp} = \frac{n_\parallel a_\parallel^2 \tau_{0\perp}}{n_\perp a_\perp^2 \tau_{0\parallel}} \exp\left(\frac{312}{T}\right). \tag{5.59}$$

At $T = 300$ K, $\sigma_\parallel/\sigma_\perp = 100$. If there is no large anisotropy in $\tau_0$ and $a$, the observed anisotropy will be due to $n$. *Nagasawa* and *Onoda* consider that $n_\perp$ is the concentration of single polarons whereas $n_\parallel$ includes that of the bipolarons on the $V_1$ site [625,630].

*Wallis* et al. analyzed the conductivity in another way. They found that $\log \sigma$ is proportional to $T^{-1/3}$ perpendicular to the $b$ axis and to $T^{-1/4}$ along the $b$ axis (Fig. 5.44) [629], which is characteristic of variable-range hopping conduction (VRH).

VRH is the conduction mechanism of one electron through intermediately localized orbitals in a random potential [636–638]. The conductivity is given by

$$\sigma \propto \exp\left[-\left(\frac{T_0}{T}\right)^{1/(n+1)}\right], \tag{5.60}$$

where $n$ is the dimension of the conduction path. A $T^{-1/2}$ dependence is reported for one-dimensional organometallic substances such as NMP-TCNQ [639] and $T^{-1/4}$ for amorphous semiconductors [638] and for amorphous $H_xWO_3$ [601]. A $T^{-1/3}$ law is found for two-dimensional polycrystalline oxide $YFe_2O_4$ [640]. In Sect. 5.9, other examples will be found. Thus this conduction mechanism seems to be a widely applicable general mechanism. The derivation was made by *Mott* as follows.

The transition probability from site $m$ to site $n$ will be

$$\Gamma_{mn} \propto \exp\left[-2\alpha|r_m - r_n| - \frac{1}{k_B T}(E_n - E_m)\right], \tag{5.61}$$

where $E_m$ is the energy at site $m$, and the wavefunction at site $m$ is expressed as

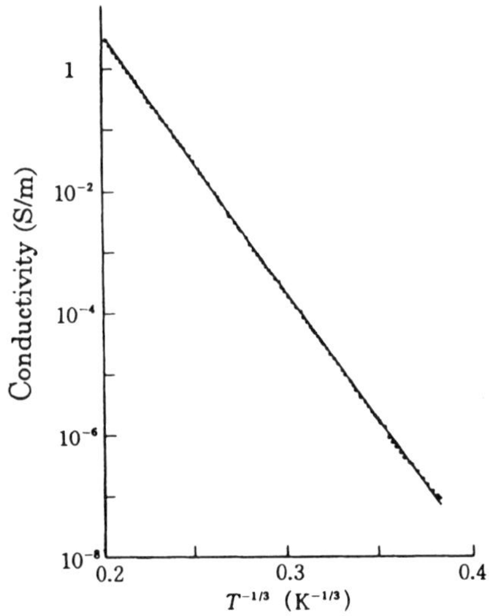


**Fig. 5.44.** Conductivity of $\beta$-$Na_{0.33}V_2O_5$ perpendicular to the $b$ axis [629]. (With permission from Pergamon, Elsevier)

$$\psi(\boldsymbol{r}) \propto \exp(-\alpha|\boldsymbol{r} - \boldsymbol{r}_m|) \,. \tag{5.62}$$

$E_m - E_n$ is taken as the average energy separation in the sphere of radius $R$ which is the maximum hopping distance. Then

$$\Delta E = \overline{E_m - E_n} = \frac{3}{4\pi R^3 N(E)} \,, \tag{5.63}$$

where $N(E)$ is the density of states in the unit volume. The transition probability becomes a function of $R$ and reaches a maximum at

$$R_0 = \left(\frac{9}{4\pi 2\alpha N(E) k_B T}\right)^{1/4} \,. \tag{5.64}$$

The dimensionality comes from the estimation of $\Delta E$. In the above example, a sphere has been postulated. In the case of two dimensions, $\Delta E$ is $1/\pi R^2 N(E)$.

*Wallis* et al. have found such a temperature dependence. However, it is $T^{-1/4}$ along the $b$ axis and not $T^{-1/2}$ which suggests a break in the one-dimensional conduction path. Thus it is not clear whether the observed temperature dependences are inherent in the perfect lattice or extrinsic, that is, originating in the defects.

If the conduction path is broken everywhere, the high frequency conductivity will be larger than the dc conductivity because the break will not be so effective as it is for dc conductivity. At the same time the high frequency conductivity will be proportional to temperature. Following *Austin* and *Mott*, the ac conductivity of the small polaron among equivalent sites will be given by [545]

$$\sigma(\omega) = \sigma(0)\exp(-\omega^2\tau^2)\frac{\sinh\left(\frac{\hbar\omega}{2k_BT}\right)}{\left(\frac{\hbar\omega}{2k_BT}\right)}, \quad \tau \sim \frac{\hbar^2}{Ek_BT}, \tag{5.65}$$

where $E$ is the activation energy. When the hopping sites are inequivalent,

$$\sigma(\omega) \propto \omega\left[\ln\left(\frac{1}{\omega\tau}\right)\right]^4 T, \tag{5.66}$$

where $\tau^{-1}$ is the probability of a downward jump.

### 5.5.3 Magnetic Properties

The magnetic susceptibility is of Curie–Weiss type as shown in Fig. 5.45. The Weiss temperature is antiferromagnetically 150 K and the effective Bohr magneton number is $1.7\mu_B$. Since $2\sqrt{S(S+1)}$ is 1.73 when $S = 1/2$, the electrons transferred from Na to $V^{5+}$ seem to localize. However, the susceptibility which was deduced from NMR is not consistent with the above result. Figure 5.46 shows the shift of the $^{51}V^{4+}$ resonance lines at the $V_1$ site, so the temperature dependence must be due to the temperature-dependent susceptibility $\chi_d$ of a $d$ electron. It was found that $\chi_d$ is expressed by

$$\chi_d = \frac{C}{T}\frac{\exp(-\Delta/T)}{1+\exp(-\Delta/T)}. \tag{5.67}$$

$\Delta$ was determined to be 170 K. Equation (5.67) is given by the bipolaron model of *Chakraverty* [573]. In this model two electrons on neighboring V ions form a spin singlet ground state, called the Chakraverty bipolaron. The triplet excited state is at $J$ and the decomposition of the bipolaron into two single polarons will occur at $\Delta$ above the ground state. If the single polaron obeys the Curie law and $J$ is much larger than $\Delta$, then (5.67) is obtained.

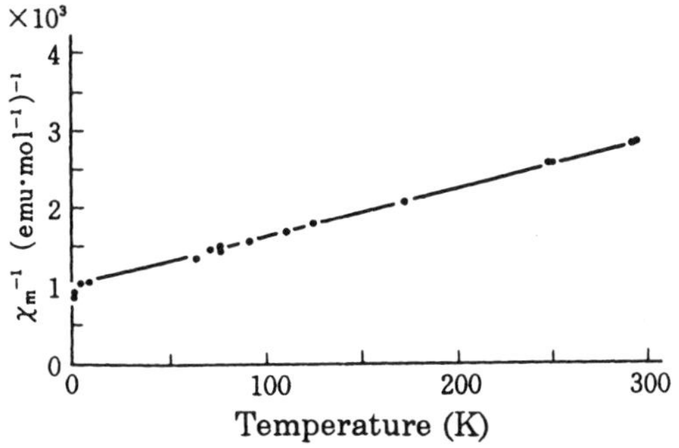


**Fig. 5.45.** Magnetic susceptibility of $\beta$-$Na_{0.33}V_2O_5$ [573]. The Weiss temperature is 150 K. (With permission from The American Physical Society)

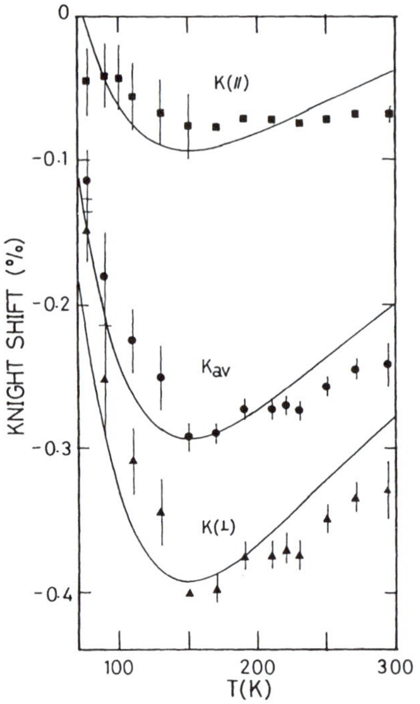


**Fig. 5.46.** Shift of the $V^{4+}$ NMR lines in $\beta$-$Na_{0.33}V_2O_5$ [641]. $\|$ and $\perp$ represent the direction of the applied field with respect to the $b$ axis. Av means the average of the two. The *solid lines* are calculated results. (With permission from The Physical Society of Japan)

*Onoda* et al. postulated that the small number of localized spins at $V_3$ whose Weiss temperature is 18 K may mask the temperature dependence of (5.67) in the macroscopic susceptibility of Fig. 5.45 [641]. The spin susceptibility given from the EPR resonance of $V^{4+}$ also supports this model [630].

*Schlenker* et al. have measured the low-temperature susceptibility and found that the magnetic transition to the antiferromagnetic-like state occurs at 18 K [642]. The easy axis is in the $a$–$c$ plane and it is not a simple antiferromagnetism but shows a magnetic cooling effect. This suggests a spin glass state which is a frozen spin state with each spin axis distributed randomly.

### 5.5.4 Specific Heat

The random potential seems to appear also in the specific heat. *Chakraverty* and *Sienko* found that the specific heat is expressed as $C = \gamma T + \beta T^3$ where $\gamma = 0.052\ \mathrm{mJg^{-1}K^{-2}} = 9.8\,\mathrm{mJmol^{-1}K^{-2}}$, $\beta = 3 \times 10^{-3}\ \mathrm{mJg^{-1}K^{-1}} = 0.57\,\mathrm{mJmol^{-1}K^{-1}}$ (Fig. 5.47) [573]. The characteristic feature is that there is a $\gamma T$ term in this insulating oxide which is also seen in the insulating $Li_{1+x}Ti_{2-x}O_4$ and in the insulating Cu oxides as will be explained in Sect. 5.9.

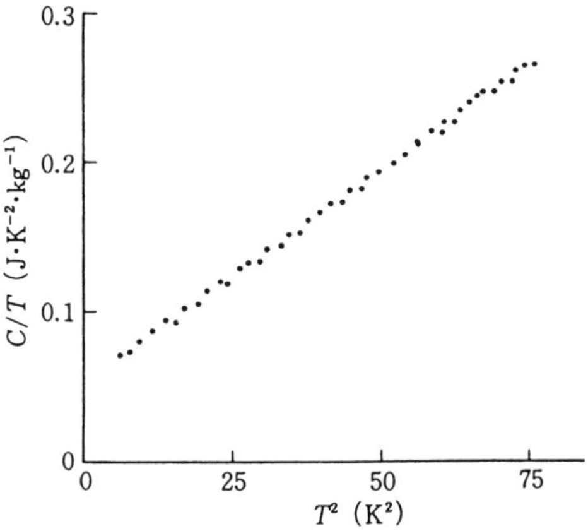


**Fig. 5.47.** Specific heat of $\beta$-$Na_{0.33}V_2O_5$. Note that the $\gamma T$ term is present [573]. (With permission from The American Physical Society)

Usually the $\gamma T$ term is attributed to metallic conduction electrons, giving a non-zero density of states at the Fermi level.

If there are two-level systems separated by an energy $E$, then the specific heat is given by [643]

$$C(E) = k_B N(E) \frac{(E/k_BT)^2 \exp(E/k_BT)}{[1+\exp(E/k_BT)]^2}, \tag{5.68}$$

where $N(E)$ is the density of the pair. Therefore, when $N(E)$ is constant over the energy range, $N_0$, the total specific heat is given after integrating over $E$ as

$$C = \frac{\pi^2}{6} N_0 k_B^2 T, \tag{5.69}$$

which is proportional to temperature. *Anderson* et al. and *Phillips* have explained the temperature-proportional specific heat of amorphous substances with this model [644, 645]. To take into account the randomness, the spatially extended states must be considered otherwise the energy difference will be finite at one site even if the environment of each site is different. If the spatially neighboring states interact with each other the randomness may produce a dense distribution of levels. Thus at least two interacting orbitals are necessary as well as the randomness. *Chakraverty* and *Sienko* postulated that the intersite bipolaron will be formed in this way [573].

### 5.5.5 EPR and NMR in $Na_xV_2O_5$

The EPR absorption spectrum has been measured at 9.3 GHz (X band) [630]. From the intensity of the absorption line, the spin susceptibility was determined and expressed as

$$\chi(T) = \frac{C_\mathrm{p}}{3k_\mathrm{B}T}\frac{\exp(-\Delta/T)}{1+\exp(-\Delta/T)} + \frac{C_i}{3k_\mathrm{B}(T+18)}\,, \tag{5.70}$$

where $C_\mathrm{p} = 0.13\,\mathrm{emuK/mole}$ and $C_i = 0.03\,\mathrm{emuK/mole}$. $\Delta$ is 169 K and is the same as that deduced from the shift of $^{51}$V resonance at the $V_1$ site. Following *Onoda* and *Nagasawa* the first term is due to the single polarons which were excited from the singlet bipolarons at $V_1$ sites, and the second term is due to the single polarons at $V_3$ sites. The *d* electrons are supposed to hop around these sites. The line width which increases with temperature is expressed as

$$\frac{1}{T_2} = \left(\frac{1}{T_2}\right)_\mathrm{dip} + \left(\frac{1}{T_2}\right)_\mathrm{hf}\,, \tag{5.71}$$

neglecting the other relaxation mechanisms (Fig. 5.48) [630]. In (5.71) $T_2$ is the transverse relaxation time of the spin. The first term is due to the magnetic dipole interaction with surrounding electron spins and the second term is due to the hyperfine interaction with the $^{51}$V nuclear moment. When the fluctuation of the effective field on the spin is characterized by the correlation time $\tau_\mathrm{c}$, and when $\omega_z\tau_\mathrm{c} \ll 1$, $1/T_2$ will be

$$\frac{1}{T_2} = [\langle\omega^2\rangle_\mathrm{dip} + \langle\omega^2\rangle_\mathrm{hf}]\tau_\mathrm{c}\,, \tag{5.72}$$

where

$$\langle\omega^2\rangle_\mathrm{dip} = \frac{3}{2}\frac{(g\mu_\mathrm{B})^4}{\hbar^2}S(S+1)\sum_{i\neq j}\frac{1+\cos^2\theta_{ij}}{r_{ij}^6}\,, \tag{5.73}$$

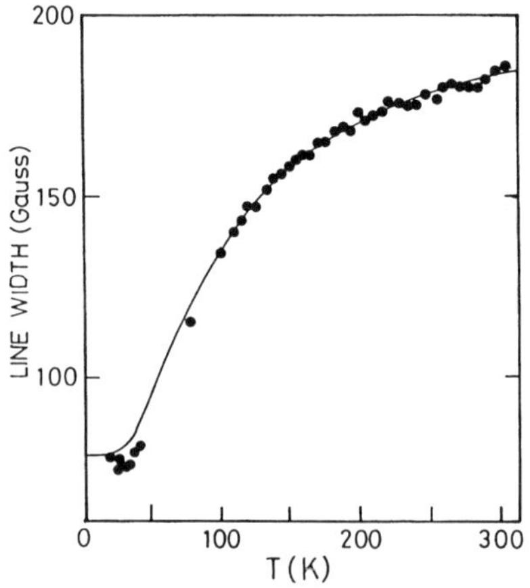


**Fig. 5.48.** Temperature dependence of the width of the EPR line in $\beta$-$Na_{0.33}V_2O_5$ [630]. The *solid line* is the calculated result. (With permission from The Physical Society of Japan)

$\theta_{ij}$ is the angle between the applied field and the $\boldsymbol{r}_{ij}$ vector which connects the $i$ and $j$ spins, $\omega_z$ is the Zeeman frequency, and $\langle\omega^2\rangle_{\rm hf}$ is the second moment of the hyperfine field. *Onoda* and *Nagasawa* have calculated the summation by assuming that the spin is on $V_1$, $O_5$, and $V_3$ sites with the probability $\frac{1}{2}\frac{\exp(-169/T)}{1+\exp(-169/T)}$, 0.025, and 0.075, respectively. $\tau_{\rm c}$ was adjusted to $1.2 \times 10^{-11}$ s. Thus broadening is due to the increase in the number of isolated spins. The solid line in Fig. 5.48 is the result calculated in this way. The agreement is good and at the same time rather curious because $\tau_{\rm c}$ was kept constant in the whole temperature range. As $\tau_{\rm c}$ is the average hopping time, it must decrease rapidly as temperature is raised. Moreover if this $\tau_{\rm c}$ is substituted in (5.58), it gives the temperature-independent mobility. To solve this contradiction *Erata* and *Nagasawa* postulated a collective motion of bipolarons [646].

As shown here the bipolaron model seems to explain various properties of $\beta$-$Na_xV_2O_5$, but, at the same time, a further study is necessary to establish whether bipolarons are formed.

### 5.5.6 Molybdenum Bronzes

In the one-dimensional 3*d* system, the pairing of localized electrons seems to occur without leading to superconductivity, which may have been the initial reason for studying the low-dimensional substance. The Bose condensation of pairs or the formation of an incommensurate charge density wave had been expected [647–651].

In the more extended 4*d* systems, metallic properties appear [652], and the molybdenum oxides especially offer a variety of low-dimensional structures, as shown in Figs. 5.49–5.52. $Pd_{1-x}Li_xO$ is also a one-dimensional metal but superconductivity is not found down to 4.2 K [653].

The blue bronzes $M_{0.3}$MoO$_3$ ($M$ = K, Rb, Tl) are quasi-one-dimensional conductors with maximum conductivity along the $b$ axis of about $2.5 \times 10^5$ Sm$^{-1}$ (K0.3) at 292 K [654]. A metal–insulator transition occurs at about 180 K. *Travaglini* et al. suggested that the transition is a Peierls transition [655] and this was confirmed later. Below the transition temperature, the $I-V$ curve becomes nonlinear and above the threshold field $E_{\rm T}$, there is excess current [656,657]. For instance $E_{\rm T}$ is 0.136 V/cm ($K_{0.3}$). In the nonlinear state, characteristic noise is generated, suggesting a depinning of the CDW. In the X-ray diffraction patterns, there is a satellite reflection line below 180 K. The satellite is at $[0, \pm q_b, 0.5]$ from the main reflection and $q_b = 0.74b^*$ at 150 K ($K_{0.3}$) which is incommensurate with the lattice [658], $b^*$ being the primitive translation in the reciprocal lattice. It becomes commensurate with the lattice at about 90 K with $q_b = 0.75b^*$. The magnetic susceptibility becomes smaller at the transition temperature due to the opening of the gap and the subsequent decrease in the Pauli paramagnetism [659].

In the one-dimensional free-electron system, the density response function, $\chi_q$, is defined by

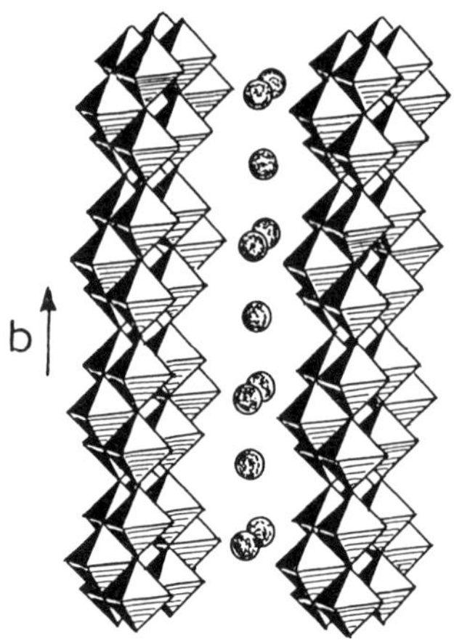


**Fig. 5.49.** Crystal structure of blue $K_{0.3}MoO_3$ [656]. Spheres stand for K and the octahedron is an $MoO_6$ unit. (With permission from Taylor & Francis)

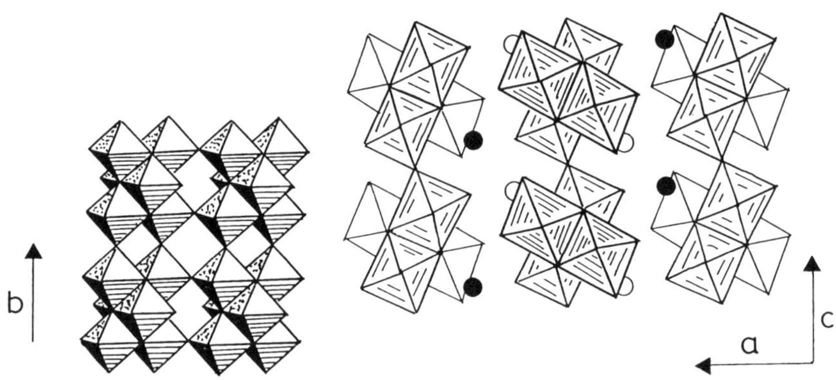


**Fig. 5.50.** Crystal structure of red $K_{0.33}MoO_3$ [680]. It is an insulator. The *right hand* is the projection onto (010). The *solid* and *open circles* stand for K at $y = 1/2$ and 0, respectively

$$\chi_q = -\frac{\rho_q}{V_q} , \tag{5.74}$$

where $V_q$ is the $q$ component of the disturbing potential and $\rho_q$ is the $q$ component of the disturbed electron density [660]. $\chi_q$ diverges logarithmically at $2k_F$ and a CDW may occur. In fact, the experimental $q_b$ value of $K_{0.3}MoO_3$ [661] coincides with the band calculation value [662].

Thus there is a CDW in $K_{0.3}MoO_3$ below 180 K and it does not show superconductivity, due to the pinning or other dissipation process. However in the quasi-one-dimensional $Li_{0.9}Mo_6O_{17}$, superconductivity appears at 1.9 K [663].

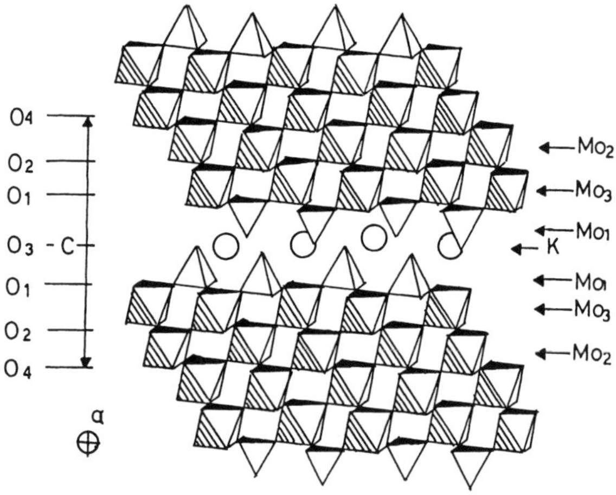


**Fig. 5.51.** Crystal structure of $K_{0.9}Mo_6O_{17}$ [656]. *Spheres* are for K and the *octahedron* is an $MoO_6$. (With permission from Taylor & Francis)

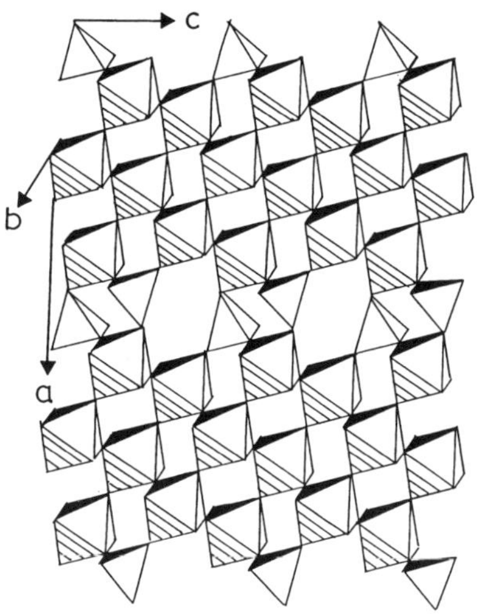


**Fig. 5.52.** Crystal structure of $\eta$-$Mo_4O_{11}$ [675]. (With permission from The Institute of Physics Publishing)

$Li_{0.9}Mo_6O_{17}$ is a quasi-one-dimensional conductor while the other $M_{0.9}Mo_6O_{17}$ ($M$ =K, Na, Tl) are two-dimensional. The conductivity ratio of the former is $\sigma_a : \sigma_b : \sigma_c = 10 : 250 : 1$ [652] where $\sigma_b$ is $1 \times 10^4\,\mathrm{Sm}^{-1}$ at 297 K [663]. The structure of the Li bronze is different from the others [664]. The $MoO_6$ octahedra are connected by $MoO_4$ tetrahedra in the $c$ direction while they are separated by K in the $c$ direction in $K_{0.3}MoO_3$. As the other $M_{0.9}Mo_6O_{17}$ are two-dimensional, this substance may be between $M_{0.9}$- and $M_{0.3}$-bronzes from the point of view of dimensionality and thus localizability.

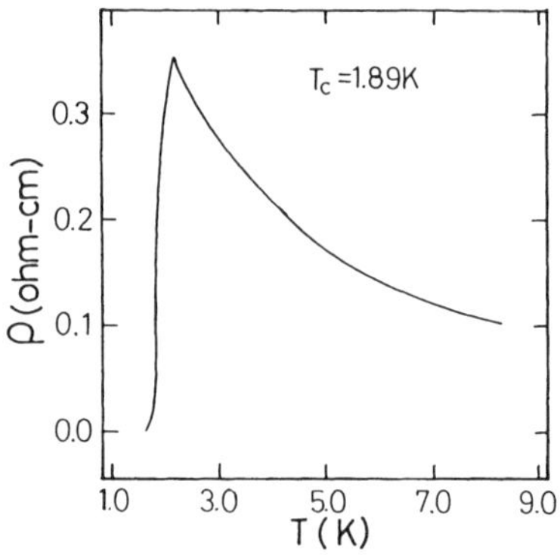


**Fig. 5.53.** Resistivity of $Li_{0.9}Mo_6O_{17}$ [663]. Superconductivity sets in at 1.9 K. $d\rho/dT$ is zero at 24.4 K and positive above that temperature. (With permission from Pergamon, Elsevier)

The resistivity becomes larger as temperature is lowered below 24 K and suddenly at 1.9 K, superconductivity sets in (Fig. 5.53) [663]. The extrapolated conductivity from above 24 K to 0 K is $1.3 \times 10^5\ Sm^{-1}$ along the $b$ axis [665]. This increase in the resistivity resembles that in $K_{0.3}MoO_3$ and suggests a Peierls transition. However, *Greenblatt* has pointed out that the $I$–$V$ curve is linear and the magnetic suceptibility does not show the anomaly below 24 K in contrast to $K_{0.3}MoO_3$ [652, 663]. *Sato* et al. consider that the resistivity increase is not due to the CDW but may be due to the electron–electron interaction in a system in which the periodicity is disturbed [665]. It should be pointed out that in the two-dimensional system, a $\log T$ anomaly may appear in the conductivity and the spin susceptibility [666, 667].

An increase in the resistivity ($d\rho/dT < 0$) is also found for the quasi-one-dimensional $La_2Mo_2O_7$ below 125 K [668, 669]. *Collins* et al. have reported that the conductivity is 1.6, 1.7, and $25 \times 10^3 Sm^{-1}$ along the $a$, $b$, and $c$ axes at 300 K. The magnetic susceptibility drops at about 125 K when the magnetic field is parallel with the $c$ axis whereas the $I$–$V$ curve is linear even below 125 K [669].

The metallic property becomes stronger in the two-dimensional Mo bronzes such as $M_{0.9}Mo_6O_{17}$ ($M$ = Na, K, Tl), and $Mo_nO_{3n-1}$ [652, 656, 670], because the band width becomes larger. CDWs appear at 120 K (K, commensurate), 80 K (Na, commensurate) [656, 671], and 120 K (Tl) [652, 672, 673], respectively. In $\eta$- and $\gamma$-$Mo_4O_{11}$, an incommensurate CDW appears at 109 K and 100 K, respectively [656, 674, 675], whereas for $Mo_nO_{3n-1}$ it is at 315 K ($n = 8$), 500 K ($n = 9$), and 610 K ($n = 10$) [652, 670, 676, 677]. The resistivity begins to increase at the onset temperature of CDW as the temperature is lowered ($d\rho/dT < 0$) but after reaching the maximum, it decreases again ($d\rho/dT > 0$). At the lowest temperature it increases again [671].

A metal–insulator transition occurs at about 103 K in $\kappa$-$Mo_{17}O_{47}$ [678, 679]. Thus in low-dimensional Mo oxides, the Fermi surface is clear and nests, and a CDW occurs suppressing the metallic conductivity, whereas in the low-dimensional high-$T_c$ copper oxides, superconductivity occurs instead of a CDW. It is not clear whether the difference is due to the marginal Fermi liquid nature of the latter or not.

## 5.6 NiO: Origin of the Band Gap and Hole Conduction

According to *Mott*, it was *Verwey* and *de Boer* [681] who first raised the question of why NiO is not a metallic conductor but rather an insulator. According to conventional band theory, which is based on the one-electron approximation, transition-metal oxides such as NiO and CoO have partially filled metal 3*d* band and would therefore be metals. Indeed, binary oxides of 3*d* transition metals from Cr to Cu are good electrical insulators except for $CrO_2$ and $Fe_3O_4$. *Peierls* answered the above question by suggesting that the insulating nature of those compounds is due to Coulomb repulsion between electrons. Following this suggestion, *Mott*, *Hubbard* and others established the notion of the localization of *d* electrons and the splitting of the *d* band caused by the Coulomb repulsion and the notion of the Mott insulator. More recently, it was pointed out on the basis of photoemission studies that the oxygen *p* band is important and that the band gap in NiO consists of the occupied oxygen *p* band and the upper part of the split Ni 3*d* band. On the other hand, much effort has been made to reproduce the band gaps of Mott insulators using band theory, which had been believed to break down in these materials. Recently, the problem of doped holes in NiO has attracted much interest in relation to superconductivity in the hole-doped Cu oxides. This problem is also closely related to the old issue of whether the electrical conduction in NiO is due to hopping conduction or due to band transport.

### 5.6.1 Optical and Magnetic Properties

3*d* transition-metal monoxides with the NaCl structure, MnO to NiO except for FeO ($Fe_xO$ with $x = 0.85 - 0.95$), as well as CuO, $Cr_2O_3$ and $Fe_2O_3$ are antiferromagnetic insulators in spite of the partially filled 3*d* orbitals and have therefore been considered to be Mott insulators. Among them, NiO has been regarded as a prototypical Mott insulator and has been most extensively studied. Since the statement by *Verwey* and *de Boer*, the most fundamental problem has been to elucidate the origin of the insulating band gap.

The magnitude of the band gap of NiO has been determined from optical studies to be about 4 eV. The optical absorption spectrum of NiO shown in Fig. 5.54 has been deduced by a Kramers–Kronig transformation of optical reflectivity spectra. The figure shows that strong absorption, which is probably

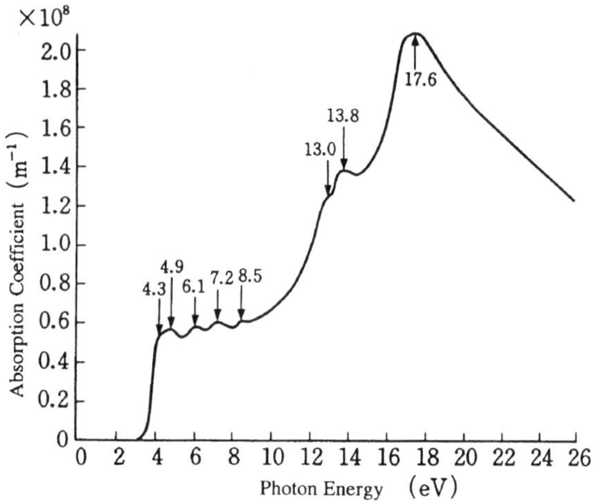


**Fig. 5.54.** Optical absorption spectrum of NiO in the ultraviolet region [682]. (With permission from The American Physical Society)

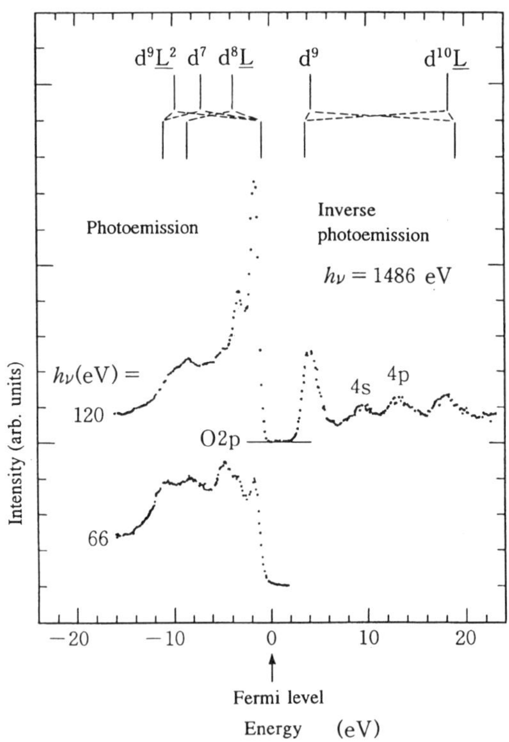


**Fig. 5.55.** Photoemission and inverse-photoemission spectra of NiO [683]. (With permission from The American Physical Society)

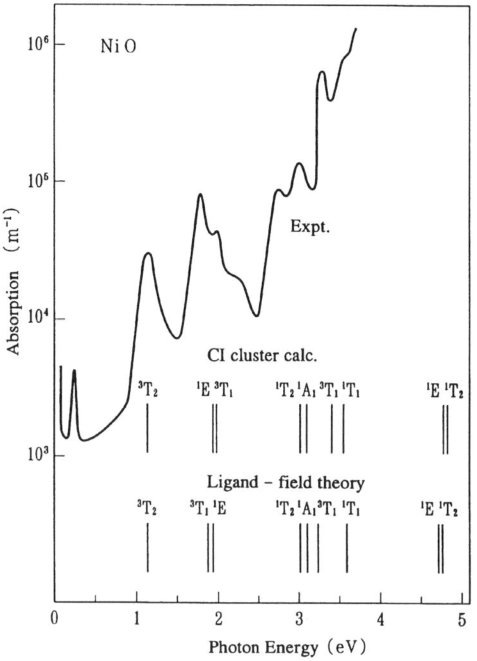


**Fig. 5.56.** Optical absorption spectrum of NiO in the infrared to visible regions [684] and its analyses [698]. (With permission from The American Physical Society)

due to interband transitions, occurs above the threshold at 4 eV. Structures above ~12 eV are commonly observed in all metal oxides and are therefore attributed to transitions from the oxygen $2p$ band to the free-electron-like conduction band (primarily of transition-metal $4s$, $4p$ atomic orbital origin). Structures below 10 eV depend on the transition-metal ions and would involve the $3d$ electrons. While the energy axis of the optical spectrum gives the energy difference between occupied and unoccupied states, the photoemission and inverse-photoemission spectra [683] shown in Fig. 5.55 give the DOS of the occupied and empty states separately. From these spectra, too, the band gap is found to be about 4 eV.

Optical absorption below 4 eV is too weak to be recognized in Fig. 5.54 but consists of sharp peaks as shown in Fig. 5.56 [684]. These peaks are assigned to transitions between the multiplet levels of the $Ni^{2+}$ ion having the $d^8$ configuration in the cubic crystal field. These intra-ionic $d \to d$ transitions are dipole-forbidden and are therefore weak. The presence of such transitions certainly indicates that the $d$ electrons are well localized, confirming that NiO is a Mott insulator. On the other hand, it should be noted that the crystal-field splitting between the $t_{2g}$ and $e_g$ levels, as large as 1.2 eV, indicates that hybridization between the $Ni_{3d}$ and $O_{2p}$ orbitals is substantial. (This crystal-

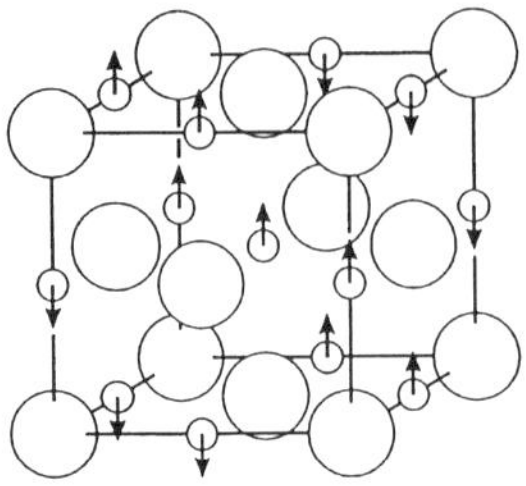

**Fig. 5.57.** Antiferromagnetic spin structure of NiO. The *large circles* are oxygen ions

field splitting is two orders of magnitude larger than that of the $4f$ level of rare-earth ions.)

Below the Néel temperature $T_{\mathrm{N}} = 525\,\mathrm{K}$, the magnetic moment of the $\mathrm{Ni}^{2+}$ ion is ordered in the antiferromagnetic structure as shown in Fig. 5.57. The spins are parallel within each (111) plane and are antiparallel between adjacent (111) planes. The magnitude of the $\mathrm{Ni}^{2+}$ magnetic moment has been obtained to be about $1.8\mu_{\mathrm{B}}$ [685]. Because the $\mathrm{Ni}^{2+}$ ion is in the high-spin ($S = 1$) state, the magnetic moment would be $2\mu_{\mathrm{B}}$ if NiO were a purely ionic insulator. The reduction of $0.2\mu_{\mathrm{B}}$ is due to the $\mathrm{Ni}_{3d}$-$\mathrm{O}_{2p}$ hybridization, which transfers part of the spin polarization of the $\mathrm{Ni}^{2+}$ ion to neighboring $\mathrm{O}^{2-}$ ions. In the magnetic structure shown in Fig. 5.57, the transferred spin density is canceled out on the oxygen atoms by the spin density transferred from $\mathrm{Ni}^{2+}$ ions with opposite spins. The finite $\mathrm{Ni}_{3d}$-$\mathrm{O}_{2p}$ hybridization has also been demonstrated by the electron momentum distribution measured using the positron annihilation technique [686].

### 5.6.2 Transport Properties

Precise control of stoichiometry is generally difficult in transition-metal oxides including NiO. A number of defects are usually present in NiO samples and their transport properties are dominated by impurity states and extra carriers introduced by the defects. It has been controversial for a long time whether the electrical conduction in NiO is due to small polaron hopping or large polaron band transport. If carriers hop from site to site accompanied by local lattice distortion, the mobility $\mu$ of the carriers is small and shows activated temperature dependence:

$$\mu \propto T^{-3/2} \exp\left(-\frac{E_{\mathrm{H}}}{k_{\mathrm{B}}T}\right), \tag{5.75}$$

where $E_{\mathrm{H}}$ is the activation energy for hopping. However, $E_{\mathrm{H}}$ is usually small (of the order 0.01 eV) and the exponential form in (5.75) does not contribute to the mobility except for low temperatures, below $\sim 100\,\mathrm{K}$. In the case of band-like transport due to large polarons, the mobility is large and decreases

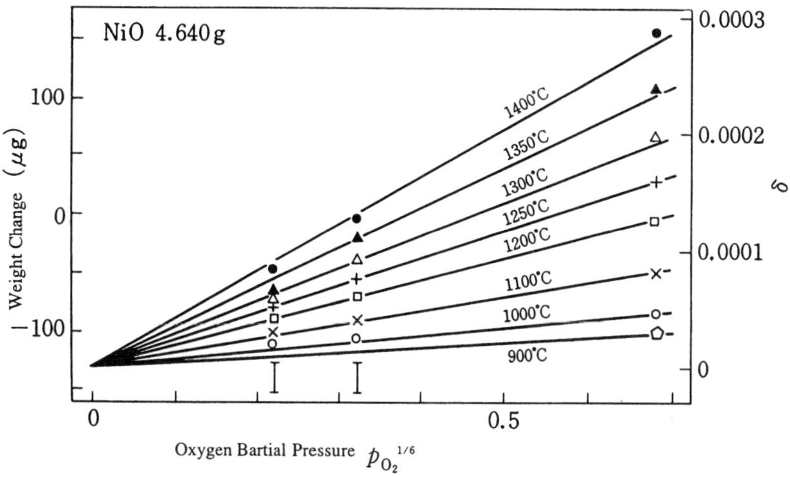


**Fig. 5.58.** Dependence of the $Ni_{1-\delta}O$ stoichiometry on temperature and oxygen partial pressure $p_{O_2}$: $\delta = 0.0107 p_{O_2}^{1/6} \exp(-9400\mathrm{K}/T)$ [687]. *Vertical bars* in the figure represent errors of the experiment. (1 atm $= 1.013 \times 10^5$ N·m$^2$). (With permission from Elsevier)

with increasing temperature. In either case, carriers are trapped by impurities at low enough temperatures and have to be thermally liberated from the impurities to become free polarons. Therefore the carrier number shows the activation-type temperature dependence

$$n \propto \exp\left(-\frac{E_0}{k_B T}\right) . \tag{5.76}$$

Here, $E_0$ is the energy needed to excite a bound polaron to a free polaron. In order to elucidate the conduction mechanism, it is necessary to study the magnitude and temperature dependence of the mobility. Since the electrical conductivity $\sigma$, which can be routinely measured, is the product of the carrier number $n$ and the mobility $\mu$, it is necessary to measure another physical quantity to deduce $n$ and $\mu$ separately.

NiO samples always have excess oxygens compared to the stoichiometric composition. Figure 5.58 gives the relationship between the equilibrium oxygen pressure and the composition of the NiO sample, showing that the amount of excess oxygen increases with temperature, a behavior opposite to that in $Fe_3O_4$ [687]. If the excess oxygens enter the sample as $O^{2-}$ ions, positive holes are introduced as charge carriers. In fact, since the NaCl structure has no space that can accommodate the excess oxygens, Ni vacancies should be created like $Ni_{1-\delta}O$. The vacancy of the $Ni^{2+}$ ion has a charge of $-2e$ compared to the ideal NiO lattice and can therefore bind two positive holes. One can also introduce hole carriers into NiO by substituting Li atoms for Ni since Li is ionized to $Li^+$. The $Li^+$ ion can bind one hole and act as an

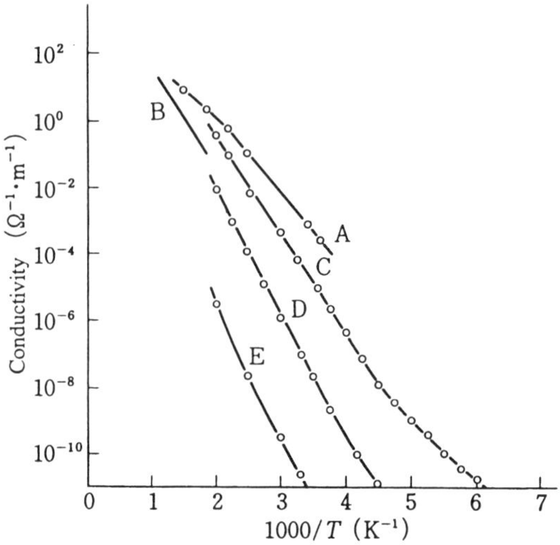


**Fig. 5.59.** Electrical conductivity of NiO single crystals. A: Bernoulli method [688]. B: Halide decomposition method [689]. C: [690]. D: Bernoulli method with annealing [690]. E: Halide decomposition method [690]

acceptor. Such samples can be doped with a high concentration of holes in a controlled way. In semiconductors, when acceptors are introduced, a small number of donors are also created and partially compensate the majority hole carriers. Compensating donors in the Ni-deficient and Li-doped NiO are metal impurities with high valences such as $Fe^{3+}$, $Cr^{3+}$ and $Al^{3+}$. The minority electron carriers will enter the acceptor levels and will shift the Fermi level to the acceptor level.

The electrical conductivity $\sigma$ of NiO single crystals shown in Fig. 5.59 follows $\log \sigma \propto -1/T$ with activation energies of 0.4–0.7 eV. Although not evident from the figure, the activation energy decreases slightly (by ~0.05 eV) above the Néel temperature $T_N = 525\,\mathrm{K}$. Li-doped samples show activation energies of 0.3–0.4 eV. Heavily Li-doped samples (over several %) show even smaller (by 0.1–0.2 eV) activation energies [689]. These activation energies are close to the estimated binding energy of a hole to the $Ni^{2+}$ vacancy or to the $Li^+$ impurity, namely, $e^2/4\pi\epsilon r$ ($\epsilon/\epsilon_0 \sim 5$: dielectric constant ratio; $r \sim 3 - 4\,\text{Å}$: the nearest neighbor distance). This supports the idea that the activated transport is due to the activation of the carrier number rather than that of the mobility [691].

The mobility of carriers in NiO was directly measured by *Spear* et al. [692]. They created electron and hole carriers by irradiating one end of a single crystal using a pulsed electron beam, applied a voltage simultaneously and measured the drift mobility of those carriers. They obtained the hole mobility of about $3.5 \times 10^{-3}\mathrm{m^2V^{-1}s^{-1}}$ at room temperature and $1.5 \times 10^{-3}\mathrm{m^2V^{-1}s^{-1}}$ at 172 K. These large values indicate that the holes enter a wide band (large

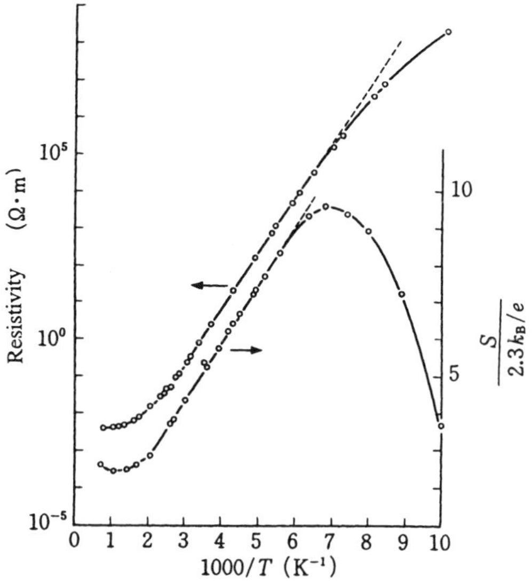


**Fig. 5.60.** Electrical resistivity and Seebeck coefficient of 0.088% Li-doped NiO polycrystal [693]. (With permission from The American Physical Society)

polarons). It is likely that the wide band is the oxygen $2p$ band but one cannot exclude from this experiment alone the possibility that the Ni $3d$ band is as large as 0.1 eV and that the holes enter the Ni $3d$ band. The decrease in mobility with temperature can be explained as due to the scattering of large polarons by optical phonons: $\mu \propto \exp(\hbar\omega_0/kT)$, where $\omega_0 \sim 0.05\,\mathrm{eV}$ is the optical phonon frequency in NiO. No electron conduction was seen in the experiment.

*Bosman* concluded from comparison between the electrical conductivity and the thermoelectric power of Li-doped polycrystalline samples that transport above 200 K is realized by band-like large polarons [693]. If minority carriers are negligible, the Seebeck coefficient is given by

$$S = \frac{k_\mathrm{B}}{e}\left(\frac{E_\mathrm{F}-E_\mathrm{V}}{k_\mathrm{B}T}+\frac{5}{2}+r\right) = \frac{k_\mathrm{B}}{e}\ln\left(\frac{N_\mathrm{c}}{n}\right) + \mathrm{const.}, \tag{5.77}$$

where $E_\mathrm{F}$ is the Fermi level, $E_\mathrm{V}$ is the valence-band maximum, $n$ is the number of hole carriers in the valence band, and $N_\mathrm{c} = 2(2\pi m_\mathrm{e}kT/h^2)^{3/2}$. Therefore, if the temperature dependence of the mobility is weak, as in band transport, $S$ should be nearly proportional to the logarithm of the resistivity $\rho \propto 1/n$. Indeed, Fig. 5.60 shows that $S$ is proportional to $\ln\rho$ in that temperature range.

More recently, however, *Keem* et al. [694] analyzed the Seebeck coefficient and the resistivity of $\mathrm{Ni}_{1-\delta}\mathrm{O}$ in the temperature range $300\,\mathrm{K} < T < 1000\,\mathrm{K}$ by taking into account the compensation of hole carriers by donors and obtained the mobility of $10^{-6}\mathrm{m}^2\mathrm{V}^{-1}\mathrm{s}^{-1}$ at 1000 K and $10^{-9}\mathrm{m}^2\mathrm{V}^{-1}\mathrm{s}^{-1}$ at 300 K. These

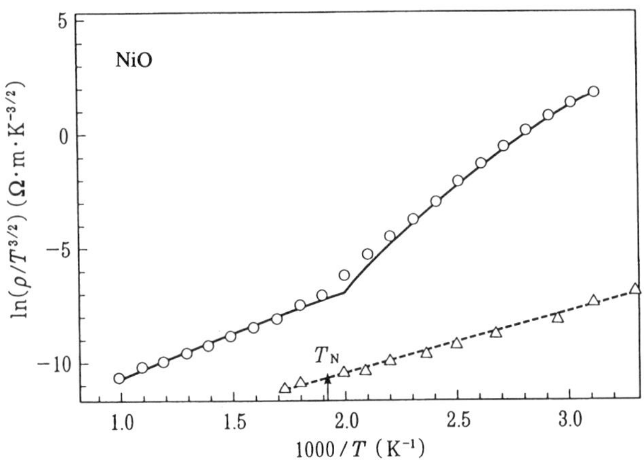


**Fig. 5.61.** Electrical resistivity (○) and Seebeck coefficient $\times e/k_B$ (Δ) of $Ni_{1-\delta}O$ single crystal [694]. (With permission from Taylor & Francis)

values are extremely small, suggesting small polaron hopping. Their results show that $S$ and $\ln\rho$ are not proportional to each other (Fig. 5.61). *Keem* et al. analyzed the conductivity using the formula for small polaron hopping of thermally activated holes:

$$\sigma = ne\mu \propto T^{-3/2} \exp\left[-\frac{E_0 + E_H(T)}{k_B T}\right] \tag{5.78}$$

[see (5.75)] as shown in Fig. 5.61. Here, it has been assumed that the hopping activation energy $E_H$ is related to the antiferromagnetic moment $M(T)$ through

$$E_H(T) = \frac{E_B}{2} - 12J\left[1 - \frac{M(T)}{M(0)}\right], \tag{5.79}$$

where $E_B = A^2/2M\omega^2$ is the small polaron binding energy ($A$: electron–phonon coupling constant [548]; $\omega$: angular frequency of the optical phonon) and $J$ is the superexchange coupling constant between $Ni^{2+}$ ions. The best fit curves shown in Fig. 5.61 have been calculated using the above equations with $E_0 = 0.2\,\text{eV}$, $A^{-1} = 2.7\,\text{eV}$, $J = 0.022\,\text{eV}$ and $\hbar\omega = 0.004\,\text{eV}$. As for the Seebeck coefficient, we set $E_F - E_V = E_0$ in (5.77) because the Fermi level is pinned by the acceptor level owing to compensation.

The temperature dependence of the electrical conductivity becomes weak below $\sim 150\,\text{K}$. Figure 5.62 shows the Arrhenius plot of the electrical conductivity multiplied by $T^{3/2}$ of Li-doped NiO polycrystals [695]. In the range $40\,\text{K} < T < 90\,\text{K}$, the plot obeys (5.78) with temperature-independent $E_H$ and becomes a straight line. From the slope of the line, one obtains $E_0 + E_H \simeq 0.06\,\text{eV}$. This together with $E_0 \sim 0.01\,\text{eV}$ yields $E_H \sim 0.05\,\text{eV}$.

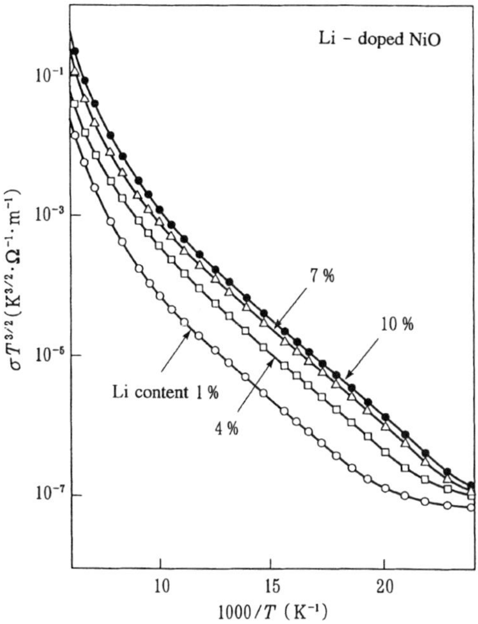


**Fig. 5.62.** Electrical conductivity of Li-doped NiO [695]. (With permission from The Physical Society of Japan)

This $E_0$ value is much reduced from the value $\sim 0.2$ eV at 200–500 K because low-temperature conduction in a compensated semiconductor occurs without activation of free polarons but within the impurity band, i.e., through impurity conduction [693]. Below $\sim 40$ K, variable-range hopping takes place, further weakening the temperature dependence of the electrical conductivity.

Thus the conduction mechanism in NiO still remains controversial as to whether it is due to band-like transport of large polarons or due to small polaron hopping. In order to make a definite conclusion, careful measurements and analyses of well characterized samples are indispensable in future studies.

### 5.6.3 Electronic Structure

Optical and magnetic studies have indicated that $Ni_{3d}$ electrons in stoichiometric NiO are essentially localized in spite of the presence of strong $d$–$p$ hybridization. If the presence of the oxygen $p$ band were ignored, the $d$ electrons would become localized under the condition that the energy $U$ required for the charge transfer $d^8 + d^8 \rightarrow d^7 + d^9$ is larger than the $d$-band width $W$. In order to elucidate the conduction mechanism in NiO, one must understand the character of holes that enter the acceptor levels or the top of the valence band. It has been controversial for a long time whether the holes enter the Ni 3$d$ orbitals or the oxygen 2$p$ orbitals. The electronic states of the top of the valence band are therefore of primary importance. Photoemission spectroscopy is expected to give important information since it probes the electronic structure of occupied states.

Following the successful interpretation of the $d \to d$ optical absorption spectra in the low energy region using ligand-field theory, photoemission spectra of NiO were initially analyzed in the same framework [696]. As shown in Fig. 5.63, emission features within $\sim$ 4 eV of the top of the valence band (called the main structure) were assigned to the final-state multiplet of the $d^8 \to d^7$ transition from the $Ni^{2+}$ ion, and the wide oxygen $p$ band is located at 4–8 eV. However, a satellite appears at an energy position deeper than the oxygen $p$ band (around $\sim$ 8 eV) and its origin could not be identified from the ligand-field analysis. Furthermore, Racah parameters ($B$ and $C$) used for the photoemission spectra had to be chosen much smaller than those used for the optical spectra. Early in the 1980s, NiO was studied using a new technique which utilizes synchrotron radiation, namely, resonant photoemission [697]. In the experiment, the photon energy was chosen in the absorption region from the Ni $3p$ core to the empty Ni $3d$ levels. Then the core absorption $3p^6 3d^8 \to 3p^5 3d^9$ followed by an autoionization process $3p^5 3d^9 \to 3p^6 3d^7 +$

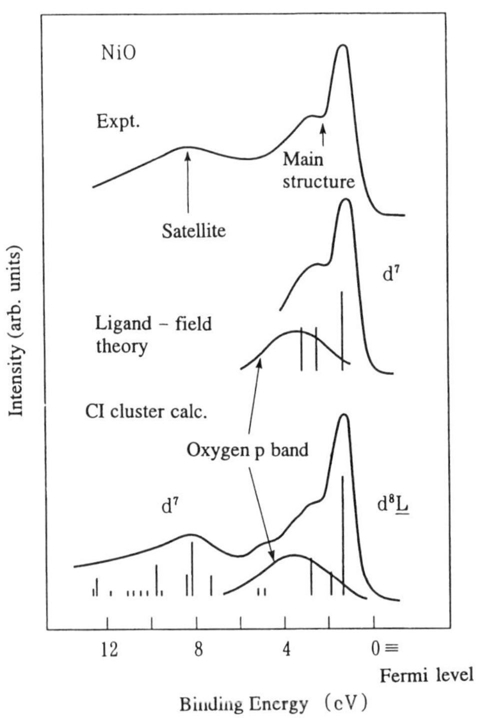


**Fig. 5.63.** Photoemission spectrum of NiO and its analyses [698]. (With permission from The American Physical Society)

photoelectron resonates quantum-mechanically with the ordinary photoemission process $d^8 \rightarrow d^7$ and enhances the cross-section of $d$-electron emission. The experimental result was surprising in that the intensity of the satellite rather than the main structure was enhanced, meaning that the satellite was indeed representative of the Ni $3d$ level, i.e., the $d^8 \rightarrow d^7$ ionization level. The intensity of the main structure, on the other hand, is suppressed at the $3p \rightarrow 3d$ resonance, meaning that the main structure largely comes from oxygen $p$ states (with substantial admixture of Ni $3d$ states). The origin of the main structure, whose line shape resembles that of the $d^8 \rightarrow d^7$ multiplet, is the strong $d$–$p$ hybridization.

In order to take into account the importance of the oxygen $p$ states, analysis of the electronic structure was made using a cluster model consisting of the central $Ni^{2+}$ ion and the surrounding six $O^{2-}$ ions by *Fujimori* and *Minami* [698]. In this model, interaction between the Ni $3d$ electrons and hybridization between the Ni $3d$ and oxygen $2p$ states were considered in a configuration interaction (CI) picture (see Sect. 4.5.2). Starting from the purely ionic configuration $[Ni^{2+}(O^{2-})_6]$, the effect of the $d$–$p$ hybridization is incorporated as a mixture of the $d^9\underline{L}$ configuration ($\underline{L}$: a hole in the $p$ orbitals), in which an electron is transferred from O to Ni, and the $d^{10}\underline{L}^2$ configuration, in which two electrons are transferred from O to Ni. If we define the energy required for the charge transfer $d^8 \rightarrow d^9\underline{L}$ by $\Delta$, charge transfer in the photoemission final state $d^7 \rightarrow d^8\underline{L}$ requires only $\Delta - U$. The above assignment that the top of the valence band is due to emission from oxygen $p$ ($d^8 \rightarrow d^8\underline{L}$) and the satellite is due to emission from Ni $d$ ($d^8 \rightarrow d^7$) means that the $d^7 \rightarrow d^8\underline{L}$ charge-transfer energy $\Delta - U$ is negative: $\Delta < U$. In inverse-photoemission spectroscopy, since there is an additional electron in the final state, we consider $d^9$ and $d^{10}\underline{L}$ configurations. The $d^9 \rightarrow d^{10}\underline{L}$ charge-transfer energy in the final state of inverse-photoemission is $\Delta + U$, which is large enough and pushes $d^{10}\underline{L}$ well above $d^9$. The band gap is defined as the minimum energy to create an electron and a hole separately, and is therefore equal to the gap between the photoemission and inverse-photoemission spectra in Fig. 5.55. This gap corresponds to the charge-transfer excitation $d^8 + d^8 \rightarrow d^9 + d^8\underline{L}$ (if the $d$–$p$ hybridization is ignored) and is given by $\Delta$. More precisely, the magnitude of the band gap is given by (4.44), in which multiplet effects, energy shifts due to the hybridization, the $d$-band width $W_d$ and the $p$-band width $W_p$ are taken into account. According to the classification of the electronic structure described in Sect. 4.5.1, NiO belongs to the charge-transfer regime and has a charge-transfer type band gap.

In Fig. 5.63, cluster-model analysis of the photoemission spectrum is also shown. Best fit parameters are $\Delta \simeq 4\,\mathrm{eV}$, $U \simeq 7\,\mathrm{eV}$ and the $d$–$p$ transfer integral $T_e \simeq 2.2\,\mathrm{eV}$. The spectral line shape has been well reproduced with the $d^8 \rightarrow d^8\underline{L}$ main structure and the $d^8 \rightarrow d^7$ satellite. Using these parameter values, mixing of the $d^9\underline{L}$ configuration (in which Ni has a magnetic moment of $1\ \mu_B$) into the $d^8$ configuration (in which Ni has a magnetic moment of

2 $\mu_B$) is calculated to be about 20%, which explains the experimental Ni moment 1.8$\mu_B$ deduced from neutron diffraction experiments. The top of the valence band has 60% $d^8\underline{L}$ character and 40% $d^7$ character. The difference between these values and values for the ground state ($d^8$: 80%, $d^8\underline{L}$: 20%) indicates that a doped hole has nearly equal weights of the Ni $d$ and oxygen $p$ character. Therefore, the classical picture that the doped holes go into the Ni 3$d$ orbital and turn $Ni^{2+}$ into $Ni^{3+}$ is not correct although they do not go exclusively into the oxygen $p$ orbitals.

If one extends the cluster model to the Anderson-impurity model, in which the oxygen $p$ band is treated as a continuum, then a discrete state of $d^8\underline{L}$ character is split off from the top of the oxygen $p$ valence band into the band-gap region due to strong hybridization with the $d^7$ state [699]. In a periodic lattice, this discrete state forms a band with $\boldsymbol{k}$-dispersion and doped holes initially enter this band. The width of the band is narrow, however, because hopping of holes between nearest-neighbor Ni ions is prohibited by the antiparallel alignment of the Ni spins and only next-nearest-neighbor hopping is allowed. Indeed, the dispersional width observed by angle-resolved photoemission spectroscopy is as narrow as $\sim$ 0.1 eV as shown in Fig. 5.64 [700]. The effective band width in the charge transport process is expected to be

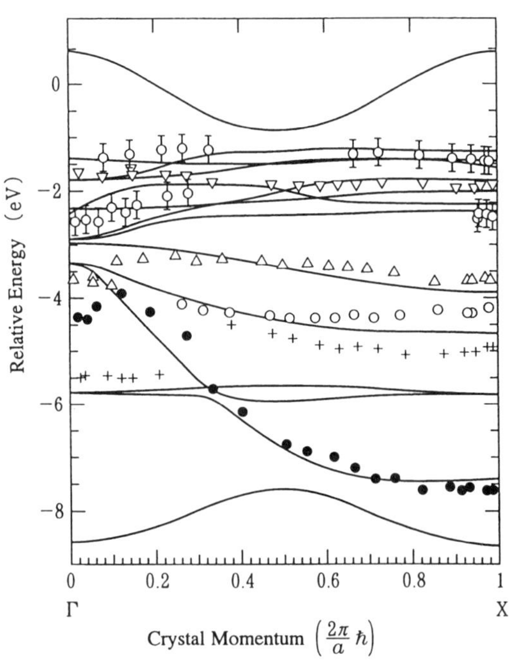


**Fig. 5.64.** Band dispersion of NiO obtained by angle-resolved photoemission spectroscopy [700]. The *solid curves* represent energy bands of the antiferromagnetic state obtained by LDA band-structure calculation. (With permission from The American Physical Society)

even narrower because phonons cannot screen the rapid ($\sim 10^{-15}$ sec) photoemission process. The acceptor levels are expected to be strongly localized because they originate from the narrow band.

The $d \to d$ optical absorption is considered as excitations within the $d^8$ configuration, which are mixed with the $d^9\underline{L}$ configuration located above $d^8$ and separated by $\sim \Delta$. The result of the cluster-model analysis shown in Fig. 5.56 reproduces the experimental data well. The widths of some absorption peaks are smaller than 0.1 eV because the $d \to d$ excited states can hop between Ni sites only via two-electron hopping, whose transfer integrals would be much smaller than those of single-electron (hole) hopping. Considering the fact that lattice vibrations contribute significantly to the spectral widths, the intrinsic widths of the $d \to d$ absorption peaks may be of order 0.01 eV. That is, as far as the $d \to d$ optical absorption is concerned, it appears as if the $d$-band width in NiO is extremely narrow.

The magnitude of the superexchange interaction $J$ ($\simeq 0.02$ eV) and hence the Néel temperature $T_{\rm N}$ has been quantitatively explained by *Zaanen* and *Sawatzky* using the parameters $\Delta$, $U$ and $T_e$ thus obtained [699]. In previous studies, however, the Hubbard model had been applied to estimate the $J$ value of NiO: $J = 4t^2/\Delta$ (where $U$ has been replaced by the charge-transfer gap $\Delta$ because NiO is indeed a charge-transfer insulator). This leads to $t \sim 0.2$ eV. The $d$-band width is then estimated to be as large as $W_d = 12t \sim 2$ eV. To summarize, the estimated $d$-band width ranges from 0.01 eV to 2 eV. The $d$-band width relevant to the transport properties is the dispersional width of a single doped hole, which may be reduced by polaron effects compared with the dispersional width observed by the angle-resolved photoemission study.

### 5.6.4 Electronic Structure of Acceptor Level

Here, we consider the acceptor level induced by Li substitution for Ni. Acceptor levels induced by Ni vacancies are expected to behave similarly to the Li acceptor level except that the Ni vacancy is capable of trapping two holes while the latter can trap only one hole. Figure 5.65 shows X-ray absorption spectra from the oxygen $1s$ core level to the unoccupied states of Li-doped NiO [701]. The absorption peak induced by Li substitution which is located 3–4 eV below the bottom of the conduction band (the $d^9$ peak) of NiO is attributed to absorption from the O $1s$ core level to the acceptor level of $d^8\underline{L}$ character ($d^8\underline{L} \to d^8$ transition). The wavefunction of the acceptor level consists of the wavefunctions of states near the top of the valence band, which have mainly $d^8\underline{L}$ character strongly hybridized with the $d^7$ states. Since the band width is narrow at the top of the valence band, the wavefunction of the acceptor level would be strongly localized around the impurity. The positive charges of the trapped hole will be mainly distributed in the $p$ orbitals of oxygen atoms neighboring the Li impurity.

It should then be clarified whether a doped hole is distributed over the six nearest-neighbor oxygen atoms around the Li impurity or localized at one of these oxygen atoms. The ac conductivity in the MHz to GHz frequency region indicates absorption which can be assigned to the hopping of holes between oxygens surrounding the Li impurities [702]. In addition, a cluster-model calculation allowing for atomic displacement has been made using the Hartree–Fock approximation and shows that a doped hole is localized at one of the oxygen atoms [703].

The experimental fact that $Li_xNi_{1-x}O$ remains semiconducting up to the solubility limit, $x \sim 0.5$, indicates that overlap between neighboring acceptor orbitals is small. $K_2NiF_4$-type $La_2NiO_4$ has $Ni^{2+}$ ions and the local electronic structure of the $NiO_6$ cluster is nearly identical to that in NiO. One can dope holes into $La_2NiO_4$ by Sr substitution for La, but $La_{2-x}Sr_xNiO_4$ remains semiconducting up to $x \sim 1$ [704]. This is contrasted with the isostructural $La_{2-x}Sr_xCuO_4$ (a high-$T_c$ superconductor with the maximum $T_c$ of $\sim 40\,K$), which already becomes metallic (superconducting) at $x \sim 0.05$.

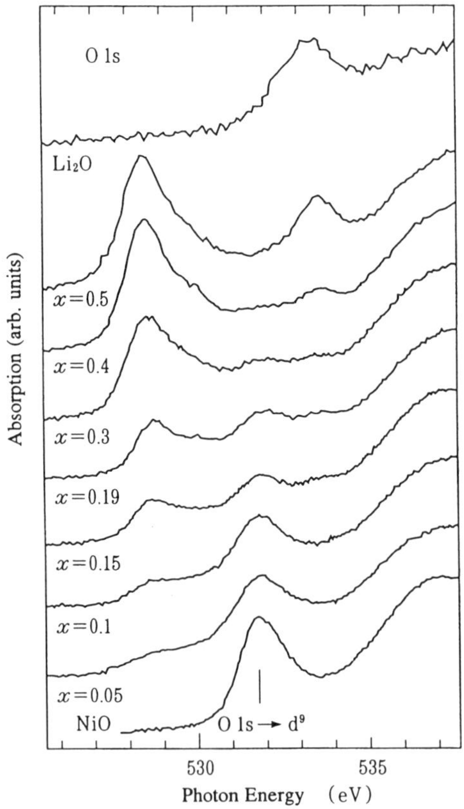


**Fig. 5.65.** Oxygen 1$s$ core-level X-ray absorption spectra of $Li_xNi_{1-x}O$ [701]. (With permission from The American Physical Society)

### 5.6.5 Band Theory of Mott Insulators

In the preceding sections, we have shown that, in order to explain the band gap, photoemission and inverse-photoemission spectra, the $d \to d$ absorption spectra and the electronic structure of doped holes in NiO, one has to treat explicitly electron–electron interaction and electron correlation using the cluster model or the Anderson-impurity model within the CI picture. Conventional band-structure calculations using the local-density approximation (LDA) to the exchange-correlation potential are unable to predict the above properties of NiO. This failure of the LDA calculations has been taken as an indication that electron correlation is important in NiO. Nevertheless, in order to correctly predict the insulating behavior and the magnitude of the band gap of NiO, various efforts have been made within band theory without resorting to an explicit treatment of electron correlation beyond the one-electron approximation.

It is well known that spin-unpolarized LDA cannot predict the insulating behavior of NiO because the Fermi level then lies within the $e_g$ band of Ni. If one introduces the spin polarization and assumes the correct antiferromagnetic structure, a band gap of $\sim 0.3\,\mathrm{eV}$ is opened between the $e_{g\uparrow}$ and $e_{g\downarrow}$ bands (see Fig. 5.66) [705]. In MnO (having $d^5$ configuration), a gap is opened between the $e_{g\uparrow}$ and $t_{2g\downarrow}$ bands. However, the same scheme is not able to make CoO and FeO ($d^6$ and $d^7$, respectively) insulating because the Fermi level necessarily lies within one of the $t_{2g\uparrow}$, $t_{2g\downarrow}$, $e_{2g\uparrow}$ and $e_{2g\downarrow}$ bands. *Terakura* et al. attributed the absence of the gap to the absence of orbital polarization (i.e., splitting of the $t_{2g\uparrow}$ band into $\xi_\uparrow$, $\eta_\uparrow$ and $\zeta_\uparrow$, see Fig. 4.1) [705]. Indeed, *Norman* performed LDA band-structure calculations using the potential which is constructed from the charge distribution of orbitally polarized states, and deduced band gaps of 0.3–0.4 eV for FeO and CoO [706].

It has thus been shown that band gaps are opened in Mott insulators within LDA, but these band gaps are one order of magnitude smaller than the experimental values (Sect. 4.3.2). In LDA, the occupied and unoccupied states both feel the same potential which has been determined by the electronic density distribution, and therefore an occupied electron should feel the potential which has been produced by that electron itself. That is, the potential includes self-interaction, resulting in an unphysical situation. If the self-interaction is removed, then the potential experienced by the occupied $d$ electron is lowered by $U$, the so-called Mott–Hubbard correction, increasing the band gap from the too small value obtained from the LDA calculations to a realistically large value.

Another method to remove the self-interaction from LDA called self-interaction-corrected LDA (SIC-LDA) has been proposed [707]. In SIC-LDA, self-interaction is subtracted from the LDA potential:

$$V_i^{\mathrm{SIC}}(\boldsymbol{r}) = V^{\mathrm{LDA}}(\boldsymbol{r}) - \frac{e\int \mathrm{d}\boldsymbol{r}' n_i(\boldsymbol{r}')}{|\boldsymbol{r}-\boldsymbol{r}'|} - E_{\mathrm{xc}}[n_i(\boldsymbol{r}),0]\,, \tag{5.80}$$

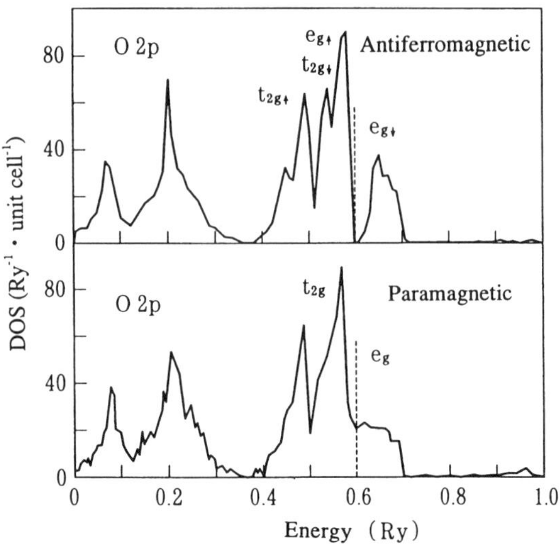


**Fig. 5.66.** Density of states of NiO derived from LDA band-structure calculations [705]. (With permission from The American Physical Society)

where $n_i(\mathbf{r})$ is the density of the electron in state $i$, and $E_{\mathrm{xc}}[n_\uparrow, n_\downarrow]$ is the exchange correlation energy for the electron densities $n_\uparrow$ and $n_\downarrow$ of spin-up and spin-down electrons, respectively. The second and third terms of (5.80) are the SIC for the Coulomb and exchange interactions, respectively. If state $i$ is a Bloch state extended over the whole crystal, then the density $n_i(\mathbf{r})$ becomes zero and hence the SIC vanishes. If $i$ is a localized state, then the SIC becomes finite and lowers the energy of the electron system (5.80). This drives the localization of the $d$ electrons and the lowering of the occupied $d$ level below the oxygen $p$ band, resulting in a large band gap of several eV between the occupied $p$ band and the empty $d$ levels. This $d$–$p$ character of the band gap is in accordance with the result of the photoemission studies combined with CI cluster model analyses. According to the SIC-LDA calculation, the magnitude of the band gap of NiO is 2.5 eV and the Ni magnetic moment is $1.5\,\mu_{\mathrm{B}}$, which are rather close to the experimental values of 4 eV and $1.8\,\mu_{\mathrm{B}}$ (cf. $1.0\,\mu_{\mathrm{B}}$ from LDA), respectively.

Another approach to eliminate the self-interaction is the LDA+$U$ method proposed by *Anisimov* and coworkers [708]. The potential for the $d$ orbital denoted by $\mu$ with spin $\sigma$ is given by

$$V_{\mu\sigma}^{\mathrm{LDA+U}}(\boldsymbol{r}) = V^{\mathrm{LDA}}(\boldsymbol{r}) + \sum_{\mu'} U(n_{\mu',-\sigma} - n_0) + \sum_{\mu' \neq \mu} (U - J)(n_{\mu',\sigma} - n_0)\,, \tag{5.81}$$

where $n_0$ is the average occupation number of the $d$ orbitals, $U$ is the intra-atomic Coulomb repulsion energy and $J$ is the intra-atomic exchange integral.

Here, the potential $V^{\mathrm{LDA}}(\boldsymbol{r})$ does not include the spin and orbital polarization. Owing to the second and third terms in (5.81), it is guaranteed that each $d$ electron interacts with electrons in the other spin-orbitals and does not interact with itself. Therefore, the energy of the electron system is lowered when the $d$ electrons are distributed in the same spin-orbitals as much as possible, which tends to enhance the spin and orbital polarization. As a result, in the case of NiO, the occupied $d$ band is pulled down below the oxygen $p$ band, and the band gap is formed between the $p$ band and the unoccupied $d$ band. The magnitude of the band gap was found to be 3.1 eV and the Ni magnetic moment to be $1.6\mu_{\mathrm{B}}$, in much better agreement with experiment than the LDA calculation. It should be noted that the $d$ states form Bloch states in the LDA+$U$ scheme, unlike the localized $d$ states in the SIC-LDA scheme.

We have shown that the band picture breaks down in the SIC-LDA approach because the $d$ electrons become localized after SIC. On the other hand, the band picture is preserved in the LDA+$U$ and Hartree–Fock methods, which correctly treat the exchange interaction and exclude the self-interaction within the one-electron approximation, thereby yielding the large band gaps (as shown in Sect. 4.4) [467]. That is, the breakdown of band-structure calculations in Mott insulators does not stem from the one-electron approximation of band theory itself but from the LDA, in which the nonlocal exchange potential is approximated by the local potential. Therefore, it is not strictly correct to claim that electrons are localized in NiO and other Mott insulators due to electron correlation. The Mott–Hubbard splitting of the $d$ band stays within the framework of the one-electron picture and should not be regarded as an electron correlation effect. However, as soon as an extra electron(s) or hole(s) is introduced in the Mott insulator, the band structure of the parent insulator is modified strongly because of electron correlation induced by the extra particle(s). In order to describe the behavior of doped carriers and the band dispersions studied by angle-resolved photoemission spectroscopy, one has to go beyond one-electron band theory and treat electron correlation explicitly.

As we have shown above, the origin of the band gap in NiO and other Mott insulators has been understood within the CI treatment of the cluster model or the impurity model as well as within the one-electron band theory, which correctly treats the nonlocal nature of the one-electron potential. However, more complicated electron correlation comes into the problem when the behavior of doped holes and their conduction mechanism are concerned. Electron–phonon interaction and randomness introduced by impurities in real materials further complicate the elucidation of the conduction mechanism in NiO. From the experimental side, studies of samples with controlled defects will be important in future.

## 5.7 Perovskite-Type Mn Oxides: Magnetoresistance

The perovskite-type antiferromagnetic insulator $A\mathrm{MnO_3}$, where $A$ = La, Pr, Nd, ... is a trivalent rare-earth ion, becomes a ferromagnetic metal when it is doped with a sufficient amount of extra holes through Ca or Sr substitution for the $A$ ions. Above the Curie temperature $T_\mathrm{C}$, the hole-doped $A\mathrm{MnO_3}$ behaves as a paramagnetic insulator, indicating a strong coupling between magnetic and transport properties. The resistivity around $T_\mathrm{C}$ is strongly reduced by an external magnetic field, exhibiting a huge negative magnetoresistance called "colossal" magnetoresistance. The origin of this magnetoresistance was originally explained by Zener's double exchange mechanism [709] but there is now growing evidence that electron–phonon interaction, that is, polaronic effects additionally play important roles. With small $A$ ions, the Mn $3d$ conduction bandwidth is reduced and conduction electrons or holes are localized in charge-ordered states, where the $d$ orbitals of Mn also form a superlattice.

### 5.7.1 Crystal Structure

The complicated crystal structure of the undoped antiferromagnetic insulator $A\mathrm{MnO_3}$ can be understood starting from the ideal cubic perovskite structure (Fig. 2.14) and considering a combination of two types of structural distortions: one is the co-operative rotation of the $\mathrm{MnO_6}$ octahedra around different axes, leading to the orthorhombic, so-called $\mathrm{GdFeO_3}$-type structure as shown in Fig. 5.67 a. The orthorhombic unit cell contains four $A\mathrm{MnO_3}$ molecules with lattice parameters $a \simeq b \simeq \sqrt{2}a_\mathrm{p}$ and $c \simeq 2a_\mathrm{p}$, where $a_\mathrm{p}$ is the lattice

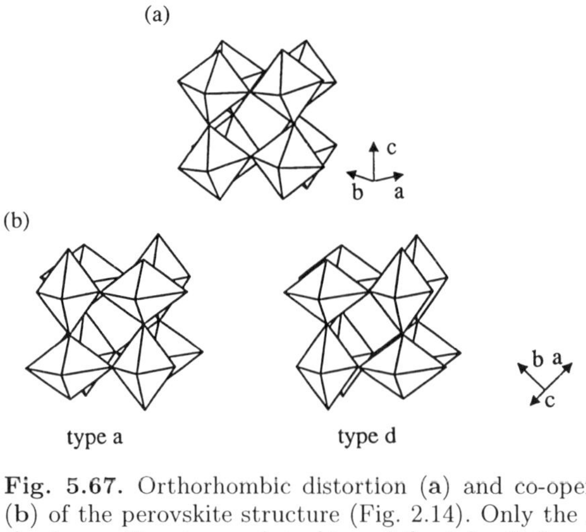


**Fig. 5.67.** Orthorhombic distortion (**a**) and co-operative Jahn–Teller distortion (**b**) of the perovskite structure (Fig. 2.14). Only the $B\mathrm{O_6}$ octahedra are shown in the figure. Two-types of stacking of the Jahn–Teller distorted $ab$-planes are shown in (**b**)

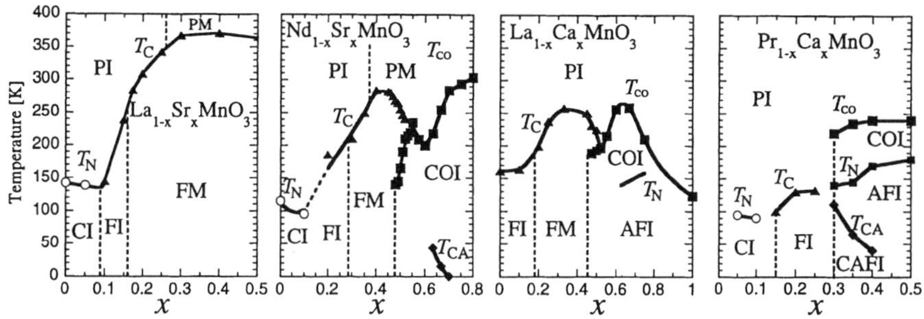


**Fig. 5.68.** Phase diagrams of $La_{1-x}Sr_xMnO_3$, $La_{1-x}Ca_xMnO_3$, $Pr_{1-x}Ca_xMnO_3$. CI: antiferromagnetic insulator with spin canting, FI: ferromagnetic insulator, FM: ferromagnetic metal, PM: paramagnetic metal, AFI: antiferromagnetic insulator, COI: charge ordered insulator [710]. (With permission from The American Physical Society)

parameter of the cubic perovskite. (Note that some literature follows the traditional convention that $a \simeq c \simeq \sqrt{2}a_p$ and $b \simeq 2a_p$.) The other distortion is the co-operative Jahn–Teller distortion, namely, an alternative elongation of the $MnO_6$ octahedra along the $x$ and $y$ directions, which are in the [110] and $[1\bar{1}0]$ directions of the orthorhombic unit cell as shown in Fig. 5.67 b.

$A$MnO$_3$ is a so-called $A$-type antiferromagnetic insulator (with slight spin canting, leading to a weak ferromagnetism), in which the Mn spins are aligned ferromagnetically within the $ab$-planes while the ferromagnetic $ab$-planes are stacked along the $c$-axis with alternating spin directions. By substituting $Sr^{2+}$ or $Ca^{2+}$ ions for the trivalent rare-earth $A^{3+}$ ions in $A$MnO$_3$, one can introduce holes into the system and make it conductive. First, it becomes a ferromagnetic insulator in a narrow composition range and then a ferromagnetic metal, as shown in Fig. 5.68. In the ferromagnetic metallic phase, the Jahn–Teller and orthorhombic distortions disappear and the structure becomes nearly cubic (with a weak rhombohedral distortion arising from the rotation of the $MnO_6$ octahedra around the [111] axis of the cubic unit cell). Note, however, that the orthorhombic-to-rhombohedral transition and the insulator-to-metal transition do not exactly coincide.

### 5.7.2 Ferromagnetic Metal–Paramagnetic Insulator Transition

Figure 5.69 shows the electrical resistivity of $La_{1-x}Sr_xMnO_3$ with various hole concentrations $x$ [711]. Resistivity decreases with increasing $x$. Above $x \simeq 0.15$, the resistivity suddenly drops and the system becomes metallic below the Curie temperature $T_C$, marked by arrows in the figure. The same materials remain "insulating" above $T_C$, i.e., $d\rho(T)/dT$ remains negative above $T_C$ up to $x <\sim 0.3$. That is, there is a temperature-dependent metal-insulator transition in the range $0.15 < x < 0.3$, as depicted in the phase diagram of Fig. 5.68 [710]. The most remarkable phenomenon among

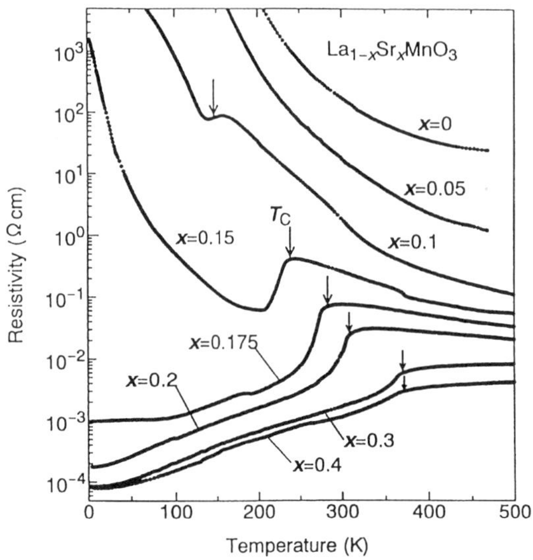


**Fig. 5.69.** Electrical resistivity of $La_{1-x}Sr_xMnO_3$ [711]. (With permission from The American Physical Society)

the transport properties of the hole-doped $AMnO_3$ is their extreme sensitivity to an external magnetic field as shown in Fig. 5.70 [710, 712]. The figure shows that the magnetoresitance is particularly pronounced around $T_C$. The saturation magnetization is close to $(4 - x)\mu_B$ per Mn atom and is equal to the $d$-band occupation number, meaning that the Mn $3d$ band is fully spin-polarized.

It is well known that, when the ionic radius of the alkali-earth or rare-earth ion at the $A$ site decreases in the perovskite-type lattice, the degree of orthorhombic distortion increases. As the average ionic radius $\langle r_A \rangle$ of the $A$-site ions decreases, the composition range of the ferromagnetic metallic phase shrinks as seen from the phase diagrams in Fig. 5.68. (Here, the ionic radii are such that $r_{Sr^{2+}} > r_{Ca^{2+}}$ and $r_{La^{3+}} > r_{Pr^{3+}} > r_{Nd^{3+}}$.) At the same time, the temperature range of the ferromagnetic metallic phase shrinks. In $Pr_{1-x}Ca_xMnO_3$, which has the smallest $\langle r_A \rangle$ in Fig. 5.68, the ferromagnetic metallic phase disappears from the phase diagram. The phase diagram of the $A_{0.7}A'_{0.3}MnO_3$ system with fixed hole concentration, where $A'$ is a divalent Ca or Sr ion, is plotted against the temperature and $\langle r_A \rangle$ in Fig. 5.71 [713]. The ferromagnetic metallic phase is realized above a critical $\langle r_A \rangle$ value. These observations already imply a great sensitivity of the magnetic and transport properties to the crystal structure. As $\langle r_A \rangle$ decreases, the orthorhombic distortion increases and hence the Mn-O-Mn bond angle $\theta$ decreases, reducing the Mn $d$ band width $W$ according to $W \propto \cos\theta$. Therefore, we may refer to perovskite-type Mn oxides with large and small $\langle r_A \rangle$ as wide-band and narrow-band compounds, respectively. As $\langle r_A \rangle$ decreases, the $T_C$ decreases,

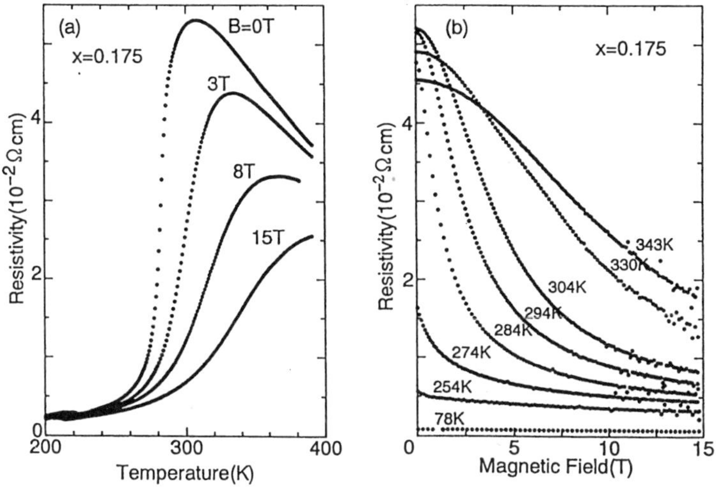


**Fig. 5.70. a, b.** Change of the electrical resistivity in $La_{1-x}Sr_xMnO_3$ under a magnetic field [712]. (With permission from The Physical Society of Japan)

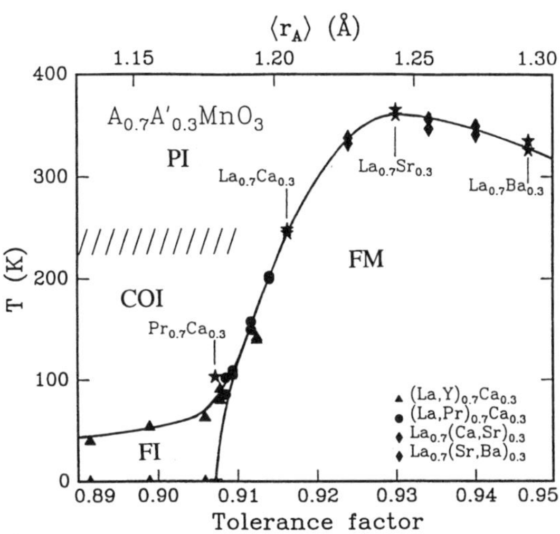


**Fig. 5.71.** Phase diagram of $A_{0.7}A'_{0.3}MnO_3$ in the temperature-$\langle r_A \rangle$ plane, where $\langle r_A \rangle$ is the average ionic radius of the $A$-site ions [713]. The tolerance factor is defined by $t = (\langle r_A \rangle + r_O)/\sqrt{2}(r_{Mn} + r_O)$ and reflects the rotation of the $MnO_6$ octahedra, which leads to orthorhombic or rhombohedral distortion. (With permission from Gordon & Breach)

electrical resistivity at $T_C$ increases and the negative magnetoresistance ratio $-[\rho(H) - \rho(0)]/\rho(0)$ increases. Therefore, the less the electrons become itinerant, the stronger the magnetoresistance effects become.

### 5.7.3 Electronic Structure

Because of the octahedral coordination by oxygen anions, the Mn $3d$ level in the Mn perovskites is split into the $t_{2g}$ and $e_g$ crystal-field levels. The magnitude of the crystal field splitting $10\,Dq$ is approximately 1 eV. In the undoped compound $A\mathrm{MnO_3}$, additional crystal field due to the Jahn–Teller distortion further splits each of the $t_{2g}$ and $e_g$ levels into two sublevels; the splitting of the $e_g$ level is $\sim 0.5\,\mathrm{eV}$. Each crystal-field level is exchange-split into the spin-up and spin-down levels separated by $J_\mathrm{H} \sim 2.5\,\mathrm{eV}$ due to the strong Hund coupling in the $\mathrm{Mn^{3+}}$ $(d^4, S = 2)$ or the $\mathrm{Mn^{4+}}$ $(d^3, S = 3/2)$ ion. Thus for the $\mathrm{Mn^{3+}}$ ion in $A\mathrm{MnO_3}$, $J_\mathrm{H} > 10\,Dq$ and therefore the $d^4$ configuration takes the high-spin state $(t^3_{2g\uparrow}e_{g\uparrow})$. Because of the Jahn-Teller distortion (Fig. 5.67 b), the two $e_{g\uparrow}$ orbitals of $3x^2-r^2$ and $3y^2-r^2$ symmetry are alternatingly occupied in undoped $A\mathrm{MnO_3}$. According to Sect. 4.6.1, the superexchange interaction between orbitals of different symmetry is ferromagnetic [(4.62)–(4.65)] and therefore the $3x^2-r^2$ and $3y^2-r^2$ orbitals are coupled ferromagnetically within the $ab$-plane. The Mn-Mn exchange coupling between the $ab$-planes is antiferromagnetic, on the other hand, because the $e_g$ orbitals are directed within the planes and have little overlap between the planes, and the exchange interaction is dominated by the (weak) antiferromagnetic interaction between the $t_{2g}$ orbitals. This exchange interaction leads to the "layered antiferromagnetic" structure called the $A$-type antiferromagnetic structure as shown in Fig. 5.72.

Holes doped into $A\mathrm{MnO_3}$ through Ca or Sr substitution for the rare-earth $A$ ions enter the $e_g$ band. Because of the Hund rule coupling within the Mn ion, the spin of a hole (electron) introduced into the $e_g$ band is aligned antiparallel (parallel) to the spin of the occupied $t_{2g}$ shell. The mobile $e_g$ holes (electrons) align the spins of the $t_{2g}$ state through the double exchange mechanism, which is described in Sect. 4.6.1, realizing the ferromagnetic metallic state. The kinetic energy of an $e_g$ electron (hole) is proportional to $t\cos(\theta_{ij}/2)$, where $\theta_{ij}$ is the angle between two neighboring Mn spins $i$ and $j$ and $t(<0)$ is the transfer integral between the neighboring $e_g$ orbitals. The kinetic energy is therefore minimized for $\theta_{ij}=0$, and the parallel spin alignment favored.

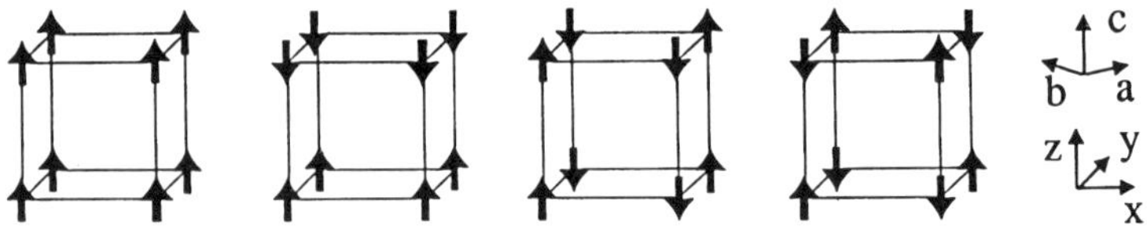


F A-AF C-AF G-AF

**Fig. 5.72.** Various magnetic structures for the perovskite-type oxides. F: ferromagnetic, A, C and G-AF: $A$-, $C$- and $G$-type antiferromagnets

The origin of the band gap in $A\mathrm{MnO_3}$ has been controversial as to whether it is due to Jahn–Teller splitting of the $e_g$ band or due to Coulomb repulsion between the $e_g$ electrons. First-principles band-structure calculations on $LaMnO_3$ using the local-density approximation (LDA) have been performed for the actual distorted crystal structure and successfully predicted the presence of the band gap between the occupied $e_{g\uparrow}$ band and the empty $t_{2g\downarrow}$ band, as shown in Fig. 5.73. This means that the Jahn–Teller distortion alone can open the band gap within the LDA band theory [714]. However, the magnitude of the gap thus obtained is of order $\sim 0.1$ eV, which is far too small compared to the experimental optical gap of $\sim 1$ eV. In order to obtain good agreement with experiment, the self-interaction inherent in the LDA should be removed as in the case of NiO, as described in detail in Sect. 5.6.5. Band-structure calculations using the LDA+$U$ method (see Sect. 5.6.5) have been made to eliminate the self-interaction and a more realistic ground-state electronic structure has been theoretically deduced [714].

Experimentally, the electronic structure of hole-doped $A\mathrm{MnO_3}$ has been studied by optical measurements and photoemission spectroscopy measurements. Optical spectra of $La_{1-x}Sr_xMnO_3$ from the infrared to visible regions show a dramatic temperature dependence over a wide energy range of a few eV, as shown in Fig. 5.74 [715]. At low temperatures, apart from the narrow Drude peak, the optical conductivity is rather flat from $\sim 0.02$ eV to $\sim 1.5$ eV. With increasing temperature, spectral weight is transferred from the low ($< 1$ eV) to high ($> 1$ eV) energies. The 1.5 eV peak, which appears in the paramagnetic insulating phase, is attributed to transition between the

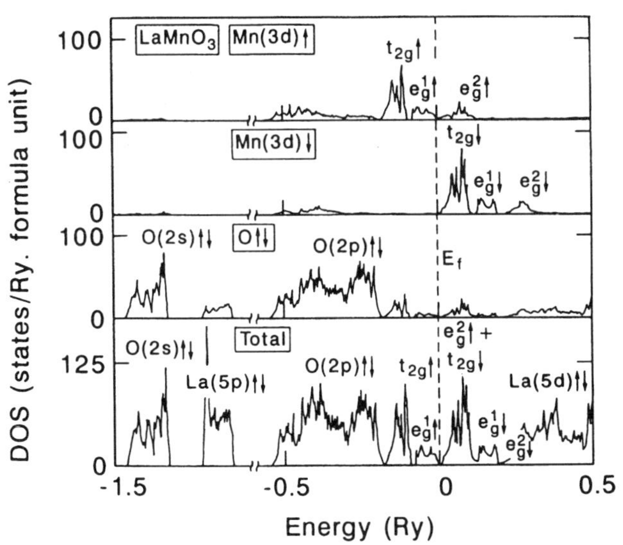


**Fig. 5.73.** Total and partial density of states of $LaMnO_3$ calculated using the LDA [714]. (With permission from The American Physical Society)

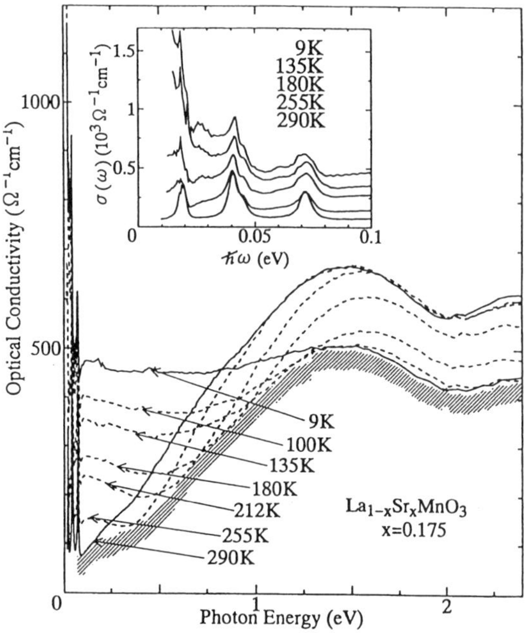


**Fig. 5.74.** Optical conductivity spectra of $La_{1-x}Sr_xMnO_3$ with $x = 0.175$ [715]. The *shaded area* is the temperature-independent part due to the O $2p$ to Mn $3d$ charge-transfer absorption. *Inset* shows the low-energy part of the spectra with a Drude component centered at zero energy appearing at low temperatures. Fine structures are due to phonons. (With permission from The American Physical Society)

exchange-split $e_g$ bands. This transition becomes spin-forbidden in the ferromagnetic state. Photoemission spectra near the Fermi level also show a dramatic temperature dependence. In $La_{0.6}Sr_{0.4}MnO_3$, the intensity at the Fermi level decreases with increasing temperature and the Fermi edge disappears already below $T_C$ [716]. This observation cannot be explained simply by the double exchange mechanism, which would predict a narrowing of the conduction band with increasing temperature and hence an increase in the density of states at the Fermi level. Angle-resolved photoemission spectra of the double-layered perovskite-type $La_{1-x}Sr_{1+x}Mn_2O_7$, which also shows a colossal negative magnetoresistance, show a "widening" rather than narrowing of the Mn $d$ band with temperature [717].

### 5.7.4 Charge and Orbital Ordering

As already described above for the undoped $AMnO_3$, the orbital degrees of freedom of the Mn $e_g$ electrons are expected to yield rich phenomena in the hole-doped $AMnO_3$, too. In 50% hole-doped "narrow-band" compounds $A_{0.5}A'_{0.5}MnO_3$ such as $Nd_{0.5}Sr_{0.5}MnO_3$ and $La_{0.5}Ca_{0.5}MnO_3$, the

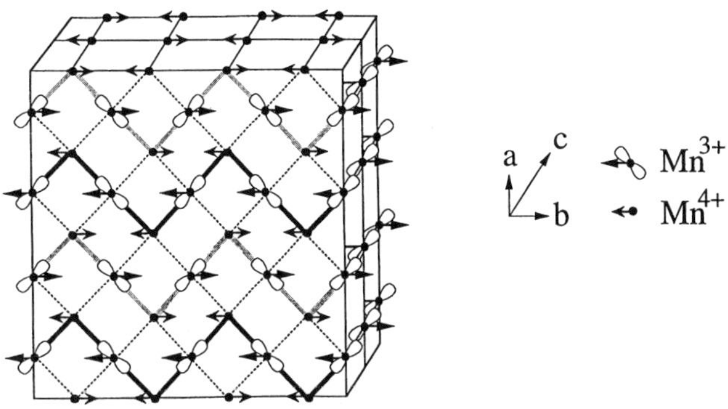


**Fig. 5.75.** Charge and orbital ordering in $A_{0.5}A'_{0.5}MnO_3$. The *arrows* indicate the spin and the *lobes* indicate the directions of the $e_g$ orbitals ($3x^2 - r^2$ or $3y^2 - r^2$) of the $Mn^{3+}$ ions [710]. (With permission from The American Physical Society)

doped holes are ordered and the $e_g$ orbitals of the $Mn^{3+}$ sites form a superlattice pattern shown in Fig. 5.75. Note that the unit cell of that orbital superlattice is twice as large as that of the charge superlattice. Its magnetic structure is different from any of those shown in Fig. 5.72 and is called the $CE$-type antiferromagnetic structure. The charge ordering is seen not only for $x = 0.5$ but also in a finite concentration range in the phase diagram of Fig. 5.68. Here it should be noted that charge ordering and orbital ordering occurs at the same temperature (denoted by $T_{CO}$ in the figure) but the magnetic ordering often occurs at a lower temperature (denoted by $T_N$). Experimental observation of the $CE$-type charge ordering was made by *Murakami* et al. for a related 50% hole-doped layered compound $La_{0.5}Sr_{1.5}MnO_4$ by X-ray diffraction using synchrotron radiation in the Mn 1$s$ core absorption region [718]. The compound consists of alternating stacks of $MnO_2$ and $(La,Sr)_2O_2$ layers, and the same ordering pattern as that in the $ab$-plane of the $CE$-type ordering (Fig. 5.75) was indeed observed.

### 5.7.5 Polaron Effects

Although the negative magnetoresistance of the hole-doped $AMnO_3$ can be qualitatively explained by the double exchange mechanism, the magnitude of the magnetoresistance is far too large compared to what is predicted by the double exchange mechanism, as first pointed out by *Millis* et al. [719]. They argued that it was necessary to take into account the effect of electron–phonon interaction, especially the effect of coupling between charge carriers and Jahn–Teller distortion.

Evidence for strong coupling between the magnetic properties and the lattice distortion has been found in temperature-dependent structural studies. When a hole-doped $AMnO_3$ undergoes a transition from the paramagnetic insulating phase to the ferromagnetic metallic phase on cooling, the lattice parameters and the unit-cell volume show anomalous decrease, as shown in Fig. 5.76 a for the case of $La_{0.75}Ca_{0.25}MnO_3$ [720, 721]. Even more dramatic structural changes occur when the system undergoes a transition into the charge- and orbital-ordered state as shown in Fig. 5.76 b for the case of $La_{0.5}Ca_{0.5}MnO_3$ [720]. Here, the crystal structure becomes more anisotropic as a result of orbital ordering while no appreciable volume change is observed. Not only the average crystal structure but also local structures exhibit unusual changes across the phase transitions. In going from the ferromagnetic metallic phase to the paramagnetic insulating phase, the Mn-O radial distribution function is split into two peaks, as revealed by extended X-ray absorption fine structure (EXAFS) studies [722], indicating the presence of dynamical, local Jahn–Teller distortion in the paramagnetic insulating phase. The dynamical Jahn–Teller distortion actually does not start exactly at the Curie temperature but from somewhat ($\sim$ 50 K) below it [723].

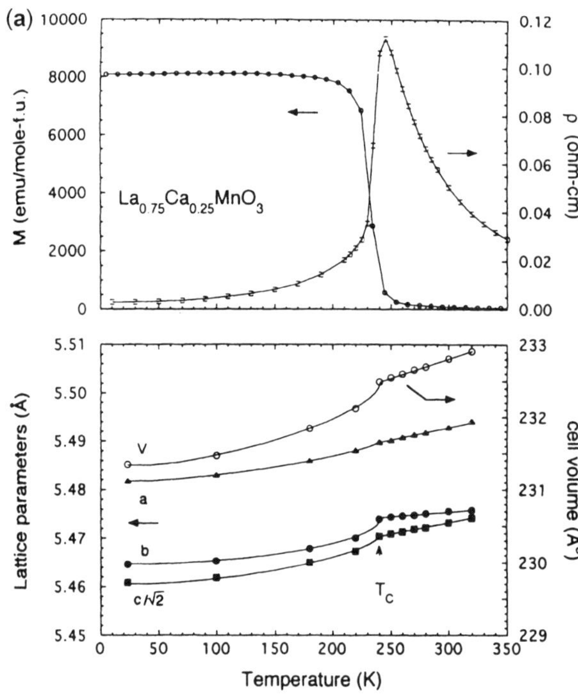


**Fig. 5.76a.** Lattice parameters and cell volumes of (**a**) $La_{0.75}Ca_{0.25}MnO_3$ and (**b**) $La_{0.5}Ca_{0.5}MnO_3$ determined by X-ray powder diffraction [720]. They have orthorhombically distorted perovskite structures. The *upper panels* show the resistivity and magnetization of each sample. The system becomes a ferromagnetic metal below $T_C$ (**a**) or a charge-ordered insulator below $T_N$ (**b**). See the phase diagram in Fig. 5.68. (With permission from The American Physical Society)

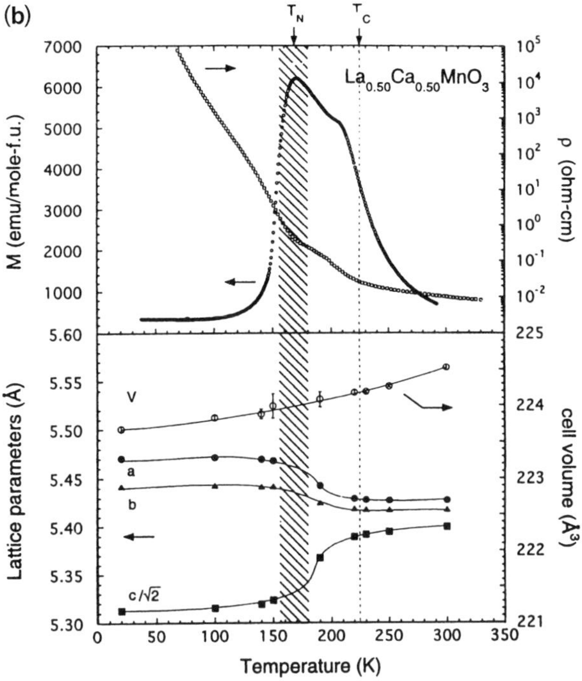


**Fig. 5.76b.** Description in caption to Fig. 5.76 a

Strong coupling between the magnetism and the local Jahn–Teller distortion would naturally lead to formation of magnetic polarons in the hole-doped $A\,MnO_3$. Doped holes would locally induce ferromagnetic spin alignment and thereby reduce the local Jahn–Teller distortion even in the paramagnetic state, whereas the other region remains paramagnetic and retains the local Jahn–Teller distortion. Such magnetic polarons have been investigated by a small-angle neutron scattering study by *De Teresa* et al. [721]. They have observed magnetic scattering signals at a small momentum transfer of $q = 0.03 - 0.6\,\text{Å}^{-1}$ from the paramagnetic insulating phase of $La_{0.33}Ca_{0.67}MnO_3$ as shown in Fig. 5.77, indicating the presence of small ferromagnetic clusters of a few tens of Å around $T_C$. The figure also shows that the temperature and magnetic field dependence of the scattering intensity correlate well with the volume change, indicating the coupling between magnetic polaron formation and lattice distortion.

Observation of local lattice distortion associated with doped holes has been made for ferromagnetic Mn oxides using pair-distribution-function (PDF) analysis of neutron powder diffraction data by *Billinge* et al. [724] and *Louca* et al. [725]. The measured PDF $g(r)$ of ferromagnetic $La_{1-x}Ca_xMnO_3$ and $La_{1-x}Sr_xMnO_3$ indeed reveals a local Jahn–Teller distortion, which is as strong as that in the undoped $LaMnO_3$, in the paramagnetic insulating

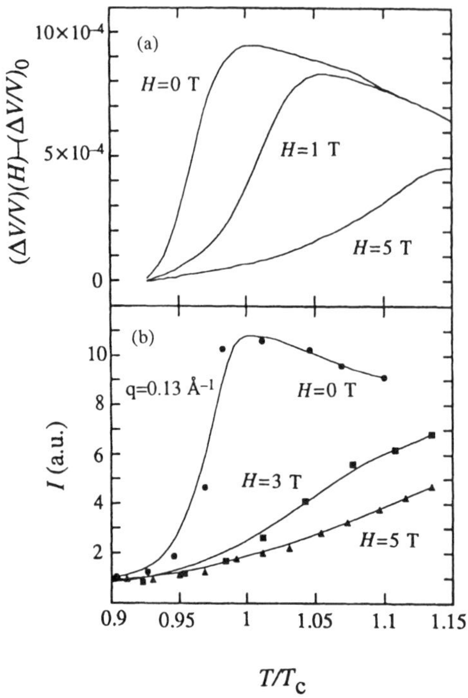


**Fig. 5.77.** Volume change induced by magnetic field (**a**) and the small angle neutron scattering intensity at $q = 0.13\,\text{Å}^{-1}$ (**b**) in $La_{0.33}Ca_{0.67}MnO_3$ as functions of temperature [721]. (With permission from MacMillan)

phase above $T_C$ although the average crystal structure does not show such a distortion. The local Jahn–Teller distortion exists at the $Mn^{3+}$ site but is absent at the hole ($Mn^{4+}$) site, where the average Mn–O distance is found to be reduced. The change in the local structure is most clearly associated with the metal–insulator transition at $T_C$, but starts already well below $T_C$ (by $\sim$ 100 K). Surprisingly, the Jahn–Teller distortion persists at low temperatures although the spatial extent of the polaron becomes larger. This implies that polaronic effects remain important and the simple double exchange mechanism is not sufficient even in the ferromagnetic metallic phase.

The most direct evidence for the importance of electron–phonon coupling in the hole-doped $AMnO_3$ was obtained from the oxygen isotope shift of the Curie temperature. The magnetization data shown in Fig. 5.78 exhibit a reproducible drop in $T_C$ in the $^{18}O$-exchanged $La_{0.8}Ca_{0.2}MnO_3$ [726]. The isotope exponent $\alpha_O \equiv -\mathrm{dln}T_C/\mathrm{d}M_O$, where $M_O$ is the oxygen isotope mass, increases from $\sim 0.1$ ($La_{0.8}Sr_{0.2}MnO_3$) to $\sim 0.9$ ($La_{0.8}Ca_{0.2}MnO_3$) as $\langle r_A \rangle$ and hence the $d$-band width decrease. That is, the stronger the isotope effect is, the larger the magnetoresistance is, implying the importance of polaronic effects in realizing the magnetoresistance. A re-examination of Fig. 5.71 re-

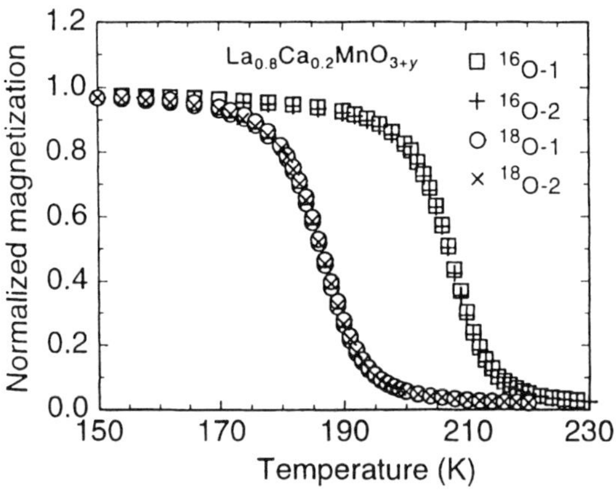


**Fig. 5.78.** Oxygen isotope shift of the Curie temperature in $La_{0.8}Ca_{0.2}MnO_3$ [726]. The $^{18}O$ samples had $\sim 95\%$ $^{18}O$ and $\sim 5\%$ $^{16}O$. (With permission from MacMillan)

veals that the change of $T_C$ with $\langle r_A \rangle$ is much stronger than the prediction of the double exchange mechanism that $T_C \propto W \propto \cos\theta$: Since the Mn-O-Mn bond angle $\theta$ changes at most between 150° and 180°, the double exchange mechanism predicts that $T_C$ should change by only $\sim$10%. If there is a polaronic effect, the effective $d$-band width $W_{\text{eff}}$ will be narrowed from the bare band width $W$ to $W_{\text{eff}} \propto W \exp(-\gamma E_B/\hbar\omega)$, where $E_B$ is the polaron binding energy, $\omega$ is the optical phonon frequency and $\gamma$ $(0 < \gamma \leq 1)$ is a dimensionless constant. Because $\gamma$ increases with decreasing $W$, $T_C \propto W_{\text{eff}}$ changes much more sensitively with the orthorhombic distortion than $T_C \propto W$ predicted by the double exchange mechanism.

The thermoelectric power of $La_{1-x}Ca_xMnO_3$ above $T_C$ displayed in Fig. 5.79 a also supports small polaron conduction [727]. The Seebeck coefficient is fitted to the form

$$S = \frac{k}{e}\frac{\Delta_S}{kT} + S_0 \tag{5.82}$$

for all hole concentrations $x$. The $x$ dependences of the energy gap $\Delta_S$ and the high-temperature limit $S_0$ as well as that of the transport gap $\Delta_\rho$ are plotted in Fig. 5.79 b. The very different $\Delta_S$ and $\Delta_\rho$ are not compatible with a simple semiconductor picture but naturally explained by small polaron transport, in which $\Delta_\rho$ is half the polaron binding energy $\frac{1}{2}E_B$ and the gap $\Delta_S$ is essentially the intersite Coulomb repulsion. The high-temperature limit $S_0 \simeq -20\mu VK^{-1}$ agrees well with $(k/e)\ln(4/5)$, the expected thermoelectric power due to positive holes with spin $S = 3/2$ ($Mn^{4+}$) in the $S = 2$ ($Mn^{3+}$) background. The Hall coefficient well above $T_C$ of the paramagnetic insulating phase shows an activation energy close to 2/3 that of the electrical conductivity, giving further support to small polaron conduction [728].

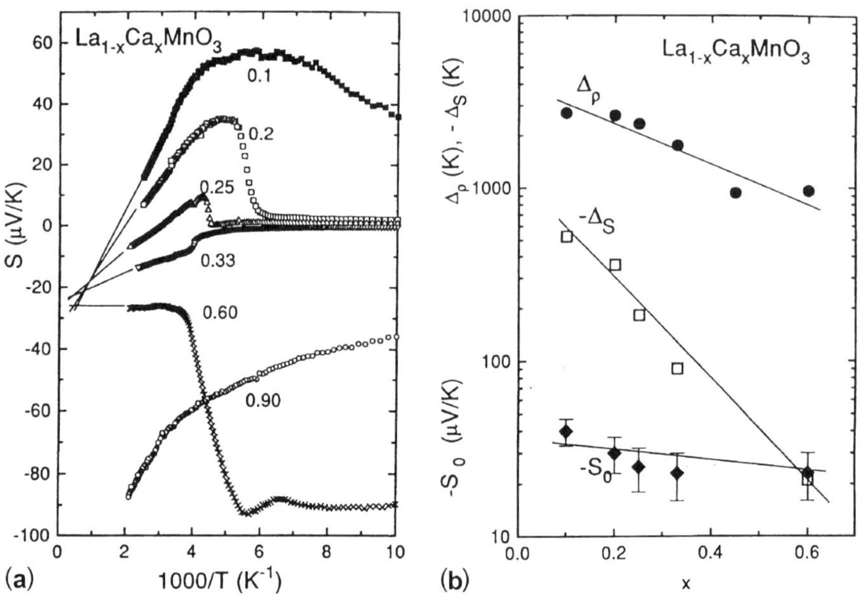


**Fig. 5.79.** Thermoelectric power of $La_{1-x}Ca_xMnO_3$ [727]. (**a**) Seebeck coefficient plotted against $1/T$. (**b**) Doping dependence of the energy gaps derived from transport $\Delta_\rho$ and thermoelectric power $\Delta_S$ and the high-temperature limit $S_0$ of the thermoelectric power. (With permission from The American Physical Society)

The optical conductivity spectra of narrow-band compounds show a broad peak at $< 1.0\,\text{eV}$ in the ferromagnetic metallic phase, implying a polaronic effect [729]. The extremely low photoemission intensity at the Fermi level in the ferromagnetic metallic phase of $La_{1-x}Sr_xMnO_3$ [716] and the vanishingly small one in the layered $La_{1-x}Sr_{1+x}Mn_2O_7$ [717] mean that the metallic carriers in these metals are highly incoherent. These phenomena are difficult to explain within a purely electronic model and may be explained by a strong polaronic effect on these metallic states.

It is by now obvious that polaronic effects play important roles in the hole-doped $AMnO_3$ and deeply influence the giant negative magnetoresistance. However, electron–electron interaction (electron correlation) should also be important in these materials owing to the narrow Mn $3d$ band. Competition/co-operation between electron–phonon and electron–electron interactions makes the problem even more difficult than the problem of only electron–phonon or of electron–electron interaction. The effects of magnetic versus lattice polarons have not yet been clarified. Further studies are necessary to elucidate the physical properties of this interesting class of materials.

## 5.8 $Fe_3O_4$, Verwey Transition

Magnetite, $Fe_3O_4$, is one of the best known minerals on the Earth and is famous for its role in the discovery of magnetism. This material is a member of the spinel type ferrites, a family of compounds important in magnetics. The formula unit can be written as $Fe^{3+}[Fe^{2+}, Fe^{3+}]O_4$, from the ionic crystal point of view. Here, [ ] indicates atoms located in the $B$ or the 16$d$ site of the spinel lattice and cations outside of the bracket are on the $A$ or the 8$a$ site (Fig. 5.83). According to this formula, 2+ and 3+ Fe ions coexist on the same crystal site. The electrical conductivity of $Fe_3O_4$ is expected to be much higher than that of $Fe_2O_3$, for example, which contains $Fe^{3+}$ ions only. In fact, the electrical conductivity of magnetite is about $2.5 \times 10^4 (\Omega\,\mathrm{m})^{-1}$ at room temperature and $Fe_2O_3$ is insulating. If the number of carriers is identical to that of $Fe^{2+}$ ions, the carrier density is about $1.5 \times 10^{28}\,\mathrm{m}^{-3}$ and the magnitude of the conductivity corresponds to a mobility of $10^{-5}\,\mathrm{m^2V^{-1}s^{-1}}$.

An anomaly in the temperature dependence of the electrical conductivity has attracted attention to the transport phenomena of magnetite. As is shown in Fig. 5.80, conductivity decreases by about two orders of magnitude at around 120 K [730]. The transition is called the Verwey transition after *E. J. Verwey*, who presented a model of the transition more than half a century ago, in the early stages of the investigation. Sometimes this terminology is extended to phase transitions in mixed valence compounds such as $Ti_4O_7$ or $YFe_2O_4$.

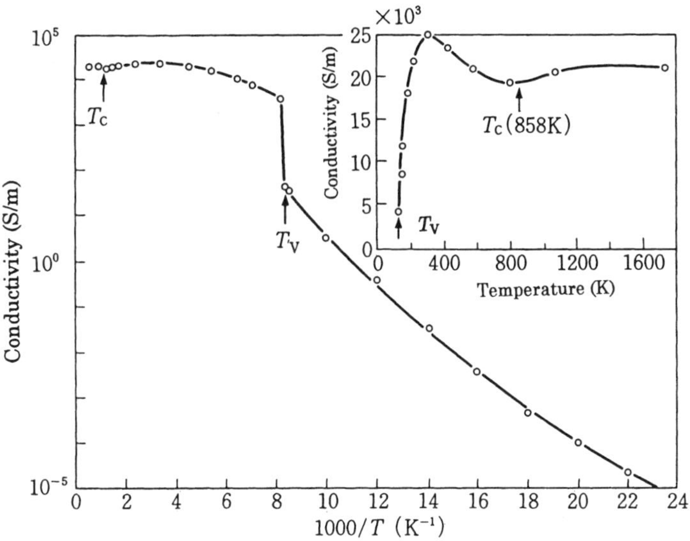


**Fig. 5.80.** Temperature dependence of the electrical conductivity of magnetite [730]. (With permission from The American Physical Society)

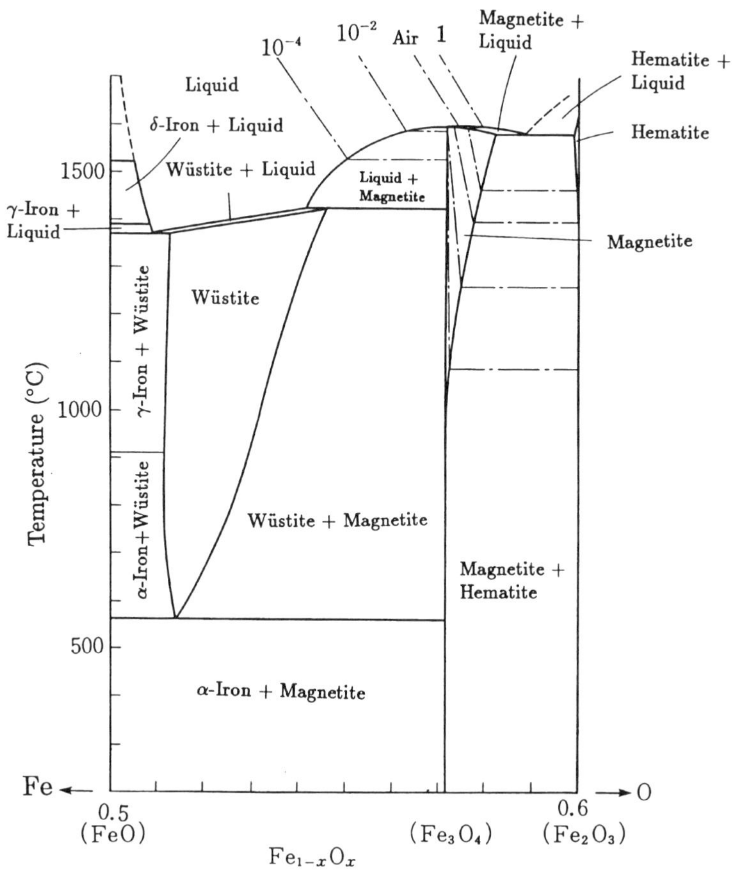


**Fig. 5.81.** Part of the Fe–O phase diagram [731]. Numbers written beside the chain lines in the figure are the partial pressures of oxygen (in atmospheric pressure) in an equilibrium atmosphere ($1\,\text{atm} = 1.013 \times 10^5$ Pa). (With permission from Addison Wesley)

### 5.8.1 Phase Diagram of the Iron–Oxygen System

A part of the phase diagram of the Fe–O system is shown in Fig. 5.81 to demonstrate the position of magnetite [731].[1] All the iron oxides, FeO (wüstite), $Fe_3O_4$ (magnetite) and $Fe_2O_3$ (hematite), are typical berthollides. Ordinarily, transport properties are sensitive to the stoichiometry or the presence of impurities. Dependence of the electrical conductivity on the stoichiometry is marked, however, only in the cases where the valence state of the elements is uniquely determined as in $Fe_2O_3$ or NiO. In a mixed valence

[1] This phase diagram is for the equilibrium state. Spinel phase magnetite can extend to $Fe_2O_3$ as a non-equilibrium state. Spinel type $Fe_2O_3$ is called $\gamma$-$Fe_2O_3$ or maghemite and is used as a primary component in magnetic recording materials.

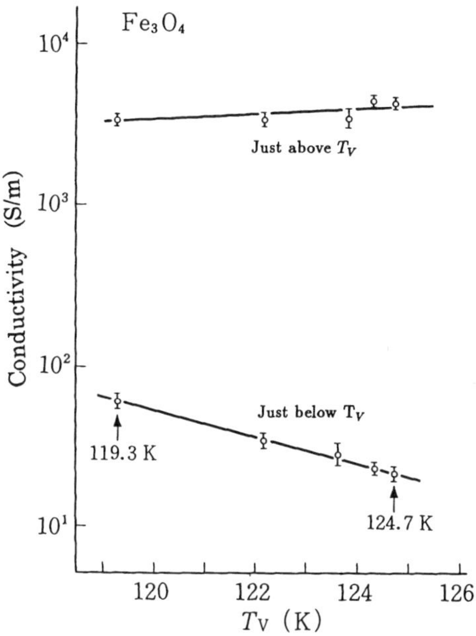


**Fig. 5.82.** Temperature of the Verwey transition and the electrical conductivity just above and below that point in several single crystals of $Fe_3O_4$, prepared by the Bridgeman method with Pt crucibles [733]. The origin of the difference is supposed to be internal strain. (With permission from The American Institute of Physics)

compound such as magnetite at room temperature, the conductivity does not depend so much on the composition of the specimen.

However, even in magnetite, the electrical resistivity in the low temperature phase as well as the transition temperature is structure sensitive. The reported temperature of the Verwey transition has increased with time by about 10 K. In the first report in 1929, it was 115 K [732]. At the same time, the magnitude of the jump in the conductivity at the transition has increased [733]. An example of the correlation is shown in Fig. 5.82 between the transition temperature and the conductivity just above and below. Besides achieving the correct stoichiometry, Fe : O = 3 : 4, the internal strain should be eliminated by appropriate annealing in order to have a high transition temperature. It has been found that the transition not only shifts to lower temperatures but becomes gradual and sometimes splits into two, in the presence of internal stress [733].

The composition of oxides is determined by the partial pressure of oxygen gas with which the material is in thermal equilibrium during growth. The chain lines in Fig. 5.81 show the composition of the compounds which are equilibrated with oxygen, indicated in units of atmospheric pressure.

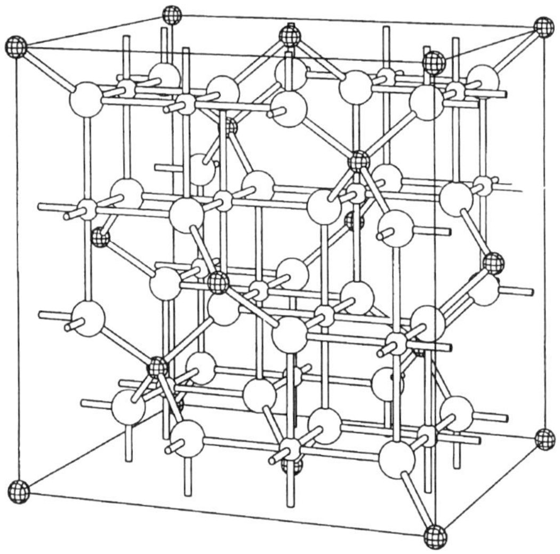

**Fig. 5.83.** Spinel structure. *Large open circles*: anions ($O^{2-}$), *small hatched circles*: cations (8$a$ site; $Fe^{3+}$), *small open circles*: cations (16$d$ site; $Fe^{2+}$+$Fe^{3+}$). (After A. Yanase)

### 5.8.2 The Spinel Structure

The spinel structure is shown in Fig. 5.83. This structure is rather complicated. The fundamental lattice can be considered as a face centered cube of $O^{2-}$ ions. Cations, $Fe^{2+}$ and $Fe^{3+}$, are located in the interstitial positions in the anion lattice. The crystal as a whole also has face centered cubic symmetry. The primitive cell is a rhombohedral parallelepiped containing two formula units, $Fe_6O_8$. To see the symmetry of the crystal explicitly, however, a cubic unit cell with the lattice constant of about 0.8 nm, four times as large as the primitive cell in volume, is better. There are two kinds of cation sites. One site is called the $A$ or 8$a$ site, tetrahedrally coordinated by four $O^{2-}$ ions composing a diamond lattice. The interstice of these coordination tetrahedra is too small for larger $Fe^{2+}$ ions and this site is occupied by $Fe^{3+}$ ions. Even the smaller $Fe^{3+}$ ions enlarge the position, shifting the anions outward. Another cation site is called the $B$ or 16$d$ site, and is coordinated by six $O^{2-}$ ions forming slightly distorted octahedra, which line up along the $\langle 110 \rangle$ axes of the cubic lattice sharing edges (Fig. 5.84). The point symmetry of the $B$ site is $\bar{3}m$, the three-fold axis of which is along one of the $\langle 111 \rangle$ axes. This site forms one half of a face centered cubic lattice. The lattice of the $B$ site can be considered as a diamond lattice of cation tetrahedra, sharing corners with each other. All the tetrahedra on the same plane, say the $xy$ plane, are isolated. This is an important point for the discussion in Sect. 5.8.4. The $B$ site is occupied by equal numbers of $Fe^{2+}$ and $Fe^{3+}$ ions, and the electron exchange between them is the topic of this section.

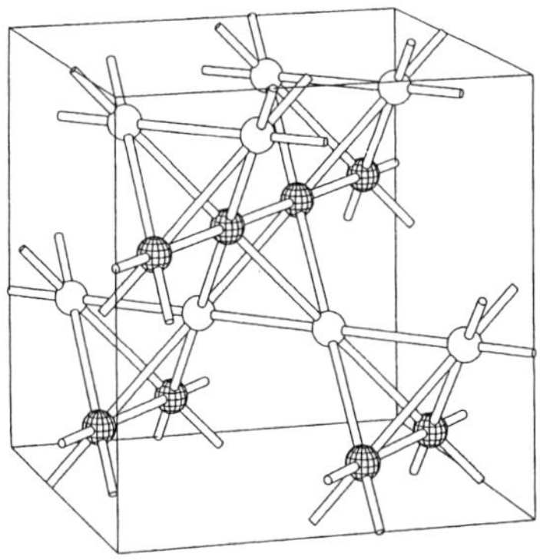

**Fig. 5.84.** 16*d* or *B* site of the spinel lattice. *Open* and *hatched circles* show $Fe^{2+}$ and $Fe^{3+}$, or vice versa, in the ordered structure, proposed by *Verwey* for the low temperature phase. (After A. Yanase)

As mentioned earlier, magnetite is a ferrimagnet caused by a strong antiferromagnetic exchange interaction between the $A$ and $B$ sites. Because of this coupling, the magnetic moments of $Fe^{3+}$ ions in the $A$ site ($\sim 5\mu_B$/Fe) and that of $Fe^{2+}$ and $Fe^{3+}$ in the $B$ site ($\sim 9\mu_B$/2Fe) order in the inverse direction. The ordering temperature of the spins, the ferromagnetic Curie point, is about 860 K. The net moment is $9-5=4\mu_B/Fe_3O_4$. Since $Fe^{2+}$ can be regarded as $Fe^{3+}+e$ and the $Fe^{3+}$ has half filled $(3d)^5-{}^6S$ electronic state, we can assume the situation of the $B$ site to be as follows: on a lattice of $Fe^{3+}$ ions there are mobile electrons, numbering just one half the number of lattice points. In this, $Fe_3O_4$ differs from NiO, etc., where the number of electrons is an integral multiple of the number of atoms. In that case, the correlation energy between two electrons on the same atom is important, whereas the correlation energy between electrons on neighboring atoms is important here. $Ti_4O_7$ [734] and $YFe_2O_4$ [735, 736] are other examples showing the same sort of transition.

At and below room temperature, $Fe^{3+}$ spins order almost perfectly and the spins of the mobile electrons are always antiparallel to the net moment. The spin degeneracy need not be taken into account, in contrast to the band theory of ordinary semiconductors, as well as $V_2O_3$ or $Ti_4O_7$ and $YFe_2O_4$. In a primitive cell, there are two $Fe^{2+}$ ions or two mobile electrons in the $B$ sites. If the second and the third band are degenerate, a Fermi surface(s) will exist and there will be appreciable electrical conductivity. If these two bands are separated for some reason, magnetite will become a semiconductor. The low temperature transition should be looked at in this way, from an itinerant electron point of view.

### 5.8.3 Verwey's Model: Order–Disorder Transformation of $Fe^{2+}$ and $Fe^{3+}$

The first model of the low temperature transition of magnetite was proposed by *Verwey* [737]. He considered the problem of how to distribute Fe ions on the $B$ sites, assuming that $Fe^{2+}$ and $Fe^{3+}$ can always be distinguished. Since a $B$ site should have an electric charge of +2.5 on average, a $Fe^{2+}$ ion can be regarded as charged by $-0.5e$ whereas $Fe^{3+}$ is charged by $+0.5e$. Then there is a Coulomb repulsion between $Fe^{2+}$–$Fe^{2+}$ or $Fe^{3+}$–$Fe^{3+}$ pairs and an attraction between $Fe^{2+}$–$Fe^{3+}$ pairs. The energy of the lattice is lowered when the ions distribute so as to make the $Fe^{2+}$–$Fe^{3+}$ pairs nearer and pairs of Fe with the same valence farther away. The state with the lowest energy is, however, not realized except at 0 K. At finite temperatures, the state of the lowest free energy ($F$), not of the energy ($U$), is realized: i.e. $F = U - TS$ should be minimum. Here, $T$ is the temperature and $S$ is the entropy of the crystal.

First, consider the ground state. Since the $B$ sites form a periodic lattice, it is natural to assume that $Fe^{2+}$ and $Fe^{3+}$ ions distribute periodically (see, however, next subsection, 5.8.4). Hence, the location of $Fe^{2+}$ can be distinguished from that of $Fe^{3+}$, both forming periodic lattices. Let us denote these lattices as $B[2+]$ and $B[3+]$, respectively. According to the definition, the energy of the crystal increases if one $Fe^{2+}$ ion (or one electron) moves from the $B[2+]$ to the $B[3+]$. The movement is accompanied by a movement of one $Fe^{3+}$ from the $B[3+]$ to the $B[2+]$. The amount by which the energy increases through this movement is determined by the difference in electric potential at $B[2+]$ and the $B[3+]$, which is produced by the extra negative charge on the $B[2+]$ and the extra positive charge on the $B[3+]$. (In this argument, only the Coulomb interaction is taken into account. The difference of the ionic radii or the shape of the electron cloud of the two ions, however, can also be considered similarly.)

At finite temperatures, some $Fe^{2+}$($Fe^{3+}$) ions are excited to $B[3+]$($B[2+]$) by thermal excitation. This excitation makes the absolute value of the averaged extra charge smaller on both $B[2+]$ and $B[3+]$ lattices and results in a decrease in excitation energy. The decrease in excitation energy makes more $Fe^{2+}$($Fe^{3+}$) move from the $B[2+]$($B[3+]$) to the $B[3+]$($B[2+]$) to increase the entropy of mixing, $S$, and thus decrease the free energy, $F = U - TS$. The number of excited ions will be larger at higher temperatures and the excitation energy is expected to be 0, or the difference between the $B[2+]$ and the $B[3+]$ is expected to vanish, at a certain temperature, $T_V$. In other words, ordering of the two kinds of ions makes ordering easier and disordering results in more disorder. Thus, two distinctly separated phases appear, the ordered and the disordered phases, though the degree of order in the former is temperature dependent. Such phenomena are called "cooperative". The order–disorder transformation in alloys is a famous example of this and has been extensively investigated since the 1930s. A quantitative discussion will be found in several books and reviews [738]. *Verwey* applied this theory of the order–disorder ransformation to the low temperature transition of magnetite. Note that the degree of order determines the carrier density directly. All $Fe^{2+}$ ions can move through the $B$ sites in the disordered phase, whereas none can move when all $Fe^{2+}$($Fe^{3+}$) are located on the B[2+] (B[3+]) sites. By this mechanism, *Verwey* explained the temperature dependence of the electrical conductivity in Fig. 5.80.

The ordered structure proposed by *Verwey* is illustrated in Fig. 5.84. The $B$ sites lined up along, say, the [110] axis are occupied by $Fe^{2+}$ and those along the $[1\bar{1}0]$ by $Fe^{3+}$. It should be stressed that, in this ordering scheme, there are always two $Fe^{2+}$ and two $Fe^{3+}$ ions in the tetrahedron formed by the neighboring $B$ sites. As seen in the figure, all the $B$ sites in a particular $c$ plane are occupied by $Fe^{2+}$ ($Fe^{3+}$) ions and 2+ planes and 3+ planes stack alternatively along the $c$ axis. This crystal structure has orthorhombic symmetry specified by [110], $[1\bar{1}0]$ and [001] of the original cubic axes. Probably, this is the simplest 1 : 1 ordered structure in the spinel $B$ lattice. Magnetic anisotropy of the low temperature phase is in accordance with this structure [739]. Because of support from several X-ray and neutron diffraction experiments as well as magnetic anisotropy, the "Verwey structure" was believed true for about 20 years. At the end of 1960s, however, electron and neutron diffraction experiments were reported which contradicted this structure. Some of the former experiments were proved to be incorrect and the Verwey structure was thrown away as a model of the structure of the low temperature phase of magnetite.[2]

In the order–disorder transformation model, the Coulomb interactions between ions play a dominant role. Transfer of electrons is considered only implicitly, as a process for altering the distributions of $Fe^{2+}$ and $Fe^{3+}$ ions. In other words, an electron is considered to localize on a $Fe^{3+}$ ion (to make it $Fe^{2+}$) and the kinetic energy of electrons is ignored. Electrical conduction in such a case is different from that of itinerant electrons, because of activated mobility. Even if the transfer integral, $t$, is finite, the energy band of the electrons splits and the material becomes an insulator when the ratio of the correlation energy $U$ to $t$ is relatively large. At high temperatures, however,

[2] The structure of the low temperature phase of magnetite is still a subject of controversy. It is generally accepted that the cubic cell in the high temperature phase is elongated a little along one $\langle 111 \rangle$ axis at the transition and the shape of the cell in the low temperature phase is nearly rhombohedral, as reported by *Rooksby* and *Willis* [740]. One $\langle 100 \rangle$ axis is selected magnetically as an easy axis. So, one $\{110\}$ plane, determined by these $\langle 111 \rangle$ and $\langle 100 \rangle$ axes, is unique and the crystal symmetry is monoclinic or lower. A point in dispute is whether the crystal is really monoclinic or minor displacements of the atoms lower the symmetry to triclinic [741]. The primitive cell in the low temperature phase is believed to be four times larger than that in the high temperature phase, i.e., it contains 24 iron and 32 oxygen atoms. The displacement of the atoms at the transition is reported to be of the order of 0.01 nm, though the atomic positions have not yet been determined [742].

thermally excited electrons will make the effective $U$ smaller and make the material metallic.

It is to be noted that the effect of compression is inverse on the relative stability of the disordered or metallic phase and the ordered or insulating phase, from the localized and the itinerant electron point of view. On the itinerant electron point of view, compression will increase the absolute value of $t$ and stabilize the metallic phase. If $t$ can be ignored and electrons can be considered localized, however, Coulomb energy between neighboring ions is increased by compression. This means an increase in the ordering energy or stabilization of the ordered phase compared with the disordered phase. The effect of compression on the Verwey transition was first studied by *Samara* [743]. The result is the lowering of $T_V$ by 4.5 K/GPa. Though the magnitude of the lowering rate is one order of magnitude smaller than that of $V_2O_3$ (38 K/GPa) [744] or $YFe_2O_4$ (26 and 105 K/GPa, each for the 2-step transition) [745], and a slightly smaller value of 2.7 K/GPa was reported later [746], the importance of electron transfer or the role of the kinetic energy in the Verwey transition in magnetite was proved by this experiment.

On the other hand, the very existence of the Verwey transition indicates the importance of the Coulomb interaction between neighboring ions. Both $U$ and $t$ should not be neglected. In this sense, Verwey's model of charge ordering is still important for metal–insulator transitions in mixed valence oxides.

### 5.8.4 Comment by Anderson: Frustration on the *B* Lattice

In the molecular field approximation, the transition temperature of the order–disorder transformation is determined by the difference in energy between low temperature ordered phase and high temperature disordered phase. If the structures are specified, this energy difference can be calculated, provided that the Coulomb energy is the main cause. According to *P. W. Anderson*, the difference of the Madelung energy is about 2 eV or 23 000 K in temperature, between the Verwey structure and the lattice of $Fe^{2.5+}$ ions: more than two orders of magnitude larger than the actual transition temperature, 125 K [747]. *Anderson* discussed this problem in 1956 and pointed out that a comparison of the Verwey structure with the totally disordered state is not adequate here, because of a rather special form of the $B$ lattice in the spinel type crystal. There are so many states in which the Fe ions distribute differently from the Verwey structure but with nearly the same energy. For this reason, he estimated the transition temperature to be as low as 500–1000 K. Moreover, screening of the crystalline electric field by the polarization of the lattice can be expected to lower the transition by a factor of about 1/5. *Anderson* concluded that Verwey's interpretation does not contradict the measured value of the transition temperature of 125 K.

As stated before, the $B$ site can be regarded as a diamond lattice of tetrahedra, sharing corners with each other (Fig. 5.84). We can consider these tetrahedra as a unit of the lattice, dividing the charge at the corners equally between the two sharing tetrahedra. It is evident that the Coulomb energy between tetrahedra is lowest when the charge of all the tetrahedra is the same, i.e. $+5e$/tetrahedron. The negative charge of neighboring $O^{2-}$ anions cancels the homogeneous positive charge of the $B$-site tetrahedra. A charge of $+5e$ on a tetrahedron means two $Fe^{2+}$ and two $Fe^{3+}$ ions occupy four corners. In this case, the Coulomb energy within the tetrahedron is also a minimum. This condition, i.e., that the charge on a $B$-site tetrahedron is constant throughout the crystal, is called Anderson's condition and has been considered as a wider basis for discussing electronic structure in spinel type oxides, not restricted to $Fe_3O_4$.

A point made by *Anderson* in the paper is that there are an infinite number of ionic configurations which meet this condition [747]. Note once more that the tetrahedra with the same $z$-coordinate are all separated and independent (Fig. 5.84). Let us consider one by one the $xy$-planes composed of the $B$ site. The lowest $B$ sites in the figure line up along the [110] direction. For any ionic distribution in this layer, we can distribute 2+ and 3+ ions in the next layer, lining up along the [110], to satify Anderson's condition for the lowest tetrahedra which are formed by these two layers of $B$ sites. If the first layer is only occupied by $Fe^{2+}$ ions, the next layer should only be occupied by $Fe^{3+}$ ions. (This is the Verwey structure.) In the same manner, tetrahedra at the center of the figure can fulfil Anderson's condition for any ionic distribution in the second layer of the $B$ sites, and so forth. Since there was no restriction on the ionic distribution in the first layer, there are an infinite number of ionic distributions which fulfil Anderson's condition.

If the interaction is restricted to the nearest neighbor pair, the energy is identical for all the configurations which fulfil Anderson's condition. This means that the long-range ordering of Fe ions, such as the Verwey structure, cannot be achieved in the spinel $B$ sites by the nearest neighbor Coulomb repulsion, $U_1$. Such a system, with its infinitely degenerate ground state, is called "frustrated". For long-range ordering, other interactions are requested, such as next nearest neighbor repulsion, $U_2$ or electron–phonon interaction to distort the crystal lattice. In the present case, finite $U_2$ can be expected since the Coulomb interaction is long range. Because the magnitude of $U_2$ is much smaller than $U_1$, the ordering temperature will be much lower than the value expected from $U_1$. At the same time, it is thought that the total charge on each $B$-site tetrahedron remains +5e even above $T_V$. In other words, ordering within a tetrahedron, or short-range ordering, develops without macroscopic long-range ordering, since the thermal energy is not enough to violate Anderson's condition.

It is to be noted here that charges cannot move in the crystal if all the $B$-site tetrahedra satisfy Anderson's condition, even though there is an infinite number of such configurations. Consider a $Fe^{2+}$–$Fe^{3+}$ pair. If these two ions belong to a tetrahedron, movement of an electron from $Fe^{2+}$ to $Fe^{3+}$ does not change the total charge of that tetrahedron. Because all $B$ sites belong to two tetrahedra, however, there should be two tetrahedra which contain only one component of the pair and the total charge in these does change. This situation can be compared with the Hubbard model, which treats a system where the number of atoms and mobile electrons is the same (Sect. 4.7.3). In that case, the ground state with the lowest Coulomb energy is realized when all atoms have one electron each. Electric current is absent. Excitation by $U$, the correlation energy, is needed to move an electron from one atom to another. In the present case, the number of electrons is one half of the number of atoms and there is an infinite number of "ground" states (i.e., satisfying Anderson's condition). However, a finite amount of energy $U_1$, the correlation energy between the nearest neighbor pair that is necessary to violate Anderson's condition, is still needed to make an electron move from one atom to the other. The motion of the electrons is governed by the short-range order, not by the long-range order of the Fe ions.

### 5.8.5 Transport Phenomena and the Fluctuation of Charge

The temperature dependence of the dc electrical conductivity of magnetite is shown in Fig. 5.80. The qualitative features are common to all reports, and the conductivity is not structure sensitive in that sense. Above $T_V$, there is a broad peak at around room temperature and a broad minimum at around 780 K, about 80 K below $T_c$, the ferromagnetic Curie point. Just at $T_c$, only a slight change of the slope in the temperature dependence can be detected. The temperature of the maximum conductivity depends on the stoichiometry or the hydrostatic pressure. The effect of the hydrostatic pressure is shown in Fig. 5.85 [746]. The maximum is shifted to lower temperature by compression and to higher by oxidation. *Kakudate* et al. estimated that the high temperature phase of magnetite becomes "metallic" under 5 Gpa or more. Ordinarily, the appearance of the maximum in the conductivity at room temperature is attributed to the correlation and/or polaron effect of carriers. On the other hand, increase of the conductivity above 800 K is considered to be due to the appearance of carriers in, or migration of $Fe^{2+}$ ions to, the $A$ sites [748].

Conductivity in the optical region, deduced from reflectivity measurements in the spectral range between 0.03 and 12 eV, was reported by *Wachter's* group [749, 750]. Reflectivity at low energy decreases with decreasing temperature, showing an anomaly at $T_V$. In a report in 1980, they analysed the difference of the low frequency conductivity at 300 K from that at 20 K, using the Drude model,

$$\sigma(\omega) = \frac{1}{4\pi}\frac{\omega_p^2\gamma}{\omega^2+\gamma^2}, \tag{5.83}$$

with $\omega_p$, the plasma frequency, between 0.5 and 1.0 eV and $\gamma$, the damping factor, between 0.3 and 1.15 eV. The effective electron concentration,

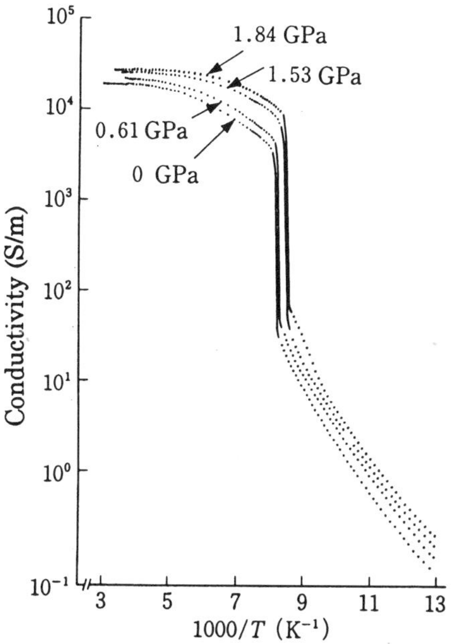


**Fig. 5.85.** Pressure dependence of the electrical resistivity of a single crystal of magnetite [746]. (With permission from Elsevier)

$N^* = N/(m^*/m) = \omega_p^2/4\pi e^2$, is estimated as 2–7 $\times\ 10^{20}\text{cm}^{-3}$ or 0.01–0.05 per formula unit.

Recently, *Park* et al. repeated optical measurements in a wider spectral range, 0.008–36 eV, between 10 and 500 K [751]. From the low energy shape of the conductivity spectrum, they argued that the charge dynamics above $T_V$ is highly diffusive. They estimated the effective number of carriers,

$$N_{\text{eff}}(\omega_c) = \frac{2m}{\pi\rho^2 N}\int_0^{\omega_c} \sigma(\omega)\mathrm{d}\omega\,, \tag{5.84}$$

as a function of temperature, as shown in Fig. 5.86. Here, $N$ is the number of $B$ site ions per unit volume and $\omega_c$ is 0.375 eV where optical conductivity is independent of temperature below 300 K. Above $T_V$, the temperature dependence of $N_{\text{eff}}$ is almost proportional to the conductivity.

The mobility of carriers above $T_V$ has not been established. Nothing is known about the drift mobility. There are several reports on the Hall effect [752–755]. These are consistent in the sign of the extraordinary coefficient, $R^{\text{ex}}$, as being negative at room temperature. The absolute value of $R^{\text{ex}}$ decreases with increasing temperature and changes sign to positive at about 650 K. This is consistent with the appearance of carriers in the $A$ site where spin directions are opposite. Below room temperature, $|R^{\text{ex}}|$ increases with decreasing temperature, proportionally to $\rho^{1/2}$ [755]. This power is quite

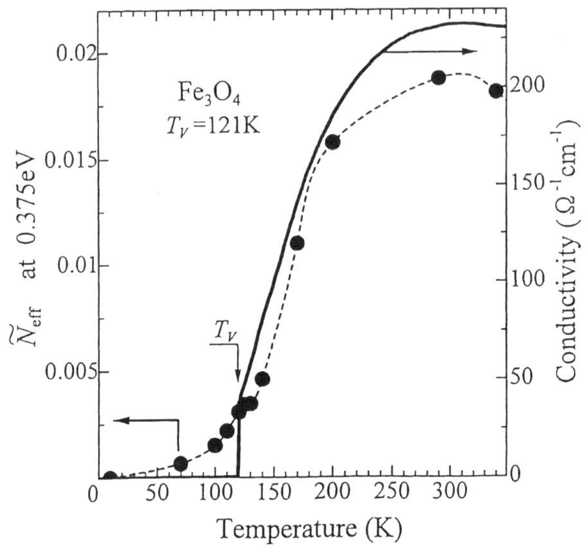


**Fig. 5.86.** Temperature dependence of the effective number of electrons, $N_{\text{eff}}$, at the frequency of 0.375 eV. Difference from the number at 10 K is shown. The *broken line* is intended to guide the eye. The *fine line* is the temperature dependence of the dc conductivity [751]. (With permission from The American Physical Society)

different from the value of 2.0, given by the phenomenology of *Karplus* and *Luttinger* [756].

In all the experiments at room temperature, the negative Hall voltage increases in absolute value with increasing external field. *Siemons* attributed this increase to an increase in the extraordinary term introduced by the increase of magnetization. He argued that the sign of the ordinary coefficient is positive, in spite of the negative slope [753]. His conclusion is inconsistent with the report that the slope remains negative at high temperatures where the extraordinary Hall coefficient is positive [755]. At any rate, difficulties still lie in the quantitative evaluation of the ordinary Hall coefficient, which is usually used to determine the character of the charge carriers, from the measured Hall voltage. No quantitative conclusion can be extracted from the Hall effect at present.

The Seebeck coefficient, $S$, is reported to be negative above $T_V$ and does not depend so much on the stoichiometry [748,757–759] (Fig. 5.87). The magnitude is about $-50\,\mu$V/K at room temperature. The absolute value increases slowly with increasing temperature to about $-80\,\mu$V/K at $T_c$, with a small jump there, and increases further up to $-120\,\mu$V/K at 1500 K. This increase in magnitude is attributed to a change of the cation distribution between the $A$ and $B$ sites in the spinel lattice, or the appearance of carriers in the $A$ site [748].

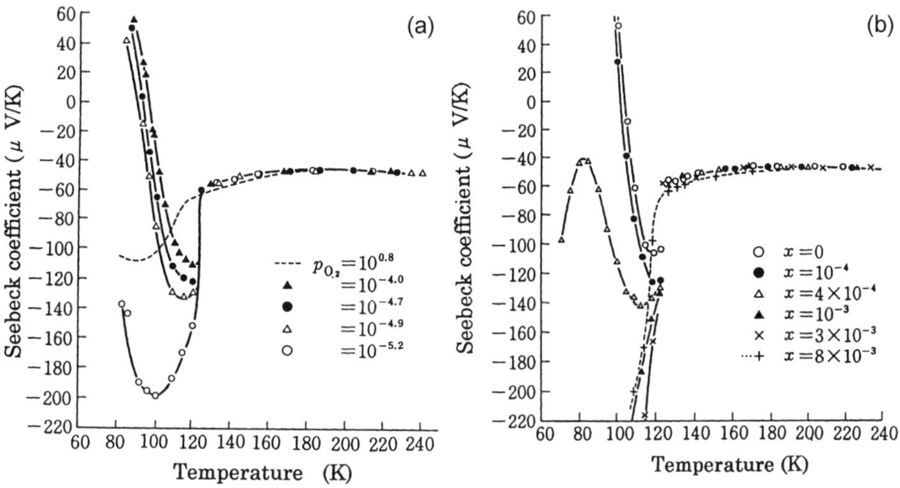


**Fig. 5.87.** Seebeck coefficient of single crystals of magnetite [757]. (**a**) Results for specimens with different stoichiometry, annealed at 1130°C for 70 h at the oxygen partial pressures indicated in the figure [762]. (**b**) Results for specimens partially substituted by Ti, as indicated in the figure [757]. (With permission from The American Physical Society)

Below $T_V$, the dc resistivity increases almost exponentially with decreasing temperature. The reported activation energies are concentrated between 0.1 and 0.15 eV. On the other hand, there exists controversy over the detailed temperature dependence and the conduction mechanism. Arguments have been made that $\log \rho$ is proportional to $T^{-1/4}$, indicating variable range hopping [760], that $\log \rho$ is proportional to $T/T_0$ rather than $1/T$ between $T_V/2$ and $T_V$, indicating incoherent tunneling of electrons induced by thermal motion of atoms [761], that there are electrons and holes both with thermal activation type mobility [762], or that $\log \rho$ is proportional to $1/T$ [763]. Differences might be due to the different state of the specimens. Experiment by *Graener* et al. were carried out on polycrystalline specimens [760]. According to *Pai* and *Honig*, $\log \rho$ of a single crystal is proportional to $1/T$ down to 77 K just after an annealing to make the specimen stoichiometric, but the proportionality is lost after an aging of 10 months in air at room temperature. Undoubtedly, electrical conduction is structure sensitive below $T_V$. The resistivity is higher below $T_V$ in the more stoichiometric specimen, whereas it is lower above $T_V$. It should be noted that no measurement has been carried out on a true "single crystal" of the low temperature phase and nothing can be said about the effect of "grain boundaries". Anisotropy of the electrical conductivity is small, measured on a specimen with magnetically aligned $a$, $b$ and $c$ axes [764]. This contradicts the Verwey structure, where conductivity along the $c$ axis is expected to be much lower than that along the $a$ or the $b$ axis.

The low frequency ac conductivity of the low temperature phase was measured by *Akishige* et al. [765, 766]. They found an anomaly in the dielectric constant between 10 and 70 K, the magnitude of which depends on the stoichiometry: smaller for the more stoichiometric sample, but could not be totally eliminated. Below this temperature region, the ac conductivity is different from the dc conductivity, similarly to NiO described in Sect. 5.6. At 6.8 K, the frequency dependence of the conductivity can be expressed as $\sigma \propto \omega^s$ with $s$ about 0.7. They attributed this to the hopping of bound carriers.

The Seebeck coefficient below $T_V$ is also very structure sensitive and depends strongly on temperature (Fig. 5.87). In a stoichiometric specimen, $S$ increases rapidly, changing sign to positive, with decreasing temperature. In contrast, the Seebeck coefficient remains negative and its absolute value increases rapidly in specimens with a higher $Fe^{2+}$ concentration, prepared by reduction or by substitution of Fe ions by $Ti^{4+}$. According to *Kuipers* et al., these data in the low temperature phase can be explained by a two-carrier model, electrons and holes, both of activation type mobility [762]. The energy gap is estimated to be about 0.1 eV.

For localized carriers changing their site by thermal motion, mobility $\mu$ is given by Einstein's relation:

$$\mu = \frac{ea^2}{k_B T \tau} . \tag{5.85}$$

Here, $e$ is the charge of the carrier, $a$ is the distance of one hop and $\tau$ is the average duration of stay. In the case of magnetite, $\tau$ can be estimated by nuclear magnetic resonance (NMR) of $^{57}Fe$ nuclei.

There is a magnetic coupling between the nuclear spin, $\boldsymbol{I}$, and the electronic spin, $\boldsymbol{S}$, proportional to their scalar product, $\boldsymbol{I} \cdot \boldsymbol{S}$. Since the mass of an electron is much smaller than that of a nucleus, the characteristic frequency is much higher for electrons than for nuclei. From the nuclear spin, the effect of this coupling can be considered as an effective field, the hyperfine field, proportional to the thermal average of the electron spin. This proportionality factor as well as the thermal average of the electron spin depends on the electronic state of the ion. Thus, the NMR frequency of the nucleus in magnetic materials reflects the electronic state of the ion.

The temperature dependence of the NMR frequency of $^{57}Fe$ in $Fe_3O_4$, without external magnetic field, is shown in Fig. 5.88 a [767]. Two resonances above $T_V$ are assigned to Fe nuclei on the $A$ (higher frequency) and the $B$ site (lower). Below $T_V$, the resonance frequency of the $B$ site Fe splits into two, with a separation of about 5 MHz or 40 kOe in the effective field at 4.2 K.[3] Apparently, in the high temperature phase, two resonances for different charge

[3] This difference of 5 MHz is very small compared with the difference between resonance frequencies of nuclei of typical $Fe^{2+}$ and $Fe^{3+}$ ions. This is further evidence of the fact that the ionic model is not adequate for magnetite. Strictly speaking, resonance points at 4.2 K are not two but are composed of many resonances corresponding to many atomic positions in the low temperature phase [768].

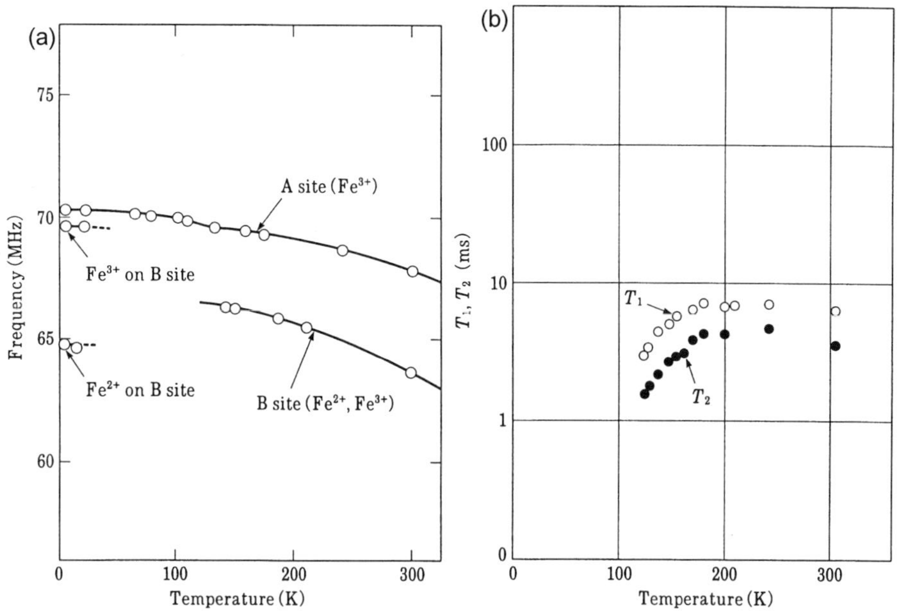


**Fig. 5.88.** (**a**) Temperature dependence of NMR frequencies of $^{57}$Fe nuclei in $Fe_3O_4$, without external magnetic field [767]. (**b**) Temperature dependence of the spin–lattice and spin–spin relaxation time, $T_1$ and $T_2$, of the Fe nuclei on the $B$ site [767]. (With permission from The Physical Society of Japan)

states of Fe ions in the $B$ site amalgamate into one because of the fluctuation of charges. This is a typical example of motional narrowing of resonance absorption where the absorption profile is determined by the relative magnitude of the difference of the resonance frequencies, $\Delta\omega$, and $1/\tau$ [769]. Here, $\tau$ is the mean duration of the state. The conclusion of the theory for two extreme cases can be understood easily: two resonances can be observed separately when $\Delta\omega$ is much larger than $1/\tau$ whereas the resonances amalgamate into one if $1/\tau$ is much larger than the frequency difference. For the latter case, the spin–spin relaxation time of nuclear spins, $T_2$, is expressed as

$$T_2^{-1} = \left(\frac{\Delta\omega}{2}\right)^2 \tau \, . \tag{5.86}$$

In the present case of magnetite, $T_2^{-1} \approx 0.35 \times 10^{15}\tau$. Figure 5.88 b shows the temperature dependence of $T_2$ for $^{57}$Fe in the $B$ site of $Fe_3O_4$, measured by the spin echo method [767]. At room temperature, $T_2$ is about 3 ms. This value gives $\tau$ about $10^{-12}$s. This magnitude corresponds to a mobility of $3 \times 10^{-6}\mathrm{m^2V^{-1}s^{-1}}$, if we assume $a \approx 0.3$ nm, the nearest neighbor distance of the $B$ lattice. Compare this value with the estimate at the beginning of this section, $10^{-5}\mathrm{m^2V^{-1}s^{-1}}$.

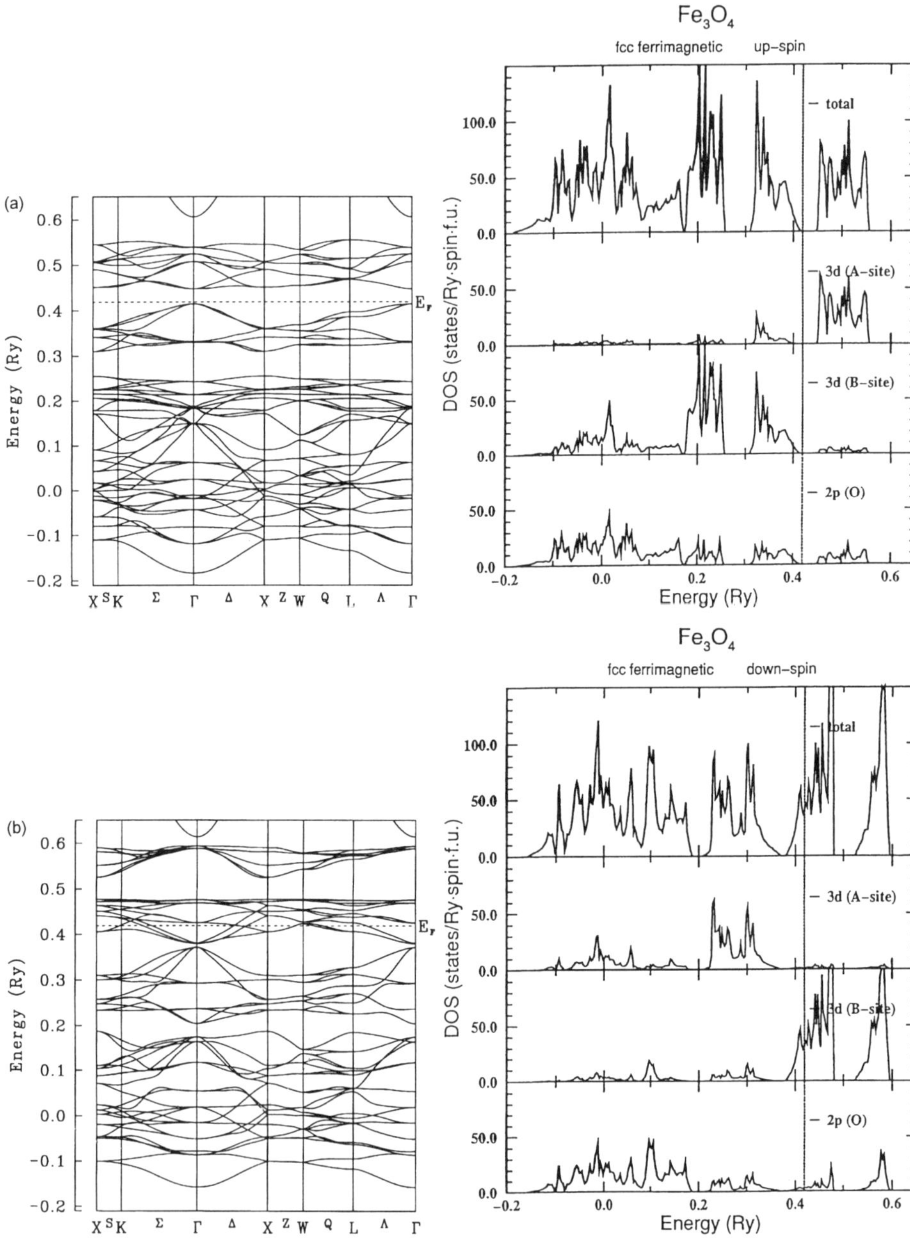


**Fig. 5.89.** Self-consistent APW energy band structure and the total and partial density of states. (**a**) Majority spin, (**b**) minority spin [770]. (With permission from The Physical Society of Japan)

An important point in this discussion is that the motional narrowing of the resonance takes place at all nuclei on the $B$ site homogeneously. If this were not so, $T_2$ would vary from nucleus to nucleus in contrast to the observation.

### 5.8.6 Band Structure

The band structure of $Fe_3O_4$ has been calculated repeatedly since 1984, for the ferromagnetic high temperature phase. In the following, the band structure is discussed according to the recent work of *Yanase* and *Hamada* [770]. Although the one-electron approximation is not sufficient in such a crystal, it is important as a basis for more detailed calculation.

The band structure has been calculated by the full potential linearized APW method, using the local spin-density approximation. The dispersion relations are shown in Fig. 5.89 with the total density of states as well as the partial density of states of Fe at $A$ and $B$ sites and of O. Though the hybridization of Fe($3d$) and O($2p$) states is rather large, the $d\epsilon$–$d\gamma$ splitting of the Fe($3d$) levels can still be recognized. As expected, the Fermi level lies within the energy gap of the majority spin band in which the $3d$ levels of Fe in the $B$ site are almost occupied whereas those in the $A$ site are almost empty. In contrast, the $3d$ states of the Fe($A$) site are almost occupied in the

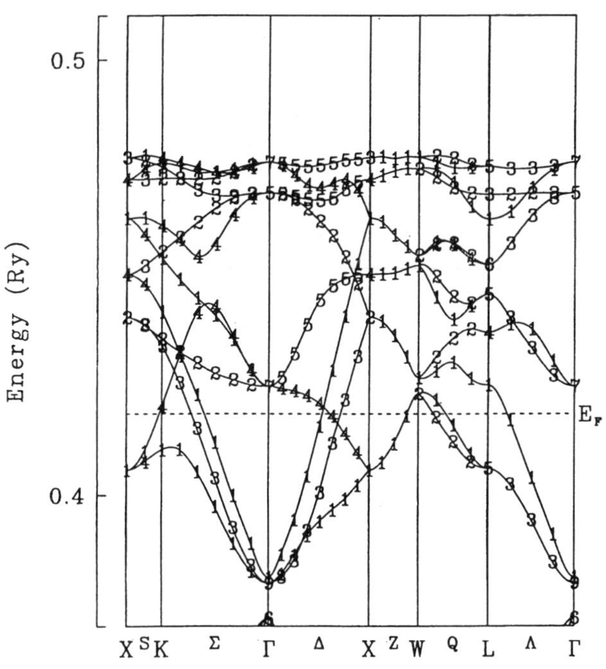


**Fig. 5.90.** Dispersion relation of minority spin electrons near the Fermi level [770]. (With permission from The Physical Society of Japan)

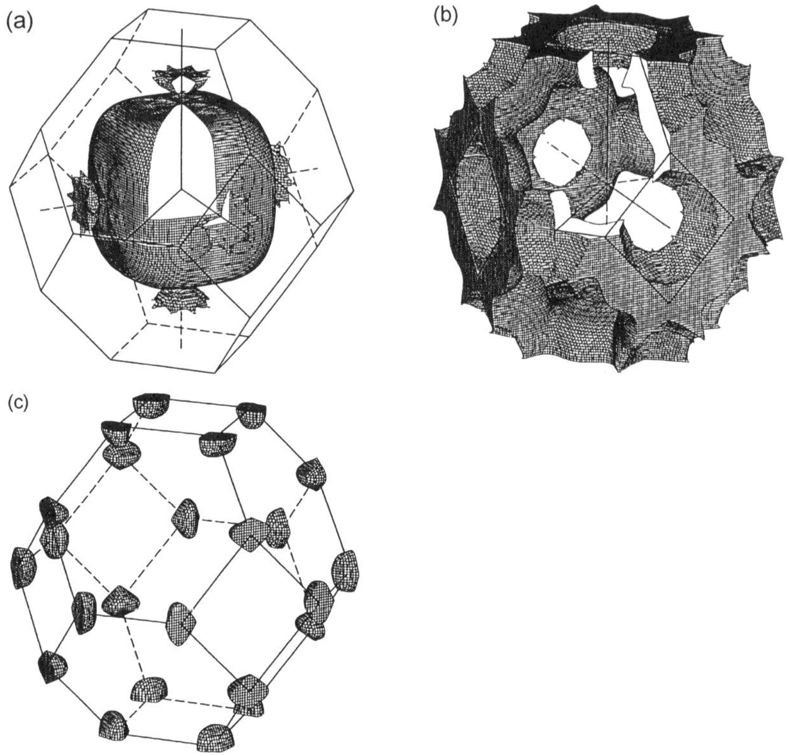


**Fig. 5.91.** Fermi surfaces of the high temperature phase of $Fe_3O_4$ [770]. (**a**) Electron surface in the 3rd band, (**b**) hole surface in the 2nd band, (**c**) hole surface in the 1st band. (With permission from The Physical Society of Japan)

minority spin band, and those of Fe($B$) site are almost empty. However, there is a finite density of states at the Fermi level, which is attributed mainly to the lowest $3d$ orbitals of Fe in the $B$ site. These are consistent with the Néel structure of spins. The total width of the valence band is about 10 eV below the Fermi level.

The minority spin band near the Fermi level is enlarged in Fig. 5.90. These states are composed mainly of $d\epsilon$ orbitals of the $B$ sites. There should be 2 electrons in a primitive cell or 4 B-site ions. This is just Anderson's condition from an itinerant electron point of view.

The Fermi surfaces are illustrated in Fig. 5.91. There is an electron surface with rounded cube shape around the $\Gamma$ point and six discs with thorns on the $\Delta$ axes. The hole surface of the 2nd band forms a multi-connected skeleton along edges of the Brillouin zone. Besides this, there are small hole pockets in the 1st zone at the $W$ point. The number of electrons in the third band is 0.245/cell.

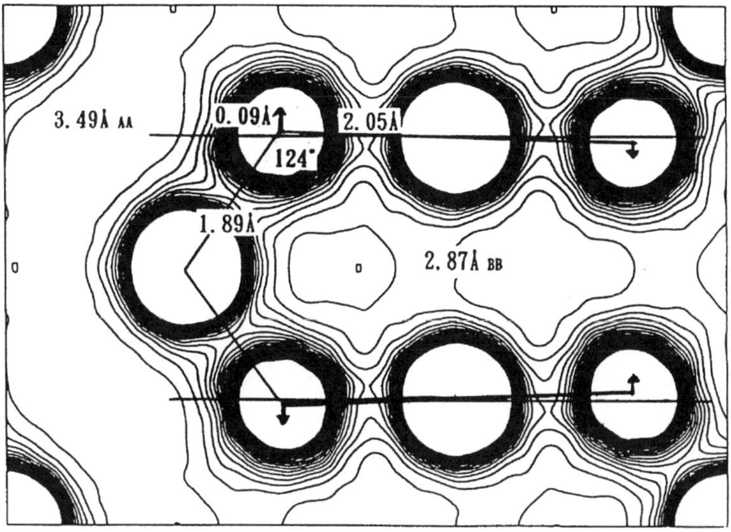


**Fig. 5.92.** The electron distribution in a (110) plane, determined by the short wavelength X-ray diffraction [772]. Four small circles are oxygen, two large circles at the center are $B$-site Fe and a large circle at the left end is the $A$-site Fe atom. *Arrows* at the oxygen indicate the direction of the shift of O ions from the face centered cubic lattice, corresponding to the difference of the $u$ parameter from 3/8. Numbers in the figure are atomic distances and the amount of the shift. (With permission from EDP Sciences)

The density of states in Fig. 5.89 shows strong hybridization between $3d$ orbitals of Fe and $2p$ of O. The calculated electron distribution in the real space is in good agreement with experiment (see below). Hybridization of orbitals reduces spin density at the Fe site: the spin density in a muffin-tin sphere at the $A$ site is $3.339\mu_B$ instead of $5\mu_B$, and is $3.469\mu_B$ at the $B$ site instead of $4.5\mu_B$. These values are to be compared with the magnetic moment of an Fe atom on the $A$ site, reported as $3.82\mu_B$, by a polarized neutron diffraction experiment [771]. The total spin density in a primitive cell is $8\mu_B$ as expected from the Néel structure.

An interesting point is that there are accidental degeneracies near the Fermi level, resulting in thorns on the Fermi surface. These degeneracies are caused by glides on the $\{100\}$ and mirrors on the $\{110\}$ plane. Lattice distortions which destroy these symmetries can remove the accidental degeneracies, resulting in band splitting. It means that the energy of the electron system can be lowered through the electron–phonon interaction or polaron formation. This point will be discussed in the next subsection. Furthermore, *Yanase* and *Hamada* suggested a possibility that the nesting of the electron Fermi surface near the middle of the $\Delta$ axis is the origin of the doubling of the unit cell in the low temperature phase along the $c$ axis (see footnote [2]).

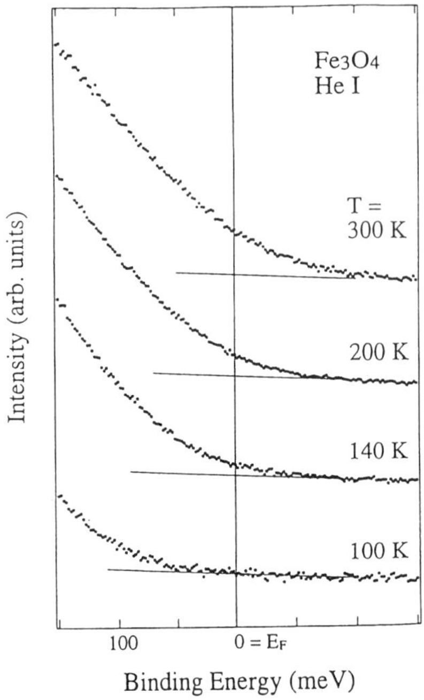


**Fig. 5.93.** Temperature dependence of the photoemission spectra near the Fermi level [773]. Energy of the incident photon (He I) is 21.2 eV. (With permission from The American Physical Society)

Several experiments are consistent with the band scheme described above. One of them is the electronic density in the real space determined by X-ray diffraction, as shown in Fig. 5.92 [772]. Hybridization of Fe($3d$) and O($2p$) orbitals is apparent, especially for the $A$ site. Compare the calculated spin density in the muffin-tin spheres at the $A$ and $B$ sites, cited above. Hybridization was also suggested by the neutron diffraction experiment, in the charge ordered low temperature phase. The difference in magnetic moment between $B$ site Fe ions was too small to be detected [742]. Another investigation to be cited here is a photoemission experiment [773]. *Chainani* et al. reported a high resolution spectrum below 300 K, indicating the existence of a finite electron density at the Fermi level. Figure 5.93 shows the spectra. The density of states decreases with decreasing temperature and becomes zero below $T_V$. It is argued that the temperature dependence of the density of states at the Fermi level is consistent with the optical study of carrier density by *Park* et al. [751].

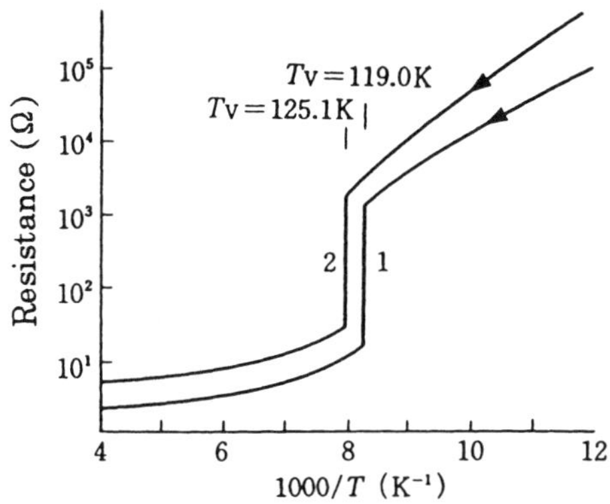


**Fig. 5.94.** Temperature dependence of the electrical resistivity of two $Fe_3O_4$ polycrystals, prepared by the same method [774]. In specimen 2, 43% of the oxygen atoms are substituted by $^{18}O$, whereas specimen 1 is composed of ordinary $^{16}O$ only. (With permission from Akademie-Verlag, Wiley–VCH)

### 5.8.7 Fluctuating Local Lattice Distortion and Electron–Phonon Coupling

As is stated in Sect. 5.8.3, the magnetite lattice is distorted almost rhombohedrally at the transition, accompanied by displacement of atoms of the order of 0.01 nm [742]. Such a distortion suggests an important role of lattice distortion or phonons in the Verwey transition. Indeed, it was discovered that the transition point depends on the mass of oxygen as shown in Fig. 5.94 [774]. When 43% of $^{16}O$ is substituted by the heavier $^{18}O$ isotope, the transition point $T_V$ shifts 6.1 K higher. Since a larger mass of atoms lowers the phonon frequency, this stabilization of the low temperature distorted phase by the substitution is quite natural.

In the high temperature phase, too, several experiments indicate the existence of a local distortion of the lattice, suggesting the formation of polarons and their relation to the Verwey transition. One example is an elastic constant, $C_{44}$, shown in Fig. 5.95 [733]. Apparently, magnetite becomes very soft against shear stress at the Verwey transition. A similar anomaly was also detected in $C_{11}$, though the magnitude is about one order of magnitude smaller [775]. It is interesting that hydrostatic pressure suppresses the anomaly in $C_{44}$ but enhances that in $C_{11}$ [775]. Another pronounced anomaly appears in the cubic magnetic anisotropy constant $K_1$, as shown in Fig. 5.96 [739].[4] Similar anomalies in the temperature dependence of the

[4] The free energy of a magnetic material is dependent on the direction of the magnetization relative to the crystal axes. In a cubic system such as $Fe_3O_4$, anisotropy energy can be expressed as

$$E = K_1(\alpha_1^2\alpha_2^2 + \alpha_1^2\alpha_3^2 + \alpha_2^2\alpha_3^2) + K_2\alpha_1^2\alpha_2^2\alpha_3^2 .$$

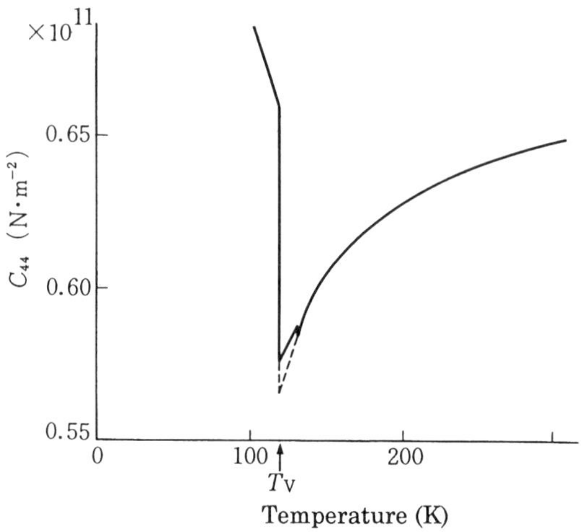


**Fig. 5.95.** Temperature dependence of the elastic constant ($C_{44}$) of magnetite, measured by Y. Kino [733]. (With permission from The American Institute of Physics)

magnetostriction constants[5] $\lambda_{100}$ and $\lambda_{111}$ indicate coupling of magnetic anisotropy and lattice distortion.

All these anomalies can be expressed by the Curie–Weiss law with a Curie temperature of about 80 K as, for example,

$$\Delta C_{44} \propto \frac{1}{(T-80)} \,. \tag{5.87}$$

Here $\Delta C_{44}$ means the difference of $C_{44}$ from an assumed value of normal temperature dependence, estimated, e.g., from $C_{44}$ of $Fe_3O_4$ partially substituted by Mn. The Verwey transition becomes much lower upon the substitution and there is no anomaly near 120 K. Using the analogy of magnetic susceptibility of a ferromagnet above the Curie point, such a temperature dependence can be interpreted quantitatively by assuming interacting units characterized by lattice distortion and magnetic anisotropy, the symmetry axes of which are, e.g., [001], [110] and [1$\bar{1}$0] [777]. The symmetry axis of the assumed unit is shown in Fig. 5.97. Note that this symmetry is incompatible with the point

Here, $\alpha_i$ is the direction cosine of the magnetization with respect to the cubic axes. $K_1$ and $K_2$ are parameters to be determined for each material and temperature, called the magnetic anisotropy constants.

[5] The lattice of a magnetic crystal distorts slightly as a function of the direction of the magnetization. In a cubic system, the strain along a direction ($\beta_1$, $\beta_2$, $\beta_3$) is

$$\frac{\delta l}{l} = \left(\frac{3}{2}\right)\lambda_{100}\left(\alpha_1^2\beta_1^2+\alpha_2^2\beta_2^2+\alpha_3^2\beta_3^2-\frac{1}{3}\right)$$
$$+3\lambda_{111}(\alpha_1\alpha_2\beta_1\beta_2+\alpha_1\alpha_3\beta_1\beta_3+\alpha_2\alpha_3\beta_2\beta_3)\,.$$

$\lambda_{100}$ and $\lambda_{111}$ are called the magnetostriction constants.

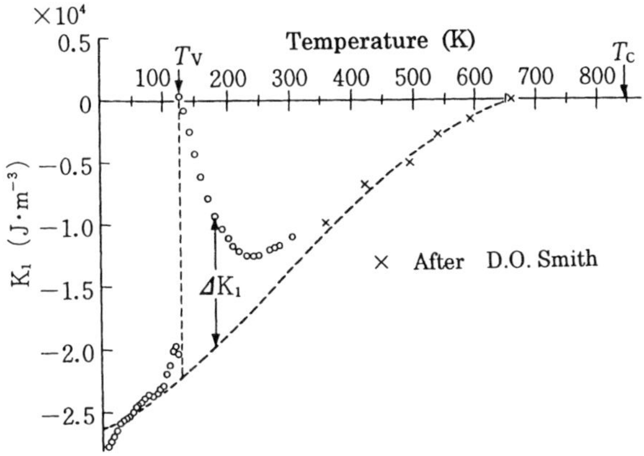


**Fig. 5.96.** The anomaly in the temperature dependence of the cubic magnetic anisotropy constant, $K_1$ [739]. (With permission from The Physical Society of Japan)

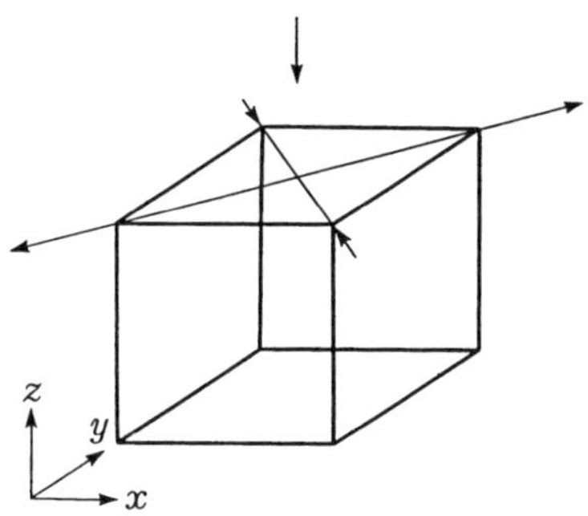


**Fig. 5.97.** Principal axes of assumed "unit", the ordering of which is the Verwey transition. $x$, $y$ and $z$ are cubic axes of the high temperature phase

symmetry of one $B$ site, which is $\bar{3}m$ with its symmetry axis along $\langle 111 \rangle$. The Verwey transition could be considered as an ordering of this unit.

Diffuse scattering of neutrons as shown in Fig. 5.98 gives microscopic information on the fluctuating local distortion of the lattice [778]. Since atomic displacements in the high temperature phase are not coherent throughout the crystal, momentum of scattered neutrons diverges in the reciprocal space. Energy analysis of the scattered neutrons gives information on the speed of the fluctuation. Figure 5.99 shows the energy distribution of neutrons scattered at $\boldsymbol{q} = (4, -0.2, -0.2)$, relative to the incident energy of 30 meV [779]. In inelastic scattering, an elementary excitation such as a phonon is created or annihilated by the scattering and the energy of the scattered neutron is changed by the process. Since the energy of the incident neutrons is the same order as that of phonons, this energy difference, i.e., the energy of a phonon, can be measured. The coordinate of Fig. 5.99 indicates this energy change.

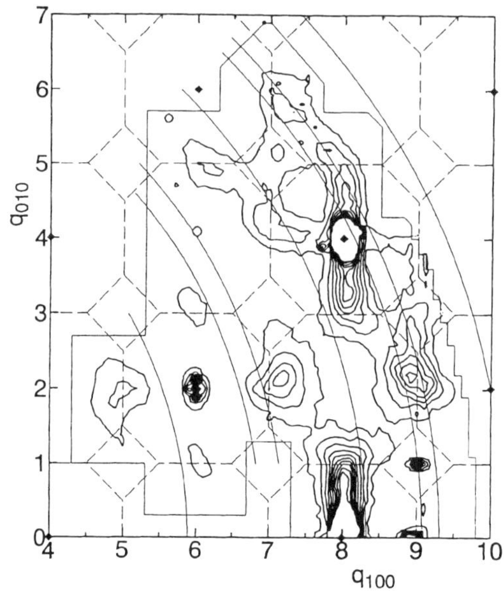


**Fig. 5.98.** Contours of neutron diffuse scattering intensity in the (001) plane in the reciprocal space, at 125 K. Rhombi indicate Bragg points of the spinel lattice [778]. (With permission from The Physical Society of Japan)

The peak at 5 meV is due to an ordinary phonon. A broad peak centered at the origin is due to the diffuse scattering we are concerned with. The theory of motional narrowing can be applied here, too, to analyse the profile of the peak. As shown in the figure by fine lines, results can be explained well if the average duration of lattice distortion is about $0.5 \times 10^{-12}$s. The coincidence between this value and that of the charge fluctuation (cf. Sect. 5.8.5) suggests strongly that the local lattice distortions are accompanied by fluctuating charges. This implies the formation of polarons.

The distribution of scattered neutrons in the reciprocal lattice (Fig. 5.98) gives information about the structure of the polaron. Strong diffuse scattering is observed at the $\Gamma$ and $X$ points in the Brillouin zone. However, the temperature dependence of the scattering suggests that the whole scattering is due to one origin. The main characteristic of the scattering is its dependence on the scattering vector. Diffuse scattering around the $\Gamma$ point, for example, is observed at (800), (840), etc., but is not detected at (620), etc. Such an existence of the "extinction rule" indicates that the symmetry of the polaron reflects symmetry in the whole lattice, instead of just a few lattice points. By analysis of the scattered neutron intensity at 19 $X$ points in the (001) and (011) planes, it was concluded that the polaron is composed of 6 phonon modes belonging to the $X_3$ symmetry. Transverse components of the atomic displacements at the $X$ point in the Brillouin zone are shown in Fig. 5.100 [778]. This rather long coherence of the atomic displacements in

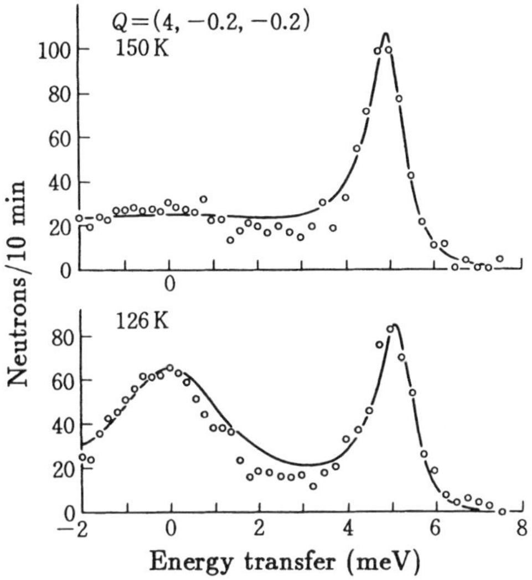


**Fig. 5.99.** Difference between energy of neutrons scattered at $q = (4, -0.2, -0.2)$ and incident energy. A peak at about 5 meV is due to an ordinary phonon. A broad peak centered at 0 meV is due to diffuse scattering. *Solid lines* are calculated line profiles for $\tau_0 = 0.44 \times 10^{-12}$ s (150 K) and $0.50 \times 10^{-12}$ s (126 K) [779]. (With permission from The American Physical Society)

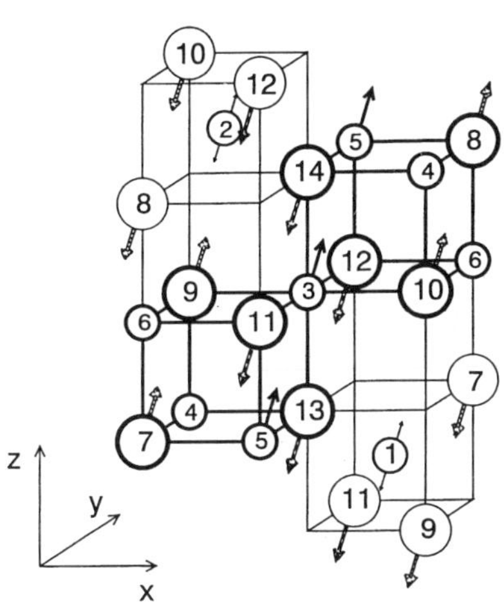

**Fig. 5.100.** Transverse displacement of atoms composing a polaron, determined at the $X$ point in the Brillouin zone [778]. (With permission from The Physical Society of Japan)

a polaron means that the electrons forming polarons do not localize on an atom but that their wave function is coherent through at least a few unit cells. This should be an indication of the itinerant character of $3d$ electrons of Fe in the $B$ site.

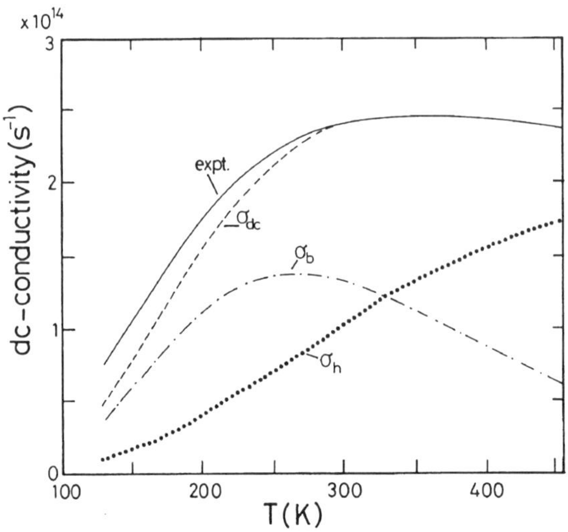


**Fig. 5.101.** Experimental and calculated temperature dependence of dc conductivity above $T_V$. The band and hopping contributions of the small polarons are shown by a *chain* and a *dotted curve*. $\sigma_{dc}$ is calculated as a superposition of these two contributions [782]. (With permission from The American Physical Society)

### 5.8.8 Itinerant Versus Localized Character of Carriers

Needless to say, the one-electron approximation is not sufficient to understand the physical properties of magnetite. The very existence of the Verwey transition shows the importance of correlation effects. Verwey's model started from here.

As for the metal–insulator transition, the ratio of correlation energy $U$ and transfer integral $t$ is important. *Ihle* and *Lorenz* estimated the critical ratio as [780]

$$\frac{U}{t} = 3, \quad U = U_1 + U_2 . \tag{5.88}$$

Here, $U_1$ is the correlation energy between electrons on nearest neighbour Fe ions and $U_2$ is that between next nearest neighbours. Note the discussion on the character of the spinel lattice, in Sects. 5.8.2 and 4.

*Ihle* and *Lorenz* later discussed small polaron conduction [781] and explained the temperature dependence of the dc conductivity as well as the frequency dependence of the optical conductivity in the range 0.001–0.65 eV, measured by *Wachter's* group [782]. The former is explained as a superposition of band and hopping conductions of small polarons whereas the latter is assigned mainly to the hopping conduction of small polarons. Examples of the comparison between their calculation and experiments are shown in Figs. 5.101 and 5.102.

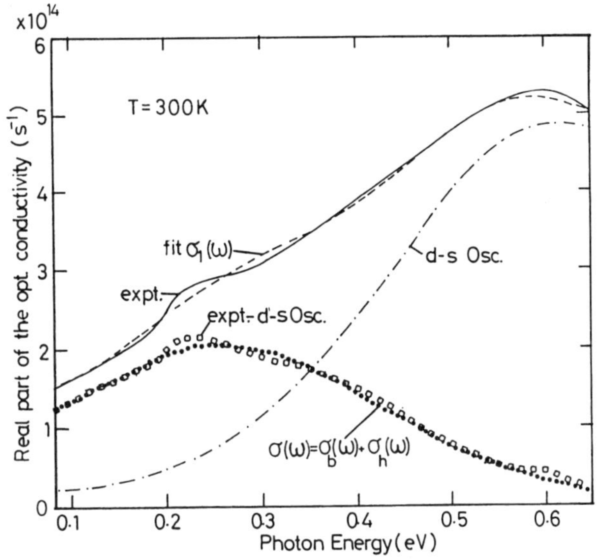


**Fig. 5.102.** Real part of the optical conductivity of $Fe_3O_4$ at 300 K, between 0.08 and 0.65 eV. The *chain line* is the estimated curve for the $3d$–$4s$ transition [782]. (With permission from The American Physical Society)

They started from the model Hamiltonian,

$$
\begin{aligned}
H &= H_0 + H_l\,, \\
H_0 &= -E_\mathrm{b}\sum n_i + \frac{1}{2}\sum U_{ij}n_in_j + \sum \omega_{q\nu}b^\dagger_{q\nu}b_{q\nu}\,, \\
H_l &= -t\sum \phi_{ij}c^\dagger_i c_j\,, \\
\phi_{ij} &= \exp\left[\sum_{q,\nu}(\Delta_{q\nu ij}b^\dagger_{q\nu} - \Delta^*_{q\nu ij}b_{q\nu})\right]\,.
\end{aligned}
\tag{5.89}
$$

Here, $i$ and $j$ refer to the $B$ sites, $b^\dagger_q$ and $c^\dagger_i$ are creation operators for a phonon and a small polaron, respectively, and $n_i = c^\dagger_i c_i$. $\phi_{ij}$ gives the coupling of small polarons to the phonon system. The electrical conductivity, $\sigma(\omega, T)$, was calculated at a frequency $\omega$ and at temperature $T$ for the high temperature phase using perturbation with respect to $t$, the nearest neighbor transfer integral, in the limit of $t \ll E_\mathrm{b}$. $E_\mathrm{b}$ is the binding energy of small polarons in the limit of $t = 0$. $\sigma(\omega)$ is expressed as a sum of $\sigma_\mathrm{b}$, diagonal part of small polaron transfer without changes in the occupation number of phonons, and $\sigma_\mathrm{h}$, off-diagonal part with changes in the phonon state. In general, the dc conductivity increases with increasing temperature by the latter, i.e., hopping conduction of small polarons, whereas by the former, i.e., band conduction, it decreases. In the present case, however, short-range ordering of small polarons due to the nearest neighbor correlation restricts the conductivity at low temperature. As pointed out in Sect. 5.8.4, no electron can move in the lattice of the $B$ site if Anderson's restriction holds perfectly, but mobile electrons should be activated by thermal energy. By this reasoning, the band conductivity of small polarons, too, should decrease with decreasing temperature below about 300 K and should result in a broad peak in the total. This temperature dependence is shown in Fig. 5.101. To explain the ac conductivity, absorption of photons by lattice vibration and $d$–$s$ transition are expressed by peaks with Lorentzian profiles characterized by the frequency, width and oscillator strength of each mode.

Several parameters had to be determined in the calculation. Drude's formula was applied to the band conduction of small polarons. The nearest neighbor correlation energy, $U_1$, was estimated as 0.11 eV from the temperature dependence of the dc conductivity below $T_V$, and the frequency of the phonon, $\omega_0$, which plays a dominant role in small polaron formation, was taken as 0.0765 eV by the highest infrared-active longitudinal optical mode. The reduced transfer integral of polarons was expressed as

$$\tilde{t} = t\exp(-S_T)\,, \quad S_T = S_0 \coth\left(\frac{\omega_0}{2k_B T}\right)\,, \tag{5.90}$$

where $t$ and $S_0$ were fixed at 0.0546 eV and 1.79, respectively, independent of temperature. Reasonable agreement was found between experiment and theory as shown in the figure.

This is a remarkable result. However, it is a phenomenological theory. The calculation is not based on a realistic band structure and parameters were set to explain optical data without any other bases. Thus, their theory can say nothing about other properties such as, e.g., the structure of the low temperature phase. The assumption of small polarons is not compatible with neutron diffuse scattering described in the last subsection. On the other hand, it is not so clear whether this assumption of the existence of small polarons is essential to their theory or not. The spread of diffuse scattering in the reciprocal space (Fig. 5.98) indicates that the coherence of polarons is rather limited in the real space: perhaps not on one atom, but probably over several unit cells. Characters of 3$d$ electrons in magnetite, with strong electron–phonon coupling as well as correlation energy, should be discussed on a realistic base. These are problems for the future.

## 5.9 High-$T_c$ Superconductors

*Bednorz* and *Müller* [783,784] first discovered the new high-$T_c$ superconductor, $La_{2-x}Ba_xCuO_4$ (LBCO), and soon after it was confirmed [785–787], a number of even higher-$T_c$ oxides were found: $YBa_2Cu_3O_7$(Y123) [788, 789], $YBa_2Cu_4O_8$(Y124), $Y_2Ba_4Cu_7O_{14}$ (Y247) [790–792], $Bi_2Sr_2Ca_{n-1}Cu_nO_{2n+4}$ ($n = 1$–3) (Bi22($n-1$)$n$) [793–795], $TlM_2Ca_{n-1}Cu_nO_{2n+2.5}$ ($n = 1$–4, $M =$

Ba, Sr) (Tl12($n-1$)$n$) [796,797], $Tl_2M_2Ca_{n-1}Cu_nO_{2n+4}$ ($n = 1$–3, $M$ = Ba or Sr) (Tl22($n-1$)$n$) [798, 799], $HgBa_2Ca_{n-1}Cu_nO_{2n+2}$ (Hg12($n-1$)$n$) [800,801] and $Nd_{2-x}Ce_xCuO_4$ (NCCO) [802,803]. NCCO is an $n$-type conductor. Still later, many other cuprates were synthesized. For instance, under high pressure: (Ag, Cu)$Ba_2Ca_{n-1}Cu_nO_{2n+4}$, (Ag, Cu12($n-1$)$n$) [804, 805], $Sr_{1-x}Nd_xCuO_2$ [806], $Sr_{1-x}Ba_xCuO_2$ [807], $La_{1.6}Sr_{0.4}CaCu_2O_6$ [808], and $La_{1.7}Ca_{1.3}Cu_2O_6$ [809]. In $R_{2-x}Ce_xMSr_2Cu_2O_{10}$ ($R$ = Eu, Gd, $M$ =Nb, Ru, Ta) [810–816], a $R_{2-x}Ce_xO_2$ layer separates the $CuO_2$ layers instead of Y in YBCO, and canted antiferromagnetism of Ru coexists with superconductivity [817]. Superconductivity was also found under a high pressure for (Sr, Ca)$_{14}Cu_{24}O_{41}$ in which a set of two lines of the corner-shared $CuO_4$ squares extends side by side with a shift of half the Cu–Cu distance as in the Magnéli phase [818]. This is called a spin ladder oxide [819, 821, 822] whose property is similar to that of $M_xV_2O_5$ (Sect. 5.5). To date, the high-$T_c$ oxides are all $d\gamma$ conductors while the other conductive oxides are $d\epsilon$ conductors except for a few examples given in Sect. 2.9. In this respect, a new field of electrically conductive oxides has been opened up and is now under development. It must be noted that the above compositions are idealized ones, and a deviation from stoichiometry is always observed in a real material.

### 5.9.1 $d\gamma$ Conductors

The first characteristic feature of the high-$T_c$ oxides is that they are the Cu oxides for which $3d$ levels are almost occupied and the intra-atomic electron–electron interaction is large. When a band picture is applicable for electrons, this means that the highest enegy level is comprised of $O_{2p}$ and Cu $d_{x^2-y^2}$ antibonding orbitals. It must be noted that the $p$ orbital is exceptionally high in energy for the high-$T_c$ oxides due to small positive charges surrounding an oxygen ion. Even when the band picture is to be modified due to the strong electron–electron interaction, the highest level is still mainly comprised of $O_{2p}$ and Cu $d_{x^2-y^2}$ orbitals. As seen from Table 2.1, most conductive oxides are composed of early elements in the periodic table, and the orbitals at the Fermi level are comprised of $O_{2p}$ and $d\epsilon$ orbitals. In this sense the high-$T_c$ oxides are rare examples of $d\gamma$ conductors for which $d\gamma$ orbitals are at the Fermi level. Compared to $d\gamma$ conductors, $T_c$ is below 30 K for $Ba_{1-x}K_xBiO_3$ (BKBO) which is an $s$–$p$ conductor [823–830], and below 14 K for $LiTi_2O_4$ which is a $d\epsilon$ conductor [552]. It must be noted that $T_c$ is as low as 0.9 K [831] for $Sr_2RuO_4((4d)^4)$ which has the same crystal structure as $La_{1-x}Sr_xCuO_4$ (LSCO) but is a $d\epsilon$ conductor and the $p$ level will not be so high as in the cuprates since Ru is +4 valent. Superconductivity of $Sr_2RuO_4$ is believed to be in a triplet $p$-wave state [1301]. The averaged Cu–O bond is a weak bond because the antibonding orbitals are nearly filled. Pd–O, Pt–O and Ag–O bonds are also weak. For instance, $PtO_2$ loses oxygens at 923 K in air [832].

The second characteristic feature of the high-$T_c$ oxides is their multilayer structure, which seems to have a large effect on the electron and phonon properties. The layer structure is also found in other oxides such as Mo bronzes [652] and $Li_xVO_2$ [833]. In the Mo bronzes, however, a charge density wave occurs and the conductivity is lost (Sect. 5.5).

The third characteristic feature is that the oxides are poor conductors in the sense that they are near the metal–insulator transition. In such a conductor, the screening of the Coulomb interaction is weak [619, 620] and a simple Fermi-liquid picture must be modified especially when the electron–electron interaction is large [834–837].

The fourth characteristic feature appears in the superconductivity itself. Besides the high critical temperature, a $d$-wave order parameter has been suggested by many experiments. In the $d$-wave state, the order parameter is supposed to have the functional form of the $d_{x^2-y^2}$ orbital.

### 5.9.2 $La_2CuO_4$

**a) Structure and Properties.** Figure 5.103 shows the $K_2NiF_4$ structure of $La_2CuO_4$ (LCO). The $LaCuO_3$ unit is shifted at every $c/2$ and it is a layered perovskite structure [838]. LCO is orthorhombic at 300 K and an orthorhombic-to-tetragonal transition occurs at about 450 K when the Néel temperature is nearly zero [839]. Metallic $La_{1.85}Sr_{0.15}CuO_4$ is tetragonal at 300 K and becomes orthorhombic near 200 K. The orthorhombic structure is

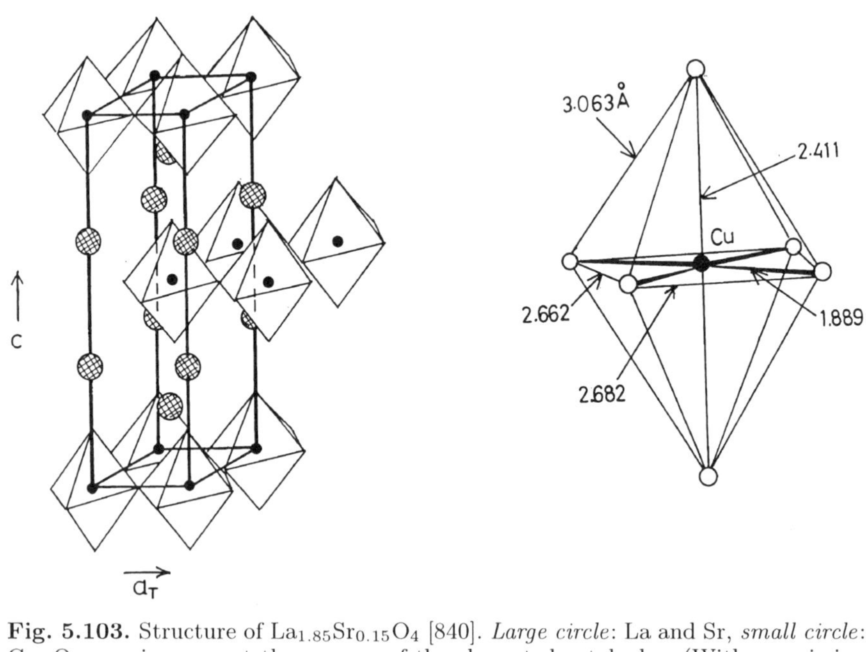


**Fig. 5.103.** Structure of $La_{1.85}Sr_{0.15}O_4$ [840]. *Large circle*: La and Sr, *small circle*: Cu. Oxygen ions are at the corners of the elongated octahedra. (With permission from The American Physical Society)

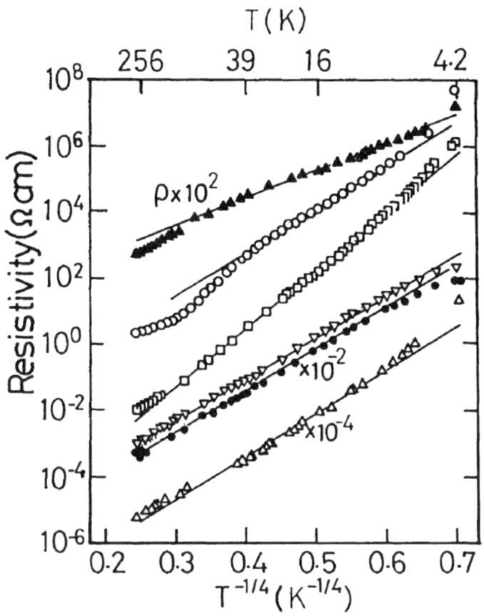


**Fig. 5.104.** Resistivity of $La_2CuO_4$ [843]. ▲: a single crystal grown by using CuO, ○: a reduced ceramic, ▽ and •: perpendicular to the $CuO_2$ plane and in the plane, respectively, for a crystal of $La_{1.98}Sr_{0.02}CuO_4$ from a Li flux, second crystal from a Li flux. (With permission from The American Physical Society)

characterized by a buckling of the Cu–O planes [840] and the buckling is common in superconducting Cu oxides except for Tl2201 [841]. *Skelton* et al. have observed a monoclinic phase below 15 K for a slowly cooled specimen [842].

LCO has a negative $d\rho/dT$ as shown in Fig. 5.104 [843]. The oxide is easily reduced or oxidized and its properties depend sensitively on the composition. LCO prepared in air shows antiferromagnetism below 220 K whereas the $T_N$ of a sample prepared at 773 K in 42 atm of $O_2$ is nearly zero [839]. *Freltoft* et al. have reported that no antiferromagnetism occurs down to 4.2 K for the sample heated in an oxygen atmosphere at 1073 K [844]. Oxygenated LCO at 1225 K in 130 atm of $O_2$ showed superconductivity below 38 K [843]. According to *Johnston* et al., the $T_N$ of $LCO_{4-y}$ changes by as much as 300 K for a change in $y$ as small as 0.03 [845]. *Saylor* et al. report that $T_N$ is maximum at 317 K for the stoichiometric sample [846]. *Reyes* et al. made a phase diagram for $LCO_{4+\delta}$ [847]. $LCO_{4+\delta}$ separates into two phases, one for $\delta$ of nearly 0.01 and the other for $\delta$ of about 0.06. The latter is metallic and shows superconductivity.

In a homologous series of $La_{n+1}Ni_nO_{3n+1}$ $(n = 1 - 3)$, $Ni^{3+}$ is observed [848]. In a normally prepared Cu oxide, the existence of $Cu^{3+}(3d^8)$ ions has not been proved, on the basis of XPS and Auger spectroscopy [849, 850].

The low-temperature-ordering magnetic moment depends on the oxygen stoichiometry and it is $0.4\mu_B$/Cu for a sample whose $T_N$ is about 300 K [844]. A spin structure is shown in Fig. 5.105 [851]. The $Cu^{2+}$ spins cant slightly

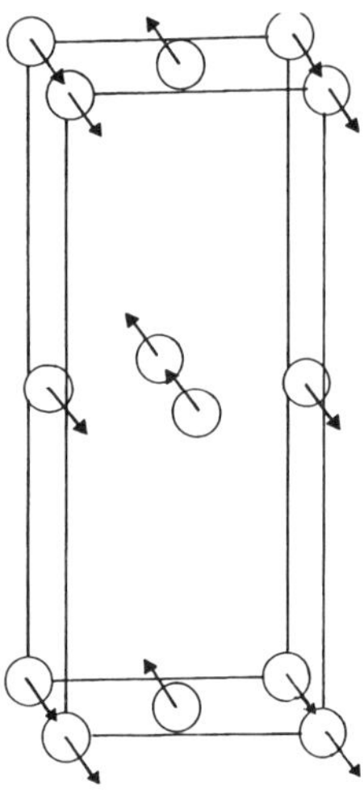

**Fig. 5.105.** Spin structure of $La_2CuO_4$ [851]. The slight canting has not been shown. (With permission from The American Physical Society)

out of the $CuO_2$ planes in an alternating way from layer to layer [852]. Under an applied field of 5.3 T (0 K), the canted component becomes ferromagnetic with a magnetic moment of $2.1 \times 10^{-3} \mu_B$ per Cu and at the same time the resistivity perpendicular to the plane decreases by about 50% while the in-plane resistivity is unaltered.

A single crystal can be prepared by using CuO or $Li_4B_2O_5$ as a flux [843] or from a melt [853]. Figure 5.104 shows the resistivity of such crystals plotted against $T^{-1/4}$ to see whether variable-range hopping conduction occurs (Sect. 5.5.2). *Brenig* et al. calculated the Seebeck coefficient of a random system [578]. In the two-dimensional lattice

$$S = \frac{W^2}{Te} \frac{2}{W} \frac{\sigma(E_F + W) - \sigma(E_F - W)}{\sigma(E_F + W) + \sigma(E_F - W)}, \tag{5.91}$$

and

$$\sigma = \frac{e^2}{k_B T} N W R^2 \gamma_0 \exp\left[-\left(\frac{T_0}{T}\right)^{1/3}\right], \tag{5.92}$$

where

$$W = \frac{k_B}{3} T_0^{1/3} T^{2/3}, \quad k_B T_0 = \delta \frac{\alpha_\perp \alpha_\parallel}{Nd}, \quad R = \frac{2}{3\alpha_\parallel} T_0^{1/3} T^{-1/3},$$

and $N$ is the density of states. $\alpha$ represents the envelope of the wavefunction as $\exp(-\alpha r)$ and $\delta$ is between 1 and 2. $d$ is the spacing of the neighboring layers. When only $N$ is energy dependent, $S$ becomes temperature-independent.

The Seebeck coefficient of LCO prepared in air is 100 μV/K and temperature-independent between 100 K and 300 K [540] which also supports the hypothesis of VRH conduction.

Here we briefly mention other non-superconducting Cu oxides. CuO is an insulator and with a doping of Li by 2% the resistivity decreases by 8 orders of magnitude. The apparent activation energy of a sintered $Li_{0.02}Cu_{0.98}O$ is 0.07 eV between 78 K and 160 K [555]. CuO is antiferromagnetic below $T_N = 230$ K and the magnetic moment is less than $0.5\mu_B$/Cu [855, 856]. PdO becomes metallic by doping of Li but does not show superconductivity down to 4.2 K [653]. $Cu_2O$ is an insulator. When prepared under an oxygen partial pressure of less than $10^{-5}$ torr, it shows $p$-type conduction with an activation energy of 0.3 eV. The magnetic susceptibility, constant between 14 and 293 K, is $\chi_m = -22.2 \times 10^{-6}$ emu/mol [857, 858]. $LaCuO_3$ has been investigated above 78 K and it is metallic [859]. $La_2SrCu_2O_{6.2}$ is metallic at least above 50 K, and it has a layer structure [860]. $La_{2-x}(Sr, Ca)_{1+x}Cu_2O_y$ is a semiconductor [861]. *Tamegai* et al. pointed out a characteristic feature similar to the weakly localized two-dimensional metal [862]. $La_4BaCu_5O_{13}$ is metallic but not superconductive down to 5 K [860]. There is a band calculation for this material [863]. $La_5SrCu_6O_{15}$ ($x = 1.33$ in $La_{8-x}Sr_xCu_8O_{20-y}$) is also a non-superconducting metal above 5 K [861]. $Bi_4Sr_8Cu_5O_{19+y}$ does not show the Meissner effect [864]. $Li_2$ (Ca, Mg)$Cu_3O_{5+y}$ is semiconducting [865].

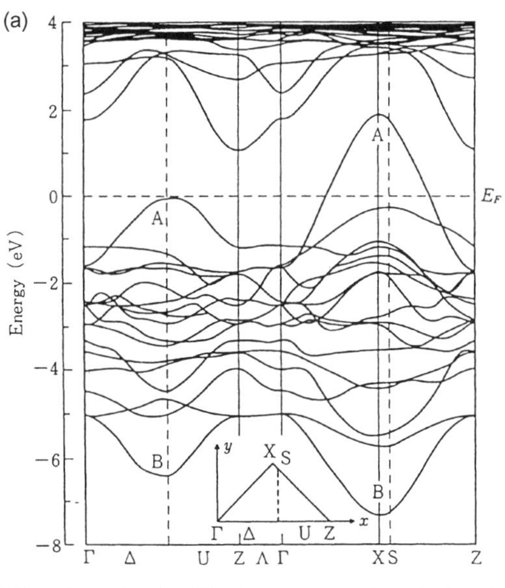


**Fig. 5.106 a, b.** Electron energy band structure of $La_2CuO_4$ in the bct (body-centered tetragonal) Brillouin zone [866]. (**b**) shows the total and projected density of states (see p. 276). (With permission from The American Physical Society)

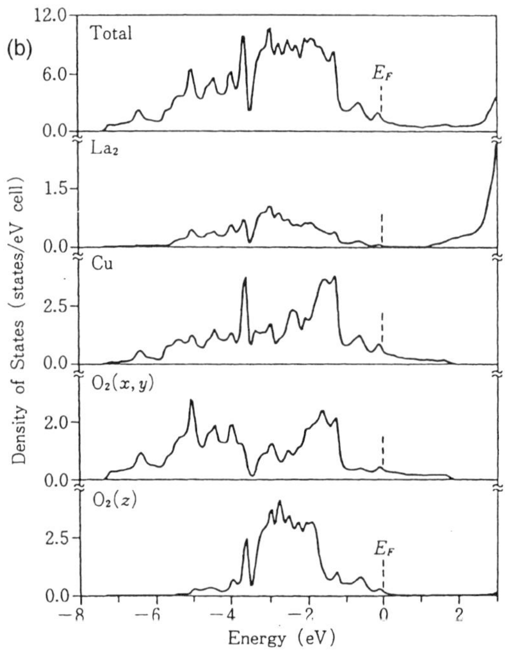


**Fig. 5.106b.** Description in caption to Fig. 5.106 a

**b) Energy Band Structure.** Band calculations were done for LCO [866–869]. Figure 5.106 shows the energy spectrum calculated by *Mattheiss* [866]. The conduction band is mainly composed of $d_{x^2-y^2}-O_{2p}$ orbitals. In the bct (body-centered tetragonal) Brillouin zone, $\Gamma - X$ is parallel with the Cu-O-Cu direction. As seen from Fig. 5.106, the calculation fails to explain the insulating property of LCO, and it is attributed to an insufficient treatment of the electron–electron interaction.

### 5.9.3 $La_{2-x}M_xCuO_4$

**a) Superconducting Properties.** *Bednorz* and *Müller* made LCO metallic by lightly substituting $Ba^{2+}$ for $La^{3+}$. They expected a strong electron–phonon (e–p) interaction at the Cu ion since it usually shows a large Jahn–Teller distortion. *Shafer* et al. determined the hole concentration in $La_{2-x}Sr_xCuO_{4-y}$ by a chemical method and found that the hole density is equal to the Sr concentration and also to $1/R_He$ up to $x = 0.15$ [870]. Above 0.15 the hole concentration decreased whereas $1/R_He$ increased. The Hall coefficient changes sign at about $x = 0.3$ [871]. *Torrance* et al. found that oxygen vacancies are generated above $x = 0.15$ and by heating the oxide in an oxygen atmosphere they extended the metallic range up to $x = 0.4$ [872]. They also determined the hole concentration, $p$, by a titration technique, with $p$ expressed as $La^{3+}_{2-x}Sr^{2+}_x[CuO_2]^{p-2}O^{2-}_{2-y}$. Figure 5.107 shows how $T_c$

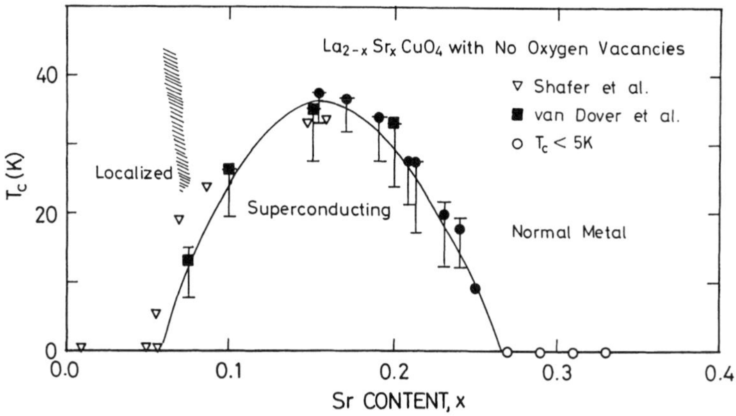


**Fig. 5.107.** $T_c$ of $La_{2-x}Sr_xCuO_4$ in relation to the hole number in the Cu-O plane per molecule [872,873]. (With permission from The American Physical Society)

depends on this hole concentration [870,872–874]. $T_c$ takes a maximum value at $x = 0.15$. Hence the terms, underdope, optimal dope and overdope come out for specifying the doping level. $T_c$ is also plotted against the Cu–O in-plane distance, becoming maximum at about 1.887 Å [875].

Many people have tried to raise $T_c$ by modifying the composition as shown in Table 5.2. As seen from the table, substitution of La by magnetic lanthanide ions changes $T_c$ very little whereas substituting Cu by Zn or Ni lowers it considerably. It must be noted that nonmagnetic Zn also reduces $T_c$ [876, 877]. The oxygen vacancy concentration will also depend on the substitution concentration and the effect of the substitution is not straightforward.

$La_{2-x}Ba_xCuO_4$ makes a transition to a low-temperature tetragonal phase at 60 K and superconductivity is severely suppressed in a narrow range of $x$ around $0.125 = 1/8$ [878]. At the same time, $d\rho/dT$ becomes negative at low temperatures and magnetic ordering occurs at $T_N = 38$ K [879, 880]. *Maeno* et el. found that the transition occurs not at $x = 0.125$ but at $x - y = 0.125$ for $La_{2-y-x}Th_yBa_xCuO_4$, and concluded that a key factor is not the Ba concentration but the hole concentration [879]. A band calculation by *Pickett* et al. shows that the density of states near the Fermi level is severely suppressed by the transition [881]. However, according to *Norman* et al., the density of states is sensitive to a tilt angle of the octahedron, and the aforementioned suppression is not so large for a real material [882]. It is widely recognized that transport properties are sensitive to the tilt angle of the $MO_6$ octahedron for perovskites. This 1/8 suppression is also observed for Zn doped Bi2212 [883]. *Tranquada* et al. observed the neutron diffraction line originating in the spin and charge stripes for $La_{1.48}Nd_{0.4}Sr_{0.12}CuO_4$ at 11 K and shed new light on this problem [884]. In this stripe structure, the doped holes are segregated at the antiphase domain boundaries between the

**Table 5.2.** Superconducting critical temperature of $La_{2-x}M_xCuO_{4-y}$ and $Nd_{2-x}Ce_xCuO_4$ type oxides. $T_{co}$: onset temperature, $T_{cm}$: mid-resistance temperature, $T_{cz}$: zero-resistance temperature

| Sample | $T_c$ | Refs. |
|---|---|---|
| $M$ = Ba | $T_{co}$ = 35 K ($x$ = 0.15) | [959] |
| | $T_{co}$ = 52 K ($x$ = 0.2, 16.8 kbar) | [960] |
| $M$ = Sr | $T_{cm}$ = 39 K ($x$ = 0.15) | [961] |
| | $T_{co}$ = 42 K ($x$ = 0.2) | [962] |
| $M$ = Ca | $T_{co}$ = 18 K ($x$ = 0.15) | [963] |
| | $T_{co}$ = 22 K ($x$ = 0.2), 25 K ($x$ = 0.3) | [964] |
| $M$ =Pb | Nonmetallic ($x = 0.1, 0.2, 0.3$) | [964] |
| $M$ = K | Nonmetallic ($x$ = 0.1) | [965] |
| $M$ = Ce | Nonmetallic ($x$ = 0.2) | [965] |
| $Ln_{0.1}Sr_{0.2}$ | $T_{co}$ = 36 K (Pr), 35 K (Nd), 32 K (Sm), 33 K (Eu), 30 K (Gd), 30 K (Tb), 32 K (Dy), 33 K (Ho), 35 K (Er), 35 K (Tm) | [966] |
| $Sr_{0.2}Cu_{1-x}M_x$ | $T_{co}$ = 23 K (Cr, $x$ = 0.01), 33 K (Mn, $x$ = 0.02) | [966] |
| $Sr_{0.15}Cu_{1-x}M_x$ | $T_{cm}$ = 39.3 K ($x$ = 0), 22.6 K ($Ni_{0.025}$), 4.2 K ($Ni_{0.05}$), no ($Ni_{0.075}$), 15 K ($Zn_{0.025}$), | [961] |
| | no ($Zn_{0.05}$), $T_{cm}$ = 13 K ($Zn_{0.025}$), ($T_{co}$ = 18 K), $T_{co}$ = 8 K $(Ni_{0.5}Zn_{0.5})_{0.025}$ | [876] |
| $Ba_{0.3}Cu_{1-x}Hg_x$ | $T_{co}$ = 35 K ($0 < x < 0.3$) | [967] |
| $Nd_2CuO_4$ type oxides.[a] | | |
| $Nd_{1.85}Ce_{0.15}CuO_{4-y}$ | $T_c$ = 24 K | [802] |
| $Nd_2Ce_{0.5}Sr_{0.5}Cu_{1.2}O_y$ | $T_{co}$ = 28 K | [803] |

[a] The crystal structure [968] is different from that of $La_2CuO_4$ and the conduction is $n$ type [802]

antiferromagnetically ordered $CuO_2$ substrate, and this charge density wave is pinned, suppressing the metallic and superconducting properties.

High pressure was applied to raise $T_c$ [885]. $dT_{co}/dp$ is 0.12–0.32 ($Sr_{0.2}$) [886, 887] and 0.64 ($Ba_{0.2}$) [787, 887] ~ 0.32 ($Ba_{0.0.15}$) [887, 888] in units of K/kbar. $T_{co}$ is the onset temperature. The pressure dependence is larger than for Y123.

*Uemura* et al. studied the spin–lattice relaxation rate of a muon and found that $T_c$ has a trapezoidal dependence on $n_s/m^*$ where $n_s$ is the density of superconducting carriers at 0 K and $m^*$ is its effective mass [889].

The isotope effect has been investigated and there appears a shift in $T_c$ suggesting a phonon contribution [890–894]. The isotope effect of a compound is expressed via a parameter $\alpha_i$ which is defined by

$$\frac{\Delta T_{ci}}{T_c} = -\alpha_i \frac{\Delta M_i}{M_i} , \tag{5.93}$$

where $\Delta$ represents the amount of shift in $T_c$ and $M_i$ which is the mass of the $i$-th ion. According to *Ronay* et al., $\alpha_{oxy} = 0.44$ for $Sr_{0.125}$ ($T_c = 28$ K) and 0.36 for $Sr_{0.15}$ ($T_c = 32$ K) [892]. The shift becomes smaller as $T_c$ increases, and is 0.5 K for $Sr_{0.15}$ ($T_c = 37$ K, $\alpha_{oxy} = 0.1$) [893, 894]. $\alpha_{oxy}$ is 0.41 for BKBO [895]. The isotope shift is also observed for Cu, and $\alpha_{Cu} = 0.93$ for LSCO ($T_c = 27.8$ K) and $\alpha_{Cu} = 0.15$ for LSCO ($T_c = 37.9$ K) [894]. The decrease in shift with an increase in $T_c$ is also observed for YBCO, suggesting the same pairing mechanism for LSCO and YBCO. As will be discussed in Sect. 5.9.4, the physical meaning of $T_c$ is disputed for underdoped (poor carrier concentration) materials, and then the isotope shift will also be different from the traditional one. The BCS theory predicts a value of 0.5 [896] which is expected when only the phonon frequency shifts. However, *Zhao* et al. point out that the effective mass of carriers also shifts suggesting a polaronic carrier which will turn into the Cooper pair [893]. *Kresin* et al. claim that the shift in $T_c$ is also affected by a number of factors not related to the lattice dynamics, for instance, a change in the carrier concentration [897].

According to the BCS theory, the superconducting energy gap at 0 K, $2\Delta(0)$, and $T_c$ obey the universal relation

**Table 5.3.** $2\Delta(0)/k_BT_c$ of $La_{2-x}Sr_xCuO_4$ and $Nd_{2-x}Ce_xCuO_4$

| $2\Delta(0)/k_BT_c$ | $2\Delta(0)$ [meV] | $T_c$ [K] | Method of measurement | Refs. |
|---|---|---|---|---|
| 2 | 6 | 35 | Reflectivity | [969] |
| 4.5 | 14 | 36 | Tunneling | [970] |
| 3.5–7 max. 12 | 10–20, max. 35 | 33 | Tunneling | [971] |
| 5.1–9.0 | 16–28 | 35.9 | Tunneling | [972] |
| 7 | 24 | 40 | Tunneling | [973] |
| 5 | | | Tunneling | [974] |
| 4.7 | 14 | 34.4 | Tunneling | [955] |
| 5.6 | 18 | 37 | Tunneling, $Sr_{0.14}$ | [975] |
| 3.9 | 7.4 | 22 | Tunneling, NCCO | [956] |
| 3.1 | 6.3 | 23 | Reflec. NCCO, ∥ *c*-axis | [976] |
| 3.9 | 6.3 | 18.5 | Reflec. NThCO, ∥ *c*-axis | [976] |
| 3.9–4.3 | 7.2–8.0 | 21.5 | Surface impedance, NCCO | [977] |

$$\frac{2\Delta(0)}{k_B T_c} = 3.52\,, \tag{5.94}$$

which is applicable for a weak coupling isotropic medium. For strong coupling superconductors, deviation from (5.94) occurs. *Schossmann* et al. calculated this quantity to be 0.53 using *Weber*'s phonon spectrum and a $\lambda$ of 2.6 [898]. The intraband Coulomb repulsion $\mu^*$ was taken to be 0.14. Table 5.3 shows experimental values of $2\Delta(0)/k_B T_c$ for $La_{2-x}M_xCuO_4$.

Since the superconducting transition is of the second kind, there appears a jump in the specific heat $\Delta C$ and the BCS theory predicts the following universal relation

$$\frac{\Delta C}{\gamma T_c} = \frac{(C_{es} - \gamma T_c)}{\gamma T_c} = 1.52\,. \tag{5.95}$$

It must be noted that $\Delta C$ is sensitive to the quality of a sample. The calculation by *Schossmann* et al. gave 2.8 [898], whereas *Nieva* et al. found experimentally that $\Delta C = 1.36\,\mathrm{J/mol\,K}$ for $La_{1.8}Sr_{0.2}CuO_4$ [899]. *Wada* et al. found that $\Delta C$ was $0.52\,\mathrm{J/mol\,K}$ for $La_{1.85}Sr_{0.15}CuO_4$ [900]. Unfortunately, $\gamma$ cannot be determined directly from the specific heat due to the large $H_{c2}$. To estimate $\gamma$, *Kwok* et al. and *Orland* et al. used $dH_{c2}/dT$ with the WHH parameter of Sect. 5.9.4 and got values of 4.9–7.3 [901] and 6.3–14.8, respectively, in units of $\mathrm{mJ/molK^2}$ [902]. Then the ratio becomes 4.4–8.4 taking $T_c$ as 42 K. *Wada* et al. measured the specific heat of a normally metallic $La_{1.7}Sr_{0.3}CuO_4$ to be $9\,\mathrm{mJ/mol\,K^2}$ [900]. If this value is applied to the $Sr_{0.15}$ sample, the ratio becomes 1.4, while *Johnston* estimated it to be larger than 2 from the magnetic susceptibility [903].

In the BCS superconductor, $\gamma$ becomes zero as temperature is lowered due to the gap opening. However, in the high-$T_c$ conductors, $\gamma$ values of $4$–$6\,\mathrm{mJ/mol\,K^2}$ were reported [899, 904–906]. As one of the RVB theories predicts gapless superconductivity [907, 908], nonzero $\gamma$ has attracted much attention. *Kato* et al. studied this term systematically for $La_{2-x}Sr_xCuO_4$ and found that it is absent only for $x = 0.01$ [909], suggesting that it cannot be attributed solely to macroscopic sample inhomogeneities. Such a nonzero term

**Table 5.4.** Critical fields of $La_{1.85}Sr_{0.15}CuO_4$. $\perp$ and $\parallel$ are with respect to the $CuO_2$-plane

| $H_{c2}(0)$ [T] | $-dH_{c2}/dT$ [T/K] | $H_{c1}(0)$ [T] | Refs. |
|---|---|---|---|
| | | $0.029(\perp)$, $0.006(\parallel)$ | [978] |
| 58 | 2.2 ($T_{cm} = 38.1\,\mathrm{K}$) | | [902] |
| | 2 ($T_{co} = 39\,\mathrm{K}$) | | [979] |
| 45 | 2.2 | | [980] |
| 50 | 2.13 ($T_c = 34.7\,\mathrm{K}$) | | [981] |

is also observed for $Li_{1+x}Ti_{2-x}O_4$ even in the insulating range [553] and it seems to be the characteristic feature of an oxide in which the Mott-Anderson type metal–insulator transition occurs. More recently a *d*-wave state has often been reported for higher-$T_c$ oxides in which the quasi-particles are excited with vanishing energy. *Momono* and *Ido* claim that the order parameter has a node based on the observation of $T^2$-dependent specific heat for LSCO far below $T_c$ [910]. On the other hand, *Ekino* et al. found an *s*-wave order parameter by the tunneling method [911].

When the magnetic field is applied, it begins to penetrate into the bulk of a superconductor at $H_{c1}$ and at $H_{c2}$ the superconductivity is destroyed except for a surface region parallel to the field. The surface superconductivity remains until $H_{c3}$, as explained in Sect. 5.3.3. In the BCS theory

$$\frac{\gamma T_c^2}{H_c^2(0)} = 0.170\,. \tag{5.96}$$

*Schossmann* et al. have calculated this to be 0.124 [898]. Unfortunately or fortunately, $H_{c2}(0)$ is so high that it cannot be determined directly and only near $T_c$ can $H_{c2}(T)$ be measured. Table 5.4 shows the results. $H_{c2}(0)$ is estimated from $dH_{c2}/dT$ at $T_c$ using the WHH calculation [912]. Note that $H_{c2}$ is not inherent to an oxide. Until these high-$T_c$ oxides were found, the Chevrel compound $PbMo_6S_8$ had the highest $H_{c2}(0)$ of 60 T [913]. As seen from (5.39–5.41), $H_{c1}$ becomes lower as $H_{c2}$ increases and the magnetic field begins to penetrate the bulk at fields as low as 300 Œ at 4.2K [914].

**b) Electronic Property in the Normal State.** Figure 5.108 shows the dependence of resistivity on the temperature and on the Sr concentration for $La_{2-x}Sr_xCuO_4$ [915]. The linear temperature dependence is observed for an optimally doped sample. *Ando* et al. applied a pulsed magnetic field to

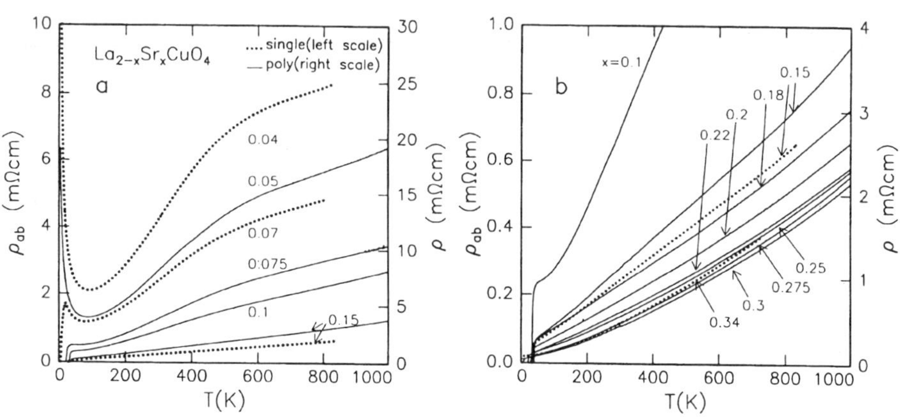


**Fig. 5.108. a, b.** The temperature dependence of the resistivity for $La_{2-x}Sr_xCuO_4$ [915]. *Dotted lines*: in-plane resistivity, *solid lines*: resistivity of polycrystals. (With permission from The American Physical Society)

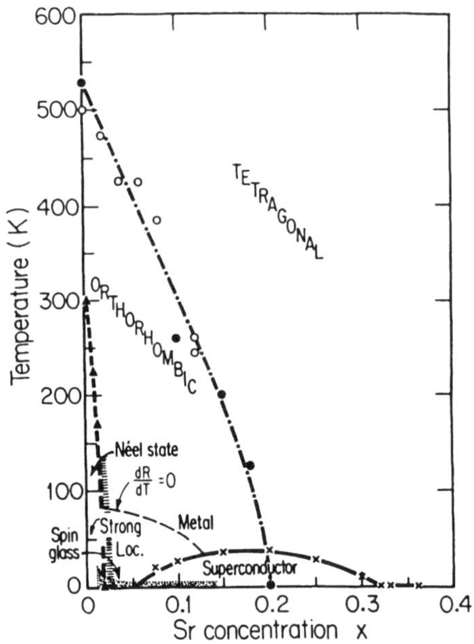


**Fig. 5.109.** Phase diagram of $La_{2-x}Sr_xCuO_4$ [920]. (With permission from The American Physical Society)

LSCO and measured the normal-state resistivity down to 2 K [916,917]. Both the in-plane and the $c$-axis resistivity showed a logarithmic temperature dependence for the underdoped (including $x = 0.15$) samples and increased as temperature decreased. For overdoped samples, in-plane resistivity showed a deviation from the higher-temperature linear dependence at lower temperatures, but it was not logarithmic. The logarithmic temperature dependence is charactersitic of the Kondo effect [918], electron localization or the electron–electron interaction in a disordered two-dimensional metallic system [919].

Figure 5.109 shows the phase diagram of $La_{2-x}Sr_xCuO_4$ [920].In general, $La_{2-x}M_xCuO_4$ becomes an insulator by either decreasing $x$ or removing oxygen ions. Both procedures reduce the hole concentration, and $p = x - 2y$ for $La_{2-x}Sr_xCuO_{4-y}$ will be a suitable parameter to express the hole concentration. In fact, *Johnston* used this parameter to scale the temperature dependence of the paramagnetic susceptibility [903]. The susceptibility shows a broad maximum at a certain temperature $T_0$, which is higher than $T_N$, as shown in Fig. 5.110 [903]. The broad peak is expected for a two-dimensional square lattice of spin $S$ in which the magnetic interaction is antiferromagnetic. $T_0$ may be expressed by [921]

$$J = \frac{k_B T_0}{2.24S(S+1) + 0.20}, \tag{5.97}$$

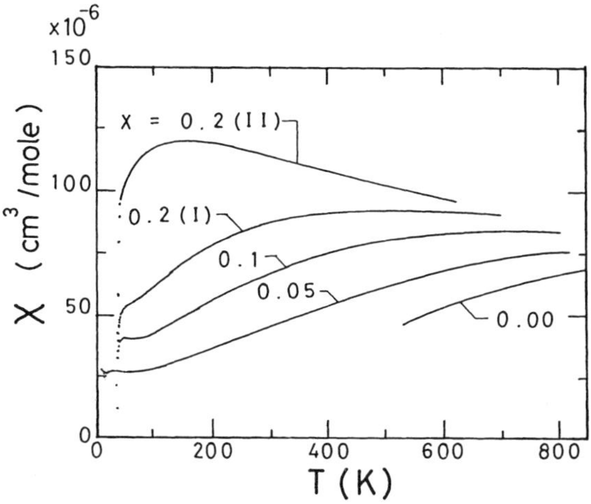


**Fig. 5.110.** Magnetic susceptibility of $La_{2-x}Sr_xCuO_{4-y}$ [903]. Curie terms have been subtracted. 0.2 (I) sample is deficient in oxygen content. (With permission from The American Physical Society)

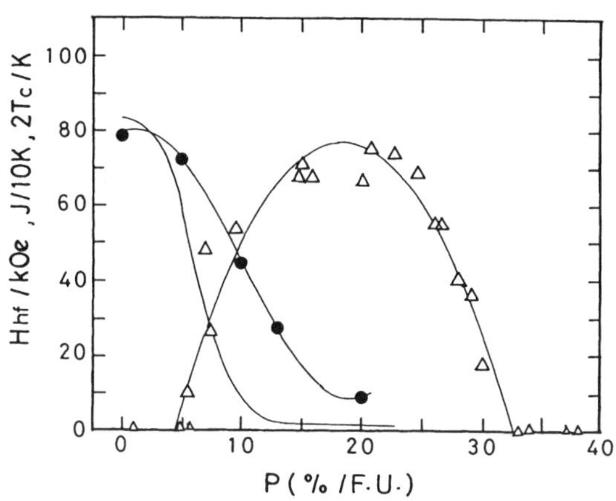


**Fig. 5.111.** Effective intralayer Cu-Cu exchange constant $J(\bullet)$ and $T_c(\triangle)$ vs $p = x-2y$ of $La_{2-x}Sr_xCuO_{4-y}$ [903]. $J$ was deduced assuming $S = 1/2$ but the magnetic moment itself decreases as $p$ increases [903]. The *solid line without a symbol* shows the hyperfine field at the Cu nucleus [900]. (With permission from The American Physical Society and Elsevier)

where the magnetic interaction between $\boldsymbol{S}_i$ and $\boldsymbol{S}_j$ is expressed as $2J\boldsymbol{S}_i \cdot \boldsymbol{S}_j$. From $T_0$, *Johnston* obtained the result shown in Fig. 5.111 [903]. If the magnetic moment were constant then $J$ would decrease as shown but *Johnston* has argued that the moment may also decrease from consideration of the intensity of the susceptibility. The decrease in the moment is consistent with a nuclear specific heat measurement by *Wada* et al. [900] which is discussed below.

*Shirane* et al. have found from inelastic neutron scattering that microscopically and in two dimensions spins are instantaneously ordered over 200 Å even at 300 K, while three-dimensional antiferromagnetism appears below 195 K for LCO [922]. When carriers are introduced, the spin correlation length is severely limited; it is $3.8/\sqrt{x}$ Å where $x$ is the number of Sr ions in one molecule [923]. They used a large single crystal of LSCO [924] and concluded that even though the spin correlation is weakened, the moment is preserved in the superconducting oxide [923, 925]. Moreover, they pointed out that the intensity of magnetic excitation becomes weaker below about 150 K when the excitation energy is 6 meV [926].

*Gutsmiedl* et al. measured the nuclear specific heat, which was proportional to $T^2$, below 1 K in both insulating and superconducting oxides and attributed it to the hyperfine field of about 30 kŒ at the Cu nuclei of a $Sr_{0.15}$ sample. They considered that the field in the superconductor was due to a fluctuating antiferromagnetism, and excluded an electric-field gradient as the origin [927]. *Wada* et al. observed a similar nuclear specific heat for LSCO and analyzed it by taking into account both the electric-field gradient and the hyperfine field at the nuclei [900]. Following them, the hyperfine field is 41 kŒ at $Sr_{0.06}$ from where superconductivity sets in. If the hyperfine constant is 122 kŒ$/\mu_B$, the magnetic moment on the Cu is 0.34 $\mu_B$ and it decreases as $x$ increases but is not zero in the superconducting compositional range. They estimated the field gradient from the nuclear quadrupole resonance (NQR) results [928]. *Kiefl* et al. reported that a muon study revealed no static or slowly-varying magnetic field in an $x = 0.15$ sample [929]. *Sugai* et al. observed the two-magnon Raman scattering and found that $JS$ depends both on the carrier density and the temperature, where $J$ and $S$ are the exchange integral and the spin. The magnon peak shifted to a lower frequency as the carrier density was increased and disappeared in the metallic compositional range [930].

The spin–lattice relaxation rate of Cu in LSCO was investigated by the NQR [931]. *Ohsugi* et al. showed that $(T_1T)^{-1}$ has a broad peak at 50 K for the $Sr_{0.15}$ sample whose $T_c$ is 38 K. Above the peak temperature, this quantity is expressed by the Curie–Weiss law for the antiferromagnetic Weiss temperature of 75 K. The Weiss temperature decreased with the Sr concentration and was nearly zero at $Sr_{0.05}$ where superconductivity disappeared. They note that these temperature dependences are well understood when the large antiferromagnetic spin fluctuation is taken into account, as in the theory due to *Moriya* et al. [932].

When the Sr concentration $x$ is increased above 0.2, the resistivity and the magnetic susceptibility become smaller until $x = 0.33$. The infrared absorption disappears for $0.25 < x < 0.50$. Superconductivity is not observed either and the oxide seems to turn into a simple metal. For larger $x$, the infrared absorption again becomes observable and the resistivity becomes large. The magnetic susceptibility follows the Curie–Weiss law. Here the randomness seems to dominate the electronic properties [933].

The electronic structure was investigated using ultraviolet photoelectron spectroscopy (UPS). According to *Fujimori* et al., the intra-atomic Coulomb interaction, $U$, of $3d$ electrons is as large as 5–6 eV and the doped hole may have mainly $O_{2p}$ character [934, 935]. Following *Shen* et al. $U$ is 6 eV and the states at or near the Fermi level have a substantial contribution from $3d$ states [936].

In many reports, the UPS density of states at the Fermi level was much smaller than the calculated one. Such a depletion of the UPS density of states was also reported for $Na_xTa_yW_{1-y}O_3$ and was attributed to the long-range Coulomb interaction as was explained in Sect. 5.4.2. The suppression of the electronic density of states and the magnetic susceptibility will be discussed in more detail for YBCO and Bi2212.

**c) Phonons and the Electron–Phonon Interaction.** The Raman spectra of single crystalline LCO show two lines near 426 cm$^{-1}$ (53 meV) and 526 cm$^{-1}$ (65 meV). The 426 line is an axial symmetric (with respect to the Cu atom) O(1, apex) stretching mode and the 526 line is a symmetric in-plane O(2, plane) stretching mode [937]. The latter should be Raman inactive in the high temperature tetragonal phase but it is observed in LCO whereas only the 427 line is observed in $La_{1.85}Sr_{0.15}CuO_4$, which is tetragonal. The authors attribute this to local symmetry breaking in the higher temperature phase. A two-phonon Raman process was observed for nonmetallic LSCO [930]. It weakened more rapidly than the single-phonon process as the oxide became metallic.

An infrared reflection study using an oriented sample of LCO and polarized light revealed TO modes at 320 (more likely 240) and 501 cm$^{-1}$, whose polarizations are parallel to the $c$ axis. The former is the vibration of all the cations against all the oxygens and the latter involves motions of oxygens with respect to copper in opposite phase in two consecutive planes. For the in-plane polarization, four TO modes were observed. They are 671 (Cu-O stretching mode), 363 (vibrations of all cations against all oxygens), 220 (motion of sheets containing Cu against intermediate sheets of oxygens), and 163 cm$^{-1}$ (motion of La against $CuO_6$) [938].

Phonon softening at the orthorhombic to tetragonal transition of LCO at 432 K was investigated [939]. The softening of the breathing mode was not found but the classic softening of the low lying (11 meV, 89 cm$^{-1}$) optical phonon involving the rotation of the $CuO_6$ octahedron was indeed observed.

The Debye temperature of $La_{1.8}Sr_{0.2}CuO_4$ is reported to be 400 K [899] or 376 K [940] from specific heat measurements. Other specific heat measurements give 439 K [905] and 356 K [941] for $La_{1.85}Sr_{0.15}CuO_4$.

Nine elastic constants of LCO were determined. For instance, $C_{11} = 1.688$ ($\parallel$ $a$-axis) and $C_{33} = 2.0$ ($\parallel$ $c$-axis) $\times 10^{12}$ dyn.cm$^{-2}$ at 44 K [942].

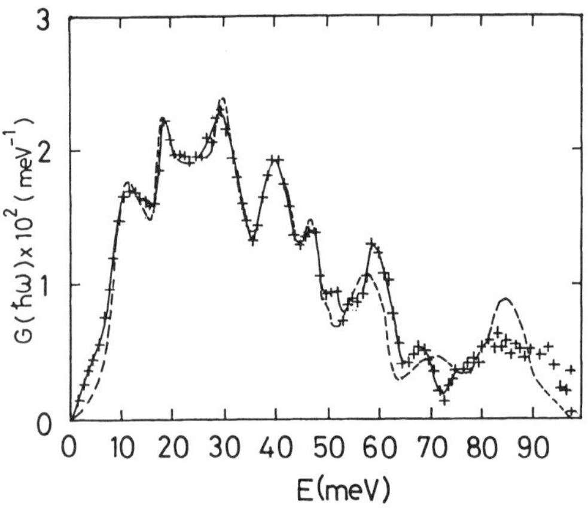


**Fig. 5.112.** Generalized phonon density of states of $La_2CuO_4$ (*crosses*) and $La_{1.85}Sr_{0.15}CuO_4$ (*dashed*) at 300 K [944]. The vibration of the *i*th atom was weighted by the ratio "scattering cross section/atomic mass". Higher frequency modes become a little softer due to the interaction with electrons

Phonon spectra of LCO and $La_{1.85}Sr_{0.15}CuO_4$ were studied using the neutron scattering [943–945]. In Fig. 5.112, the generalized phonon density of states at 300 K is shown for both oxides following *Renker* et al. [944]. Nearly the same results were obtained at 6 K. By the doping of Sr, the softening of the phonon was noted along the $\Delta$ axis for the planar breathing mode which involves the apical O(1) and the plane O(2) stretching motions at 20.4 THz (84 meV, 681 cm$^{-1}$) [945]. They also noted that a LA phonon in the $Sr_{0.1}$ sample is strongly anharmonic. Figure 5.113 shows the calculated phonon density of states and their isotope shifts for LBCO and YBCO after *de Wette* et al. [946]. Figure 5.114 shows the infrared active normal modes of the $La_2CuO_4$ lattice [947–949].

According to *McMillan*, $T_c$ is given by (5.38) [496]. Usually one speaks of a strong-coupling superconductor when the electron-phonon, e–p, interaction $\lambda$ is larger than about 1, as in $V_3Si$ (1.1), Pb (1.12), and Hg (1.0). $LiTi_2O_4$ on the other hand is a weak-coupling superconductor ($\lambda = 0.3-0.7$) as explained in Sect. 5.3.2 though $T_c$ is as high as 13 K.

*Weber* has calculated the strength of e–p interaction in a hypothetical LCO using the band calculation of *Mattheiss*, and has found that a $T_c$ as high as 35 K is attainable [950]. $\lambda$ was 2.5. *Cohen* et al. have compared the calculated phonon frequencies with the experimental ones and have suggested that the phonons are not heavily dressed by magnetic or excitonic fluctuations in LCO [869]. *Krakauer* et al. calculated $\lambda$ by the frozen-phonon model based on the LDA band calculation, and got a value of 1.37 which gives a $T_c$ of 41 K [951,952]. The e–p interaction strength was largest for an apical-oxygen oscillation along the *c* axis. *Motizuki* et al. pointed out that the interaction depends sensitively on the wavevector of phonons, and also got a large e–p

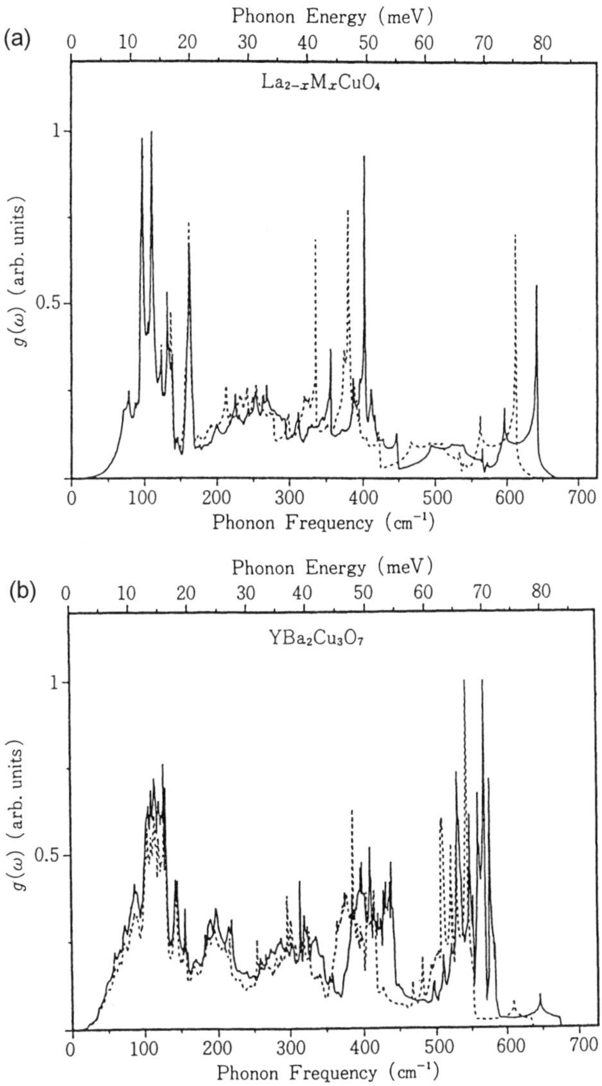


**Fig. 5.113. a, b.** Calculated density of states of phonons of LBCO and YBCO [946]. The *solid line* is for $^{16}O$ and the *dashed* for $^{18}O$. (With permission from The American Physical Society)

interaction strength [953]. *Bulaevskii* et al. and *Ekino* et al. observed the phonon structures in the tunneling conductance and thus proved that the quasi-particle states are heavily modulated by the e–p interaction [954, 955, 1171]. The spectral function of the e–p interaction was also determined for BKBO and NCCO by the tunneling method [828, 956, 957]. Using the thus determined spectral function, $T_c$ was calculated as 20 K for BKBO and 22 K

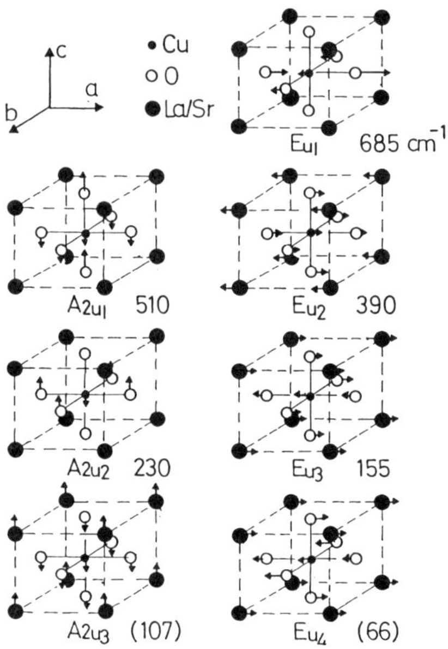


**Fig. 5.114.** Schematic representation of the infrared-active normal modes of $La_{1.9}Sr_{0.1}CuO_4$ [948, 949]. The observed and estimated frequencies are shown in units of cm$^{-1}$. (With permission from The American Physical Society)

for NCCO, and the interaction parameter $\lambda$ was 1.3 and 1.0, respectively. The experimental $T_c$ was 25 K and 22 K, respectively. *Shirai* et al. calculated $T_c$ for BKBO, and got a large value for both [830, 958].

*Tralshawala* et al. calculated the resistivity by using the spectral function of the e–p interaction which was determined by the tunneling experiment for NCCO [616]. The calculated resistivity was proportional to temperature. Subtracting it from the observed resistivity, they found that the remaining resistivity was proportional to $T^2$ which indicates electron–electron scattering. The $T^2$ term had almost the same magnitude as the $T$-linear term at 300 K.

*Salchow* et al. have shown that the $T_c$ of $Na_xWO_3$ is very sensitive to the carrier concentration [619, 620] and it increases toward the metal–insulator boundary (Sect. 5.4.3). In the cuprates, the e–p interaction will also be strengthened toward the boundary. Nevertheless, the experimental $T_c$ decreases toward the boundary, and instead, the spin effects become noticeable.

### 5.9.4 $YBa_2Cu_3O_{7-x}$

*Wu, Chu,* [787,788] and *Hikami* et al. [789] discovered the YBCO oxide which has an even higher $T_c$ than LBCO. Its properties sensitively depend on the oxygen vacancy concentration and the vacancies are easily produced due to the weaker Cu–O bond especially in a linear chain. A loss of oxygen has been noted even at 50 K in a vacuum, especially at the surface region [982–984].

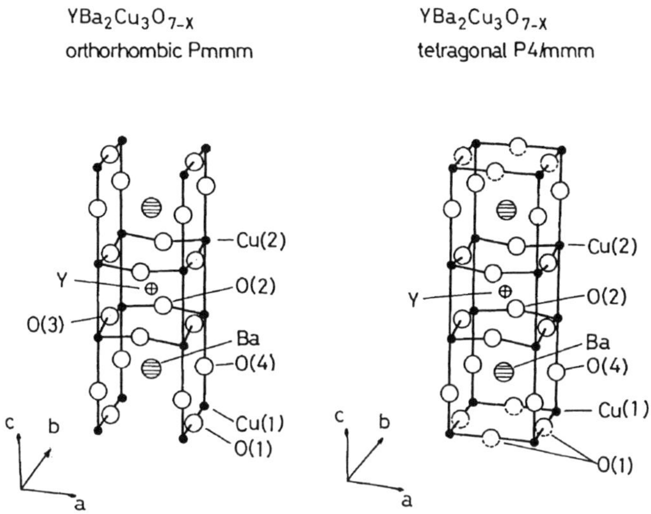


**Fig. 5.115.** Crystal structure of $YBa_2Cu_3O_{7-x}$ [985]. In the tetragonal phase, O ions occupy O(1) sites randomly. (With permission from The American Physical Society)

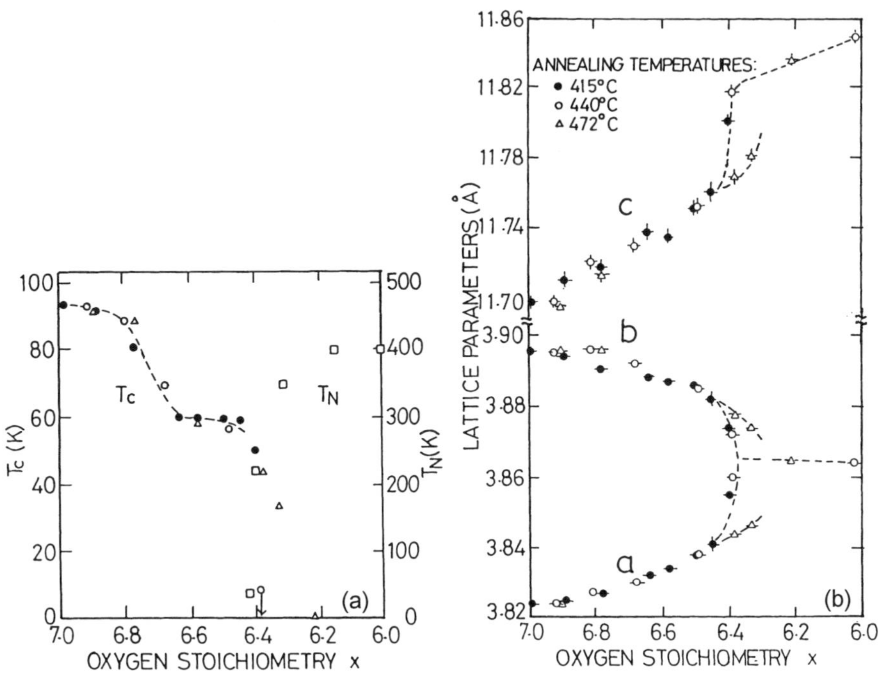


**Fig. 5.116. a, b.** $T_c$, $T_N$ and the lattice parameters in relation to $x$ of $YBa_2Cu_3O_x$ [986, 1087]. (With permissions from Elsevier and The American Physical Society)

**a) Crystal Structure.** The crystal structure is shown in Fig. 5.115 [985]. As oxygen is removed, the 60 K-orthorhombic II phase appears and finally the lattice becomes tetragonal with the disappearance of superconductivity as shown in Fig. 5.116 [986–988]. At elevated temperatures oxygen vacancies distribute randomly in the chain, and the orthorhombic lattice transforms to a tetragonal one [988–991].

*Karpinski* et al. and *Santro* et al. have synthesized two other 81 K superconductors $YBa_2Cu_4O_8$ and 40 K $Y_2Ba_4Cu_7O_{14+x}$, which contain double linear Cu-O chains [790–792].

In many BCS superconductors, a phonon softening is often observed above $T_c$ due to a strong e–p interaction. For instance a phase transition occurs at 43 K in $Nb_3Sn$ whose $T_c$ is 18 K [992]. In YBCO, such a phase transition has not been observed but a gradual crossover to a pseudogap state occurs in the underdoped region as explained in Sect. 5.9.4.

**b) Superconducting Transition Temperature.** The $T_c$ depends on the hole concentration on the $CuO_2$ layers. This was determined by *Tokura* et al. who measured the hole concentration with a titration technique for variously substituted oxides [993]. They proposed that a hole enters first into the Cu-O linear chain and after reaching a critical concentration, it then begins to enter the $CuO_2$ layer and the oxide becomes metallic and superconducting. Figure 5.117 shows how $T_c$ changes with the estimated hole number on one $CuO_2$ layer per molecule. For $YBa_2Cu_3O_y$, it is $(y - 6.5)/2$. Figure 5.118 shows a relation between $T_c$ and the Hall carrier density per molecule after *Wang* et al. [994]. The Hall carrier density is defined by $n_H = 1/R_He$ where $R_H$ is the Hall coefficient.

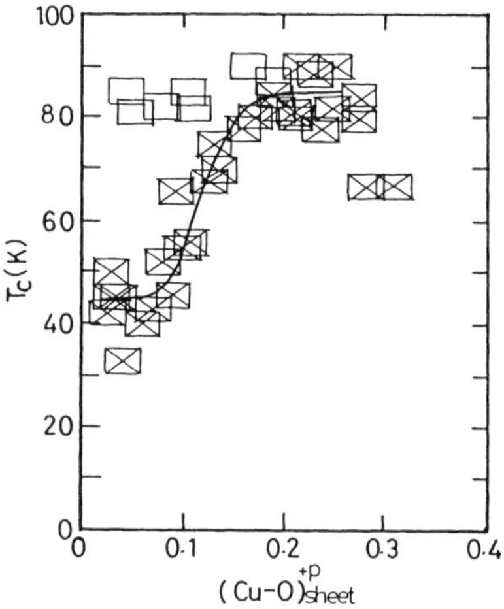


**Fig. 5.117.** $T_c$ in relation to the estimated hole number $p$ on one $CuO_2$ sheet per molecule [993]. (With permission from The American Physical Society)

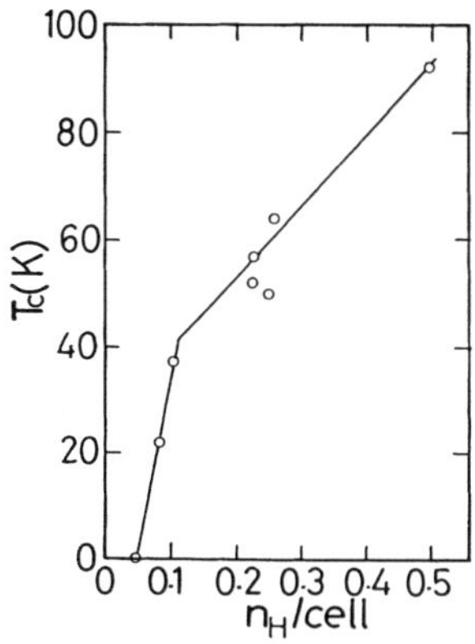


**Fig. 5.118.** $T_c$ in relation to the Hall carrier density per molecule $n_H$ [994]. (With permission from The American Physical Society)

Various elements have been substituted and the $T_c$ measured as shown in Table 5.5. Divalent elements such as $Ni^{2+}$, $Zn^{2+}$, and $Cd^{2+}$ tend to enter the $CuO_2$ plane and trivalent elements such as $Al^{3+}$, $Fe^{3+}$, and $Ga^{3+}$ enter into the Cu–O linear chain [995, 996]. It is not clear in many cases whether the variation in $T_c$ is simply due to the substituted elements or to the accompanying oxygen vacancies.

The isotope effect has been intensively investigated [997–1006]. It is very small when $T_c$ is high. For instance, $\Delta T_c/T_c$ was 1.65 K/30.6 K ($\alpha_i$ = 0.45), 1.42/54.3 (0.23), and 0.80/75.6 (0.09) for $Y_{1-x}Pr_xBa_2Cu_3O_{7-y}$ when 72% to 85% oxygen was substituted [1001]. Similar shifts were found for $YBa_{1-x}La_xCu_3O_{7-y}$ such as 1.75 K/38.3 K ($\alpha_i$ = 0.38), 1.93/60 (0.27), 1.29/77.3 (0.14), 0.41/91.9 (0.039), and 0.28/92.3 (0.025) when more than 94% of oxygens were substituted [1002]. *Zech* et al. investigated the shift in $T_c$ by selectively substituting oxygens and concluded that the plane oxygens contribute more than 80% to the shift and the apical oxygen contributes little [1003]. *Müller* suggested that the apex oxygen greatly contributes to $T_c$ for its anharmonic vibration whose isotope effect will be small [1004]. The anharmonicity of the apex oxygen vibration, is inferred from the isotope shift of the 490 $cm^{-1}$ mode [1005]. *Zhao* et al. observed a large Cu isotope effect for underdoped YBCO [1006]. The small isotope shift at higher $T_c$ was discussed by many people. According to *Carbotte*, it indicates a coexistence of the e–p interaction with other non-phonon interactions which give a higher $T_c$ [1007]. The van Hove singularity [1008–1010], anharmonicity [1011–1013], strong e–p interaction [999] and non-adiabatic effects [1014, 1015] may also reduce the shift. *Zech* et al. claim that the shift in $T_c$ occurs not only by the frequency shift but also by the isotope-dependent condensate density (carrier density/carrier mass), and that the former is always small while the latter is large for underdoped regions, thus giving the larger isotope effect for lower $T_c$

**Table 5.5.** Superconducting transition temperature of $YBa_2Cu_{3-x}M_xO_{7-y}$ and related materials. $T_{co}$: onset temperature, $T_{cm}$: mid-resistance temperature, $T_{cz}$: zero-resistance temperature, $n_H$: Hall carrier density per molecule

| Sample | $T_c$ | Refs. |
|---|---|---|
| YBCO | $T_{co} = 93$ K, | [788] |
| | 85 K | [789] |
| | 96 K | [1149] |
| | 22 K ($n_H = 0.08$), 37–43 K ($n_H = 0.1$), 50 – 65 K ($n_H = 0.22$–$0.25$), 92 K ($n_H = 0.5$) | [994] |
| $H_xYBa_2Cu_3O_{7-y}$ | $T_{co} = 92$ K→ 94.8 K ($x = 0 \rightarrow 0.58$) | [1150] |
| $O_7 \rightarrow O_6S$ | $T_{co} = 90$ K | [1151] |
| $O_7 \rightarrow O_6Cl$ | $T_{co} = 72$ K | [1151] |
| $M$ = Al | $T_{co} = 92$ K ($x = 0$), 82 K (0.11), 45 K (0.13), 20 K (0.18), < 4.2 K (0.22) | [1152] |
| $M$ = Fe | $T_{co} = 90$ K ($x = 0$), 82 K (0.03), 60 K (0.06), 50 K (0.15), 33 K (0.3) | [1151] |
| $M$ = Co | $T_{co} = 92$ K ($x = 0$), 90 K (0.1), 68 K (0.2), 15 K (0.4) | [1153] |
| $M$ = Ni | $T_{co} = 91$ K ($x = 0$), 75 K (0.09), 54 K (0.15) | [1154] |
| | $T_{co} = 92$ K ($x = 0$), 64 K (0.25), 52 K (0.5) | [961] |
| $M$ = Zn | $T_{co} = 92$ K ($x = 0$), 59 K (0.1), 50 K (0.2), 15 K (0.3) | [1153] |
| $M$ = Ga | $T_{co} = 92$ K ($x = 0$), 75 K (0.15), 66 K (0.3) | [1155] |
| $Ba_2 \rightarrow Ba_1Ca_1$ | $T_{co} = 83$ K | [1156] |
| $Ba_2 \rightarrow Ba_{2-x}Sr_x$ | 92 K→ 80 K ($x = 0 \rightarrow 1.2$, stability limit) | [1157] |
| $Ba_2 \rightarrow Ba_{1.9}La_{0.1}$ | $T_{cm} = 92.5$ K→ 94.0 K | [1158] |
| $Ba_2 \rightarrow Ba_{2-x}La_x$ | 97 K→ 99 K ($x = 0 \rightarrow 0.05$, $y_{min} = 0.06 - 0.5\,x$), 95 K→ 38 K ($x = 0.1 \rightarrow 0.5$, $y_{min} = 0.15 - 0.5\,x$, tetragonal when $x > 0.3$) | [1159] |
| $Sr_2$ | $T_{cz} = 81$ K (tetragonal) | [1160] |
| Y→Ln | $T_{co} = 93$ K (Y), 81 K (La), $X$ (Ce), $X$ (Pr), 94 K (Nd), 89 K (Sm), 94 K (Eu), 94 K (Gd), $X$ (Tb), 93 K (Dy), 93 K (Ho), 93 K (Er), 92 K (Tm), 93 K (Yb), ($X$ nonsuper.,) | [1161] |
| $Y \rightarrow Y_{0.75}Sc_{0.25}$ | $T_{co} = 92$ K | [1156] |
| $Y \rightarrow Y_{0.9}Ca_{0.1}$ | $T_c = 89$ K | [1162] |
| $YBa_2Cu_4O_8$ | $T_{co} = 81$ K | [790] |
| $Y_2Ba_4Cu_7O_{14+y}$ | $T_{co} = 40$ K | [791] |
| | $T_{cz} = 85.5$ K | [1163] |
| | $T_c = 92$ K | [1164] |
| $Tl_{0.5}Pb_{0.5}Ca_{0.8}Y_{0.2}Sr_2Cu_2O_y$ | $T_{cz} = 108$ K | [1165] |

substances [1016]. On the other hand, *Schüttler* and *Pao* point out that cooperation of the electron-anharmonic phonon interactions with the electron-spin fluctuation interaction can account for the observed isotope effect, while harmonic phonons cannot [1017]. Some people consider that $T_c$ in the underdoped region is a temperature which separates a coherent pairing state from an incoherent state and not a temperature at which the amplitude of the order parameter becomes zero. Then the isotope effect will have different physical meaning, and the traditional isotope effect would have to be considered to be very small, as found for the optimally doped highest $T_c$ substances.

High-pressure experiments show that $dT_{co}/dp$ is 0.043 K/kbar, for YBCO up to 170 kbar where $T_{co}$ is an onset temperature, and is 91 K at 1 atm [1018, 1019]. This value is much smaller than the 0.64 K/kbar of LBCO. For a single crystal with a $T_c$ of about 91 K, a negative pressure effect, $dT_c/dp = -0.085$ K/kbar, has been reported [1020]. For the double chain oxide Y248, as large a value as 0.55 K/kbar was reported for both polycrystalline and single crystalline specimens [1021].

**c) Superconducting Properties.** When an alloy becomes superconducting, superconductivity is usually present throughout the whole specimen. However, in the new oxides the volume fraction of the superconducting region is not one and volume quantities may be somewhat ambiguous.

The superconducting energy gap has been estimated by various methods and it is shown in Table 5.6 in the form $2\Delta(0)/k_B T_c$ which is 3.52 for weak-coupling BCS superconductivity. *Marsiglio* et al. have calculated $2\Delta(0)/k_B T_c$ in the extremely strong coupling limit, and found that it increases with $T_c/\omega_0$, where $\omega_0$ is a phonon frequency, and reaches a value as large as 10 [1022].

One of the most characteristic features of superconductivity appears in the order parameter. For usual BCS superconductors, the energy gap is isotropic and it is an $s$-wave. However, when the electron–electron (e–e) Coulomb repulsion is large the attractive interaction may become effective for carriers on different sites and then the order parameter may become a $d$-wave for the tetragonal lattice [1023–1028]. For the $d$-wave of $E_g$ symmetry, the absolute value of the gap becomes largest along the Cu-O-Cu direction as for a $d_{x^2-y^2}$ orbital. The $d$-wave state is possible even for phonon-mediated superconductivity [1029–1032]. Experimentally the situation is still controversial, but results for the $d$-wave are growing. *Wollman* et al. reported that the order parameter changes sign between the $a$ and $b$ directions by using the YBCO-Au-Pb Josephson junction [1033]. *Kirtley* et al. found a half-integer flux quantization by using specially designed grain-boundary Josephson junctions, and analyzed the result by the $d$-wave order parameter [1034]. *Alff* et al. observed a pronounced zero-bias conductance peak for a sputtered YBCO/vacuum gap junction along the Cu-Cu direction which had been predicted by their theory for the $d$-wave junction [1035, 1036]. The peak to peak gap $2\Delta_{pp}$ was 40 meV and $2\Delta_{pp}/k_B T_c$ was 5.7. *Zhang* et al. measured a microwave surface impedance for 93.4 K-$YBCO_{6.95}$, and found that the penetration depth was

**Table 5.6.** $2\Delta(0)/k_BT_c$ of $YBa_2Cu_3O_{7-x}$ and related materials (see also Sect. 5.9.4)

| $2\Delta(0)/k_BT_c$ | $2\Delta(0)$ [meV] | $T_c$ [K] | Method of measurement | Refs. |
|---|---|---|---|---|
| 3.2 | 23 | 84 | Tunneling | [1166] |
| 3.6 | 28 | 90 | Tunneling | [1167] |
| 4.8 | 39 | 92.4 | Tunneling | [1168] |
| 5.8 | 33 | 83 | Tunneling, *c*-plane | [1169] |
| 3.6 | 20 | 83 | Tunneling, *c*-axis | [1169] |
| 3-4 | 22–29 | 85 | Tunneling | [1170] |
| 10–11.6 (Y→Eu) | 73–91 | 85–91 | Tunneling | [1171] |
| 3.3 (Y→Eu) | 26 | 90 | Thermal conduc. | [1172] |
| 4.7 | | | Infrared reflec. | [1173] |
| 8 | 62 | 90 | Infrared reflec. | [1174] |
| 8 | 63 | 92 | Infrared reflec., *c*-plane | [1175] |
| 3 | 24 | 92 | Infrared reflec., *c*-axis | [1175] |
| 3.5 (chain) | 27 | 90 | NMR | [1047] |
| 3.8–6.2 (plane) | 29–48 | 90 | NMR | [1047] |
| 2.9 (Fe 1%) | 23 | 92 | Reflec. | [1176] |
| 1.1 (Fe 6%) | 6.8 | 72 | Reflec. | [1176] |
| 9.0-3.9 (Y→Gd) | $\Delta_{max}35$, $\Delta_{min}15$ | 90 | tunneling | [1177] |
| 6.5 (Y124 plane?) | | 72 | Raman | [1110] |
| 2.3 (Y124 chain?) | | 72 | Raman | [1110] |

proportional to temperature below 15 K [1037]. This also indicates a *d*-wave order parameter. At the same time, there are experiments for an *s*-wave component or a large anisotropy [1038–1041] and it is not yet clear whether any high-$T_c$ superconductivity is characterized by a pure *d*-wave. It must be noted that the origin of the attractive interaction cannot be identified only from the symmetry of the order parameter when both the e–e and the e–p interactions are large.

The nuclear spin–lattice relaxation rate $T_1^{-1}$ of plane Cu for 93 K-YBCO ($T_c = 93$ K) was as large as $1\,s^{-1}$ at 10 K and showed a gradual decrease even below this temperature [1042–1044]. $T_1^{-1}$ of plane O was smaller than that of Cu by one order but showed a similar temperature dependence [1044]. Since the quadrupolar relaxation cannot easily explain such a large and nearly temperature-proportional rate [1045], these phenomena suggest low-lying excitation states to exist as in the *d*-wave superconductor [1046]. The Knight shift of the plane Cu fits to both the *s*- and *d*-wave as well [1047]. For a usual BCS superconductor, the relaxation rate of a phenomenon in which the spin flips increases just below $T_c$, and it is known as the Hebel–Slichter peak for the nuclear-spin relaxation [1048]. For 91 K-YBCO such a peak was very weak or absent [1042–1044], suggesting a *d*-wave state [1049,1050]. However, *Allen* and *Rainer* showed that the peak may be absent even for an *s*-wave when the e–p interaction is very large [1051].

The jump in the specific heat $\Delta C$ at the transition temperature was measured. For instance, 3.91 J/Kmol (93.8 K-GdBCO) [1052], 4.13 J/Kmol [1053], 5.1 J/Kmol (90 K-YBCO) [1054], 3.7 J/Kmol (92.5 K-DyBCO) [1055], and 5.3 J/Kmol (92.5K-YBCO) [1056]. It is assumed that 1 mol=13 gatom. According to *Loram* et al. $\Delta C/C$ at $T_c$ is estimated to be 2.5 for 90 K-YBCO and 0.5 for 60 K-YBCO [1056–1058]. *Marsiglio* et al. have shown by calculation that $\Delta C/C$ at $T_c$ increases with $T_c/\omega_0$ at first and, after reaching a maximum, decreases to even smaller values than the BCS value of 1.43 in the extremely large e–p coupling limit, where $\omega_0$ is a phonon frequency [1022].

The low-energy excitation within the gap is observed even in a superconducting state [905, 1052, 1056, 1059, 1060], and the density of states of excitations increases with Zn doping [1060] suggesting a *d*-wave state.

The thermal conductivity increases below $T_c$ as temperature is lowered, and after reaching a maximum it decreases to zero, suggesting that the heat transport is predominantly by phonons which are strongly scattered by normal electrons. Above $T_c$, the thermal conductivity is nearly constant up to 140 K while the electrical resistivity increases in proportion to temperature. This also suggests that the main carriers of heat are the phonons [1062].

The diamagnetic properties have been investigated as shown in Table 5.7. $H_{c2}(0)$ is given by solving the WHH equation in which a parameter $\alpha$ is given by

$$\alpha = \frac{3e^2\hbar\gamma\rho_n}{2m\pi^2 k_B^2},$$

or equivalently

$$\alpha = 5.2758 \times 10^{-5}\left(-\frac{dH_{c2}}{dT}\right)_{T_c},$$

where the tangent is in units of gauss/K [912]. From $H_{c2}(0)$, the effective coherence length is obtained using [1062]

$$H_{c2\perp}(0) = \frac{\phi_0}{2\pi\xi_\parallel^2}, \quad H_{c2\parallel}(0) = \frac{\phi_0}{2\pi\xi_\parallel\xi_\perp},$$

where $\phi_0$ is the flux quantum $= h/2e = 2.068 \times 10^{-15}$ Tm$^2$, $\xi_\parallel(0)$ is 27 Å and $\xi_\perp(0)$ is 6 Å from 190 T and 45 T, respectively [1063]. *Umezawa* et al. have reported 22–31 Å for $\xi_\parallel(0)$ and 4.8–6.7 Å for $\xi_\perp(0)$ [1064]. From $H_{c1}$ and $H_{c2}$, $\kappa$ has been calculated using (5.39,40), as 130 (YBCO) [1065] and 92 or 42 (LaBCO) [1066]. Using the single crystal values, $\kappa_\perp$ is calculated to be 230 and $\kappa_\parallel$ is 29–44 [1064].

**Table 5.7.** Critical field of Y(Eu, La)$Ba_2Cu_3O_{7-y}$. $\parallel$ and $\perp$ are with respect to the $CuO_2$-plane

| $H_{c2}(0)$ [T] | $-dH_{c2}/dT$ [T/K] | $H_{c1}(0)$ [T] | Refs. |
|---|---|---|---|
| $\parallel$ 222 | 0.7 at 89.7 K, 3.6 at 85 K | | [1070] |
| $\perp$61 | 0.2 at 89.7 K, 1.1 at 67 K | | [1070] |
| $\parallel$ 210 | 3.3 | | [1179] |
| $\perp$35 | 0.56 | | [1179] |
| $\parallel$ | | 0.02 | [1067] |
| $\perp$ | | 0.4 | [1067] |
| $\parallel$ 674 | 10.5, $T_c = 92.5$ K | | [1180] |
| $\perp$122 | 1.9 | | [1180] |
| $\parallel$ 230 | | 0.012 (11 K) | [1064] |
| $\perp$35–70 | | 0.069 (11 K) | [1064] |
| $\parallel$ | | 0.018 | [1181] |
| $\perp$ | | 0.053 | [1181] |
| $\parallel$ 190 (Eu) | 3.0 at 94.8 K | | [1063] |
| $\perp$45 (Eu) | 0.7 at 94.8 K | | [1063] |
| 75 (La), poly. | 1.2 at 83.5 K | 0.02 | [1066] |

**d) Electronic Properties in the Normal States.** Since reduction occurs at higher temperatures in air, the oxygen content can be varied by controlling the quenching temperature [1068, 1069].

The resistivity of a single crystal of 89.7 K-YBCO is about 140$\mu\Omega$ cm at 100 K in the $c$ plane. It increases linearly with temperature and is about 400$\mu\Omega$ cm at 280 K [1070]. For a single crystal 94.8 K-EuBCO, $\rho$ is 410$\mu\Omega$ cm at 100 K. The linear temperature dependence is lost when the oxygen concentration is decreased from 6.9 of 90 K-YBCO as shown in Fig. 5.119 [1071]. The deviation from linearity occurs around $T^*$, below which the pseudogap states of spin and charge are observed as explained below. This deviation indicates that at least the non-linearly increasing part of the resistivity is not due to phonon scattering but due to spin or electron scattering. It is important to compare the scattering rate by phonons and by spins or electrons to clarify the pairing mechanism, but this has not yet been done qualitatively except for a few cases: examples are the analysis by *Tralshwala* et al. for NCCO [616] and by *Shimada* et al. for Bi2212 [617] (Sect. 5.9.3).

The Hall coefficients $R_H$ of high-$T_c$ YBCO and GdBCO are proportional to $T^{-1}$ [1072]. When Ni or Co is substituted for Cu, $R_H$ becomes larger and decreases as temperature increases but is not proportional to $T^{-1}$ [1073]. Similarity of $R_H$ with that of a dilute magnetic alloy was pointed out [1074]. The Hall angle $\Theta_H$ ($= \tan^{-1} \sigma_{xy}/\sigma_{xx}$), where $\sigma_{ij}$ is the conductivity tensor, shows

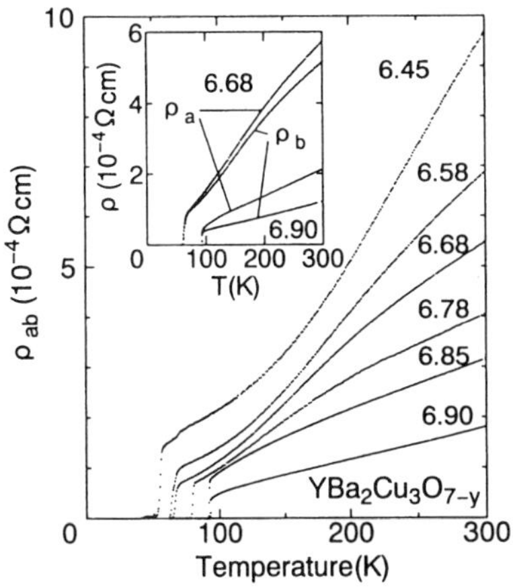


**Fig. 5.119.** The temperature dependence of in-plane resistivity for twinned $YBa_2Cu_3O_{7-y}$ crystals [1071]. *Inset*: for detwinned crystals of $T_c = 90$ K and 60 K. (With permission from The American Physical Society)

a temperature dependence of $\cot\Theta_H = aT^2 + b$ for 60 K-90 K-YBCO [1075]. *Anderson* has pointed out that the $T^2$ dependence is explainable by the spinon–spinon and spinon-holon scattering [1076]. *Stojković* and *Pines* tried to explain the temperature dependence of the Hall angle by spin fluctuation scattering [1077]. *Timusk* et al. noted that the plasma frequency, which depends on the carrier density, was not temperature dependent, from the analysis of the infrared-frequency conductivity [1078]. For a simple single-band metal, $\cot\Theta_H$ is proportional to $T$, while when there are many Fermi surfaces as for NbO, $R_H$ may depend on temperature (Sect. 2.9.1).

The magnetic susceptibility is also sensitive to oxygen content and shows a characteristic feature of a two-dimensional antiferromagnetic lattice as shown in Fig. 5.120 [1079, 1080]. However, there is no peak in the susceptibility in the measured temperature range in contrast to LCAO indicating that the exchange interaction is stronger in YBCO than in LSCO. *Westerholt* and *Bach* estimated that the superexchange interaction $J$ between in-plane Cu ions is larger than 1000 K in the insulating oxide [1080]. *Pennington* et al. studied the transverse NMR relaxation of Cu nuclei in a 90 K single crystal and suggested that there might be an electronic magnetic moment on Cu and that $J$ is about 1100 cm$^{-1}$ (140 meV) [1081]. *Tranquada* et al. have estimated $J$ of $T_N = 200$ K oxide as 80 meV from neutron scattering experiments [1082]. As shown in Fig. 5.120 b, the susceptibility becomes smaller below a certain temperature when the sample is reduced, and this has been attributed to a fluctuation in the order parameter [1079]. However, *Schneider* claims that the fluctuation occurs only for a sample whose width is thinner than the 4-unit-cell length [1083]. Then the smaller susceptibility suggests that the short-range antiferromagnetic coupling suppresses spin freedom in these temperature ranges. Sometimes the Curie–Weiss paramagnetism is observed but it has been attributed to an impurity oxide such as $Y_2BaCuO_5$ [1084]. Anisotropy in the susceptibility was attributed to that in the orbital nature of electrons [1085].

The muon experiment shows that superconducting YBCO does not manifest a long-range magnetic ordering while the insulating one ($O_{6.2}$) is antiferromagnetic with $T_N$ about 300 K [1086]. *Tranquada* et al. constructed a phase diagram using the neutron diffraction results as shown in Fig. 5.116 [1087]. For $y = 1$ in $YBCO_{7-y}$, the Néel temperature is about 500 K and the magnetic moment is 0.66 $\mu_B$ per Cu ion on the plane. When $y = 0.6$, $T_N$ is about 30 K and the moment is 0.16 $\mu_B$. The easy axis of the spins is perpendicular to the $c$ axis and they are antiparallel both in the plane and in the $c$ axis direction.

The spin–lattice relaxation rate of plane Cu showed a similar temperature dependence as for LSCO, and $(T_1T)^{-1}$ has a maximum at $T_c$ for $YBCO_7$ [1088], at about 100 K for 92 K-YBCO, and at about 150 K for 62 K-$YBCO_{6.63}$ [1089]. Since the Knight shift is temperature independent above $T_c$ for $YBCO_7$ [1090], the Korringa relation $T_1TK^2 = \hbar\gamma_e^2/4\pi k_B(\gamma_N)^2$ [1091] does not hold, indicating that the relaxation is not so simple as in a nonmagnetic metal. Since the antiferromagnetic fluctuation was observed for 91 K-, 92.5 K-YBCO by neutron experiments [1092, 1093], the nuclear relaxation may partly occur through the fluctuation of magnetically correlated electron spins [932, 1094, 1095]. As for the magnetic peaks observed by neutron scattering, *Fong* et al. claim that the broad 30 meV peak and the sharp 41 meV peak in the normal state are nonmagnetic peaks, and that the magnetic peak

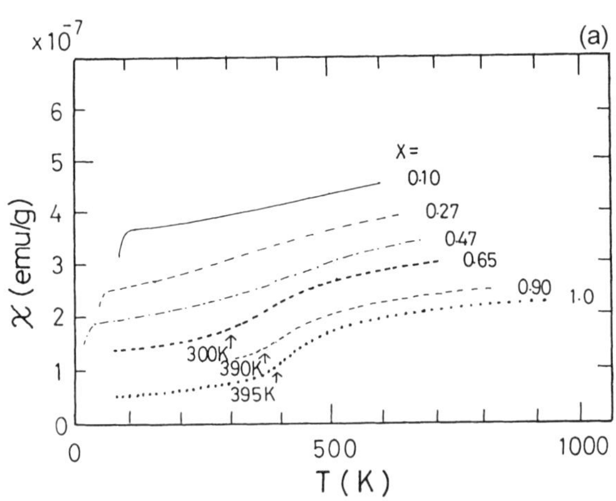


**Fig. 5.120a.** Magnetic susceptibility of $YBa_2Cu_3O_{7-x}$ [1079, 1080]. (**a**) Curie term subtracted. The *arrows* indicate Néel temperatures [1080]. (**b**) No increase in $\chi$ was observed at lower temperatures (see p. 299) [1079]. (With permissions from The American Physical Society and The Physical Society of Japan)

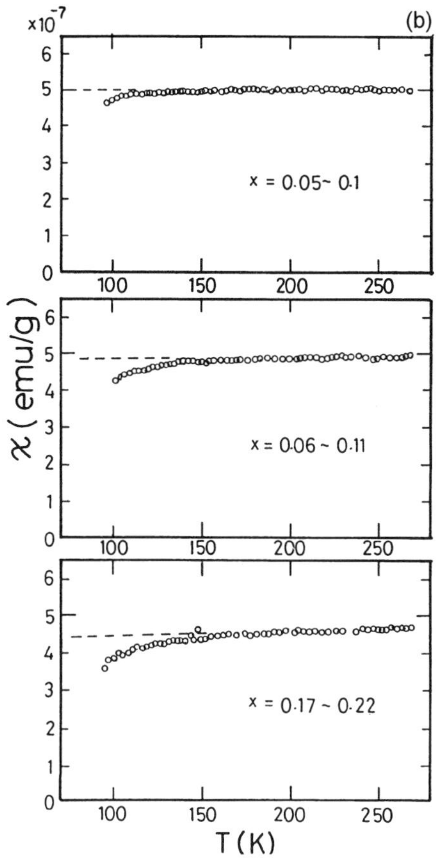


**Fig. 5.120b.** Description in caption to Fig. 5.120 a

is at 40 meV and vanishes at $T_c$ of 93 K [1096]. For reduced 62 K-YBCO$_{6.63}$ the Knight shifts of Cu and O showed identical temperature dependence in the normal state, and they decreased as temperature decreased [1089]. The intensity of spin susceptibility is reduced to 30% of that at 300 K at $T_c$. Thus the antiferromagnetic correlation becomes stronger as YBCO is reduced. In contrast to $(T_1T)^{-1}$ of Cu, that of plane O is temperature-independent for 93 K-YBCO above $T_c$ indicating that it is not sensitive to the spin correlation [1097].

In a strong-coupling superconductor, there often occurs a phonon softening above $T_c$. For high-$T_c$ superconductors, there also occur certain lattice anomalies. For instance, some weakening of the bond strength near 170 K [1098], a slight phonon softening [1099,1100], and an anomaly in the acoustic velocity around 120 K [1101]. The phonon anomalies were optically investigated by systematically changing the doping level, and a certain phonon softening was noted above $T_c$ for underdoped materials [1102–1105]. However, the

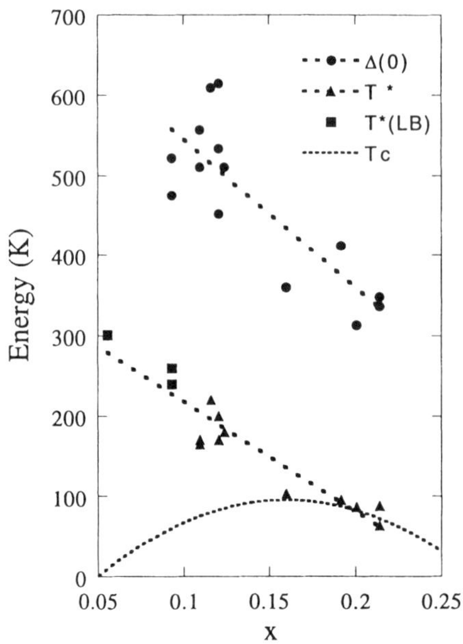


**Fig. 5.121.** $T^*$ and $\Delta(0)$ of $Bi_2Sr_2CaCu_2O_{8+x}$ determined by the ARPES study [1107]. LB means a lower bound. $\Delta(0)$: the superconducting energy gap near $(\pi, 0)$, where $(0,0)-(\pi,0)$ is parallel with the Cu-O-Cu direction. Note that this is not for YBCO but for Bi2212. The situation is similar between the two materials. (With permission from Elsevier)

anomalous change is not restricted to the phonon property but also observed for various electronic properties around $T^*$, which depends on the carrier concentration as shown in Fig. 5.121 [1107]. Figure 5.121 is not for YBCO but for Bi2212, but the situation is quite similar between the two. What occurs below $T^*$ is characterized by the depression of a spin or a charge freedom, that is, an ordering of some kind, and is known as a pseudogap phenomenon. It is noted that $T^*$ is not a phase transition temperature but a gradual crossover temperature, and is a measure of an energy scale for what occurs. *Williams* et al. reported that there was no isotope effect in the pseudogap from the NMR result of Y [1108]. The anomalous phenomena appear above $T_c$ in the nuclear spin–lattice relaxation rate $T_1^{-1}$, Knight shift, dc and optical conductivity [1109], Seebeck coefficient [1106], magnetic susceptibility, electronic specific heat [1056–1058], electron lifetime, magnetic excitation spectrum, electronic density of states, phonon frequency [1102–1105, 1110, 1111], etc. Anomalies in $T_1^{-1}$, dc resistivity, Hall coefficient, the low frequency magnetic excitation spectrum [1112], and the magnetic susceptibility were already described. High-frequency conductivity of $CuO_2$ planes at $500\,cm^{-1}$ begins to decrease even above $T_c$ as temperature decreases, and the temperature dependence is very similar to that of $T_1^{-1}$ [1109]. *Tallon* et al. attribute an

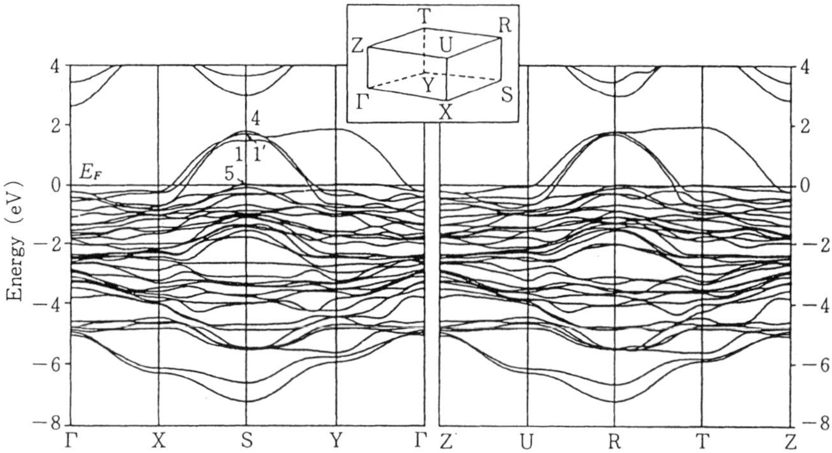


**Fig. 5.122.** Electron energy band structure of $YBa_2Cu_3O_7$ in the $k_z = 0$ and $\pi/c$ planes of the orthorhombic Brillouin zone [1114]. $S_1'$ and $S_4$: $d_{x^2-y^2} - O_{p_x}, O_{p_y}$ in the $CuO_2$ plane, $S_1$: $d_{z^2-y^2} - O_{p_y}, O_{p_z}$ antibonding orbitals in the chain. $S_5$: $d_{zy} - O_{p_z}, O_{p_y}\pi$ bonding orbitals in the chain. (With permission from Elsevier)

enhancement of the thermoelectric power and its dependence on Zn concentration to the opening of a gap [1106]. *Loram* et al. estimated the electronic contribution to the specific heat, and found a depression in $S/T$, where $S$ is the entropy, above $T_c$ for underdoped YBCO [1057,1058]. The depression in the electronic spectrum by ARPES study is explained after a description of the band structure. These phenomena are observed in the low concentration region of carriers, a so-called underdoped region. However, *Itoh* et al. claim that $T^*$ is above $T_c$ even in the overdoped region for Hg1201 [1113].

The band calculation was done as shown in Fig. 5.122 [1114–1118]. The states at the Fermi surface are mainly composed of $d_{x^2-y^2}$ and the corresponding $O_{2p}$ orbitals (Figs. 2.2 and 2.3). A characteristic feature appears in the position of the $O_{2p}$ orbitals which is sensitive to the distribution of surrounding cations and is exceptionally high in this compound especially in the linear chain. An inverse photoemission study revealed that there are unoccupied $O_{2p}$ hole states above the Fermi level [1119]. About 20% mixing of $Cu_{3d}$ orbitals was noted at $E_F$ [1120]. According to *Pennington* et al., the electric field gradient tensor indicates that the hole in the Cu $d$-shell is in the $(x^2 - y^2)$ orbital [1121].

The calculated Fermi surface consists of four sheets, and most of them were observed by the photoemission experiments for 92 K-$YBCO_{6.9}$ as shown in Fig. 5.123 [1122]. In the orthorhombic BZ, the $\Gamma - Y$ is parallel with both the chain and plane Cu-O-Cu directions. *Andersen* et al. emphasized that not only the $O_x$, $O_y$, and $Cu_{x^2-y^2}$ orbitals but also the $O_z$ orbitals contribute to the Fermi surface formation due to the dimpling, and they predicted an

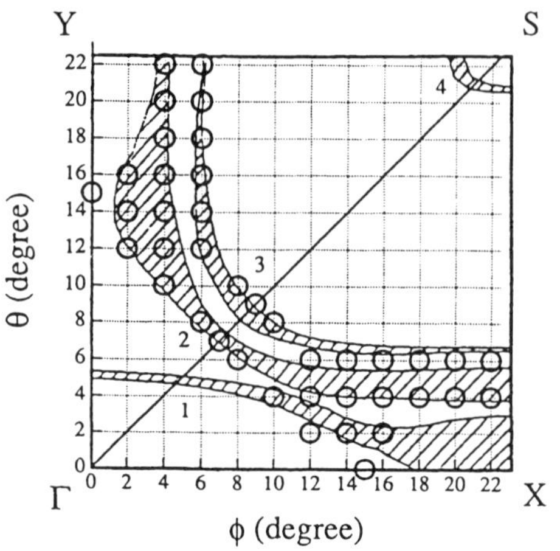


**Fig. 5.123.** The Fermi surface of a twinned $YBa_2Cu_3O_{6.9}$ crystal [1122]. *Open circles*: APRES results; *shaded area*: the calculated Fermi surface of YBCO projected along the $k_z$ direction [1126]. (With permission from The American Physical Society)

anomalously flat saddle point. The flat dispersion relation along the $\Gamma - Y$ and a van Hove singularity in the density of states were experimentally observed for $YBCO_{0.69}$ by *Campuzano* et al. [1123]. *Abrikosov* et al. pointed out the importance of the divergent singularity in the density of states for the high-$T_c$ phenomena [1010]. However, the singularity is also found for low-$T_c$ $Sr_2RuO_4$ [1250]. De Haas–van Alphen oscillations (dHvA) were observed for 93 K-YBCO [1124]. The cyclotron masses of the chain band extremals are $7.0\,m_0$ ($k_z = 0$) and $7.2\,m_0$ ($k_z = \pi/c$) while the calculated band masses are: 2.5 and 3.0 [1125], 2.1 and 2.3 [1115, 1124] and 1.6 and 1.7 [1124, 1126], respectively, in units of the free electron mass $m_0$. Damping of electrons by a phonon excitation was observed, and this gives a value of $7.5 \times 10^5\,\mathrm{ms}^{-1}$ for the maximum Fermi velocity [1127].

*Loram* et al. determined the electronic specific heat for $YBCO_{6+x}$ ($0.16 \leq x \leq 0.97$) between 1.8 K and 300 K [1056]. The temperature coefficient $\gamma$ is about 18 mJ/mol K$^2$ at 280 K for $x > 0.92$, which is as small as the band structure value 13 mJ/molK$^2$ [1128], and barely temperature-dependent. Therefore, they claim that superconductivity is due to $k$-space pairing of charge carriers which obey Fermi statistics.

**e) Phonons.** The Raman spectra of a single crystal YBCO has been investigated [1129–1132]. In Fig. 5.124, the 36 eigenvectors of the optical mode are shown following *Liu* et al. [1132]. The phonon frequency depends on the oxygen content. Particularly the frequencies of the 340 cm$^{-1}$ bend of planar Cu(2)-O(2,3) with O(2), O(3) out of phase and the 437 cm$^{-1}$ bend with O(2), O(3) in phase increase as the oxygen content is reduced from $O_7$ to $O_6$, to 341cm$^{-1}$ and 455cm$^{-1}$, respectively [1133]. With this reduc-

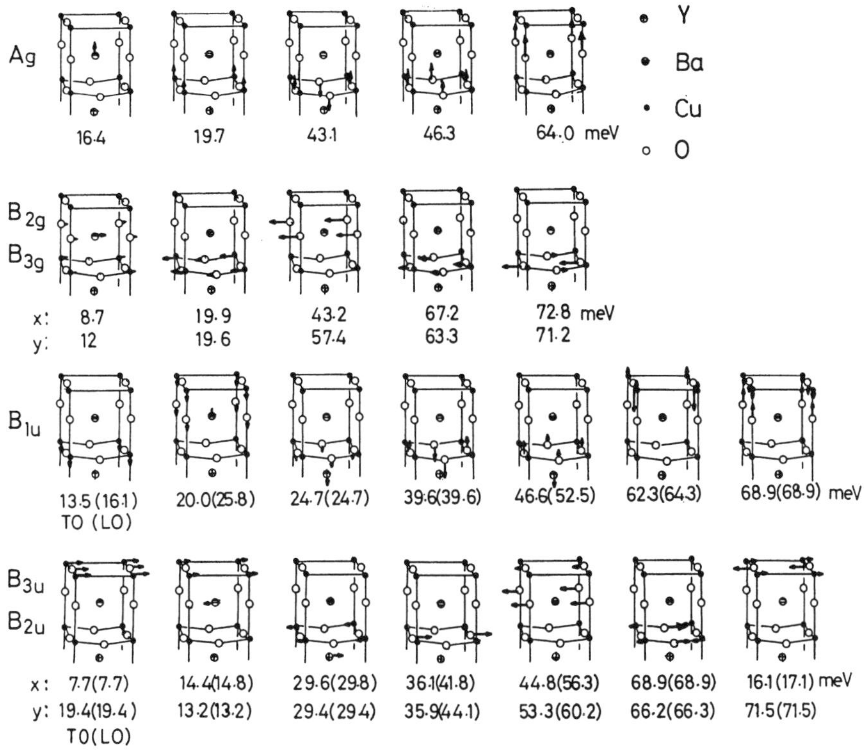


**Fig. 5.124.** The 36 eigenvectors of the optical modes in the upper half of the $YBa_2Cu_3O_7$ unit cell [1132]. Frequencies are in meV. (With permission from The American Physical Society)

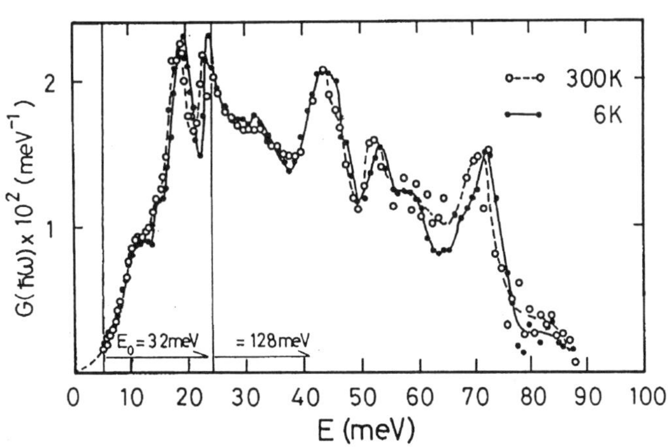


**Fig. 5.125.** Generalized phonon density of states of $YBa_2Cu_3O_7$ at 6 K and 300 K by a neutron scattering method [1136]

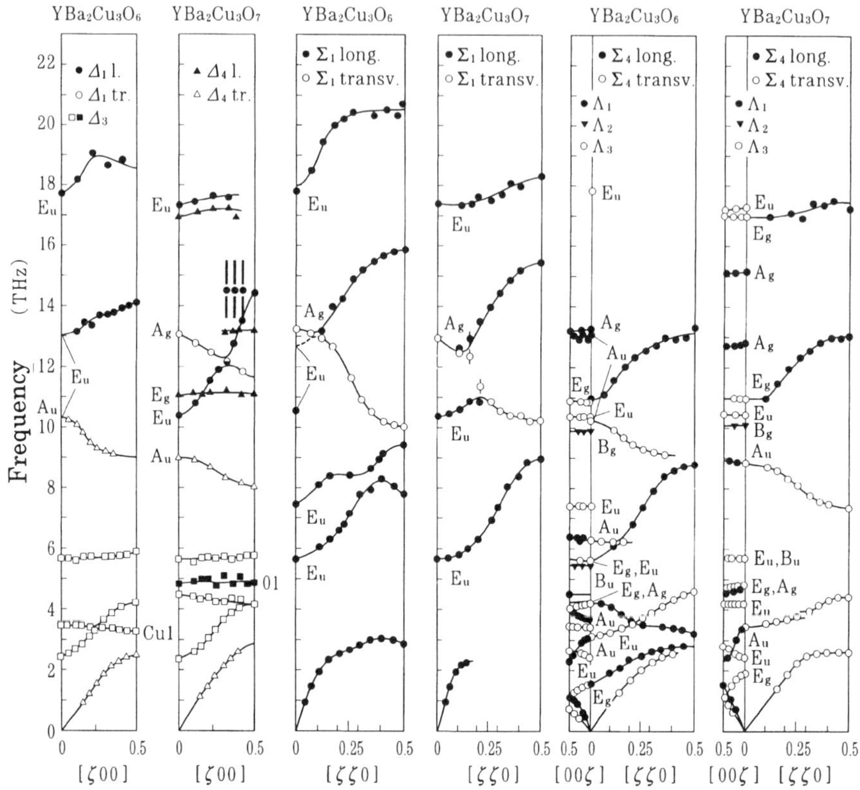


**Fig. 5.126.** The dispersion relation of phonons for $YBa_2Cu_2O_7$ and $YBa_2Cu_3O_6$ [945]. *Lines* are a guide to the eye. (With permission from World Scientific)

tion the carrier concentration decreases and it induces a hardening of the high-frequency phonons [945,1134–1136]. In Fig. 5.113, the calculated phonon density of states is shown [946]. Experimentally the phonon density of states was investigated by neutron scattering [1134–1137], and Fig. 5.125 shows the generalized phonon density of states for $YBCO_6$ and $YBCO_7$ after *Renker* et al. [1136]. They are not corrected for the different scattering cross section of each nucleus. The phonon dispersion relations are shown in Fig. 5.126 [945]. As shown there, the Cu-O stretching mode at 20 THz (83 meV) shows a softening of 2.3 THz upon oxidization at [110], and it is due to the e–p interaction. *Heyen* et al. studied the phonon structure of Y124 [1110, 1178].

The Debye temperature $\theta_D$ of YBCO:Fe is calculated to be 455 K from a fraction of the recoil-free Mössbauer absorption peak [1139]. 432 K has been obtained from the heat capacity of YBCO [905].

The thermal expansion coefficient is about $2 \times 10^{-5}$/K along the $c$ axis and $6 \times 10^{-6}$/K along the $a$ axis at 300 K [1139]. For $TiO_2$, it is $8.8 \times 10^{-6}$/K along the $c$ axis and $6.9 \times 10^{-6}$/K perpendicular to the $c$ axis at 300 K.

**f) Electron–Phonon Interaction.** *Zeyher* calculated the e–p interaction strength for a hypothetical cylindrical Fermi surface using a rigid ion harmonic oscillator approximation, and got a value of $\lambda$ as 3.2 which gives $T_c$ as 95 K for *s*-wave superconductivity. The frequency and the width of some phonons change at $T_c$ as expected from the theory [1140]. In the calculation, the phonon frequency was assumed to be much less than the electron band width. It was emphasized that the remote oscillations do interact with electrons by varying the chemical potential in a particular layer structure in which the screening of the long-range Coulomb interaction is incomplete. This long-range interaction along the *c*-axis was also noted by *Cohen* et al. in their frozen-phonon calculation [1141]. According to *Cohen* et al., $\lambda$ of the $A_g$ mode decreases as 0.049, 0.043, 0.026, 0.021, and 0.005 as the frequency increases. The total $\lambda$ is estimated to be 1, which is not large enough to give the high value of $T_c$ in YBCO. *Mazin* et al. got a value of 1.7 [1142], and *Zhao* and *Callaway* [1143] a value of 1.95. The latter emphasize that the e–p interaction is strong enough to explain $T_c$ of *s*-wave superconductivity. It must be noted that $\lambda$ is not a good parameter for determining $T_c$ since $T_c$ depends on a functional form of the spectral function of the e–p interaction $\alpha^2 F$ while $\lambda$ is a low-frequency weighted integrated value $2\int d\omega\alpha^2 F/\omega$ [1007, 1144]. *Savrasov* and *Andersen* calculated $\lambda$ for a hypothetical infinite layer compound $ACuO_2$ as 0.4, and concluded that the e–p interaction is too weak to give high $T_c$ [1031]. For these calculations, it is always disputed whether the band picture holds good for the cuprates. The frozen phonon calculation is also a kind of approximation. Moreover, these calculations use the Migdal–Eliashberg equations where the velocity of phonons is assumed to be much smaller than that of electrons [1145]. *Pietronero* et al. extended the calculation beyond Migdal's theorem and found that $T_c$ is considerably enhanced for a small momentum exchange [1146]. *Zeyher* and *Kulić* noted that the small momentum transfer is enhanced when the electron correlation is large [1029]. They also point out that a *d*-wave state is possible even for a homogeneous e–p interaction when the electron correlation is large. *Song* and *Annett*, and *Kamimura* et al. also claim that *d*-wave superconductivity may occur by the e–p interaction for a correlated electron system [1030, 1032]. *Bishop* et al. suggested that the charge transfer excitation perpendicular to the $CuO_2$ plane may enhance the attractive interaction of the pair in the $CuO_2$ plane and such an electronic excitation mechanism will benefit from the e–p interaction [1147]. For instance, the 155 cm$^{-1}$ and 515 cm$^{-1}$ modes seem to be influenced by such an excitation [1148].

### 5.9.5 $Bi_2Sr_2CaCu_2O_8$

After the discovery of the lanthanum and yttrium oxides there was a big search for other high-$T_c$ oxides. *Sheng* et al., and *Maeda* et al., then discovered the Tl-Ba-Ca-Cu oxides and the Bi-Sr-Ca-Cu oxides, respectively [793, 798, 799]. Compared with Y and La oxides, Bi oxides are more stable in

**Table 5.8:** Superconducting transition temperature of Bi-, Tl- and some other oxides. $T_{co}$: onset, $T_{cm}$: mid-resistance, $T_{cz}$: zero-resistance (see also Sect. 5.9.4)

| Sample | $T_c$ | Refs. |
|---|---|---|
| $Bi_2Sr_2Ca_{n-1}Cu_nO_{2n+4}$ | | |
| (Bi22$(n-1)n$) | | |
| Bi2201 | $T_{cm} = 22$ K (nominally 2202) | [794] |
| | $T_{co} = 8$ K | [795] |
| $Bi_2Sr_{1.5}Nd_{0.5}CuO_y$ | $T_{co} = 28$ K | [1260] |
| $Bi_2Sr_{1.5}La_{0.2}CuO_y$ | $T_{co} = 27$ K | [1260] |
| Bi2212 | $T_{co} = 94.4$ K | [1203] |
| Bi2223 | $T_c = 105$ K | [1261] |
| $Bi_2Sr_2YCu_2O_{8.5}$ | Semiconductor, $T_N = 13$ K | [1262] |
| $Bi_2Sr_2Ca_{1-x}Y_xCu_2O_y$ | $T_c = 85$ K ($x < 0.4$) | [1208] |
| (M-I at $x = 0.4$ or 0.5) | $T_c$ decreases to zero as $x$ increases | [1263] |
| $Bi_2Sr_{3-x}Y_xCu_2O_y$ | $T_c = 72$–65 K ($x = 0.2$–0.4) | [1264] |
| (M-I at $x = 0.5$) | | |
| $Bi_2Sr_{3-x}Ca_xCu_2O_y$ | $T_{co} = 70$–89 K and 108–118 K? | [1265] |
| (single phase $1.25 < x < 2.0$) | | |
| $Bi_{1.6}Pb_{0.4}Sr_2Ca_2Cu_3O_y$ | $T_{cz} = 120$ K | [1266] |
| $Bi_{1.73}Pb_{0.35}Sr_{1.89}Ca_{1.86}Cu_3O_{9.91}$ | $T_{cz} = 105$ K | [1267] |
| $Pb_2Sr_2Y_{0.5}Ca_{0.5}Cu_3O_y$ | $T_c = 75$–25 K | [1268] |
| $Pb_2Sr_2(Dy,Ca)Cu_3O_y$ | $T_{cm} = 51.5$ K | [1269] |
| $TlBa_2Ca_{n-1}Cu_nO_{2n+2.5}$ | | |
| Tl1201 | No superconductivity | [796] |
| Tl1212 | $T_{co} = 85$ K, $T_{cm} = 78$ K | [796, 1270] |
| Tl1223 | $T_{cm} = 120$ K | [796, 1270] |
| Tl1234 | $T_{co} = 122$ K, $T_{cm} = 121$ K | [1270, 127 |
| Tl1245 | $T_{co} < 120$ K, $T_{cm} = 107$ K | [1270, 127 |
| $Tl(Ba, Ca)_2Ca_{n-1}Cu_nO_{2n+2.5}$ | | |
| 1212 | $T_{co} = 98$ K | [797] |
| 1223 | $T_{co} = 114$ K | [797] |
| 1234 | $T_{co} = 125$ K | [797] |
| $(Tl,Bi)(Sr,Ca)_2Ca_{n-1}Cu_nO_{2n+2.5}$ | | |
| 1201 | $T_{co} = 50$ K | [797] |
| 1212 | $T_{co} = 90$ K | [797] |
| $Tl_2Ba_2Ca_{n-1}Cu_nO_{2n+4}$ | | |
| Tl2201 | $T_{co} = 90$ K | [841] |
| Tl2212 | $T_{co} = 112$ K | [1192] |
| Tl2223 | $T_{co} = 120$ K | [1272] |

**Table 5.8:** (continued)

| Sample | $T_c$ | Refs. |
|---|---|---|
| $Tl_{1.7}Ba_2Ca_{1.06}Cu_{2.32}O_y(2212)$ | $T_{cz} = 108\,K$ | [1273] |
| $Tl_{1.64}Ba_2Ca_{1.87}Cu_{3.11}O_y(2223)$ | $T_{cz} = 125\,K$ | [1273] |
| $Tl_{0.75}Pb_{0.25}Sr_2CaCu_2O_7$ | $T_{co} = 70\,K$ | [1274] |
| $Tl_{0.5}Pb_{0.5}Sr_2CaCu_2O_7$ | $T_{cz} = 90\,K$ | [1275] |
| $Tl_{0.5}Pb_{0.5}SrCa_3Cu_3O_y$ | $T_{co} = 115\,K$ | [1276] |
| $Tl_2Ba_{2-x}Sr_xCaCu_2O_8$ | $T_{co} = 116\,K\,(x=0)$, 90 K $(x=1)$, 61 K $(x=2)$ | [1277] |
| $HgBa_2Ca_{n-1}Cu_nO_{2n+2+y}(n=2,3)$ | | |
| Hg1212 | $T_{co} = 93\,K\,(O_{6.10})$, 127 K $(O_{6.21})$, 98 K $(O_{6.31})$ | [1278] |
| Hg1223 | $T_{co} = 98\,K\,(O_{8.17})$, 135 K $(O_{8.29})$, 134 K $(O_{8.30})$ | [1278] |
| $Ag_{1-x}Cu_xBa_2Ca_{n-1}Cu_nO_{2n+3-y}$ $(n=4)$ | | |
| Ag1234 | $T_c = 117\,K$ | [804] |
| $CuBa_2Ca_{n-1}Cu_nO_{2n+4-y}$ $(n=4)$ | | |
| Cu1234 | $T_c = 116\,K$ | [804] |
| $R_{1.4}Ce_{0.6}RuSr_2Cu_2O_{10-y}$ $(R =$ Eu, Gd) | | [817] |
| $R$ =Eu | $T_c = 32\,K$, $T_N = 122\,K$ | |
| $R$ =Gd | $T_c = 42\,K$, $T_N = 180\,K$ | |
| (Ln,Ce)$RuSr_2Cu_2O_{10}$ (Ln = Sm, Eu, Gd) | $T_c = 42\,K$ | [816] |
| $Sr_2Nd_{1.5}Ce_{0.5}NbCu_2O_{10-y}$ | $T_c = 28\,K$ | [811] |
| $Sr_2(Nd,Ce)_2NbCu_2O_{10}$ | $T_c = 28\,K$ | [812] |
| $Sr_{0.4}Ca_{13.6}Cu_{24}O_{41.84}$ (ladder) | $T_{co} = 12\,K$ (3 Gpa) | [818] |

air and in a vacuum, therefore the properties of the surface region reflect the bulk properties more reliably, and studies such as photoelectron emission and tunneling have been advanced. However, it is true that surface properties are very sensitive to the atmosphere even for these materials.

*Michel* et al. [794] and *Akimitsu* et al. [795] discovered the low-$T_c$ Bi oxides, and *Halder* et al. [797] and *Parkin* et al. [796] added other oxides. The $T_c$ values are shown in Table 5.8. *Halder* et al. expressed these Bi and Tl oxides as the following homologous compounds:

$$TlB_2Ca_{n-1}Cu_nO_{2n+2.5+x}(n = 1-4)\,, B = \mathrm{Ba\,or\,Sr}\,, \mathrm{Tl}12n-1\,,\, n\,,$$
$$Tl_2B_2Ca_{n-1}Cu_nO_{2n+4+x}(n = 1-3)\,, B = \mathrm{Ba\,or\,Sr}\,, \mathrm{Tl}22n-1\,,\, n\,,$$
$$Bi_2Sr_2Ca_{n-1}Cu_nO_{2n+4+x}(n = 1-3)\,, \mathrm{Bi}22n-1\,,\, n\,.$$

These are idealized expressions and Cu is +2 valent when $x = 0$. In the real oxides, the ratios of elements are different from the idealized ones depending

**Table 5.9.** Lattice parameters of Bi-, Tl- and some other cuprates

| | $a$ [Å] | $b$ [Å] | $c$ [Å] | | Refs. |
|---|---|---|---|---|---|
| Bi2201 | 5.37 ($= 3.8 \times \sqrt{2}$) | 5.36 | 24.60 | Ammaa, $T_c = 9\,K$ | [841] |
| Bi2212 | 5.414 | 5.418 | 30.89 | | [1188] |
| Bi2223 | | | 37.08 | $T_c = 110\,K$ | [1279] |
| Tl1201 | 3.745 | | 9.00 | $T_{co} = 50\,K$ | [797] |
| Tl1212 | 3.847 | | 12.74 | $T_{co} = 98\,K$ | [797] |
| Tl1223 | 3.8429 | | 15.871 | $T_c = 110\,K$ | [796] |
| Tl1234 | 3.850 | | 19.01 | $T_{co} = 125\,K$ | [797] |
| Tl2201 | 3.87 | | 23.24 | $T_{co} = 90\,K$ | [841] |
| Tl2212 | 3.8500 | | 29.318 | $T_{co} = 112\,K$ | [1192] |
| Tl2223 | 5.40 | | 36.25 | | [1272] |
| Ag1234 | 3.8635 | | 18.111 | | [804] |
| Cu1234 | 3.8557 | | 17.993 | | [804,805] |
| $Pb_2Sr_2YCu_3O_8$ | 5.4019 | 5.4333 | 15.7388 | Magnetic | [1269] |
| $Eu_{1.4}Ce_{0.6}RuSr_2Cu_2O_{10-y}$ | 3.846 | | 28.50 | | [817] |

on the preparation conditions. For instance, *Hibble* et al. analyzed the compositions with an analytic electron microscope as $Tl_{1.76}Ba_{2.0}Ca_{0.70}Cu_2O_8$ and $Tl_{1.83}Ba_{2.0}Ca_{1.44}Cu_3O_{10}$ [1182]. Bi and Tl may be substituted by Pb, Cu and Hg [800,801,804,805] and many new compounds have been prepared as shown in Tables 5.8 and 5.9. Single crystals of the Bi oxides have been grown from a CuO flux [1183], an alkali chloride flux [1184, 1185] or by the Bridgeman method [1186, 1187]. It must be noted that $Tl_2O_3$ has a high vapor pressure and is poisonous.

**a) Crystal Structure.** The structures are shown schematically in Fig. 5.127 [792, 796, 797, 841, 1188–1194], and the lattice parameters are shown in Table 5.9. Sometimes these idealized structures are modulated by a density wave of Bi whose dense band extends along the $a$ axis and appears periodically along the $b$ axis, and thus the lattice parameters of this orthorhombic cell are $a$, $\sim 5b$ and $c$ where $a$ is about 5.41 Å (= Cu–Cu separation, not Cu–O–Cu distance). The modulation is incommensurate with the primitive cell [1195, 1197].

*Torardi* et al. have identified the superconducting phases 9 K-$Bi_2Sr_2CuO_6$ (Bi2201) and 90 K-$Tl_2Ba_2CuO_6$ (Tl2201) and compared their structures [841]. The Tl2201 structure is tetragonal and the Cu-O sheet is strictly flat while Bi2201 structure is orthorhombic and the Cu-O sheet is not flat, as in the other Cu oxides.

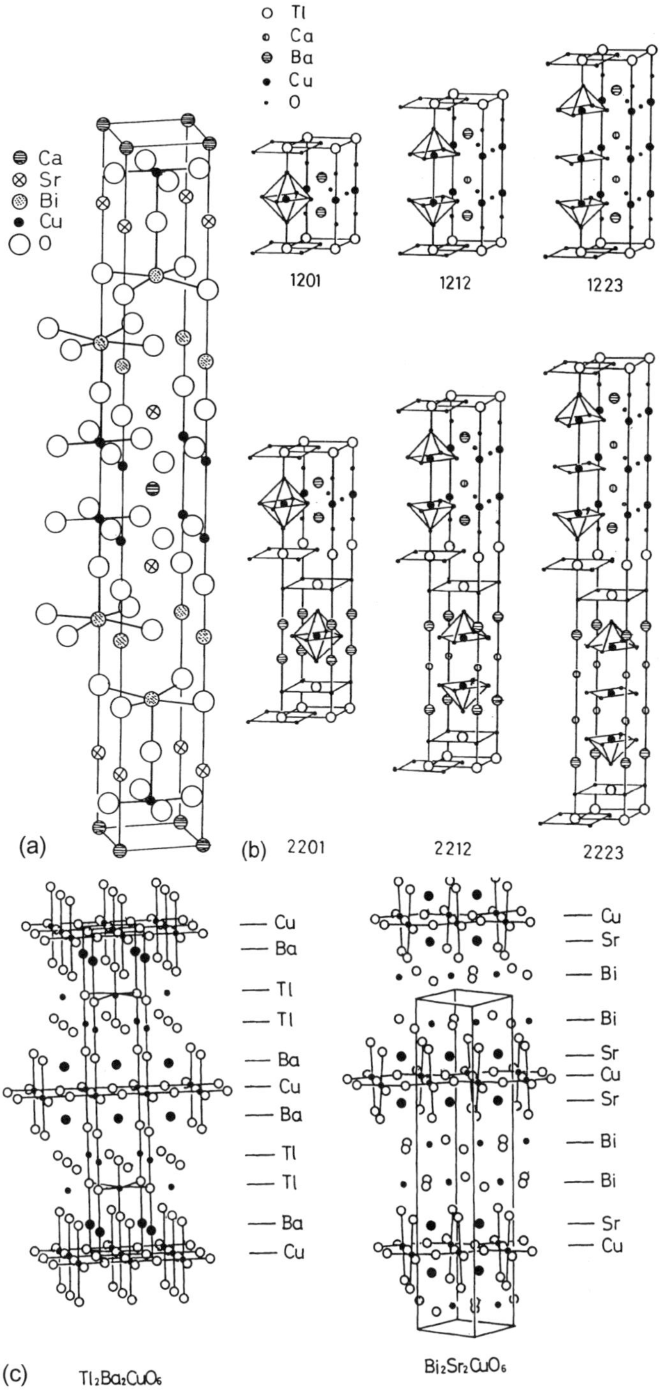


**Fig. 5.127.** Crystal structure of Tl and Bi containing oxides. (**a**) $Bi_2Sr_2CaCu_2O_8$ [1189]. (**b**) Schematic representation of Tl12$(n-1)n$ and Tl22$(n-1)n$ [796]. (**c**) Flat and buckling planes of $CuO_2$ in Tl2201 and Bi2201, respectively [841]. (With permission from The American Physical Society)

It was found by the EXAFS study on Tl1234 that the apex oxygen is in the double well potential whose minima are at 2.57 Å and 2.74 Å from Cu [1198]. The double well potential becomes single near $T_c$.

**b) Superconducting Properties.** In the Bi oxides, $T_c$ increases as the number of $CuO_2$ layers between $Bi_2O_2$ increases, at least until $n = 3$. For Tl oxides, $T_c$ increases at least until $n = 4$ and the extrapolated value to $n = \infty$ is 180 K [797]. $T_c$ is as high as 90 K for Tl2201 but the subsequent increment in $T_c$ with the layer number is as small as that for Bi oxides. $T_c$ of Tl2201 decreases with oxydization down to 4 K with a contraction of the $c$-axis lattice parameter [1199, 1200]. $T_c$ values of Bi2201 and Bi2212 also depend on the carrier concentration [1201–1203]. According to *Watanabe* et al., the optimal oxygen content is about $O_{8.25}$ for Bi2212, and $T_c$ decreases on both sides. According to *Triscone* et al., the optimal carrier concentration for Bi2212 is about 0.22 hole per Cu atom where the onset $T_c$ is 94.4 K [1203]. It must be noted that the dependence of $T_c$ on oxygen partial pressure and the annealing temperature is influenced by the starting metal composition [1204].

$T_c$ of Bi2212 decreases when Cu is substituted with other metals. For instance, $T_c$ of $Bi_2Sr_{1.8}Ca_{1.2}Cu_2O_y$ decreased from 78 K to 52 K (Fe), 58 K (Co), 67 K (Ni), and 67 K (Zn) for the 4% substitution [1205]. In this experiment, Hall density was kept constant. The metal–insulator transition occurs when Ca is substituted with Y. $Bi_2Sr_2Ca_{1-x}Y_xCu_2O_y$ becomes an insulator for $x$ larger than about 0.5 [1206–1208]. $T_c$ seems to be constant at least until $x = 0.3$ and rapidly decreases near the boundary while the Hall carrier density decreases gradually [1206]. *Jayaram* et al. found that $T_c$ was constant for $0 < x < 0.35$ while hole number per subcell decreased from 0.3 to 0.21, and then $T_c$ decreased rapidly to zero at $x = 0.6$ where the hole number was 0.7 per subcell [1209]. The situation is similar to the case of $Li_{1+x}Ti_{2-x}O_4$ (Sect. 5.3).

The isotope shift was investigated by partially substituting the $^{16}O$ in a mixture of Bi2212 and Bi2223. It was 0.32 K for 110 K-Bi2223 and 0.34 K for 75 K-Bi2212 [1210].

The magnetic properties are shown in Table 5.10. From $H_{c2\parallel}(0) = 400$ T and $H_{c2\perp}(0) = 21$ T, $\xi_\parallel(0)$ and $\xi_\perp(0)$ are 2.1 Å and 40 Å, respectively, for Bi2212, where $\parallel$ and $\perp$ are parallel and perpendicular to the $CuO_2$ plane [1183].

The energy gap $2\Delta$ was determined by tunneling, UPS and Raman scattering (Table 5.11). As for the symmetry of the order parameter, *Ding* et al. observed a $d_{x^2-y^2}$ type superconducting gap at 13 K for a lightly overdoped 87 K-Bi2212 with the ARPES. The gap was largest along the Cu-O-Cu direction and $\Delta$ was 33 meV [1211]. *Shen* et al. also observed such a $d$-wave like order parameter. $\Delta$ was about 20 meV along the Cu-O-Cu direction near $\bar{M}$, which is the notation for the face-centered orthorhombic lattice, and was very small along the Cu–Cu direction for 88 K-Bi2212. For the overdoped 78 K-Bi2212, the maximum value was about 12 meV [1212, 1213]. *Ma* et al.

**Table 5.10.** $H_{ci}$ of Bi-cuprates. $\parallel$ and $\perp$ mean parallel and perpendicular to the $CuO_2$ plane, respectively

| Sample | $H_{c2}(0)$ [T] | $-dH_{c2}/dT$ at $T_c$ [T/K] | Refs. |
|---|---|---|---|
| Bi2212 | $\parallel$ 440 | 7.3 | [1280] |
| | $\perp$ 20 | 0.36 | [1280] |
| | $\parallel$ 400 | 7.3 | [1183] |
| | $\perp$ 21 | 0.39 | [1183] |
| | $\parallel$ 812 | 14 | [1281] |
| | $\perp$ 19 | 0.32 | [1281] |
| Bi2223 | $\parallel$ 500 | 7 | [1280] |
| | $\perp$ 25 | 0.34 | [1280] |

observed an anisotropic gap for overdoped 83 K-Bi2212 by ARPES. $\Delta$ was 16 meV along the $\Gamma$–$\bar{M}$ (Cu–O–Cu) direction and 10 meV along the $\Gamma$–$X$ (Cu–Cu) direction at 40 K, and the latter vanished at 66 K while the former remained until 83 K. They suggest a mixing of a $d$- and an $s$-wave state [1214]. *Kelly* et al. observed a gap even along the Cu-Cu direction for the overdoped sample [1215].

Compared with ARPES, tunneling results are more complex. *Renner* and *Fisher* found nonzero conductance within a gap of the $c$-axis conductance for 92.3 K–Bi2212 by the vacuum tunneling method, and suggested an anisotropic $s$- or a $d$-wave state [1216]. *Ichimura* and *Nomura* reported an anisotropic $s$-state by using the same method [1217]. *Barbiellini* et al. analyzed the vacuum tunneling conductance and found that $\Delta$ distributed from 0 meV to 30 meV like a $d$-wave and also had a weak component at about 90 meV [1218]. On the other hand, *Kane* and *Ng* reported that $\Delta$ was largest along the Cu–Cu direction with a value of about 40 meV while it was about 25 meV along the Cu-O-Cu direction [1219].

When a normal electron enters into a superconductor the Cooper pair is formed, and as a result a hole is emitted back to the normal metal. This is known as an Andreev reflection [1220]. *Blonder*, *Tinkham* and *Klapijwk* (BTK) calculated the conductance for a metal–$s$ wave superconductor contact [1221], and *Kashiwaya* et al. extended it to a $d$-wave case [1036]. For the $s$-wave, the zero-bias conductance is at largest twice that of the normal-state conductance, but for the $d$-wave it may become much larger along the gap suppressed direction. *Alff* et al. observed such a large conductance along the Cu–Cu direction [1035] for YBCO, but there are at the same time many examples of a small zero-bias conductance along this direction.

The symmetry of the order parameter does not necessarily identify the pairing mechanism. The most direct method to identify the mechanism is to study the energy and the momentum distribution of electrons in a supercon-

**Table 5.11.** $2\Delta(0)/k_BT_c$ of Bi- and Tl-cuprates. $\Delta(0)$ and $T_c$ depend on the carrier concentration, and even an anisotropic or $d$-wave gap has been reported. The present $2\Delta$ is a separation between the two edge peaks in the effective density of states. See also Sect. 5.9.4

| Sample | $2\Delta(0)/k_BT_c$ | $2\Delta$ [meV] | $T_c$ [K] | Method | Refs. |
|---|---|---|---|---|---|
| Bi2201 poly. | 5.4–6.2 | 3.0–3.5 | 6.5 | Tunnel. | [1282] |
| Bi2201 poly. | 6.5–7.4 | 5.0–5.7 | 8.9 | Tunnel. | [1282] |
| Bi2212 | 8.7 | 60 | 80 | Tunnel. | [1283] |
| Bi2212 | 9.6–11 | 73 | 76–88 | Tunnel. | [1282] |
| Bi2212 | 4.5–6.5 | 30–46 | 80 | Tunnel. | [1284] |
| Bi2212 | 5.9–6.7 | 44 | 76–87? | Tunnel. | [1285] |
| Bi2212 | 7 | 50 | 81–85 | Tunnel. | [1286] |
| Bi2212 | 6.8 | 50 | 86 | Raman | [1287] |
| Bi2212 | 4–7 | 32–56 | 90 | Raman | [1288] |
| Bi2212 | 7 | 49 | 82 | UPS | [1289] |
| Bi2212 | 8 | 60 | 88 | UPS | [1290] |
| Bi2212 | 7.8–9.3 | 54–64 | 80 | Tunnel. | [1291] |
| Bi2212 | 7.1 | 52 | 85 | Tunnel. | [1292] |
| Bi2212 | 6.5 | 48 | 86 | Tunnel. | [1226] |
| Bi2212 | 12.6 → 9.2 → 7.3 | 90 → 75 → 52 | 83 → 95 → 82 | Tunnel. | [1236] |
| BiSrCaCuO | 7.1 | 40 | 65 | Tunnel. | [1293] |
| Bi2212, $c$-axis | 3.3 | 24 | 85 | Tunnel. | [1294] |
| Bi2212, $c$-plane | 6.2 | 46 | 85 | Tunnel. | [1294] |
| Bi2223 | 7.3 | 60 | 95 | Tunnel. | [1295, 1296] |
| Bi2223 film (in plane) | 9–10 | 80–88 | 103–119 | Tunnel. | [1282] |
| Bi2223 film ($\perp$ to the plane) | 5.5–6.1 | 49–54 | 103–119 | Tunnel. | [1282] |
| Tl2212 | 6.1–7.7 | 60 | 90–115 | Tunnel. | [1297] |
| Tl2223 | 4 | 38 | 110 | Tunnel. | [1298] |
| $Bi_2Pb_{0.4}Sr_2$-$Ca_{2.5}Cu_{3.5}O_y$ | 5.5 | 50 | 105 | Tunnel. | [1299] |

ductor. For instance, the phonon at $\omega_0$ makes a structure at $\Delta + \hbar\omega_0$ in the quasiparticle density of states [1145]. When the mediator is the electron, there occurs a gap $2\Delta$ in the spectral function of the interaction [1222] and there will appear a structure at the corresponding energy in the density of states.

The spectral function of the interaction is deduced from the experimental tunneling conductance by the McMillan–Rowell inversion procedure of the

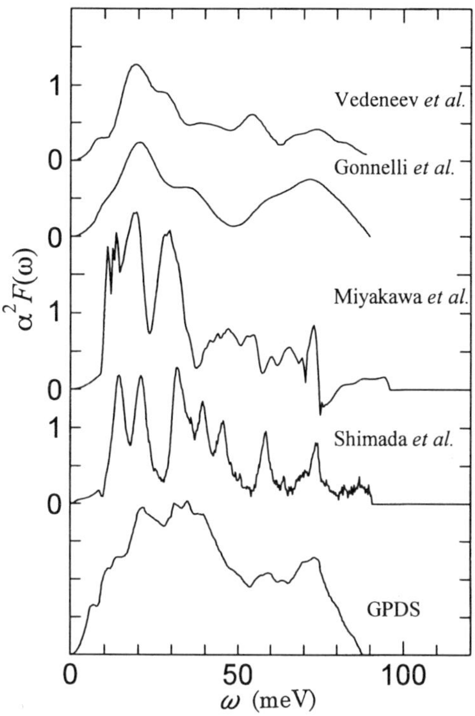


**Fig. 5.128.** The spectral function $\alpha^2 F$ of the electron–phonon interaction and the generalized phonon density of states for Bi2212 [1226]. Data in [1225, 1226, 1228, 1230, 1231] have been collected. (With permission from The American Physical Society)

Eliashberg equations for an $s$-wave superconductor [1223–1230]. Figure 5.128 shows spectral functions thus obtained, together with the generalized phonon density of states determined by neutron scattering [1225, 1226, 1228, 1230, 1231], and Fig. 5.129 shows the good correspondence between the spectral function and the calculated phonon spectrum [1232]. The sharp spectral function by *Simada* et al. suggests an isotropic $s$-wave, and it gives a value of 87 K for $T_c$. *Shimada* et al. emphasized the importance of the large density of phonon modes which interact with electrons in the layer structure for raising $T_c$ and for explaining the rise in $T_c$ with the layer number [1232, 1233]. If the order parameter is a $d$-wave, their tunneling current must have flowed from a very limited portion of the $k$-space. It must be noted, however, that there are many studies which do not observe the phonon structure in the elastic tunneling process. For inelastic tunneling in which phonons are emitted in the barrier, the structure may also appear at $\Delta + \hbar\omega_0$, as *Koltun* et al. observed for Bi2212 and Bi2223 [1234]. In this context, however, the shape of the structure is different from that mentioned above.

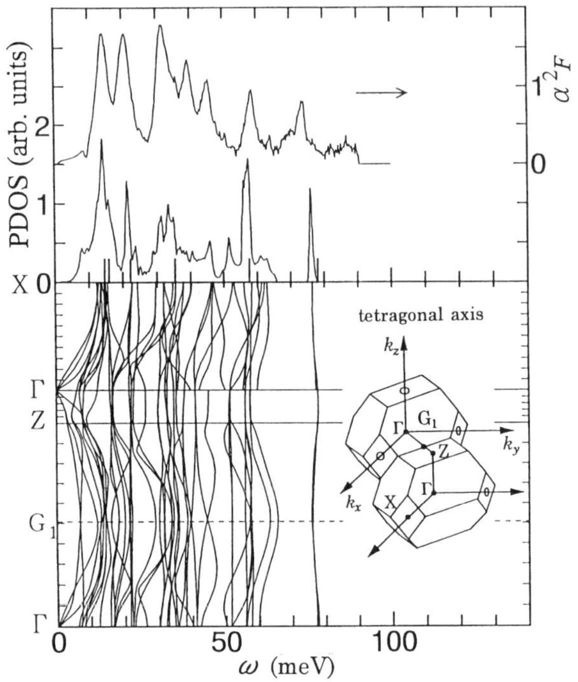


**Fig. 5.129.** The spectral function $\alpha^2 F$ of the electron–phonon interaction [1226] is compared with the calculated phonon density of states and the calculated phonon dispersion relations for Bi2212 [1232]. (With permission from Elsevier)

*Renner* and *Fisher* noted another feature of the tunneling conductance [1216]. They did not observe the phonon structure but a broad dip around 80 mV in the vacuum tunneling conductance, and the dip was larger in the bias range corresponding to the states below the Fermi level than above the Fermi level. *DeWilde* et al. found that the dip position scales with the gap value suggesting an electron–electron interaction mechanism for the pairing [1222, 1235–1237]. The scaling feature is observed at all doping levels. The gap increases as the carrier density decreases, and the ratio $2\Delta/k_\mathrm{B}T_\mathrm{c}$ becomes as large as 14 [1236, 1239, 1240]. It must be noted that the increase in the gap value itself will not be against the electron–phonon mechanism since the interaction becomes stronger as the screening of the Coulomb interaction becomes weaker [619]. *Loram* et al. [1057, 1106] conjectured that the quasiparticle energy is given by

$$E_k = (e_k^2 + E_\mathrm{g}^2 + \Delta_\mathrm{s}^2)^{1/2}, \tag{5.98}$$

where $e_k$ is the kinetic energy, $E_\mathrm{g}$ is the pseudogap, and $\Delta_\mathrm{s}$ is the superconducting gap, and claim that $E_\mathrm{g}$ increases as the hole concentration decreases while $\Delta_\mathrm{s}$ is relatively independent of the concentration. *Miyakawa* et al. claim that the large gap in the underdoped region is of superconducting origin based on their tunneling study, and it decreases toward $T_c$ but is not zero at that temperature [1300].

**c) Normal State Properties.** The temperature dependence of in-plane resistivity depends on the carrier concentration as shown in Fig. 5.130 [1202]. The deviation from the linear temperature dependence becomes noteworthy below $T^*$ as for YBCO. The in-plane resistivity is 300 μΩ cm along the $a$ axis at 300 K for 84 K-$Bi2212O_{8.27}$. This value is typical of metallic oxides and is not exceptionally large as shown in Table 2.1. The $c$ axis resistivity has an activation energy of about 17 meV, independently of the carrier concentration.

*Ando* et al. measured the normal-state in-plane and out-of-plane resistivity for 13 K-$Bi_2Sr_{1.95}La_{0.05}CuO_x$ by suppressing superconductivity [1241]. The in-plane resistivity was metallic while out-of-plane resistivity was semiconducting even at a temperature as low as 0.66 K, where the phonons will be barely excited. Therefore, they claim that this is evidence for the non-Fermi liquid nature of Bi2201 [834, 836, 837].

*Kubo* et al. studied the Hall mobility ($R_H/\rho$) of Tl2201 and Tl1212 and found that it behaves as $(a+bT^2)^{-1}$ [1242]. The value of the Hall mobility is 5.4 cm$^2$/Vs at 300 K for 50 K-Tl2201. Compared with this, the Hall mobility of $ReO_3$, which has the maximum normal-state conductivity among oxides, is 30 cm$^2$/Vs at 300 K (Sect. 5.1.2).

For LSCO, anomalous suppression of superconductivity occurs at the hole concentration of 1/8 per Cu. This suppression also occurs for Zn doped $Bi_2Sr_2Ca_{1-x}Y_xO_{8+\delta}$ [883], and it is a common feature of high-$T_c$ cuprates.

The Fermi surface has been studied by using ARPES [1243–1247]. The experimental Fermi surface is similar to the band calculation [1248] as shown in Fig. 5.131 [1247]. This Fermi surface is usually labeled as a large Fermi surface. The $\Gamma$–$\bar{M}$ is parallel with the Cu–O–Cu direction at the basal plane $k_z = 0$ and the $\Gamma$–$X(Y)$ is parallel with the Cu–Cu direction. *Aebi* et al. reported nearly rectangular pieces around $X(Y)$ which are repeated weakly around $\Gamma$. They claim that the repeat is due to the antiferromagnetic alignment of Cu spins [1245]. However, this was criticized by *Chakravarty* because of the absence of the averaged staggered moments [1249]. *Ma* et al. emphasized the orthorhombic nature of the Fermi surface [1246]. *Ding* et al. observed a large Fermi surface centered at $Y$ for heavily underdoped 15 K-, moderately underdoped 83 K-, and slightly overdoped 87 K-Bi2212, but found no pocket expected for the antiferromagnetic case [1247].

The electron dispersion relation is very flat near $\bar{M}$ [1243, 1246] as for YBCO. A similar van Hove singularity also exists either for low-$T_c$ $Sr_2RuO_4$ [1250]. Therefore, large density of states alone cannot be a cause for high $T_c$.

Another interesting feature of the ARPES spectra is the pseudogap phenomenon [1211, 1244, 1251, 1252]. *Ding* et al. observed the suppression of the spectral weight in the ARPES spectra near the Fermi level for under-

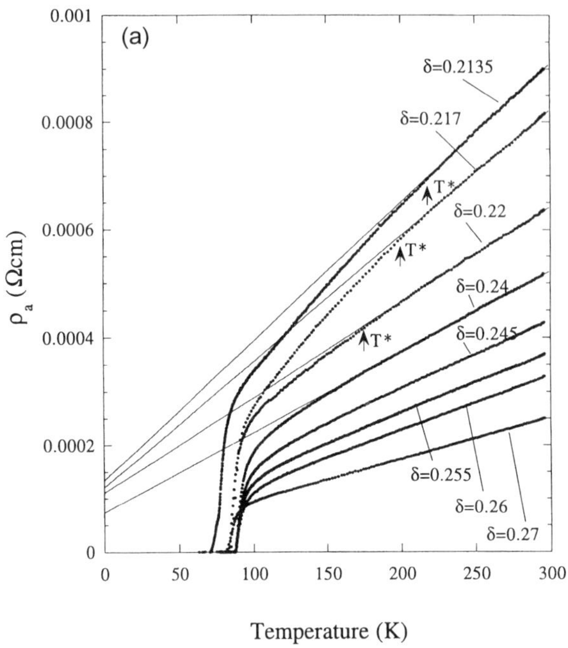


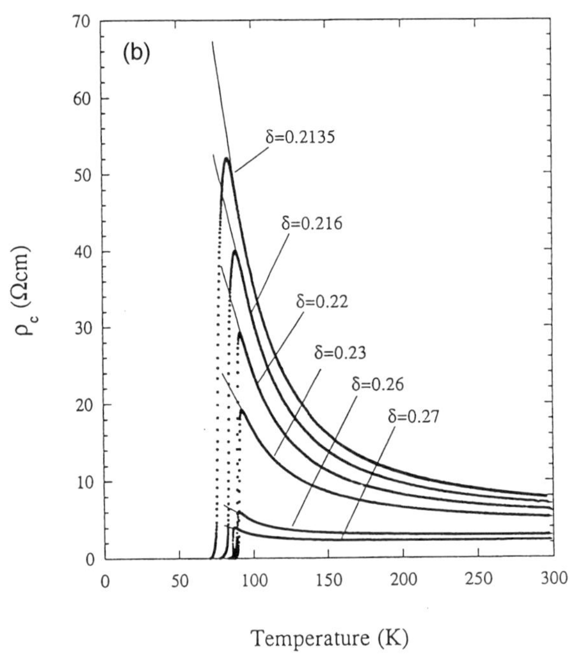


**Fig. 5.130.** In-plane (**a**) and out-of-plane (**b**) resistivity for $Bi_2Sr_2CaCu_2O_{8+\delta}$ [1202]. $T_c$ is highest (89–90 K) for $\delta = 0.24$–$0.25$. $T^*$ is shown by the *arrow*. (With permission from The American Physical Society)

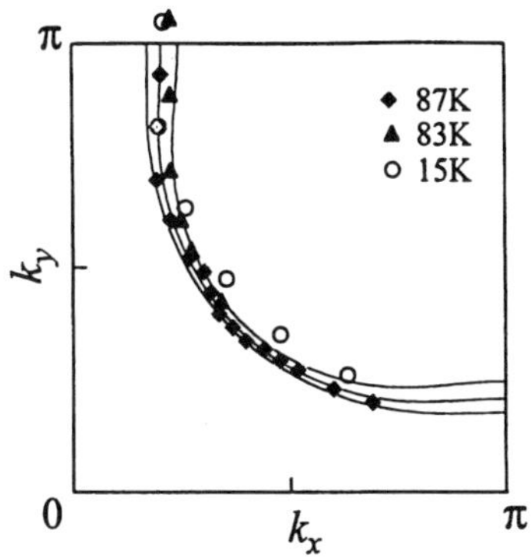


**Fig. 5.131.** Fermi surfaces of Bi2212 samples whose $T_c$ values are 87 K, 83 K, and 15 K [1247]. The *solid lines* are tight binding estimates of the Fermi surface at 18%, 13%, and 6% doping assuming rigid band behavior [1248]. $(\pi, 0)$ is $\bar{M}$ and $(\pi, \pi)$ is $Y$. $(0, 0) - (\pi, 0)$ is parallel with the Cu-O-Cu direction. (With permission from The American Physical Society)

doped Bi2212 crystals. This suppression is called a pseudogap. The pseudogap opened at 170 K and smoothly turned into a superconducting gap for moderately underdoped 83 K-Bi2212, while the gap opened very near to $T_c$ for slightly overdoped 87 K-Bi2212. The pseudogap had a *d*-wave-like momentum dependence [1211,1251]. They also found that a minimum gap locus (locus of points in the momentum space with minimum excitation energy) in the pseudogap phase coincides with the Fermi surface (locus of points with gapless excitations) for 83 K-Bi2212 [1211]. Therefore, they conjecture that the pseudogap is of a pairing origin and not of magnetic origin. *Marshall* et al. observed a large Fermi surface which is consistent with the band calculations for slightly overdoped 85 K-Bi2212, but the Fermi surface was not observed at 110 K along the $Y$–$\bar{M}$ for underdoped 67 K-Bi2212 suggesting an opening of a gap [1252]. The suppression of the spectral weight is compensated by the enhancement of the weight at another region of the *k*-space. *Shen* et al. noted that the transfer of the spectral weight occurs from $(0.36\pi/a, 0)$ to $(0.8\pi/a, 0)$ by comparing the spectra at 100 K and 20 K [1253]. Here $(\pi/a, \pi/a)$ is parallel with the Cu–Cu (not Cu–O–Cu) direction. The suppression occurs over an energy range as wide as 300 meV whereas the maximum phonon energy is about 80 meV.

*Renner* et al. observed the pseudogap-like depression in the tunneling conductance measured along the *c* axis with an STM. It was present both for underdoped and overdoped Bi2212, and seems to merge into the superconducting gap structure. The superconducting gap was claimed to be independent of temperature, and increased with decreasing carrier density [1239]. On the other hand, *Suzuki* et al. observed a superconducting gap structure within the pseudogap structure using the *c* axis Josephson junction [1254].

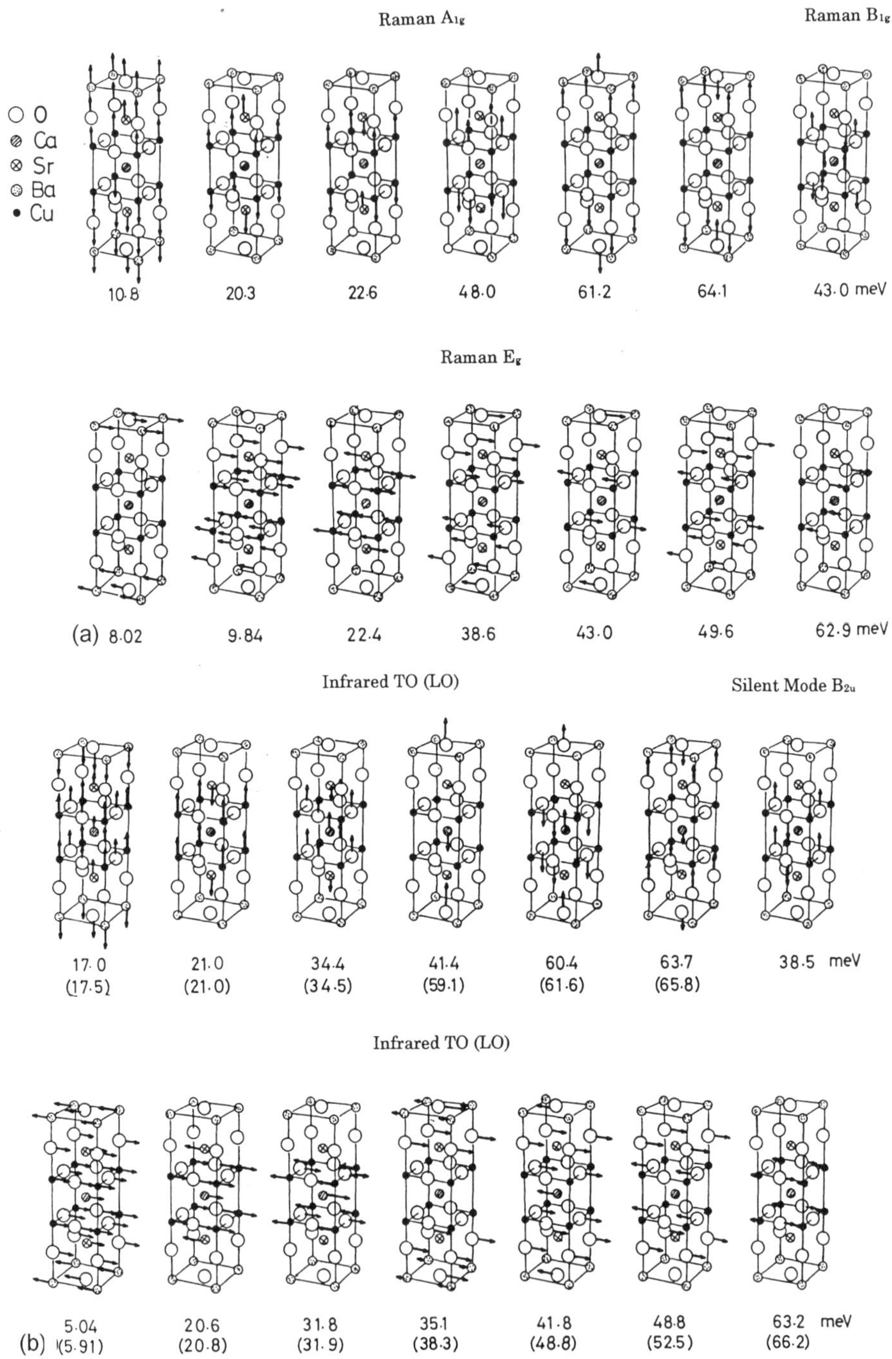


**Fig. 5.132.** Raman (**a**) and IR (**b**) modes of $Bi_2Sr_2CaCu_2O_8$. Amplitudes are schematic and not to scale [1257]. (With permission from The American Physical Society)

Thus it has been disputed whether the pseudogap phenomenon is a precursor of high-$T_c$ superconductivity or whether it is a phenomenon peculiar to the nearly antiferromagnetic poor metal and not directly related with high-$T_c$ superconductivity [1255].

**d) Phonons.** Phonon density of states (PDOS) are shown in Figs. 5.128 and 5.129. The former is the generalized PDOS experimentally determined by *Renker* et al. and has not been corrected for the neutron scattering cross section [1231], while the latter was calculated by *Paltzer* and *de Wette* fitting to the Raman phonon frequencies [1232, 1256]. In Fig. 5.132, normal modes in Bi2212 are shown following *Prade* et al. [1257] Those of Tl oxides were calculated by *Kurkarni* et al. [1258, 1259].

# References

## Chapter 1

1. C. Kittel: *Introduction to Solid State Physics*, 6th ed. (Wiley, New York 1986)
2. H. Ibach, H. Lüth: *Solid-State Physics*, 2nd ed. (Springer, Berlin, Heidelberg 1995)
3. A. Lakar, S. Chandra (Eds.): *Superionic Solids and Solid Electrolytes* (Academic, New York 1989)

## Chapter 2

4. H. Kamimura, S. Sugano, Y. Tanabe: *Ligand Field Theory and Its Applications* (Syokabo, Tokyo 1969) in Japanese
5. S. Sugano, Y. Tanabe, H. Kamimura: *Multiples of Transition-Metal Ions in Crystals* (Academic, New York 1970)
6. J.H. Slater: *Quantum Theory of Molecules and Solids*, Vol. 2 (McGraw-Hill, New York 1965)
7. J.F. Annett: Adv. in Phys. **39**, 83 (1990)
8. H. Ibach, H. Lüth: *Solid-State Physics*, 2nd edn. (Springer, Berlin, Heidelberg 1995)
9. J.C. Slater, G.F. Koster: Phys. Rev. **94**, 1498 (1954)
10. C. Kittel: *Quantum Theory of Solids* (Wiley, New York, 1963)
11. L.F. Mattheiss: Phys. Rev. **181**, 987 (1969)
12. T. Wolfram, E.A. Kraut, F.J. Morin: Phys. Rev. B **7**, 1677 (1973)
13. J. Kondo: In *Solid State Phys.* Vol. **23**, ed. by F. Seitz, D. Turnbull, H. Ehrenreich (Academic Press, New York 1969) p. 183
14. A.J. Heeger: In *Solid State Phys.* Vol. **23**, ed. by F. Seitz, D. Turnbull, H. Ehrenreich (Academic Press, New York 1969) p. 283
15. D.R. Hamann: Phys. Rev. **158**, 570 (1967)
16. T. Sugiyama and N Tsuda: J. Phys. Soc. Jpn. **68**, 1306 (1999)
17. M. Born, J.R. Oppenheimer: Ann. Phys. **84**, 457 (1927)
18. D. Adler: In *Handbook on Semiconductors* Vol. 1, ed. by W. Paul (North-Holland, Amsterdam 1982) p. 805
19. H. Fröhlich: Adv. Phys. **3**, 325 (1954)
20. D.J. Howarth, E.H. Sondheimer: Proc. Roy. Soc. London A **219**, 53 (1953)
21. J. Bardeen, W. Shockley: Phys. Rev. **80**, 72 (1950)
22. A.H. Wilson: *The Theory of Metals*, 2nd edn. (Cambridge University Press 1953)
23. J.M. Ziman: *Electrons and Phonons* (Oxford University Press, Oxford 1963)
24. W.L. McMillan, J.M. Rowell: Phys. Rev. Lett. **14**, 108 (1965)
25. H. Fröhlich, H. Pelzer, S. Ziemann: Philos. Mag. **41**, 221 (1950)
26. T.D. Lee, F.E. Low, D. Pines: Phys. Rev. **90**, 297 (1953)

27. W.L. McMillan: Phys. Rev. **167**, 331 (1968)
28. P.B. Allen: Phys. Rev. B **17**, 3725 (1978)
29. F.J. Pinski, P.B. Allen, W.H. Butler: Phys. Rev. B **23**, 5080 (1981)
30. T. Holstein: Ann. of Phys. **8**, 325, 343 (1959)
31. H.Böttger, V.V. Bryksin: Phys. Stat. Sol. (B) **78**, 9, 415 (1976)
32. T. Kasuya, A. Yanase: Rev. Mod. Phys. **40**, 684 (1968)
33. P.W. Anderson: Phys. Rev. **109**, 1492 (1958)
34. N.F. Mott: Phil. Mag. **22**, 7 (1970)
35. J.T. Edwards, D.J. Thouless: J. Phys. C **5**, 807 (1972)
36. D.C. Licciardello, D.J. Thouless: J. Phys. C **8**, 4157 (1975)
37. P.A. Lee, T.V. Ramakrishnan: Rev. Mod. Phys. **57**, 287 (1985)
38. H. Fukuyama: J. Phys. Soc. Jpn. **50**, 407 (1981)
39. H. Fukuyama: In *Anderson Localization*, Ed. Y. Ohtsuki (Kyouritu, Tokyo 1982) p. 59 (A review paper in Japanese)
40. H. Fukuyama: In *Anderson Localization*, ed. by Y. Nagaoka, H. Fukuyama, Springer Series in Solid-State Sci., Vol. **39** (Springer, Berlin, Heidelberg 1982)
41. E. Abrahams, P.W. Anderson, D.C. Licciardello, T.V. Ramakrishnan: Phys. Rev. Lett. **42**, 673 (1979)
42. N.F. Mott: Phil. Mag. **19**, 835 (1969)
43. W. Brenig, G.H. Döhler, P. Wolfle: Z. Physik **258**, 381 (1973)
44. V. Ambegaokar, B.I. Halperin, J.S. Langer: Phys. Rev. **4**, 2612 (1971)
45. A.L. Efros: J. Phys. C **9**, 2021 (1976)
46. R.P. Huebener: Phys. Rev. **146**, 490 (1966)
47. L. Friedman, M. Pollak: In *Disordered Semiconductors*, ed. by M.A. Kastner, G.A. Thomas, S.R. Ovshinsky (Plenum Press, New York 1987) p. 745
48. C. Wood, D. Emin: Phys. Rev. B **29**, 4582 (1984)
49. K.-D. Schotte: Z. Phys. **196**, 393 (1966)
50. T. Holstein: Phys. Rev. **124**, 1329 (1961)
51. H. Böttger, V. Bryksin: Phys. Stat. Sol. (B) **80**, 569 (1977)
52. P.N. Butcher, J. Clark, A.A. Kumar, J.A. McInnes: J. Non-Cryst. Solids **35–36**, 89 (1980)
53. L. Friedman, M. Pollak: Philos. Mag. B **44**, 487 (1981)
54. B. Movaghar, B. Pohlmann, D. Wurtz: J. Phys. C **14**, 5127 (1981)
55. D. Emin: In *The Hall Effect and Its Applications*, ed. by C.L. Chien, C.R. Westgate (Plenum, New York 1980) p. 281
56. P.W. Anderson: In *Solid State Physics*, Vol. 14, ed. by F. Seitz, D. Turnbull (Academic, New York 1963) p. 99
57. C. Kittel: In *Solid State Physics*, Vol. 22, ed. by F. Seitz, D. Turnbull, H. Ehrenreich (Academic, New York 1968) p. 199
58. C. Zener: Phys. Rev. **82**, 403 (1951)
59. P.-G. DeGennes: Phys. Rev. **118**, 141 (1960)
60. U. Löw, V.J. Emery, K. Fabricius, S.A. Kivelson: Phys. Rev. Lett. **72**, 1918 (1994)
61. Y. Nagaoka: Phys. Rev. **147**, 392 (1966)
62. W.F. Brinkman, T.M. Rice: Phys. Rev. B**2**, 4302 (1970)
63. K.-F. Berggren: Phil. Mag. **30**, 1 (1974)
64. B.E. Sernelius, K.-F. Berggren: Philos. Mag. B **43**, 115 (1981)
65. T. Moriya, H. Hasegawa: J. Phys. Soc. Jpn. **48**, 1490 (1980)
66. C.Kitte: *Introduction to Solid State Physics*, 6th edn. (Wiley, New York 1986)
67. J.H. Van Vleck: *The Theory of Electric and Magnetic Susceptibilities* (Oxford University Press, Oxford 1932)
68. P.W. Selwood: *Magnetochemistry*, 2nd edn. (Interscience, New York 1956) p. 78

69. N.F. Mott: *Metal–Insulator Transition*, 2nd edn. (Taylor & Francis, London 1990)
70. N.F.Mott: Proc. R. Soc. London A **382**, 1 (1982)
71. F.J. Rogers, H.C. Graboske, Jr., D.J. Harwood: Phys. Rev. A **1**, 1577 (1970)
72. P. Leroux Hugon, A.Ghazali: Phys. Rev. B **14**, 602 (1976)
73. J.B. Krieger, M. Nightingale: Phys. Rev. B **4**, 1266 (1971)
74. P.P. Edwards, M.J. Sienko: Phys. Rev. B **17**, 2575 (1978)
75. J.B. Goodenough: Mater. Res. Bull. **5**, 621 (1970)
76. B.K. Chakraverty: J. Solid State Chem. **12**, 376 (1975)
77. D.B. McWhan, M. Marezio, J.P. Remeika, P.P. Dernier: Phys. Rev. B **10**, 490 (1974)
78. A. Zylbersztejn, N.F. Mott: Phys. Rev. B **11**, 4383 (1975)
79. A. Bianconi, S. Stizza, R. Bernardini: Phys. Rev. B **24**, 4406 (1981)
80. M. Hase, I. Terasaki, K. Uchinokura: Phys. Rev. Lett. **70**, 3651 (1993)
81. H. Winkelmann, E. Gamper, B. Büchner, M. Braden, A. Revcolevschi, G. Dhalenne: Phys. Rev. B **51**, 12884 (1995-II)
82. C. Schlenker, J. Dumas, C. Escribe-Fillipini, H. Guyot, J. Marcus, G. Fourcaudot: Philos. Mag. B **52**, 643 (1985)
83. M. Imada, A. Fujimori, Y. Tokura: Rev. Mod. Phys. **70**, 1039 (1998)
84. D. Adler: In *Solid State Physics*, Vol. 21, edn. by F. Seitz, D. Turnbull, H. Ehrenreich (Academic, New York 1968) p. 1
85. K. Suzuki: In *Nonstoichiometric Metal Compounds*, ed. by S. Takeuchi (Maruzen, Tokyo 1975) p. 1
86. S. Takeuchi, K. Suzuki: Trans. Jpn. Inst. Metal. **33**, 279 (1969)
87. S. Takeuchi, K. Suzuki: Trans. Jpn. Inst. Metal. **33**, 284 (1969)
88. S. Takeuchi, K. Suzuki: Trans. Jpn. Inst. Metal. **33**, 409 (1969)
89. A.D. Pearson: Phys. Chem. Solids **5**, 316 (1958)
90. J.M. Honig, W.E. Wahnsiedler, P.C. Eklund: J. Solid State Chem. **6**, 203 (1973)
91. W.W. Schulz, L. Forro, C. Kendziora, R. Wentzcovitch, D. Mandrus, L. Mihaly, P.B. Allen: Phys. Rev. B **46**, 14001 (1992)
92. H. Aoki, Y. Asada, T. Hatano, K. Ogawa, A. Yanase, M. Koiwa: J. Low Temp. Phys. **81**, 19 (1990)
93. E. Wimmer, K. Schwarz, R. Podloucky, P. Herzig, A. Neckel: J. Phys. Chem. Solids **43**, 439 (1982)
94. J.M. Leger, P. Aimonino, J. Loriers, P. Dordor, B. Coqblin: Phys. Lett. **80A**, 325 (1980)
95. C. Llinares, J.P. Desfours, J.P. Nadai, C. Gordat, A. Percheron, J.C. Achard: Phys. Stat. Sol. (A) **25**, 185 (1974)
96. T.A. Hewston, B.L. Chamberland: J. Solid State Chem. **65**, 100 (1986)
97. T.A. Hewston, B.L. Chamberland: J. Phys. Chem. Solids **48**, 97 (1987)
98. M. Onoda, T. Naka, H. Nagasawa: J. Phys. Soc. Jpn. **60**, 2550 (1991)
99. D.C. Johnston: J. Low Temp. Phys. **25**, 145 (1976)
100. R.W. McCallum, D.C. Johnston, C.A. Luengo, M.B. Maple: J. Low Temp. Phys. **25**, 177 (1976)
101. T. Inukai, T. Murakami, T. Inamura: Thin Solid Films **94**, 47 (1982)
102. U. Roy, A. DasGupta, C.C. Kock: IEEE Trans. Magn. **13**, 836 (1977)
103. I. Maekawa, F. Takagi, Y. Sakai, N. Tsuda: J. Phys. Soc. Jpn. **56**, 2119 (1987)
104. See references cited in Sect. 5.3
105. P.M. Lambert, M.R. Harrison, D.E. Logan, P.P. Edwards: In *Disordered Semiconductors*, ed. by M.A. Kastner, G.A. Thomas, S.R. Ovshinsky (Plenum, New York 1987) p. 135

106. P.M.Lambert, M.R. Harrison, P.P. Edwards: J. Solid State Chem. **75**, 332 (1988)
107. D.B. Rogers, J.L. Gillson, T.E. Gier: Solid State Commun. **5**, 263 (1967) and references in Sect. 5.3
108. P.A. Miles, W.B. Westphal, A. von Hippel: Rev. Mod. Phys. **29**, 279 (1957) and references in Sect. 5.6
109. Muhtar, F. Takagi, K. Kawakami, N. Tsuda: J. Phys. Soc. Jpn. **57**, 3119 (1988)
110. T. Omata, N. Ueda, K. Ueda, H. Kawazoe: Appl. Phys. Lett. **64**, 1077 (1994)
111. A.W. Sleight, J.L. Gillson: Solid State Commun. **4**, 601 (1966) and references in Sect. 5.1
112. S. Kachi, K. Kosuge: In *Nonstoichiometric Metal Compounds*, ed. by S. Takeuchi (Maruzen, Tokyo 1975) p. 59
113. D.R. Wanlass, M.J. Sienko: J. Solid State Chem: **12**, 362 (1975)
114. L.D. Ellerbeck, H.R. Shanks, P.H. Sidles, G.C. Danielson: J. Chem. Phys. **35**, 298 (1961)
115. T.A. Bither, J.L. Gillson, H.S. Young: Inorg. Chem. **5**, 1559 (1966) and references in Sect. 5.4
116. J.P. Giroult, M. Goreaud, P. Labbe, J. Provost, B. Raveau: Mater. Res. Bull. **16**, 811 (1981)
117. J.P. Doumerc, P. Dordor, E. Marquestant, M. Pouchard, P. Hagenmuller: Philos. Mag. B **42**, 487 (1980)
118. M.A. Dubson, D.F. Holcomb: Phys. Rev. B **34**, 25 (1986)
119. D. Ridgley, R. Ward: J. Am. Chem. Soc. **77**, 6132 (1955)
120. K. Ishikawa, G. Adachi, M. Hasegawa, K. Sato, J. Shiokawa: J. Electrochem. Soc. **128**, 1374 (1981)
121. B. Ellis, J.P. Doumerc, P. Dordor, M. Pouchard, P. Hagenmuller: Solid State Commun. **51**, 913 (1984)
122. J.B. Goodenough: J. Appl. Phys. **37**, 1415 (1966)
123. J.E. Greedan: J. Less-Common Met. **111**, 335 (1985)
124. D.A. Crandes, T. Timusk, J.D. Garrett, J.E. Greedan: Physica C **201**, 407 (1992)
125. F. Lichtenberg, D. Widmer, J.B. Bednorz, A. Reller: Z. Phys. B **82**, 211 (1991)
126. Y. Fujishima, Y. Tokura, T. Arima, S. Uchida: Phys. Rev. B **46**, 11167 (1992)
127. I.S. Kim. T. Nakamura, Y. Inaguma, M. Itoh: J. Solid State Chem: **113**, 281 (1994)
128. G.V. Bazuev, G.P. Shveikin: Izv. Akad. Nauk SSSR Neorg. Mater. **22**, 1354 (1986)
129. I. Kim, M. Itoh, T. Nakamura: J. Solid State Chem. **101**, 77 (1992)
130. B.L. Chamberland, P.S. Danielson: J. Solid State Chem. **3**, 243 (1971)
131. I.H. Inoue, K. Morikawa, H. Fukuchi, T. Tsujii, F. Iga, Y. Nishihara: Jpn. J. Appl. Phys. Suppl. **32**, 451 (1993)
132. P. Dougier, J.C.C. Fan, J.B. Goodenough: J. Solid State Chem: **14**, 247 (1975)
133. M. Onoda, H. Ohta, H. Nagasawa: Solid State Commun. **79**, 281 (1991)
134. G.V. Bazuev, G.D. Milova, O.V. Makarova: Izv. Akad. Nauk SSSR Neorg. Mater. **17**, 342 (1981)
135. G.V. Bazuev, G.D. Milova, G.P. Shveikin: Izv. Akad. Nauk SSSR Neorg. Mater. **15**, 1268 (1979)
136. G.V. Bazuev, O.V. Makarova, G.P. Shveikin: Izv. Akad. Nauk SSSR Neorg. Mater. **18**, 859 (1982)
137. P. Dougier, P. Hagenmuller: J. Solid State Chem. **15**, 158 (1975)

138. P. Dougier, A. Casalot: J. Solid State Chem. **2**, 396 (1970)
139. G.D. Milova, G.V. Bazuev, N.A. Kirsanov: Izv. Akad. Nauk SSSR Neorg. Mater. **24**, 278 (1988)
140. M. Itoh, M. Shikano, H. Kawaji, T. Nakamura: Solid State Commun. **80**, 545 (1991)
141. M. Cyrot, B. Lambert-Andron, J.L. Soubeyroux, M.J. Rey, Ph. Dehauht, F. Cyrot-Lacjmann G. Fourcaudt, J. Bellie, J.L. Tholence: J. Solid State Chem. **85**, 321 (1990)
142. A. Nozaki, H. Yoshikawa, T. Wada, H. Yamauchi, S. Tanaka: Phys. Rev. B **43**, 181 (1991)
143. S. Tanaka, N. Fukushima, R. Yoshimura, T. Tada: J. Phys. Soc. Jpn. **61**, 2399 (1992)
144. J.B. Goodenough, J.M. Longo, J.A. Kafalas: Mater. Res. Bull. **3**, 471 (1968)
145. J.F. Weiher, B.L. Chamberland, J.L. Gillson: J. Solid State Chem. **3**, 529 (1971)
146. B.L. Chamberland: Solid State Commun. **5**, 663 (1967)
147. H. Taguchi, M. Shimada: J. Solid State Chem. **63**, 290 (1986)
148. K. Chahara, T. Ohno, M. Kasai, Y. Kozono: Appl. Phys. Lett. **63**, 1990 (1993)
149. S. Jin, T.H. Tiefel, M. McCormack, R.A. Fastnacht, R. Ramesh, L.H. Chen: Science **264**, 413 (1994)
150. J.M. De Teresa, M.R. Ibarra, P.A. Algarabel, C. Ritter, C. Marquina, J. Blasco, J. Garcia, A. del Moral, Z. Arnold: Nature **386**, 256 (1997)
151. J.H. van Santen, G.H. Jonker: Physica **16**, 599 (1950)
152. J.B. Goodenough: In *Progress in Solid State Chemistry*, Vol. 5, ed. by H. Reiss (Pergamon, London 1971) p. 145
153. S. Methfessel, D.C. Mattis: *Magnetism*, Handbuch der Physik, Vol. 18, Part 1, ed. by H.P. Wijn (Springer, Berlin, Heidlberg 1963) p. 389
154. T.M. Perekalina, I.E. Lipin'ski, V.A. Timofeeva, S.A. Cherkezyan: Fiz. Tverd. Tela **32**, 3146 (1990)
155. V.I. Pavlov, A.K. Bogush, G.L. Bychkov: Izv. Akad. Nauk SSSR Neorg. Mater. **20**, 863 (1984)
156. R. von Helmolt, J. Wecker, B. Holtzapfel, L. Schultz, K. Samwer: Phys. Rev. Lett. **71**, 2331 (1993)
157. Z. Jirak, E. Pollert, A.F. Andersen, J.-C. Grenier, P. Hagenmuller: Eur. J. Solid State Inorg. Chem. **27**, 421 (1990)
158. S.B. Ogale, V. Talyansky, C.H. Chen, R. Ramesh, R.L. Greene, T. Venkatesan: Phys. Rev. Lett. **77**, 1159 (1996)
159. R.M. Kusters, J. Singleton, D.A. Keen, R. McGreevy, W. Hayes: Physica B **155**, 362 (1989)
160. Y. Takeda, S. Naka, M. Takano, T. Shinjo, T. Takeda, M. Shimada: Mater. Res. Bull. **13**, 61 (1978)
161. J.B. MacChesney, R.C. Sherwood, J.F. Potter: J. Chem. Phys. **43**, 1907 (1965)
162. T. Takeda, H. Watanabe: J. Phys. Soc. Jpn. **33**, 973 (1972)
163. T. Takeda, Y. Yamaguchi, H. Watanabe: J. Phys. Soc. Jpn. **33**, 967 (1972)
164. A.E. Bocquet, A. Fujimori, T. Mizokawa, T. Saitoh, H. Namatame, S. Suga, N. Kimizuka, Y. Takeda, M. Takano: Phys. Rev. B **45**, 1561 (1992)
165. V.G. Jadhao, R.M. Singru, G. Rama Rao, D. Bahadur, C.N.R. Rao: J. Chem. Soc., Faraday Trans. **2**, 71, 1885 (1975)
166. P.M. Raccah, J.B. Goodenough: Phys. Rev. **155**, 932 (1967)
167. G. Thornton, B.C. Tofield, D.E. Williams: Solid State Commun. **44**, 1213 (1982)

168. K. Asai, P. Gehring, H. Chou, G. Shirane: Phys. Rev. B **40**, 10982 (1989)
169. I.F. Kononyuk, S.P. Tolochko, V.A. Lutsko, V.M. Anishchik: J. Solid State Chem. **48**, 209 (1983)
170. P. Ganguly, N.Y. Vasanthacharya, C.N.R. Rao, P.P. Edwards: J. Solid State Chem. **54**, 400 (1984)
171. P.M. Raccah, J.B. Goodenough: J. Appl. Phys. **39**, 1209 (1968)
172. J.B. Goodenough: J. Phys. Chem. Solids **6**, 287 (1958)
173. A. Minesige, M. Inaba, T. Yao, Z. Oguchi, K. Kikuchi, M. Kawase: J. Solid State Chem. **121**, 423 (1996)
174. K.P. Rajeev, A.K. Raychaudhuri: Phys. Rev. B **46**, 1309 (1992)
175. H.E. Hofer, R. Schmidberger: J. Electrochem. Soc. **141**, 782 (1994)
176. J.B. Goodenough: Phys. Rev. **164**, 785 (1967)
177. J.B. Goodenough, P.M. Raccah: J. Appl. Phys. **36**, 1031 (1965)
178. K. Asai, H. Sekizawa: J. Phys. Soc. Jpn. **49**, 90 (1980)
179. J.B. Goodenough, N.F. Mott, M. Pouchard, G. Demazeau, P. Hagenmuller: Mater. Res. Bull. **8**, 647 (1973)
180. R.M. Mohan Ram, L. Ganapathi, P. Ganguly, C.N.R. Rao: J. Solid State Chem. **63**, 139 (1986)
181. J. Blasco, J. Garcia: J. Phys. Condens. Matter **6**, 10759 (1994)
182. J.B. Torrance, P. Lacorre, A.I. Nazzal, E.J. Ansaldo, Ch. Niedermayer: Phys. Rev. B **45**, 8209 (1992)
183. P. Lacorre, J.B. Torrance, J. Pannetier, A.I. Nazzal, P.W. Wang, T.C. Huang: J. Solid State Chem. **91**, 225 (1991)
184. J.K. Vassiliou, M. Hornbostel, R. Ziebarth, F.J. Disalvo: J. Solid State Chem. **81**, 208 (1989)
185. G. Demazeau, A. Marbeuf, M. Pouchard, P. Hagenmuller, J.B. Goodenough: C. R. Acad. Sci. C **272**, 2163 (1971)
186. G. Demazeau, C. Parent, M. Pouchard, P. Hagenmuller: Mater. Res. Bull. **7**, 913 (1972)
187. C. Michel, L. Er-Rakho, B. Raveau: Mater. Res. Bull. **20**, 667 (1985)
188. C. Michel, L. Er-Rakho, M. Hervieu, J. Pannetier, B. Raveau: J. Solid State Chem. **68**, 143 (1987)
189. J.B. Torrance, Y. Tokura, A. Nazzal, S.S. Parkin: Phys. Rev. Lett. **60**, 542 (1988)
190. R. Desfeux, J.F. Hamet, B. Mercey, C. Simon, M. Hervieu, B. Raveau: Physica C **221**, 205 (1994)
191. R. Genouel, C. Michel, N. Nguyen, M. Hervieu, B. Raveau: J. Solid State Chem. **115**, 469 (1995)
192. N. Murayama, Y. Torii: Mater. Sci. Eng. A, A **109**, 307 (1989)
193. K. Ishikawa, G. Adachi, J.Shiokawa: Mater. Res. Bull. **16**, 419 (1981)
194. R.J. Cava, B. Batlogg, J.J. Krajewski, P. Gammel, H.F. Poulsen, W.F. Peck, Jr., L.W. Rupp, Jr.: Nature **350**, 598 (1991)
195. J.M. Longo, P.M. Raccah, J.B. Goodenough: J. Appl. Phys. **39**, 1327 (1968)
196. A. Callaghan, C.W. Moeller, R. Ward: Inorg. Chem. **5**, 1572 (1966)
197. R.J. Bouchard, J.L. Gillson: Mater. Res. Bull. **7**, 873 (1972)
198. T.C. Gibb, R. Greatrex, N.N. Greenwood, D.C. Puxley, K.G. Snowdon: J. Solid State Chem. **11**, 17 (1974)
199. P.A. Cox, R.G. Egdell, J.B. Goodenough, A. Hamnett, C.C. Naish: J. Phys. C **16**, 6221 (1983)
200. T. Kiyama, K. Yoshimura, K. Kosuge, H. Michor, G. Hilscher: J. Phys. Soc. Jpn. **67**, 307 (1998)
201. J.M. Longo, J.A. Kafalas, R.J. Arnott: J. Solid State Chem. **3**, 174 (1971)

202. M. Shikano, T.K. Huang, Y. Inaguma, M. Itoh, T. Nakamura: Solid State Commun. **90**, 115 (1994)
203. M. Shepard, G. Cao, S. McCall, F. Frebert, J.E. Crow: J. Appl. Phys. **79**, 4821 (1996)
204. A. Kanbayashi: J. Phys. Soc. Jpn. **44**, 108 (1978)
205. G. Cao, S. McCall, H. Shepard, J.E. Crow, R.P. Guertin: Phys. Rev. B **56**, 321 (1997)
206. F. Fukunaga, N. Tsuda: J. Phys. Soc. Jpn. **63**, 3798 (1994)
207. J.B. Goodenough, J.M. Longo: In *Landolt-Boernstein*, New Series III/4a (Springer, Berlin, Heidelberg 1970) p. 224
208. F.M. Da Costa, R. Greatrex, N.N. Greenwood: J. Solid State Chem. **20**, 381 (1977)
209. R.J. Bouchard, J.F. Weiher: J. Solid State Chem. **4**, 80 (1972)
210. M. Jansen, B. Klinkert, S. Elschner: Mater. Res. Bull. **25**, 1415 (1990)
211. R.J. Bouchard, J.F. Weiher, J.L. Gillson: J. Solid State Chem. **21**, 135 (1977)
212. B. Szymanik, A. Edger: Solid State Commun. **79**, 355 (1991)
213. T. Nakamura, T. Shimura, M. Itoh, Y. Takeda: J. Solid State Chem. **103**, 523 (1993)
214. T. Shimura, M. Itoh, T. Nakamura: Physica C **190**, 104 (1991)
215. B.L. Chamberland: J. Less-Common Metals **171**, 377 (1991)
216. R.D. Shannon, P.E. Bierstedt: J. Amer. Ceram. Soc. **53**, 635 (1970)
217. V.V. Bogatko, Yu.N. Venevtsev: Sov. Phys.-Solid State **22**, 705 (1980)
218. M. Itoh, T. Sawada, I.S. Kim, Y. Inaguma, T. Nakamura: Solid State Commun. **83**, 33 (1992)
219. A.W. Sleight, J.L. Gillson, P.E. Bierstedt: Solid State Commun. **17**, 27 (1975)
220. L.F. Mattheiss, E.M. Gyorgy, D.W. Johnston, Jr.: Phys. Rev. B **37**, 3745 (1988)
221. Q. Huang, J.F. Zasadzinski, N. Tralshawala, K.E. Gray, D.G. Hinks, J.L. Peng, R.L. Greene: Nature **347**, 369 (1990)
222. D.G. Hinks, D.R. Richards, B. Dabrowski, D.T. Marx, A.W. Mitchell: Nature **335**, 419 (1988)
223. Z. Schlesinger, R.T. Collins, J.A. Calise, D.G. Hinks, A.W. Mitchell, Y. Zheng, B. Dabrowski, N.E. Bickers, D.J. Scalapino: Phys. Rev. B **40**, 6862 (1989)
224. D.N. Manh, D. Mayou, F. Cyrot-Lackmann: Europhys. Lett. **13**, 167 (1990)
225. R.J. Cava, B. Batlogg, G.P. Espinosa, A.P. Ramirez, J.J. Krajewski, W.F. Peck, Jr., L.W. Rupp, Jt., A.S. Cooper: Nature **339**, 291 (1989)
226. M. Itoh, T. Sawada, I. Kim, Y. Inaguma, T. Nakamura: Physica C **204**, 194 (1991)
227. G.H. Bouchard, Jr., M.J. Sienko: Inorg. Chem. **7**, 441 (1968)
228. L.H. Brixner: J. Inorg. Nucl. Chem: **14**, 225 (1960)
229. J.B. Goodenough, J.M. Longo, J.A. Kafalas: Mater. Res. Bull. **3**, 471 (1968)
230. S.Hayashi, R. Aoki, T. Nakamura: Mater. Res. Bull. **14**, 409 (1979)
231. G. Liu, X. Zhao, H.A. Eick: J. Alloys Compd. **187**, 145 (1992)
232. V.A. Fotiev, G.V. Bazuev, V.G. Zubkov: Izv. Akad. Nauk SSSR Neorg. Mater. **23**, 1005 (1987)
233. F.S. Galasso: *Structure, Properties and Preparation of Perovskite-type Compounds* (Pergamon, Oxford 1969)
234. J.B. Goodenough: *Magnetism and Chemical Bond* (Wiley-Interscience, New York 1963)
235. J.F. Schooley, W.R. Hosler, M.L. Cohen: Phys. Rev. Lett. **12**, 474 (1964)
236. A.R. Mackintosh: J. Chem. Phys. **38**, 1991 (1963)

237. F. Abraham, J. Trehoux, D. Thomas, F.E. Wagner: J. Less-Common Met. **84**, 245 (1982)
238. J.M. Longo, A.W. Sleight: Inorg. Chem. **7**, 108 (1968)
239. G.A. Thomas, D.H. Rapkine, S.A. Carter, T.F. Rosenbaum, P. Metcalf, D.F. Honig: J. Low Temp. Phys. **95**, 33 (1994)
240. See references cited in Sect. 5.8 of the 1st edition
241. S.C. Abraham: Phys. Rev. **130**, 2230 (1963)
242. J.M. Honig, T.B. Reed: Phys. Rev. **174**, 1020 (1968)
243. H. Leiva, R. Kershaw, K. Dwight, A. Wold: Mater. Res. Bull. **17**, 1539 (1982)
244. Y. Kachi, K. Kosuge: In *Nonstoichiometric Compounds*, ed. by S. Takeuchi ( Maruzen, Tokyo 1975) p. 233
245. D.H. Henler: J. Appl. Phys. **39**, 2354 (1968)
246. A.Bianconi, S. Stizza, R. Bernardini: Phys. Rev. B **24**, 4406 (1981)
247. R.M. Wentzcovitch, W.W. Schluz, P.B. Allen: Phys. Rev. Lett. **72**, 3389 (1994)
248. T.M. Rice, H. Launois, J.P. Pouget: Phys. Rev. Lett. **73**, 3042 (1994)
249. R.M. Wentzcovitch, W.W. Schluz, P.B. Allen: Phys. Rev. Lett. **73**, 3043 (1994)
250. D.S. Rodbell, J.M. Lommel, R.C. Devries: J. Phys. Soc. Jpn. **21**, 2430 (1966)
251. G. Belanger, J. Destry, G. Perluzzo: Can. J. Phys. **52**, 2272 (1974)
252. Y. Sakai, N. Tsuda, T. Sakata: J. Phys. Soc. Jpn. **54**, 1514 (1985)
253. D.B. Rogers, R.D. Shannon, A.W. Sleight, J.L. Gillson: Inorg. Chem. **8**, 841 (1969)
254. R.C. Vickery, J.C. Hipp: J. Appl. Phys. **37**, 2926 (1966)
255. S.R. Butler, J.L. Gillson: Mater. Res. Bull. **6**, 81 (1971)
256. W.D. Ryden, A.W. Lawson, C.C. Sartain: Phys. Rev. B **1**, 1494 (1970)
257. J.B. Graebner, E.S. Greiner, W.D. Ryden: Phys. Rev. B **13**, 2426 (1976)
258. W.D. Ryden, A.W. Lawson: J. Chem. Phys. **52**, 6058 (1970)
259. L.F. Mattheiss: Phys. Rev. B **13**, 2433 (1976)
260. Y.S. Huang, H.L. Park, F.H. Pollak: Mater. Res. Bull. **17**, 1305 (1982)
261. K.M. Glassford, J.R. Chelikowsky: Phys. Rev. B **47**, 1732 (1993)
262. R.D. Shannon: Solid State Commun. **6**, 139 (1968)
263. J.M. Honig: J. Solid State Chem. **45**, 1 (1982)
264. B.F. Griffin, S.P. Faile, J.M. Honig: Phys. Rev. B **21**, 154 (1980)
265. E.I. Terkov. D.I. Khomskii, F.A. Chundnovskii: Sov. Phys.-JETP **46**, 1160 (1977)
266. S. Kachi: AIP Conf. Proc. **10**, 714 (1972)
267. S. Kachi, Y. Ueda, K. Kosuge, Y. Kitaoka, H. Yasuoka: *Ferrites*, Proc. of the Intern. Conf., September–October 1980, Japan, p. 131.
268. P. Dordor, A. Erragh, G. Villeneuve, J.C. Launay, P. Hagenmuller: Mater. Res. Bull. **16**, 817 (1981)
269. A.D. Inglis, C.M. Hurd, P.Strobel: J. Phys. C **17**, 6801 (1984)
270. A.C. Gossard, J.P. Remeika, T.M. Rice, H. Yasuoka: Phys. Rev. B **9**, 1230 (1974)
271. D.B. McWhan, J.P. Remeika, J.P. Maita, H. Okinaka, K. Kosuge, S. Kachi: Phys. Rev. B **7**, 326 (1973)
272. S. Nagata, P.H. Keesom, S.P. Faile: Phys. Rev. B **20**, 2886 (1979)
273. S. Nagata, B.F. Griff D.R. Frenkl: Phys. Rev. **187**, 828 (1969)
274. B.F. Griffing, S.A. Shivashankar, S. Nagata, S.P.Faile, J.M. Honig: Phys. Rev. B **25**, 1703 (1982)
275. B.F. Griffing, S.A. Shivashankar, S.P. Faile, J.M. Honig: Phys. Rev. B **31**, 8143 (1985)
276. R.F. Bartholomew, D.R. Frenkl: Phys. Rev. **187**, 828 (1969)

277. A.D. Inglis, Y. Le Page, P. Strobe, C.M. Hurd: J. Phys. C **16**, 317 (1983)
278. S. Lakkis, C. Schlenker, B.K. Chakraverty, R. Buder, M. Marezio: Phys. Rev. B **14**, 1429 (1976)
279. See references cited in Sect. 5.2
280. A.F. Reid, M.J. Sienko: Inorg. Chem. **6**, 321 (1967)
281. P. Ganguly, C.N.R. Rao: Mater. Res. Bull. **8**, 405 (1973)
282. C.N.R. Rao, D.J. Bua, P. Ganguly, H.R. Harrison, C.J. Sandberg, J.M. Honig: J. Solid State Chem. **52**, 266 (1984)
283. K.K. Singh, P. Ganguly, J.B. Goodenough: J. Solid State Chem. **52**, 254 (1984)
284. D.J. Buttery, J.M. Honig, C.N.R. Rao: J. Solid State Chem. **64**, 287 (1986)
285. M. Sayer, P. Odier: J. Solid State Chem. **67**, 26 (1987)
286. H. Eisaki, S. Uchida, T. Mizokawa, H. Namatame, A. Fujimori, J. van Elp, P. Kuiper, G.A. Sawatzky, S. Hosoya, H. Katayama-Yoshida: Phys. Rev. B **45**, 12513 (1992)
287. Y. Takeda, R. Kanno, M. Sakano, O. Yamamoto, M. Takano, Y. Bando, H. Akinaga, K. Takit, J.B. Goodenough: Mater. Res. Bull. **25**, 293 (1990)
288. J. Gopalakrishnan, G. Colsmann, B. Reuter: J. Solid State Chem. **22**, 145 (1977)
289. R.J. Cava, B. Batlogg, T.T. Palstra, J.J. Krajewski, W.F. Peck, Jr., A.P. Ramirez, L.W. Rupp, Jr.: Phys. Rev. B **43**, 1229 (1991)
290. A.M. George, I.K. Gopalakrishnan, K.D. Karkhanavala: Mater. Res. Bull: **9**, 721 (1974)
291. C.N.R. Rao, P. Ganguly: In *The metallic and Nonmetallic States of Matter*, ed. by D.P. Edwards, C.N.R. Rao (Taylor & Francis, London 1985)
292. See references cited in Sect. 5.9
293. K.K. Singh, P. Ganguly, C.N.R. Rao: Mater. Res. Bull. **17**, 493 (1982)
294. A. Callaghan, C.W. Moeller, R. Ward: Inorg. Chem. **5**, 1572 (1966)
295. T. Williams, F. Lichtenberg, A. Reller, J.G. Bednorz: Mater. Res. Bull. **26**, 763 (1991)
296. F. Lichtenberg, A. Catana, J. Mannhart, D.G. Schlom: Appl. Phys. Lett. **60**, 1138 (1992)
297. Y. Maeno, H. Hashimoto, K. Yoshida, S. Nishizaki, T. Fujita, J.G. Bednorz, F. Lichtenberg: Nature **372**, 532 (1994)
298. K. Ishida, H. Mukuda, Y. Kitaoka, K. Asayama, Z.Q. Mao, Y. Mori, Y. Maeno: Nature **396**, 658 (1998)
299. R.J. Cava, B. Batlogg, K. Kiyono, H. Takagi, J.J. Krajewski, W.F. Peck, Jr,, L.W. Rupp, Jr., C.H. Chen: Phys. Rev. B **49**, 11890 (1994)
300. T. Shimura, M. Itoh, T. Nakamuta: J. Solid State Chem. **98**, 198 (1992)
301. T. Shimura, M. Itoh, Y. Inaguma, T. Nakamuta: Phys. Rev. B **49**, 5591 (1994)
302. See references cited in Sect. 2.9.7
303. See references cited in Sect. 5.5.6
304. J. Grannec, J. Portier, R. Von der Muhll, G. Demazeau, P. Hagenmuller: Mater. Res. Bull. **5**, 185 (1970)
305. A.W. Sleight, J.L. Gillson, B.L. Chamberland: Mater. Res. Bull. **5**, 807 (1970)
306. G.P. Wirtz, C.J. Yu, R.W. Doser: J. Am. Ceram. Soc. **64**, 269 (1981)
307. J.M. Phillips, R.J. Cava, G.A. Thomas, S.A. Carter, J. Kwo, T. Siegrist, J.J. Krajewski, J.H. Marshall, W.F. Peck, Jr., D.H. Rapkine: Appl. Phys. Lett. **67**, 2246 (1995)
308. A.C.W.P. James, J.B. Goodenough: J. Solid State Chem. **74**, 287 (1988)
309. K. Bushida, I. Yazawa, G. Zouganelis, T. Conard, N. Terada, K. Kaneko, M. Hirabayashi, H. Ihara: Physica C **185–189**, 2727 (1991)

310. M.B. Robin, K. Andres, T.H. Geballe, N.A. Kuebler, D.B. McWhan: Phys. Rev. Lett. **17**, 917 (1966)
311. J.J. Sheer, A.E. van Arkelu, R. Heyding: Cand. J. Chem. **33**, 683 (1955)
312. K. Itoh, N. Tsuda: Solid State Commun. **109**, 715 (1999)
313. K. Machida, M. Enyo, G. Adachi, J. Shiokawa: Bull. Chem. Soc. Jpn. **60**, 411 (1987)
314. R. Uriu, D. Shimada, N. Tsuda: J. Phys. Soc. Jpn. **60**, 2479 (1991)
315. See references cited in Sect. 5.5
316. A.Taylor, N.J. Doyle: In *The Chemistry of Extended Defects in Nonmetallic Solids*, ed. by L. Eyring, M.O'Keefe (North-Holland, Amsterdam 1970) p. 523
317. N.J. Doyle, J.K. Hulm, C.K. Jones, R.C. Miller, A. Taylor: Phys. Letters **26A** A604 (1968)
318. M.D. Banus, T.B. Reed, A.J. Strauss: Phys. Rev. B **5**, 2775 (1972)
319. A. Neckel, P. Rastl, R. Eibler, P. Weinberger, K. Schwarz: J. Phys. C **9**, 579 (1976)
320. L.F. Mattheiss: Phys. Rev. B **5**, 290 (1972)
321. R.D. Shannon, C.T. Prewitt: Acta Crystallogr. B **25**, 925 (1969)
322. A.B. Kunz: J. Phys. C **14**, L455 (1981)
323. L.F. Mattheiss: Phys. Rev. B **13**, 2433 (1976)
324. Y. Tokura, A. Urushibara, Y. Moritomo, T. Arima, A. Asamitsu, G. Kido, N. Furukawa: J. Phys. Soc. Jpn. **63**, 3931 (1994)
325 A.J. Millis, P.B. Littlewood, B.I. Shraiman: Phys. Rev. Lett. **74**, 5144 (1995)
326. L.F. Mattheiss: Phys. Rev. B **6**, 4718 (1972)
327. L.F. Mattheiss: Phys. Rev. B **6**, 4740 (1972)
328. J.K. Hulm, M. Ashkin, D.W. Deis, C.K. Jones: In *Progress in Low Temperature Physics*, Vol. 6, ed. by C.J. Gorter (North-Holland, Amsterdam 1970) p. 205
329. R.W.G. Wyckoff: *Crystal Structures*, 2nd edn., (Robert E. Krieger, Malabar, 1986) p. 411
330. T. Takeda, S. Ohara: J. Phys. Soc. Jpn **37**, 275 (1974)
331. E.O. Wollan, W.C. Koehler: Phys. Rev. **100**, 545 (1955)
332. C.M. Varma: Rev. Mod. Phys. **48**, 219 (1976)
333. M. Takano, N. Nakanishi, Y. Takeda, S. Naka, T. Takeda: Mater. Res. Bull. **12**, 923 (1977)
334. K. Asai, O. Yokokura, N. Nishimori, H. Chou, J.M. Tranquada, G. Shirane, S. Higuchi, Y. Okajima, K. Kohn: Phys. Rev. B **50**, 3025 (1994)
335. J.B. Goodenough: Phys. Rev. **164**, 785 (1967)
336. T. Moriya: J. Magn. Magn. Mat. **14**, 1 (1979)
337. T.C. Gibb, R. Greatrex, N.N. Greenwood, P. Kaspi: J. Chem. Soc., Dalton Trans. 1253 (1973)
338. T. Katano, N. Tsuda: J. Phys. Soc. Jpn. **65**, 207 (1995)
339. K. Kamata, T. Nakamura, T. Sata: Chem. Lett. No.1, 81 (1975)
340. B. Batlogg, J.P. Remeika, R.C. Dynes, H. Barz, A.S. Cooper, J.P. Garno: In *Superconductivity in d- and f-band Metals*, ed. by W. Buckel, W. Weber (Kernforschungszentrum Karlsruhe 1982) p. 401
341. C. Methfessel, S. Methfessel: In *Superconductivity in d- and f-band Metals*, ed. by W. Buckel, W. Weber (Kernforschungszentrum Karlsruhe 1982) p. 393
342. L.F. Mattheiss, D.R. Hamann: Phys.Rev. B **28**, 4227 (1983)
343. D.E. Cox, A.W. Sleight: Solid State Commun. **19**, 969 (1976)
344. T.M. Rice, L. Sneddon: Phys. Rev. Lett. **47**, 689 (1981)

345. J.F. Zasadzinski, N. Tralshawala, D.G. Hinks, B. Dabrowski, A.W. Mitchell, D.R. Richards: Physica C **158**, 519 (1989)
346. P. Samuely, N.L. Bobrov, A.G.M. Jansen, P. Wyder, S.N. Barilo, S. Shryaev: Phys. Rev. B **48**, 13904 (1993)
347. N. Tralshawala, J.F. Zasadzinski, L. Coffey, W. Gai, M. Romalis : Phys. Rev. B **51**, 3812 (1995)
348. J.W. Lynn, I.W. Smarlin, D.A. Neumann, J.J. Rush, J.L. Peng, Z.Y. Li: Phys. Rev. Lett. **66**, 919 (1991)
349. C.-K. Loong, P. Vashishta, R.K. Kalia, Wei Jin, M.H. Degani, D.G. Hinks, D.L. Price, J.D. Jorgensen, B. Dabrowski, A.W. Mitchell, D.R. Richards, Y. Zheng: Phys. Rev. B **45**, 8052 (1992)
350. M. Shirai, N. Suzuki, K. Motizuki: J. Phys. C **2**, 3553 (1990)
351. M. Shirai, N. Suzuki, K. Motizuki: Solid State Commun. **73**, 633 (1990)
352. W. Jin, M.H. Degani, R.K. Kalia, P. Vashishta: Phys. Rev. B **45**, 5535 (1992)
353. T. Sugihara, K. Siratori, I. Sindo, T.Katsura: J. Phys. Soc. Jpn. **45**, 1191 (1978)
354. T. Thio, T.R. Thurston, N.W. Preyer, P.J. Picone, M.A. Kastner, H.P. Jenssen, D.R. Gabbe, C.Y. Chen, R.J. Birgeneau, A. Aharony: Phys. Rev. B **38**, 905 (1988)
355. J. Fontcuberta, G. Longworth, J.B. Goodenough: Phys. Rev. **30**, 6320 (1984)
356. H.W. Blöte, R.F. Wielinga, W.J. Huiskamp: Physica **43**, 549 (1969)
357. T. Shin-ike, G. Adachi, J. Shiokawa: Mat. Res. Bull. **12**, 1149 (1977)
358. G.V. Bazuev, A.A. Samokhvalov, Yu.N. Morozov, I.I. Matveenko, V.S. Babushkin, T.I. Arbuzova, G.P. Shveikin: Sov. Phys.-Solid State **19**, 1913 (1977)
359. J.E. Greedan: Mat. Res. Bull. **14**, 13 (1979)
360. R.A. McCauley: J. Appl. Phys. **51**, 290 (1980)
361. H. Fujinaka, N. Kinomura, M. Koizumi, Y. Miyamoto, S. Kume: Mat. Res. Bull. **14**, 1133 (1979)
362. W.R. Cook, H. Jaffe: Phys. Rev. **89**, 1297 (1953)
363. M.A. Subramanian, G. Aravamudan, G.V. Subba Rao: Mat. Res. Bull. **15**, 1401 (1980)
364. M. Sato, J.E. Greedan: J. Solid State Chem. **67**, 248 (1987)
365. M. Sato, Xu Yan, J. Greedan: Z. anorg. allg. Chem. **540/541**, 177 (1986)
366. J.E. Greedan, M. Sato, Xu Yan: Solid State Commun. **59**, 895 (1986)
367. J.E. Greedan, M. Sato, Naushad Ali, W.R. Datars: J. Solid State Chem. **68**, 300 (1987)
368. V. Vasudeva Rao, G. Rangarajan, R. Srinivasan: J. Phys. Chem. Solids **47**, 395 (1986)
369. A.W. Sleight: Mat. Res. Bull. **3**, 699 (1968)
370. J.M. Longo, P.M. Raccah, J.B. Goodenough: Mat. Res. Bull. **4**, 191 (1969)
371. P.C. Donohue, J.M. Longo, R.D. Rosenstein, L. Katz: Inorg. Chem. **4**, 1152 (1965)
372. J.A. Kafalas, J.M. Longo: Mat. Res. Bull. **5**, 193 (1970)
373. A.W. Sleight: Mat. Res. Bull. **6**, 775 (1971)
374. A.W. Sleight, J.L. Gillson: Mat. Res. Bull. **6**, 781 (1971)
375. P.A. Cox, J.B. Goodenough, P.J. Tavener, D. Telles, R.G. Egdel: J. Solid State Chem. **62**, 360 (1986)
376. R.J. Bouchard, J.L. Gillson: Mat. Res. Bull. **6**, 669 (1971)
377. J.M. Longo, P.M. Raccah, J.A. Kafalas, J.W. Pierce: Mat. Res. Bull. **7**, 137 (1972)
378. I.S. Shaplygin, V.B. Lazarev: Mat. Res. Bull. **8**, 761 (1973)
379. A.F. Reid, C. Li, A.E. Ringwood: J. Solid State Chem. **20**, 219 (1977)

380. P.A. Cox, R.G. Egdell, J.B. Goodenough, A. Hamnett, C.C. Naish: J. Phys. C **16**, 6221 (1983)

## Chapter 3

381. Y. Toyozawa: *International School of Physics 'Enrico Fermi', Highlights of Condensed-Matter Theory*, ed. by F. Bassani, F. Fumi, M. Tosi (North-Holland, Amsterdam, 1985) p. 798
382. M. Ueta, H. Kanzaki, K. Kobayashi, Y. Toyozawa, E. Hanamura: *Excitonic Processes in Solids* (Springer-Verlag, Berlin, 1984)
383. P. Peierls: *Quantum Theory of Solids* (Oxford, 1955) p. 108
384. B. Johanson, K. Berggren: Phys. Rev. **181**, 855 (1969)
385. K. Nasu: J. Phys. Soc. Jpn. **52**, 3865 (1983); **53**, 301 (1984); **53**, 427 (1984)
386. J. Schrieffer: *Theory of Superconductivity* (Benjamin, New York, 1964)
387. F. Gautier: *Magnetism of Metals and Alloys*, ed. by M. Cyrot (North-Holl, Amsterdam, 1982) p. 1
388. T. Moriya: *Spin Fluctuations in Itinerant Electron Magnetism* (Springer-Verlag, Berlin, Heidelberg 1985)
389. E. Goryachev, E. Kuzmin: Sov. Phys.-JETP **64**, 531 (1986)
390. K. Nasu: J. Phys. Soc. Jpn. **65**, 2285 (1996)
391. W. McMillan: Phys. Rev. **167**, 331 (1968)
392. H. Fröhlich: In *Polarons and Excitons*, ed. by C. Kuper, G. Whitfield (Oliver and Boyd, Edinburgh, 1963) p. 1
393. Y. Shinozuka, Y. Toyozawa: J. Phys. Soc. Jpn. **46**, 505 (1989)
394. Y. Toyozawa: In *Polarons and Excitons*, ed. by C. Kuper, G. Whitfield (Oliver and Boyd, Edinburgh, 1963) p. 211
395. K. Cho, Y. Toyozawa: J. Phys. Soc. Jpn. **30**, 155 (1971)
396. A. Sumi, Y. Toyozawa: J. Phys. Soc. Jpn. **35**, 137 (1973)
397. M. Matsuura: Solid State Commun. **44**, 1471 (1982)
398. I. Kuusmann, P. Liblik, C. Lushchik: Sov. Phys.-JETP **21**, 72 (1975)
399. A. Matsui, K. Mizuno, N. Tamai, I. Yamazaki: Chem. Phys. **113**, 111 (1987)
400. K. Nasu, Y. Toyozawa: J. Phys. Soc. Jpn. **50**, 235 (1981); K. Nasu: J. of Lumin. **38**, 90 (1987)
401. E. Rashba: J. Mol. Electron. **4**, 149 (1988)
402. V. Vinetskii: Sov. Phys.-JETP **13**, 1023 (1961)
403. C. Schlenker: *Physics of Disordered Materials* ed. by D. Adler, H. Fritzsche, S. Ovshinsky (Plenum, New York, 1985) p. 369
404. M. Onoda, H. Nagasawa: J. Phys. Soc. Jpn. **52**, 223 (1983)
405. J. Scott, P. Pfluger, M. Krounbi, G. Street: Phys. Rev. B **28**, 2140 (1983)
406. A. Sumi: J. Phys. Soc. Jpn. **43**,1286 (1977)
407. S. Nagasaka, K. Nasu: Proceedings of U.S.–Japan Joint Seminar on Atomic Processes Induced by Electronic Excitation in Non-Metallic Solids, ed. by N. Itoh (September 11–15, 1989, Nagoya, Japan) p. 42
408. K. Song, C. Leung: Proceedings of U.S.–Japan Joint Seminar on Atomic Processes Induced by Electronic Excitation in Non-Metallic Solids, ed. by N. Itoh (September 11–15, 1989, Nagoya, Japan), p. 32
409. K. Kanno: Proceedings of U.S.–Japan Joint Seminar on Atomic Processes Induced by Electronic Excitation in Non-Metallic Solids, ed. by N. Itoh (September 11–15, 1989, Nagoya, Japan) p. 38
410. K. Tanimura: *Relaxations of Excited States and Photo-Induced Structural Phase Transitions*, ed. by K. Nasu (Springer, 1997, Berlin) p. 193
411. S. Nagasaka, K. Nasu: J. Phys. Soc. Jpn. **65**, 1826 (1996)

412. P. Day: *Mixed Valence Chemisty and Metal Chain Compounds*, in *Low-Dimensional Cooperative Phenomena*, ed. by H.J. Keller (Plenum, New York 1974) p. 191
413. M. Yamashita, Y. Nonaka, S. Kida, Y. Hamaue, R. Aoki: Inorg. Chim. Acta **52**, 43 (1981)
414. M. Yamashita, K. Toriumi, T. Ito: Acta Crystallogr. C **41**, 876 (1985)
415. M. Yamashita, N. Matsumoto, S. Kida: Inorg. Chim. Acta **31**, L381 (1978)
416. R. Aoki,Y. Hamaue, S. Kida, M. Yamashita, T. Takemura, Y. Furuta, A. Kawamori: Mol. Cryst. Liq. Cryst. **81**, 301 (1982)
417. K. Toriumi, Y. Wada, T. Mitani, S. Bando, M. Yamashita, Y. Fujii: J. Am. Chem. Soc. **111**, 2341 (1989)
418. H. Okamoto: Phys. Rev. B**54**, 8438 (1996)
419. S. Roth, H. Bleier: Adv. in Phys. **36**, 385 (1987); A. Heeger, S. Kivelson, J. Schrieffer, W. Su: Rev. Mod. Phys. **60**, 781 (1988)
420. H. Hayashi, K. Nasu: Phys. Rev. B **32**, 5295 (1985)
421. K. Nasu, Y. Toyozawa: J. Phys. Soc. Jpn. **51**, 2098 (1982); **51**, 3111 (1982)
422. J. Takimoto, Y. Toyozawa: J. Phys. Soc. Jpn. **52**, 4331 (1983)
423. Y. Wada, T. Mitani, M. Yamashita, T. Koda: J. Phys. Soc. Jpn. **54**, 3143 (1985)
424. K. Iwano, K. Nasu: J. Phys. Soc. Jpn. **61**, 1380 (1992)
425. A. Mishima, K. Nasu: Phys. Rev. B **39**, 5758 (1989)
426. M. Suzuki, K. Nasu: Phys. Rev. B**45**, 1605 (1992)
427. S. Tajima, S. Uchida, A. Masaki, H. Takagi, K. Kitazawa, S. Tanaka, S. Sugai: Phys. Rev. B**35**, 696 (1987)
428. A. Inoue, A. Iyo, H. Uwe, T. Sakudo, Y. Tanaka, M. Tokumoto: Advances in Superconductivity IV, Proc. of 4th Internat. Symp. Superconduct. (ISS'91), Tokyo, (1991) p. 139
429. H. Uwe, K. Tachibana, Advances in Superconductivity VII, Proc. of 7th Internat. Symp. Superconduct. (ISS'94), Tokyo, (1994) p. 94; H. Uwe: *Relaxations of Excited States and Photo-Induced Structural Phase Transitions* ed. by K. Nasu (Springer, Berlin, Heidelberg 1997) p. 55
430. S. Hasanuzzarman, K. Iwano, K. Nasu: J. Phys. Soc. Jpn. **68**, 1376 (1999); S. Hasanuzzarman: Doctoral Thesis, The School of Mathematical and Physical Studies, The Graduate Universty for Advanced Studies, March 1999, Hayama, Kanagawa, Japan
431. L. Matteiss, D. Hamann: Phys. Rev. B **28**, 4227 (1983)
432. D. Jerome, C. Beithier, P. Molinie, J. Rouxel: J. de Phys. C **4**, 125 (1976)
433. J. Jorgensen, D. Hinks, G. Flecher: Phys. Rev. B **35**, 5365 (1987)
434. S. Tajima, S. Uchida, A. Masaki, H. Takagi, K. Kitazawa, S. Tanaka, A. Katsui: Phys. Rev. B **32**, 6302 (1985)
435. H. Sato, S. Tajima, H. Takagi, S. Uchida: Nature **338**, 241 (1989)
436. S. Kondo, M. Sera, Y. Ando, M. Sato: Physica C **157**, 469 (1989)
437. G. Bilbro, W. McMillan: Phys. Rev. B **14**, 1887 (1976)
438. B. Chakraverty: J de Phys. **42**, 1351 (1981)
439. A. Alexandrov, J. Ranninger: Phys. Rev. B **23**, 1796 (1981)
440. K. Nasu: Phys. Rev. B **35**, 1748 (1987)
441. P. Anderson: Phys. Rev. **112**, 1900 (1958)
442. P. Pincus, P. Charkin, C. Coll III: Solid State Commun. **12**, 1265 (1973)
443. A. Leggett: J. de Phys. **41**, C7–19 (1980)
444. P. Nozieres, S. Schmitt-Rink: J. Low Temp. Phys. **59**, 195 (1985)
445. K. Nasu: Phys. Rev. B **42**, 6076 (1990)
446. P. Soven: Phys. Rev. **156**, 809 (1967)
447. W. Evenson, J. Schrieffer, S. Wang: J. Appl. Phys. **41**, 1199 (1970)

448. J. Bednorz, K. Müller: Z. Phys. B **64**, 199 (1986)
449. V. Ginzburg, D. Kirzhnit: *High-Temperature Superconductivity* (Consultant Bureau, New York, 1982)
450. P. Morel, P.W. Anderson: Phys. Rev. **125**, 1263 (1962)
451. B. Batlogg, G. Kourouklis, W. Weber, R.J. Cava, A. Jayaraman, A.E. White, K.T. Short, L.W. Rupp, E.A. Rietmann: Phys. Rev. Lett. **55**, 912 (1987)
452. H.J. Bornemann, D.E. Morris, H.B. Liu: Physica C **182**, 132 (1991); G. Zhao, A. Inyushkin: Phys. Rev. B **54**, 14956 (1996)
453. M.K. Crawford, M.N. Kunchur, W.E. Farneth, E.M. Nccarron III, S.J. Poon: Phys. Rev. B **41**, 282 (1990); J.P. Frank, J. Jung, M.A.K. Mohamed, S. Gygax, G.I. Sproule: Phys. Rev. B **44**, 5318 (1991)
454. V. Postnikov, V. Postnikov, M. Fedorov: Phys. Status Solidi (B) **87**, 11 (1978)
455. K. Nasu: Phys. Rev. B **44**, 7625 (1991); J. Phys. Condens. Matter **4**, 3793 (1992)
456. A.B. Migdal: Sov. Phys.–JETP **7**, 996 (1958); G.M. Eliashiberg: Sov. Phys.–JETP **11**, 696 (1960); P.B. Allen, B. Mitrovic, in *Solid State Physics*, ed. by H. Ehrenreich, F. Seitz, D. Turnbull (Academic, New York 1982) Vol. 37, p. 1
457. Y. Inada, K. Nasu: J. Phys. Soc. Jpn. **61**, 4511 (1992)
458. D. Khandekar, S. Lawande , K. Bhagwat: *Path-Integral Methods and Their Applications* (World Scientific, Singapore 1993)
459. J. Hubbard: Phys. Rev. B **19**, 2626 (1979)
460. D. Scalapino: *Frontiers and Borderline in Many-Particle Physics*, ed. R. Broglia, J. Schrieffer (North-Holland, Amsterdam 1988)
461. C. Bloch, C.D. Dominicis: Nuclear Physics **7**, 459 (1958)
462. M. Gaudin: Nuclear Physics **15**, 89 (1960)
463. A. Fetter, J. Walecka: *Quantum Theory of Many-Particle Systems* (McGraw–Hill, New York 1971)
464. E. Fradkin: *Field Theories of Condensed Matter Systems* (Addison-Wesley, New York, 1991)
465. F. Berezin: *The Methods of Second Quantization* (Academic, New York,1966)
466. N. Tomita, K. Nasu: Phys Rev B **56**, 3779 (1997)

## Chapter 4

467. B.H. Brandow: Adv. Phys. **26**, 651 (1977)
468. S. Sugano, Y. Tanabe, H. Kamimura: *Multiplets of Transition-Metal Ions in Crystals* (Academic, New York, 1970)
469. J. Zaanen, G.A. Sawatzky, J.W. Allen: Phys. Rev. Lett. **55**, 418 (1985)
470. T. Mizokawa, A. Fujimori, H. Namatame, K. Akeyama, N. Kosugi: Phys. Rev. B **49**, 7193 (1994).
471. A.E. Bocquet, T. Mizokawa, T. Saitoh, H. Namatame, A. Fujimori: Phys. Rev. B **46**, 3771 (1992); A.E. Bocquet, T. Mizokawa, K. Morikawa, A. Fujimori, K.B. Maiti, S.R. Barman, D.D. Sarma, Y. Tokura, M. Onoda: Phys. Rev. B. **53**, 1161 (1996)
472. J. Zaanen, G.A. Sawatzky: Can. J. Phys. **65**, 1292 (1987)
473. M. Cyrot, C. Lyon-Caen: J. Phys. (Paris) **36**, 253 (1975)
474. H. Ichikawa, J. Akimitsu, M. Nishi, K. Kakurai: Physica B **281–282**, 482 (2000)
475. J.H. van Santen, G.H. Jonker: Physica (Amsterdam) **16**, 599 (1950)
476. W.F. Brinkman, T.M. Rice: Phys. Rev. B **2**, 4302 (1970)
477. T. Moriya, H. Hasegawa: J. Phys. Soc. Jpn. **48**, 1490 (1980)

478. L. Hedin, S. Lundqvist: In *Solid State Physics*, Vol. 23, Eds. F. Seitz, D. Turnbull, E. Ehrenreich (Academic, New York, 1996) p. 1
479. Y. Tokura, Y. Taguchi, Y. Okada, Y. Fujishima, T. Arima, K. Kumagai, Y. Iye: Phys. Rev. Lett. **70**, 2126 (1993)
480. A. George, G. Kotliar, W. Krauth, M. J. Rosenberg: Rev. Mod. Phys. **68**, 13 (1996)
481. N.F. Mott: *Metal-Insulator Transitions* (Tayor & Francis, London, 1990)
482. P.P. Edwards, M.J. Sienko: Phys. Rev. B **17**, 2575 (1978)
483. A.L. Efros, B.I. Shklovskii: J. Phys. C **8**, L49 (1975)
484. J.H. Davies, J.R. Franz: Phys. Rev. Lett. **57**, 475 (1986)

## Chapter 5

485. N. Tsuda, Y. Sumino, I. Ohno, T. Akahane: J. Phys. Soc. Jpn. **41**, 1153 (1976)
486. M. Ishii, T. Tanaka, T. Akahane, N. Tsuda: J. Phys. Soc. Jpn. **41**, 908 (1976)
487. T.P. Pearsall, L.A. Coldren: Solid State Commun. **18**, 1093 (1976)
488. R.E. Benner, E.M. Brody, H.R.Shanks: J. Solid State Chem. **22**, 361 (1977)
489. F.C. Zumsteg, T. Perasall: Solid State Commun. **16**, 751 (1975)
490. T. Tanaka, T. Akahane, E. Bannai, S. Kawai, N. Tsuda, Y. Ishizawa: J. Phys. C **9**, 1235 (1976)
491. A.H. Wilson: *The Theory of Metals*, 2nd ed. (Cambridge University Press, Cambridge 1965) p. 278
492. D.J. Howarth, E.H. Sondheimer: Proc. Roy. Soc. London A **219**, 53 (1953)
493. H. Fröhlich: Proc. Roy. Soc. A **160**, 230 (1937)
494. T.P. Perarsall, C.A. Lee: Phys. Rev. B **10**, 2190 (1974)
495. C.N. King, H.C. Kirsch, T.H. Geball: Solid State Commun. **9**, 907 (1971)
496. W.L. McMillan: Phys. Rev. **167**, 331 (1968)
497. P.W. Selwood: *Magnetochemistry*, 2nd. edn. (Interscience, New York 1956)
498. E. König: *Magnetic Propertiey of Transition Metal Compounds*, Landolt-Börnstein Numerical Data and Functional Relationships, New Series, Group II, Vol. 2, ed. by K.H. Hellwege (Springer, Berlin, Heidelberg 1966)
499. L.F. Mattheiss: Phys. Rev. **181**, 987 (1969)
500. L.F. Mattheiss: Phys. Rev. B **2**, 3918 (1970)
501. F.S. Razari, W.R. Datas: Can. J. Phys. **54**, 845 (1976)
502. A.H. Kahn, A.J. Leyendecker: Phys. Rev. **135**, A1321 (1964)
503. T. Wolfram, E.A. Kraut, F.J. Morin: Phys. Rev. B **7**, 1677 (1973)
504. J.C. Slater, G.F. Koster: Phys. Rev. **94**, 1498 (1954)
505. S. Ellialtioglu, T. Wolfram: Phys. Rev. B **18**, 4509 (1978)
506. F.J. Morin, T.Wolfram: Phys. Rev. Lett. **30**, 1214 (1973)
507. M. Tsukada, N. Tsuda, F. Minami: J. Phys. Soc. Jpn. **49**, 1115 (1980)
508. T. Chiba, T. Akahane, N. Tsuda: Proc. of the Fifth Int. Conf. on Positron Annihilation, ed. by R.Hashiguti, K. Fujiwara (Japan Institute of Metals, Sendai 1979) p. 653
509. J. Robertson: J. Phys. C **12**, 4767 (1979)
510. V.T. Agekyan: Phys. Stat. Sol. (a) **43**, 11 (1977)
511. K.J. Button, D.G. Fonstad, W. Dreybradt: Phys. Rev. B **4**, 4539 (1971)
512. Z.M. Jarzebsky, J.P. Marton: J. Electrochem. Soc. **123**, 199C (1976) (A review paper)
513. Z.M. Jarzebsky, J.P. Marton: J. Electrochem. Soc. **123**, 299C (1976) (A review paper)

514. Z.M. Jarzebsky, J.P. Marton: J. Electrochem. Soc. **123**, 333C (1976) (A review paper)
515. M. Nagasawa, S. Shionoya: J. Phys. Soc. Jpn. **30**, 1213 (1971)
516. M. Nagasawa, S. Shionoya: Jpn. J. Appl. Phys. **10**, 472 (1971)
517. M. Nagasawa, S. Shionoya: Jpn. J. Appl. Phys. **10**, 727 (1971) and refs. in [515–517]
518. C.G. Fonstad, R.H. Rediker: J. Appl. Phys. **42**, 2911 (1971)
519. P.P. Debye, E.M. Conwell: Phys. Rev. **93**, 693 (1954)
520. J. Bardeen, W. Shockley: Phys. Rev. **80**, 72 (1950)
521. T.D. Lee, F.E. Low, D. Pines: Phys. Rev. **90**, 297 (1953)
522. F.E. Low, D. Pines: Phys. Rev. **91**, 193 (1953)
523. F.E. Low, D. Pines: Phys. Rev. **98**, 414 (1955)
524. S. Samson, C.G. Fonstad: J. Appl. Phys. **44**, 4618 (1973)
525. H. Koch: Phys. Status Solidi **3**, 1059 (1963)
526. N. Yamazoe, J. Fuchigami, M.Kishikawa, T. Seiyama: Surf. Science **86**, 335 (1979)
527. M. Nitta, M. Haradome: J. Electron. Mater. **8**, 571 (1979)
528. T. Arai: J. Phys. Soc. Jpn. **15**, 916 (1960)
529. M. Nagasawa, S. Shionoya: J. Phys. Soc. Jpn. **30**, 158 (1971)
530. D. Fröhlich, R. Kenklies, R. Helbig: Phys. Rev. Lett. **41**, 1750 (1978)
531. V.N. Bogomolov, E.K. Kudinov, Yu.A. Firsov: Sov. Phys.-Solid State **9**, 2502 (1968)
532. A. Fujishima, K. Honda: Nature **238**, 37 (1972)
533. R.G. Breckenridge, W.R. Hosler: Phys. Rev. **91**, 793 (1953)
534. D.C. Kronemeyer: Phys. Rev. **87**, 876 (1952)
535. H. Fröhlich: Adv. Phys. **3**, 325 (1954)
536. J.L. Jourdan, C. Gout, J.P. Albert: Solid State Commun. **31**, 1023 (1979)
537. N. Daude, C. Gout, C. Jouanin: Phys. Rev. B **15**, 3229 (1977)
538. H.P.R. Frederikse: J. Appl. Phys. **32**, 2211 (1961)
539. R. Summitt: J. Appl. Phys. **39**, 3762 (1968)
540. W.G. Spitzer, R.C. Miller, D.A. Kleinman, L.E. Howarth: Phys. Rev. **126**, 1710 (1962)
541. D.M. Eagles: J. Phys. Chem. Solids **25**, 1243 (1964)
542. F.A. Grant: Rev. Mod. Phys. **31**, 646 (1959)
543. Y. Toyozawa: Prog. Theor. Phys. **26**, 29 (1961)
544. E. Chang, E.K. Graham: J. Geophys. Res. **80**, 2595 (1975)
545. I.G. Austin, N.F. Mott: Adv. Phys. **18**, 41 (1969)
546. A.J. Bosman, H.J. van Daal: Adv. Phys. **19**, 1 (1970)
547. T. Holstein: Ann. of Phys. **8**, 325 (1959)
548. T. Holstein: Ann. of Phys. **8**, 343 (1959)
549. J. Yamashita, T. Kurosawa: J. Phys. Chem. Solids **5**, 34 (1958)
550. H. Böttger, V.V. Bryksin: Phys. Stat. Sol. (B) **78**, 9 (1976)
551. H.Böttger, V.V. Bryksin: Phys. Stat. Sol. (B) **78**, 415 (1976)
552. D.C. Johnston: J. Low. Temp. Phys. **25**, 145 (1976)
553. R.W. McCallum, D.C. Johnston, C.A. Luengo, M.B. Maples: J. Low Temp. Phys. **25**, 177 (1976)
554. L.A. de Picciotto, M.M. Thackeray: Mater. Res. Bull. **20**, 187 (1985)
555. R.G. Goodrich: Phys. Rev. B **44**, 5326 (1991)
556. M.R. Harrison, P.P. Edwards, J.B. Goodenough: Philos. Mag. B **52**, 679 (1985)
557. M. Watanabe, K. Kaneda, H. Takeda, N. Tsuda: J. Phys. Soc. Jpn. **53**, 2437 (1984)

558. M.R. Harrison, P.P. Edwards, J.B. Goodenough: J. Solid State Chem. **54**, 136 (1984)
559. W.L. McMillan: Phys. Rev. **167**, 331 (1968)
560. P.B. Allen, R.C. Dynes: Phys. Rev. B **12**, 905 (1975)
561. F. Gompf, B. Renker, H. Mukta: Physica B **180–181**, 459 (1992)
562. S. Satpathy, R.M. Martin: Phys. Rev. B **36**, 7269 (1987)
563. W.F. Brinkman, T.M. Rice: Phys. Rev. B **2**, 1324 (1970)
564. W.F. Brinkman, T.M. Rice: Phys. Rev. B **2**, 4302 (1970)
565. P.P. Edwards, R.G. Edgell, I. Fragala, J.B. Goodenough, M.R. Harrison, A.F. Orchard, E.G. Scott: J. Solid State Chem. **54**, 127 (1984)
566. A.L. Efros: J. Phys. C **9**, 2021 (1976)
567. A.L. Efros, B.I. Shklovskii: J. Phys. C **8**, L49 (1975)
568. J.H. Davies, P.A. Lee, T.M. Rice: Phys. Rev. **29**, 4260 (1984)
569. J.H. Davies, J.R. Franz: Phys. Rev. Lett. **57**, 475 (1986)
570. A. Fujimori, K. Kawakami, N. Tsuda: Phys. Rev. B **38**, 7889 (1988)
571. P.G. de Gennes: *Superconductivity of Metals and Alloys* (Benjamin, New York 1966)
572. J. Bardeen, L.N. Cooper, J.R. Schrieffer: Phys. Rev. **108**, 1175 (1957)
573. B.K. Chakraverty, M.J. Sienko: Phys. Rev. B **17**, 3781 (1978)
574. P.W. Anderson: Phys. Rev. Lett. **34**, 953 (1975)
575. T. Takahashi, H. Nagasawa: Solid State Commun. **39**, 1125 (1981)
576. L.N. Mulay, W.J. Danley: J. Appl. Phys. **41**, 877 (1970)
577. I. Maekawa, F. Takagi, Y. Sakai, N. Tsuda: J. Phys. Soc. Jpn. **56**, 2119 (1987)
578. W. Brenig, G.H. Döhler, P. Wölfle: Z. Phys. **258**, 381 (1974)
579. D.B. Rogers, J.L. Gillson, T.E. Gier: Solid State Commun. **5**, 263 (1967)
580. A. Sawaoka, S. Miyahara, H. Nagasaki, S. Minomura: Solid State Commun. **3**, 155 (1965)
581. D.B. Rogers, R.J. Arnott, A. Wold, J.B. Goodenough: J. Phys. Chem. Solids **24**, 347 (1963)
582. K. Kawakami, Y. Sakai, N. Tsuda: J. Phys. Soc. Jpn. **55**, 3174 (1986)
583. B. Reuter, K. Müller: Naturwissenschaften **54**, 164 (1967)
584. Muhtar, F. Takagi, K. Kawakami, N. Tsuda: J. Phys. Soc. Jpn. **57**, 3119 (1988)
585. S. Kondo, D.C. Johnston, C.A. Swenson, F. Borsa, A.V. Mahajan, L.L. Miller, T. Gu, A.I. Goldman, M.B. Maple, D.A. Gajewski, E.J. Freeman, N.R. Dilley, R.P. Dickey, J. Merrin, K. Kojima, G.M. Luke, Y.J. Uemura, O. Chamaissem, J.D. Jorgensen: Phy. Rev. Lett. **78**, 3729 (1997)
586. A.S. Ribnick, B. Post, E. Banks: Advances in Chemistry Series, Vol. 39, *Non-stoichiometric Compounds*, ed. by R.F. Gould (American Chemical Society, Washington, DC 1963) p. 246
587. S. Kachi, K. Kosuge: In *Hikagakuryouronteki Kinzokukagoubutsu*, ed. by S. Takeuchi (Maruzen, Tokyo 1975) p. 65 (in Japanese)
588. R. Clarke: Phys. Rev. Lett. **39**, 1550 (1977)
589. M. Sato, B.H. Grier, G. Shirane: Phys. Rev. B **25**, 6876 (1982)
590. S. Tanisaki: J. Phys. Soc. Jpn. **15**, 566 (1960)
591. A. Fujimori, N. Tsuda: Solid State Commun. **34**, 433 (1980)
592. S. Nagasawa: Denki Kagaku **16**, 13 (1948)
593. J.M. Berak, M.J. Sienko: J. Solid State Chem. **2**, 109 (1970)
594. D.C. Langreth: Phys. Rev. **159**, 717 (1967)
595. L.D. Mühlestein, G.C. Danielson: Phys. Rev. **158**, 825 (1967)
596. W. McNeill, L.E.Convoy: J. Chem. Phys. **36**, 87 (1962)
597. P.A. Lightsey: Phys. Rev. B **8**, 3586 (1973)
598. P.A. Lightsey: Phys. Rev. B **14**, 4730 (1976)

599. Data of Refs. [595–598] are accumulated
600. B.W. Faughnan, R.S. Crandall, P.M. Heyman: RCA Rev. **36**, 177 (1975)
601. R.S. Crandall, B.W. Faughnan: Phys. Rev. Lett. **39**, 232 (1977)
602. S. Kirkpatrick; Rev. Mod. Phys. **45**, 574 (1973)
603. F.C. Zumsteg: Phys. Rev. B **14**, 1406 (1976)
604. D.P. Tunstall: Phys. Rev. B **11**, 2821 (1975)
605. M.A. Langell, S.L. Bernasek: J. Vac. Sci. Technol. **17**, 1287 (1980)
606. R. Fuchs: J. Chem. Phys. **92**, 3781 (1965)
607. N.F. Mott: Philos. Mag. **35**, 111 (1977)
608. G. Hollinger, P. Petrosa: Chem. Phys. Lett. **74**, 341 (1980)
609. G. Hollinger, P. Petrosa, J.P. Doumerc, F.J. Himpsel, B. Reihl: Phys. Rev. B **32**, 1987 (1985)
610. M.D. Hill, R.G. Egdell: J. Phys. C **16**, 6205 (1983)
611. M.A. Dubson and D.F. Holcomb: Phys. Rev. B **34**, 25 (1986)
612. J.P. Doumerc, P. Dordor, E. Marquestaut, M. Pouchard, P. Hagenmuller: Philos. Mag. B **42**, 487 (1980)
613. B.I. Shklovskii, A.L. Efros: Sov. Phys.-Semicond. **14**, 487 (1980)
614. L.D. Ellerbeck, H.R. Shanks, P.H. Sidles, C. Danielson: J. Chem. Phys. **35**, 298 (1961)
615. (4.132) A. Narath, D.C. Barham: Phys. Rev. 176, 479 (1968)
616. N. Tralshawala, J.F. Zasadzinski, L. Coffey, Q. Huang: Phys. Rev. B **44**, 12102 (1991)
617. D. Shimada, Y. Shiina, N. Miyakawa, N. Tsuda: Physica B **219–220**, 192 (1996)
618. H.R. Shanks: Solid State Commun. **15**, 753 (1974)
619. R. Salchow, R. Liebmann, J. Appel: J. Phys. Chem. Solids **44**, 245 (1983)
620. R. Salchow, R. Liebmann, J. Appel: Physica B **107**, 523 (1981)
621. A.D. Wasley: Acta Crystallogr. **8**, 695 (1955)
622. J.B. Goodenough: Phys. Rev. **117**, 1442 (1960)
623. A. Casalot, P. Hagenmuller: J. Phys. Chem. Solids **30**, 1341 (1969)
624. P. Hagenmuller, J. Galy, M. Pouchard, A. Casalot: Mat. Res. Bull. **1**, 45 (1966)
625. H. Nagasawa: Kotai Buturi **21**, 187 (1986)
626. Y. Kanai, S. Kagoshima, H. Nagasawa: J. Phys. Soc. Jpn. **51**, 697 (1982)
627. Y. Kanai, S. Kagoshima, H. Nagasawa: Synthetic Metals: **9**, 369 (1984)
628. J.H. Perlstein, M.J. Sienko: J. Chem. Phys. **48**, 174 (1968)
629. R.H. Wallis, N. Sol, A. Zylberstejn: Solid State Commun. **23**, 539 (1977)
630. M. Onoda, H. Nagasawa: J. Phys. Soc. Jpn. **52**, 2231 (1983)
631. J.B. Goodenough: J. Solid State Chem. **1**, 349 (1970)
632. J. Galy, J. Darriet, A. Casalot, J.B. Goodenough: J. Solid. State Chem. **1**, 339 (1970)
633. K. Maruyama, H. Nagasawa: J. Phys. Soc. Jpn. **48**, 2159 (1980)
634. J. Gendell, R.M. Cotts, M.J. Sienko: J. Chem. Phys. **37**, 220 (1962)
635. W.J. Gunning, A.J. Heeger, R.H. Wallis, N. Sol, A. Zylbersztejn: Solid State Commun. **26**, 155 (1978)
636. N.F. Mott: Philos. Mag. **19**, 835 (1969)
637. W. Brenig, G.H. Döhler, H. Heyszenau: Philos. Mag. **27**, 1093 (1973)
638. V. Ambegaokar, B.I. Halperin, J.S. Langer: Phys. Rev. **4**, 2612 (1971)
639. A.N. Bloch, R.B. Weisman, C.M. Varma: Phys. Rev. Lett. **28**, 753 (1972)
640. Y. Sakai, K. Kawakami, N. Tsuda, M. Tanaka: J. Phys. Soc. Jpn. **55**, 3181 (1986)
641. M. Onoda, T. Takahashi, H. Nagasawa: J. Phys. Soc. Jpn. **51**, 3868 (1982)

642. C. Schlenker, R. Buder, V.D. Nguyen, J. Dumas: J. Appl. Phys. **50**, 1720 (1979)
643. C. Kittel: *Introduction to Solid State Physics*, 2nd edn. (Wiley, New York 1956) p. 137
644. P.W. Anderson, B.I. Halperin, C.M. Varma: Philos. Mag. **25**, 1 (1972)
645. W.A. Phillips: J. Low Temp. Phys. **7**, 351 (1972)
646. T. Erata, H. Nagasawa: J. Phys. Soc. Jpn. **52**, 3652 (1983)
647. B.K. Chakraverty, J. Ranninger: Philos. Mag. B **52**, 669 (1985)
648. A. Alexandrov, J. Ranninger: Phys. Rev. B **23**, 1796 (1981)
649. K. Nasu: J. Phys. Soc. Jpn. **54**, 1933 (1985)
650. K. Kubo, S. Takada: J. Phys. Soc. Jpn. **52**, 2108 (1983)
651. H. Fröhlich: Proc. R. Soc. London A **223**, 296 (1954)
652. M. Greenblatt: Chem. Rev. **88**, 31 (1988) (review paper)
653. R. Uriu, D. Shimada, N. Tsuda: J. Phys. Soc. Jpn. **60**, 2479 (1991)
654. G.H. Bouchard, Jr., J. Perlstein, M.J. Sienko: Inorg. Chem. **6**, 1683 (1967)
655. G. Travaglini, P. Wachter, J. Marcus, C. Schlenker: Solid State Commun. **37**, 599 (1981)
656. C. Schlenker, J. Dumas, C. Escribe-Filippini, H.Guyot, J. Marcus, G. Fourcaudot: Philos. Mag. B **52**, 643 (1985) (review paper)
657. J. Dumas, C. Schlenker, J. Marcus, R. Buder: Phys. Rev. Lett. **50**, 757 (1983)
658. J.P. Pouget, S. Kagoshima, J. Marcus, C. Schlenker: J. de Phys. Lett. **44**, L113 (1983)
659. L.F. Schneemeyer, F.J. DiSalvo, R.M. Fleming, J.V. Waszczak: J. Solid State Chem. **54**, 358 (1984)
660. S. Kagoshima, T. Sambongi, H. Nagasawa: *One-Dimensional Conductors*, Springer Ser. Solid-State Sci., Vol. 72 (Springer, Berlin, Heidelberg 1988)
661. J. Dumas, C. Schlenker, J.Y. Veuillen, R. Chevalier, J. Marcus, R. Cinti, E. Al Khoury Nemeh: Synth. Met. **19**, 937 (1987)
662. M.H. Whangbo, L.F. Schneemeyer: Inorg. Chem. **25**, 2424 (1986)
663. M. Greenblatt, W.H. McCarroll, R. Neifeld, M. Croft, J.V. Waszczak: Solid State Commun. **51**, 671 (1984)
664. M. Onoda, K. Toriumi, Y. Matsuda, M. Sato: J. Solid State Chem. **66**, 163 (1987)
665. M. Sato, Y. Matsuda, H. Fukuyama: J. Phys. C **20**, L137 (1987)
666. H. Fukuyama: J. Phys. Soc. Jpn. **48**, 2169 (1980)
667. H. Fukuyama: In *Anderson Localization*, ed. by Y. Ohtsuki (Kyoritu, Tokyo 1982) p. 99
668. A. Moini, M.A. Subramanian, A. Clearfield, F.J. DiSalvo, W.H. McCarrol: J. Solid State Chem. **66**, 136 (1987)
669. B.T. Collins, M. Greenblatt, W.H. McCarroll, G.W. Hull: J. Solid State Chem: **73**, 507 (1988)
670. M. Sato: Nihon Kesshou Gakkaisi **29**, 190 (1987) (review paper in Japanese)
671. R. Buder, J. Devenyi, J. Dumas, J. Marcus, J. Mercier, C. Schlenker, H. Vincent: J. de Phys. Lett. **43**, L59 (1982)
672. K.V. Ramanujachary, B.T. Collins, M. Greenblatt, J.V. Waszczak: Solid State Commun. **59**, 647 (1986)
673. M.Ganne, M. Dion, A. Boumaza, M. Tournoux: Solid State Commun. **59**, 137 (1986)
674. L.F. Schneemeyer, S.E. Spengler, F.J. DiSalvo, J.V. Warzczack, C.E. Rice: J. Solid State Chem. **55**, 158 (1984)
675. H. Guyot, C. Escribe-Flippini, G. Fourcaudot, K. Konate, C. Schlenker: J. Phys. C **16**, L1227 (1983)
676. M.Onoda, H. Fujisita, Y. Matsuda, M. Sato: Synth. Met. **19**, 947 (1987)

677. M. Sato, H. Fujusita, S. Sato, S. Hoshino: J. Phys. C **19**, 3059 (1986)
678. L. Kihlborg: Acta Chem. Scand. **13**, 954 (1959)
679. H. Gruber, E. Krautz, H.P. Fritzer: Phys. Stat. Sol. (A) **65**, 589 (1981)
680. N.C. Stephenson, A.D. Wasley: Acta Crystallogr. **19**, 241 (1965)
681. J.H. de Boer, E.J.W. Verway: Proc. Phys. Soc. **49**, 59 (1937)
682. R.J. Powell, W.E. Spicer: Phys. Rev. B**2**, 2182 (1970)
683. G.A. Sawatzky, J.W. Allen: Phys. Rev. Lett. **53**, 2339 (1984)
684. R. Newman, R.M. Chrenko: Phys. Rev. **114**, 1507 (1959)
685. B.E.F. Fender, A.J. Jacobson, F.A. Wedgewood: Phys. Rev. **43**, 990 (1968)
686. T. Chiba, N. Tsuda: Appl. Phys. **5**, 37 (1974)
687. C.M. Osborn, R.W. Vest: J. Phys. Chem. Solids **32**, 1331 (1971)
688. V.P. Zhuze, A.I. Shelkh: Fiz. Tverd. Tela **5**, 1576 (1963) [Engl. transl.: Sov. Phys.-Solid State **5** 1278 (1963)]
689. S. Koide: J. Phys. Soc. Jpn. **20**, 123 (1965)
690. J.G. Aiken, A.G. Jordan: J. Phys. Chem. Solids **29**, 2153 (1968)
691. D. Adler, J. Feinleib: Phys. Rev. B **2**, 3112 (1970)
692. W.E. Spear, D.S. Tannhauser: Phys. Rev. **144**, 763 (1966)
693. A.J. Bosman, C. Crevecoeur: Phys. Rev. **144**, 763 (1966)
694. J.E. Keem, J.M. Honig, L.L. van Zandt: Philos. Mag. B **37**, 537 (1978)
695. E. Iguchi, K. Akashi: J. Phys. Soc. Jpn. **61**, 3385 (1992)
696. D.E. Eastman, J.L. Freeouf: Phys. Rev. Lett. **34**, 395 (1975)
697. M.R. Thuler, R.L. Benbow, Z. Hurych: Phys. Rev. B **27**, 2082 (1983)
698. A. Fujimori, F. Minami: Phys. Rev. B **30**, 957 (1984)
699. J. Zaanen, G.A. Sawatzky: Can. J. Phys. **65**, 1262 (1987)
700. Z.-X. Shen, R.S. List, D.S. Dessau, B.O. Wells, O. Jepsen, A.J. Arko, R. Bartlet, C.K. Shin, F. Parmigiani, J.C. Huang, P.A. Lindberg: Phys. Rev. B **44**, 3604 (1991)
701. J. van Elp, H. Eskes, P. Kuiper, G.A. Sawatzky: Phys. Rev. B **45**, 1612 (1992)
702. M.A. Kobler, R.K. MacCrone: Phys. Rev. Lett. **29**, 1457 (1972)
703. J. Meng, P. Jena, J.M. Vail: J. Phys. Cond. Matt. **2**, 1037 (1990)
704. H. Eisaki, S. Uchida, T. Mizokawa, H. Namatame, A. Fujimori, J. van Elp, P. Kuiper, G.A. Sawatzky, S. Hosoya, H. Katayama-Yoshida: Phys. Rev. B **45**, 12513 (1992)
705. T. Oguchi, K. Terakura, A.R. Williams: Phys. Rev. B **28**, 6443 (1983)
706. M.R. Norman: Phys. Rev. B **44**, 1364 (1991)
707. A. Svane, O. Gunnarsson: Phys. Rev. Lett. **65**, 1148 (1990)
708. V.I. Anisimov, J. Zaanen, O.K. Andersen: Phys. Rev. B **44**, 943 (1991)
709. C. Zener: Phys. Rev. **82**, 403 (1951)
710. M. Imada, A. Fujimori, Y. Tokura: Rev. Mod. Phys. **70**, 1039 (1998)
711. A. Urushibara, Y. Moritomo, T. Arima, A. Asamitsu, G. Kido, Y. Tokura: Phys. Rev. B **51**, 14103 (1995)
712. Y. Tokura, A. Urushibara, Y. Moritomo, T. Arima, A. Asamitsu, G. Kido, N. Furukawa: J. Phys. Soc. Jpn. **63**, 3931 (1994) x
713. S.-W. Cheong and H.Y. Hwang: In *Colossal Magnetoresistive Oxides*, edited by Y. Tokura (Gordon & Breach, 2000)
714. S. Satpathy, Z. S. Popović, F. R. Vukajlović: Phys. Rev. Lett. **76**, 960 (1996)
715. Y. Okimoto, T. Katsufuji, T. Ishikawa, A. Urushibara, Y. Tokura: Phys. Rev. Lett. **75**, 109 (1995)
716. D.D. Sarma, N. Shanthi, S.R. Krishnakumar, T. Saitoh, T. Mizokawa, A. Sekiyama, K. Kobayashi, A. Fujimori, E. Weschke, R. Meier, G. Kaindl, Y. Takeda, M. Takano: Phys. Rev. B **53**, 6873 (1996)
717. D.S. Dessau, C.-H. Park, T. Saitoh, Z.-X. Shen, P. Villela, N. Hamada, Y. Moritomo, Y. Tokura: Phys. Rev. Lett. **81**, 192 (1998)

718. Y. Murakami, H. Kawada, H. Kawata, M. Tanaka, T. Arima, Y. Morimoto, Y. Tokura: Phys. Rev. Lett. **80**, 1932 (1998)
719. A.J. Millis, P.B. Littlewood, B.I. Shraiman: Phys. Rev. Lett. **74**, 5144 (1995)
720. P.G. Radaelli, D.E. Cox, N.Marezio, S-W. Cheong, P.E. Schiffer, A.P. Ramirez: Phys. Rev. Lett. **75**, 4488 (1995)
721. J.M. De Teresa, M.R. Ibarra, P.A. Algarabel, C. Ritter, C. Marquina, J. Blasco, J. García, A. del Moral, Z. Arnold: Nature **386**, 256 (1997)
722. T.A. Tyson, J. Mustre, de Leon, S.D. Conradson, A.R. Bishop, J.J. Neumeier, J. Zang: Phys. Rev. B **53**, 13985 (1996)
723. C.H. Booth, F. Bridges, G.J. Snyder, T.H. Geball: Phys. Rev. B **54**, R15606 (1966)
724. S.J.L. Billinge, R.G. DiFrancesco, G.H. Kwei, J.J. Neumeier, J.D. Thompson: Phys. Rev. Lett. **77**, 715 (1996)
725. D. Louca, T. Egami, E.L. Brosha, H. Röder, A.R. Bishop: Phys. Rev. B **56**, R8475 (1997)
726. G.-M. Zhao, K. Conder, H. Keller and K.A. Müller: Nature **381**, 676 (1996)
727. T.T.M. Palstra, A.P. Ramirez, S-W. Cheong, B.R. Zegarski, P. Shiffer, J. Zaanen: Phys. Rev. B **56**, 5104 (1997)
728. M. Jamie, H.T. Hardner, M.B. Salamon, M. Rubinstein, P. Dorsey, D. Emin: Phys. Rev. Lett. **78**, 951 (1997)
729. S.G. Kaplan, M. Quijada, H.D. Drew, D.B. Tanner, G.C. Xiong, R. Ramesh, C. Kwon, T. Venkatesan: Phys. Rev. Lett. **77**, 2081 (1996)
730. P.A. Miles, W.B. Westphal, A. von Hippel: Rev. Mod. Phys. **29**, 279 (1957)
731. A. Muan, E.F. Osborn: *Phase Equilibria among Oxides for Steel Making* (Addison Wesley, Reading, MA 1965)
732. R. W. Millar: J. Amer. Chem. Soc. **51**, 215 (1929)
733. S. Chikazumi: AIP Conf. Proc. **29**, 382 (1976)
734. C. Schlenker, M. Marezio: Phil. Mag. **42**, 453 (1980)
735. M. Tanaka, J. Akimitsu, Y. Inada, N. Kimizuka, I. Shindo, K. Siratori: Solid State Commun. **44**, 687 (1982)
736. N. Kimizuka, E. Muromachi, K. Siratori: In *Handbook on the Physics and Chemistry of Rare Earths*, Vol. 13, ed. by K.A. Gschneidner Jr., L. Eyring, (North-Holland, Amsterdam 1990) p. 283
737. E.J.W. Verwey, P.W. Haaymann: Physica **8**, 979 (1941); E.J.W. Verwey, P.W. Haaymann, F.C. Romeijn: J. Chem. Phys. **15**, 181 (1947)
738. See, for example, F.C. Nix, W. Shockley: Rev. Mod. Phys. **10**, 1 (1938); T. Muto, Y. Takagi: In *Solid State Physics*, Vol, 1, ed. by F. Seitz, D. Turnbull (Academic, New York 1955) p. 193
739. K. Abe, Y. Miyamoto, S. Chikazumi: J. Phys. Soc. Jpn. **41**, 1894 (1976)
740. H.P. Rooksby, B.T.M. Willis: Acta Cryst. **6**, 565 (1953)
741. See, for example, N. Otsuka, H. Sato: J. Solid State Chem. **61** 212 (1986) and references therein
742. M. Iizumi, T.F. Koetzle, G. Shirane, S. Chikazumi, M. Matsui, S. Todo: Acta Cryst. B **38**, 2121 (1982)
743. G.A. Samara: Phys. Rev. Lett. **21**, 795 (1968)
744. D.B. McWhan, J.P. Remeika: Phys. Rev. B **2**, 3734 (1970)
745. K. Siratori, N. Môri, H. Takahashi, G. Oomi, J. Iida, M. Tanaka, M. Kishi, Y. Nakagawa, N. Kimizuka: J. Phys. Soc. Jpn. **59**, 631 (1990)
746. Y. Kakudate, N. Môri, Y. Kino: J. Magn. Magn. Mat. **12**, 22 (1979)
747. P.W. Anderson: Phys. Rev. **102**, 1008 (1956)
748. C.C. Wu, T.O. Mason: J. Amer. Cer. Soc. **64**, 520 (1981)
749. A. Schlegel, S.F. Alvarado, P. Wachter: J. Phys. C **12**, 1157 (1979)
750. A. Schlegel, P. Wachter: J. de Phys. **41**, C5–19 (1980)

751. S.K. Park, T. Ishikawa, Y. Tokura: Phys. Rev. B **58**, 3717 (1998)
752. J.M. Lavine: Phys. Rev. **114**, 482 (1959); **123**, 1273 (1961)
753. W.J. Siemons: IBM J. R. & D. **14**, 245 (1970)
754. I. Shiozaki, C.M. Hurd, S.P. McAlister, W.R. McKinnon, P. Strobel: J. Phys. C **14**, 4641 (1981)
755. S. Todo, K. Siratori, S. Kimura: J. Phys. Soc. Jpn. **64**, 2118 (1995)
756. R. Karplus, J.M. Luttinger: Phys. Rev. **95**, 1154 (1954)
757. A.J.M. Kuipers, V.A.M. Brabers: Phys. Rev. B **14**, 1401 (1976), B **20**, 594 (1979)
758. J.M. Lavine: Phys. Rev. **114**, 482 (1959)
759. B.A. Griffiths, D. Elwell, R. Parker: Phil. Mag. **22**, 163 (1970)
760. H. Graener, M. Rosenberg, T.E. Whall, M.R.B. Jones: Phil. Mag. B **40**, 389 (1979)
761. W.R. McKinnon, C.M. Hurd, I. Shiozaki: J. Phys. C **14**, L877 (1981)
762. A.J.M. Kuipers, V.A.M. Brabers: Phys. Rev. B **20**, 594 (1979)
763. M. Pai, J.M. Honig: J. Phys. C **16**, L35 (1983)
764. M. Matsui, S. Todo, S. Chikazumi: J. Phys. Soc. Jpn. **42**, 1517 (1977)
765. Y. Akishige, T. Fukatsu, M. Kobayashi, E. Sawaguchi: J. Phys. Soc. Jpn. **54**, 2323 (1985)
766. M. Kobayashi, Y. Akishige, E. Sawaguchi: J. Phys. Soc. Jpn. **55**, 4044 (1986)
767. T. Mizoguchi, M. Inoue: J. Phys. Soc. Jpn. **21**, 1310 (1966)
768. M. Mizogushi: J. Phys. Soc. Jpn. **44**, 1501, 1512 (1978)
769. P.W. Anderson: J. Phys. Soc. Jpn. **9**, 316 (1954)
770. A. Yanase, N. Hamada: J. Phys. Soc. Jpn. **68**, 1607 (1999)
771. V.C. Rakhecha, N.S. Satya-Murthy: J. Phys. C **11**, 4389 (1978)
772. Y. Kino, K. Yamamoto, F. Okamura, K. Siratori: J. de Phys. **IV–7, C1** 587 (1997)
773. A. Chainani, T. Yokota, T. Morimoto, T. Takahashi, S. Todo: Phys. Rev. B **51**, 17976 (1995)
774. E.I. Terukov, W. Reichelt, D. Ihle, H. Oppermann: Phys. Stat. Solidi B **95**, 491 (1979)
775. S. Ishida, M. Suzuki, S. Todo, N. Môri, K. Sirataori: J. Phys. Soc. Jpn. **67**, 3125 (1998)
776. K.I. Arai, K. Ohmori, N. Tsuya, S. Iida: Phys. Stat. Solidi A **34**, 325 (1976)
777. K. Siratori, Y. Kino: J. Magn. Magn. Mat. **20**, 87 (1980)
778. K. Siratori, Y. Ishii, Y. Morii, S. Funahashi, S. Todo, A. Yanase: J. Phys. Soc. Jpn. **67**, 2818 (1998)
779. S.M. Shapiro, M. Iizumi, G. Shirane: Phys. Rev. B **14**, 200 (1976)
780. D. Ihle, B. Lorenz: Phil. Mag. B **42**, 337 (1980)
781. D. Ihle, B. Lorenz: J. Phys. C **19**, 5239 (1986)
782. L. Degiorgi, P. Wachter, D. Ihle: Phys. Rev. B **35**, 9259 (1987)
783. J.G. Bednorz, K.A. Müller: Z. Phys. B **64**, 189 (1986)
784. J.G. Bednorz, M. Takashige, K.A. Müller: Europhys. Lett. **3**, 379 (1987)
785. S. Uchida, H. Takagi, S. Tanaka, K. Kitazawa: Jpn. J. Appl. Phys. **26**, L1 (1987)
786. H. Takagi, S. Uchida, K. Kitazawa, S. Tanaka: Jpn. J. Appl. Phys. **26**, L123 (1987)
787. C.W. Chu, P.H. Hor, R.L. Meng, L. Gao, Z.J. Huang, Y.Q. Wang: Phys. Rev. Lett. **58**, 405 (1987)
788. M.K. Wu, J.R. Ashburn, C.J. Torng, P.H. Hor, R.L. Meng, L. Gao, Z.J. Huang, Y.Q. Wang, C.W. Chu: Phys. Rev. Lett. **58**, 908 (1987)
789. S. Hikami, T. Hirai, S. Kagoshima: Jpn. J. Appl. Phys. **26**, L314 (1987)

790. J. Karpinski, E. Kaldis, E. Jilek, S. Rusiecki, B. Bucher: Nature **336**, 660 (1988)
791. P. Bordet, C. Challiout, J. Chenavas, J.L. Hodeau, M. Marezio, J. Karpinski, E. Kaldis: Nature **334**, 596 (1988)
792. A. Santro, F. Beech, M. Marezio, R.J. Cava: Physica C **156**, 693 (1988)
793. H. Maeda, Y. Tanaka, M. Fukutomi, T. Asano: Jpn. J. Appl. Phys. **27**, L209 (1988)
794. C. Mitchel, H. Herview, M.M. Borel, A. Grandin, F. Deslanges, J. Provost, B. Raveau: Z. Phys. B **68**, 421 (1987)
795. J. Akimitsu, A. Yamazaki, H. Sawa, H. Fujiki: Jpn. J. Appl. Phys. **26**, L2080 (1987)
796. S.S. Parkin, V.Y. Lee, A.I. Nazzal, R. Savoy, R. Beyers: Phys. Rev. Lett. **61**, 750 (1988)
797. P. Haldar, K. Chen, B. Maheswaran, A. Roig-Janicki, N.K. Jaggi, R.S. Markiewics, B.C. Giessen: Science **241**, 1198 (1988)
798. Z.Z. Sheng, H. Hermann: Nature **332**, 55 (1988)
799. Z.Z. Sheng, A.M. Hermann: Nature **332**, 138 (1988)
800. A Schilling, M. Cantoni, J.D. Guo, H.R. Otto: Nature **363**, 56 (1993)
801. S.N. Putilin, E.V. Antipov, O. Schmaissem, M. Marezio: Nature **362**, 226 (1993)
802. Y. Tokura, H. Takagi, S. Uchida: Nature **337**, 345 (1989)
803. J. Akimitsu, S. Suzuki, M. Watanabe, H. Sawa: Jpn. J. Appl. Phys. **27**, L1859 (1988)
804. H. Ihara, K. Tokiwa, A. Iyo, M. Hirabayashi, N. Terada, M. Tokumoto, Y.S. Song, Physica C **235–240**, 981 (1994)
805. J. Akimoto, Y. Oosawa, K. Tokiwa, M. Hirabayashi, H. Ihara, Physica C **242**, 360 (1995)
806. M.G. Smith, A. Manthiram, J. Zhou, J.B. Goodenough, J.T. Markert: Nature **351**, 549 (1991)
807. M. Takano, M. Azuma, Z. Hiroi, Y. Bando, Y. Takeda: Physica C **176**, 441 (1991)
808. R.J. Cava, B. Batlogg, R.B. van Dover, J.J. Krajewski, J.V. Waszczak, R.L. Fleming, W.F. Peck, Jr., L.W. Rupp, Jr., P. Marsh, A.C.W.P. James, L.F. Schneemeyer: Nature **345**, 602 (1990)
809. B. Okai: Jpn. J. Appl. Phys. **30**, L179 (1991)
810. L. Rukang, Z. Yingjie, Q. Yitai, C. Zuyao: Physica C **176**, 19 (1991)
811. R.J. Cava, J.J. Krajewski, H. Takagi, H.W. Zandbergen, R.B. van Dover, W.F. Peck Jr., B. Hessen: Physica C **191**, 237 (1992)
812. R.-K. Li, Y.-J. Zhu, C. Xu, Z.-Y. Chen: Physica C **176**, 19 (1991)
813. I. Felner, U. Yaron, U. Asaf, T. Kröner, V. Breit: Phys. Rev. B **49**, 6903 (1994)
814. I. Felner, D. Sprinzak, U. Asaf, T. Kröner: Phys. Rev. B **51**, 3120 (1995)
815. L. Bauerfeind, W. Widder, H.F. Braun: J. Low. Temp. Phys. **105**, 1605 (1996)
816. L. Bauerfeind, W. Widder, H.F. Braun: Physica C **254**, 151 (1995)
817. I. Felner, U. Asaf, Y. Levi, O. Millo: Phys. Rev. B **55**, R3374 (1997)
818. M. Uehara, T. Nagata, J. Akimitsu, H. Takahashi, N. Mori, K. Kinosita: J. Phys. Soc. Jpn. **65**, 2764 (1996)
819. M. Sigrist, T.M. Rice, F.C. Zhang: Phys. Rev. B **49**, 12058 (1994)
820. Z. Hiroi, M. Azuma, M. Takano, Y. Bando: J. Solid State Chem. **95**, 230 (1991)
821. E.M. McCarron, III, M.A. Subramanian, J.C. Calabrese, R.L. Harlow: Mat. Res. Bull. **23**, 1355 (1988)

822. K. Ishida, Y. Kitaoka, K. Asayama, M. Azuma, Z. Hiroi, M. Takano: J. Phys. Soc. Jpn. **63**, 3222 (1994)
823. S. Pei, J.D. Jorgensen, B. Dabrowski, D.G. Hinks, D.R. Richards, A.W. Mitchell, J.M. Newsam, S.K. Sinha, D. Vaknin, A.J. Jacobson: Phys. Rev. B **41**, 4126 (1990)
824. L.F. Mattheiss, E.M. Gyorgy, D.W. Johnson, Jr.: Phys. Rev. B **37**, 3745 (1988)
825. D.G. Hinks, D.R. Richards, B. Dabrowski, D.T. Marx, A.W. Mitchell: Nature **335**, 419 (1988)
826. L.F. Schneemeyer, J.K. Thomas, T. Siegrist, B. Batlogg, L.W. Rupp, Jr., R.L. Opila, R.J. Cava, D.W. Murphy: Nature **335**, 421 (1988)
827. B. Batlogg, R.J. Cava, L.W. Rupp, Jr., A.M. Mujsce, J.J. Krajewski, J.P. Remeika, W.F. Peck, Jr., A.S. Cooper, G.P. Espinosa: Phys. Rev. Lett. **61**, 1670 (1988)
828. J.F. Zasazdinski, N. Tralshawala, D.G. Hinks, B. Dabrowski, A.M. Mitchell, D.R. Richards: Physica C **158**, 519 (1989)
829. T.J. Wagner, H.M. Meyer III, D.M. Hill, Y. Hu, M.B. Jost, J.H. Weaver, D.G. Hinks, B. Dabrowski, D.R. Richards: Phys. Rev. B **40**, 4532 (1989)
830. M. Shirai, N. Suzuki, K. Motizuki: J. Phys. C **2**, 3553 (1990)
831. Y. Maeno, H. Hashimoto, K. Yoshida, S. Nishizaki, T. Fujita, J. D. Bednorz, F. Lichtenberg: Nature **372**, 532 (1994)
832. R.D. Shannon: Solid State Commun. **6**, 139 (1968)
833. T.A. Hewston, B.L. Chamberland: J. Solid State Chem. **65**, 100 (1986)
834. P.W. Anderson, Z. Zou: Phys. Rev. Lett. **60**, 132 (1988)
835. P.W. Anderson: Science **256**, 1526 (1992)
836. N. Nagaosa: Phys. Rev. B **52**, 10561 (1995)
837. D.G. Clarke, S.P. Strong, P.W. Anderson: Phys. Rev. Lett. **74**, 4499 (1995)
838. J.M. Delgado, R.K. McMullan, G. Diaz de Delgado, B.J. Wuensch, P.J. Picone, H.P. Jenssen, D.R. Gabbe: Phys. Rev. B **37**, 9343 (1988)
839. D.C. Johnston, J.P. Stokes, D.P. Goshorn, J.T. Lewandowski: Phys. Rev. B **36**, 4007 (1987)
840. R.J. Cava, A. Santro, D.W. Johnson, Jr., W.W. Rhodes: Phys. Rev. B **35**, 6716 (1987)
841. C.C. Torardi, M.A. Subramanian, J.C. Calabrese, J. Gopalakrishnan, E.M. McCarron, K.J. Morrissey, T.R. Askew, R.B. Flippen, U. Chowdhry, A.W. Sleight: Phys. Rev. B **38**, 225 (1988)
842. E.F. Skelton, W.T. Elam, D.U. Gubser, V. Letourneau, M.S. Osofsky, S.B. Qadri, L.E. Toth, S.A. Wolf: Phys. Rev. B **36**, 5713 (1987)
843. M.A. Kastner, R.J. Birgeneau, C.Y. Chen, Y.M. Chiang, D.R. Gabbe, H.P. Jenssen, T. Junk, C.J. Peters, P.J. Picone, T. Thio, T.R. Thurston, H.L. Tuller: Phys. Rev. B **37**, 111 (1988)
844. T. Freltoft, G. Shirane, S. Mituda, J.P. Remeika, A.S. Cooper: Phys. Rev. B **37**, 137 (1988)
845. D.C. Johnston, S.K. Sinha, A.J. Jacobson, J.M. Newsam: Physica C **153–155**, 572 (1988)
846. J. Saylor, L. Takacs, C. Hohenemser, J.I. Budnick, B. Chamberland: Phys. Rev. B **40**, 6854 (1989)
847. A.P. Reyes, P.C. Hammel, E.T. Ahrens, J.D. Thompson, P.C. Canfield, Z. Fisk, J.E. Schirber: J. Phys. Chem. Solids **54**, 1393 (1993)
848. R.A. Mohan Ram, L. Ganapathi, P. Ganguly, C.N.R. Rao: J. Solid State Chem. **63**, 139 (1986)
849. J.C. Fuggle, P.J.W. Weijs, R. Schoorl, G.A. Sawatzky, J. Fink, N. Nucker, P.J. Durham, W.M. Temmerman: Phys. Rev. B **37**, 123 (1988)

850. P. Steiner, V. Kinsinger, I. Sander, B. Siegwart, S. Hufner, C. Politis, R. Hoppe, H.P. Muller: Z. Phys. B **67**, 497 (1987)
851. D. Vaknin, S.K. Sinha, D.E. Moncton, D.C. Johnston, J.M. Newsam, C.R. Safinya, H.E. King, Jr.: Phys. Rev. Lett. **58**, 2802 (1987)
852. T. Thio, T.R. Thurston, N.W. Preyer, P.J. Picone, M.A. Kastner, H.P. Jenssen, D.R. Gabbe, C.Y. Chen, R.J. Birgeneau, A. Aharony: Phys. Rev. B **38**, 905 (1988)
853. I. Tanaka, K. Yamane, H. Kojima: J. Cryst. Growth **96**, 711 (1989)
854. R.R. Heikes, W.D. Johnston: J. Chem. Phys. **26**, 582 (1957)
855. M. O'Keeffe, F.S. Stone: J. Phys. Chem. Solids **23**, 261 (1962)
856. B.N. Brockhouse: Phys. Rev. **94**, 781 (1954)
857. M. O'Keeffe, F.S. Stone: Proc. Roy. Soc. A **267**, 501 (1962)
858. M. O'Keeffe, W.J. Moore: J. Chem. Phys. **35**, 1324 (1961)
859. G. Demazeau, C. Parent, M. Pouchard, P. Hagenmuller: Mater. Res. Bull. **7**, 913 (1972)
860. J.B. Torrance, Y. Tokura, A. Nazzal, S.S.P. Parkin: Phys. Rev. Lett. **60**, 542 (1988)
861. N. Nguyen, C. Michel, F. Studer, B. Raveau: Mater. Chem. **7**, 413 (1982)
862. T. Tamegai, Y. Iye: Physica C **159**, 181 (1989)
863. F. Herman, R.V. Kasowski, W.Y. Hsu: Phys. Rev. B **37**, 2309 (1988)
864. A. Fuertes, C. Miravitlles, J. Gonzalez-Calbet, M. Vallet-Regi, X. Obradors, J. Rodriguez-Carvajal: Physica C **157**, 525 (1989)
865. J. Hauck, K. Bickmann, B. Bishof, S. Ipta: J. Solid State Chem. **76**, 416 (1988)
866. L.F. Mattheiss: Phys. Rev. Lett. **58**, 1028 (1987)
867. K. Takegahara, H. Harima, A. Yanase: Jpn. J. Appl. Phys. **26**, L352 (1987)
868. T.C. Leung, X.W. Wang, B.N. Harmon: Phys. Rev. B **37**, 384 (1988)
869. R.E. Cohen, W.E. Pickett, H. Krakauer: Phys. Rev. Lett. **62**, 831 (1989)
870. M.W. Shafer, T. Penney, B.L. Olson: Phys. Rev. B **36**, 4047 (1987)
871. H. Takagi, T. Ido, S. Ishibashi, M. Uota, S. Uchida, Y. Tokura: Phys. Rev. B **40**, 2254 (1989)
872. J.B. Torrance, Y. Tokura, A.I. Nazzal, A. Bezinge, T.C. Huang, S.S.P. Parkin: Phys. Rev. Lett. **61**, 1127 (1988)
873. J.B. Torrance, A. Bezinge, A.I. Nazzal, T.C. Huang, S.S.P. Parkin, D.T. Keane, S.J. LaPlaca, P.M. Horn, G.A. Held: Phys. Rev. B **40**, 8872 (1989)
874. R.B. van Dover, R.J. Cava, B. Batlogg, E.A. Reitman: Phys. Rev. B **35**, 5337 (1987)
875. M.-H. Whangbo, D.B. Kang, C.C. Torardi: Physica C **158**, 371 (1989)
876. G. Hilsher, N. Pillmayr, R. Eibler, E. Bauer, K. Remshnig, P. Rogl: Z. Phys. B **72**, 461 (1988)
877. Y. Oi Nakamura, N. Matsuda, Y. Shiina: Solid State Commun. **81**, 923 (1992)
878. A.R. Moodenbaugh, Y. Xu, M. Suenaga, T.J. Folkerts, R.N. Shelton: Phys. Rev. B **38**, 4596 (1988)
879. Y. Maeno, F. Nakamura, T. Suzuki, T. Fujita: JJAP Series 7, Mechanism of Superconductivity, (1992) p. 91
880. G.M. Luke, L.P. Le, B.J. Sternlieb, W.D. Wu, Y.J. Uemura, J.H. Brewer, T.M. Riseman, S. Ishibashi, S. Uchida: Physica C **185–189**, 1175 (1991)
881. W.E. Pickett, R.E. Cohen, H. Krakauer: Phys. Rev. Lett. **67**, 228 (1991)
882. M.R. Norman, G.J. Mamullan, D.L. Novikov, A.J. Freeman: Phys. Rev. B **48**, 9935 (1993)
883. M. Akoshima, T. Noji, Y. Ono, Y. Koike: Phys. Rev. B **57**, 7491 (1998)
884. J.M. Tranquada, B.J. Sternlieb, J.D. Axe, Y. Nakamura, S. Uchida: Nature **375**, 561 (1995)

885. N. Mori, C. Murayama, H. Takahashi, H. Kaneko, K. Kawabata, Y. Iye, S. Uchida, H. Takagi, Y. Tokura, Y. Kubo, H. Sasakura, K. Yamaya: Physica C **185–189**, 40 (1991)
886. S. Yomo, C. Maruyama, H. Takahashi, N. Mori, K. Kishio, K. Kitazawa, K. Fueki: Jpn. J. Appl. Phys. **26**, L602 (1987)
887. R. Griessen: Phys. Rev. B **36**, 5284 (1987)
888. M. Kurisu, H. Kadomatsu, H. Fujiwara, Y. Maeno, T.Fujita: Jpn. J. Appl. Phys. **26**, L361 (1987)
889. Y.J. Uemura, G.M. Luke, B.J. Sternlieb, J.H. Brewer, J.F. Carolan, W.N. Hardy, R. Kadono, J.R. Kepton, R.F. Kiefl, S.R. Kreitzman, P. Mulhern, T.M. Riseman, D. L. Williams, B.X. Yang, S. Uchida, H. Takagi, J. Gopalakrishnan, A.W. Sleight, M.A. Subramanian, C.L. Chien, M.Z. Cieplak, G. Xiao, V.Y. Lee, B.W. Statt, C.E. Stronach, W.J. Kossler, X.H. Yu: Phys. Rev. Lett. **62**, 2317 (1989)
890. B. Batlogg, G. Kourouklis, W. Weber, R.J. Cava, A. Jayaraman, A.E. White, K.T. Short, L.W. Rupp, E.A. Rietman: Phys. Rev. Lett. **59**, 912 (1987)
891. T.A. Faltens, W.K. Ham, S.W. Keller, K.J. Leary, J.N. Michaels, A.M. Stacy, H.-C zur Loye, D.E. Morris, T.W. Barbee III, L.C. Bourne, M.L. Cohen, S. Hoen, A. Zettl: Phys. Rev. Lett. **59**, 915 (1987)
892. M. Ronay, M.A. Frisch, T.R. McGuire: Phys. Rev. B **45**, 355 (1992)
893. G. Zhao, M.B. Hunt, H. Keller, K.A. Müller: Nature **385**, 236 (1997)
894. J.P. Franck, S. Harker, J.H. Brewer: Phys. Rev. Lett. **71**, 283 (1993)
895. D.G. Hinks, D.R. Richards, B. Dabrowski, D.T. Marx, A.W. Mitchell: Nature **335**, 419 (1988)
896. J. Bardeen, L.N. Cooper, J.R. Schrieffer: Phys. Rev. **108**, 1175 (1957)
897. V.Z. Kresin, A. Bill, S. A. Wolf, Y.N. Ovchinnikov: Phys. Rev. B **56**, 107 (1997)
898. M. Schossmann, F. Marsiglio, J.P. Carbotte: Phys. Rev. B **36**, 3627 (1987)
899. G. Nieva, E.N. Martinez, F. de la Cruz, D.A. Esparza, C.A. D'Ovidio: Phys. Rev. B **36**, 8780 (1987)
900. N. Wada, H. Muro-oka, Y. Nakamura, K. Kumagai: Physica C **157**, 453 (1989)
901. W.K. Kwok, G.W. Crabtree, D.G. Hinks, D.W. Capone, J.D. Jorgensen, K. Zhang: Phys. Rev. B **35**, 5343 (1987)
902. T.P. Orland, K.A. Delin, S. Foner, E.J. McNiff, Jr., J.M. Tarascon, L.H. Greene, W.R. McKinnon, G.W. Hull: Phys. Rev. B **35**, 5347 (1987)
903. D.C. Johnston: Phys. Rev. Lett. **62**, 957 (1989)
904. M.E. Reeves, D.S. Citrin, B.G. Pazol, T.A. Friedmann, D.M. Ginsberg: Phys. Rev. B **36**, 6915 (1987)
905. S.J. Collocott, G.K. White, S.X. Dou, R.K. Williams: Phys. Rev. B **36**, 5684 (1987)
906. B. Batlogg, A.P. Ramirez, R.J. Cava, R.B. van Dover, E.A. Rietman: Phys. Rev. B **35**, 5340 (1987)
907. P.W. Anderson, G. Baskaran, Z. Zou, T. Hsu: Phys. Rev. Lett. **58**, 2790 (1987)
908. Z. Zou, P.W. Anderson: Phys. Rev. B **37**, 627 (1988)
909. M. Kato, Y. Maeno, T. Fujita: Physica C **152**, 116 (1988)
910. N. Momono, M. Ido: Physica C **264**, 311 (1996)
911. T. Ekino, T. Minami, H. Fujii, J. Akimitsu: Physica C **235–240**, 1899 (1994)
912. N.R. Werthamer, E. Helfand, P.C. Hohenberg: Phys. Rev. **147**, 295 (1966)
913. S. Foner: IEEE Trans. MAG-**11**, 155 (1975)
914. S. Senoussi, M. Oussena, M. Ribault: Phys. Rev. B **36**, 4003 (1987)

915. H. Takagi, B. Batlogg, H.L. Kao, J. Kwo, R.J. Cava, J.J. Krajewski, W.F. Peck, Jr.: Phys. Rev. Lett. **69**, 2975 (1992)
916. Y. Ando, G.S. Boebinger, A. Passner, T. Kimura, K. Kishio: Phys. Rev. Lett. **75**, 4662 (1995)
917. Y. Ando, G.S. Boebinger, A. Passner, K. Tamasaku, N. Ichikawa, S. Uchida, M. Okuya, T. Kimura, J. Shimiyama, K. kishio: J. Low Temp. Phys. **105**, 867 (1996)
918. J. Kondo in *Solid State Phys.*, Vol. 23 ed. by F. Seitz, D. Turnbull, H. Ehrenreich (Academic, New York, 1969) p. 183
919. P.A. Lee. T.V. Ramakrishnan: Rev. Mod. Phys. **57**, 287 (1985)
920. B. Keimer, N. Belk, R.J. Birgeneau, A. Cassanho, C.Y. Chen, M. Greven, M.A. Kastner, A. Aharony, Y. Endoh, R.W. Erwin, G. Shirane: Phys. Rev. B **46**, 14034 (1992)
921. M.E. Lines: J. Phys. Chem. Solids **31**, 101 (1970)
922. G. Shirane, Y. Endoh, R.J. Birgeneau, M.A. Kastner, Y. Hidaka, M. Oda, M. Suzuki, T. Murakami: Phys. Rev. Lett. **59**, 1613 (1987)
923. R.J. Birgeneau, D.R. Gabbe, H.P. Jenssen, M.A. Kastner, P.J. Picone, T.R. Thurston, G. Shirane, Y. Endoh, M. Sato, K. Yamada, Y. Hidaka, M. Oda, Y. Enomoto, M. Suzuki, T. Murakami: Phys. Rev. B **38**, 6614 (1988)
924. I. Tanaka, H. Kojima: Nature **337**, 21 (1989)
925. G. Shirane, R.J. Birgeneau, Y. Endoh, P. Gehring, M.A. Kastner, K. Kitazawa, H. Kojima, I. Tanaka, T.R. Thurston, K. Yamada: Phys. Rev. Lett. **63**, 330 (1989)
926. T.R. Thurston, R.J. Birgeneau, M.A. Kastner, N.W. Preyer, G. Shirane, Y. Fujii, K. Yamada, Y. Endoh, K. Kakurai, M. Matsuda, Y. Hidaka, T. Murakami: Phys. Rev. B **40**, 4585 (1989)
927. P. Gutsmiedl, G. Wolff, K. Angres: Phys. Rev. B **36**, 4043 (1987)
928. K. Kumagai, Y. Nakamura: Physica C **152**, 286 (1988)
929. R.F. Kiefl, J.H. Brewer, J. Carolan, P. Dosanjh, W.N. Hardy, R. Kadono, J.R. Kempton, R. Krahn, P. Schleger, B.X. Yang, H. Zhou, G.M. Luke, B. Sternlieb, Y.J. Uemura, W.J. Kossler, X.H. Yu, E.J. Ansaldo, H. Takagi, S. Uchida, C.L. Seaman: Phys. Rev. Lett. **63**, 2136 (1989)
930. S. Sugai, S. Shamoto, M. Sato: Phys. Rev. B **38**, 6436 (1988)
931. S. Ohsugi, Y. Kitaoka, K. Ishida, K. Asayama: J. Phys. Soc. Jpn. **60**, 2351 (1991)
932. T. Moriya, T. Takahashi, K. Ueda: J. Phys. Soc. Jpn. **59**, 2905 (1990)
933. K. Sreedhar, P. Ganguly: Phys. Rev. B **41**, 371 (1990)
934. A. Fujimori, E. Takayama-Muromachi, Y. Uchida, B. Okai: Phys. Rev. B **35**, 8814 (1987)
935. A. Fujimori, E. Takayama-Muromachi, Y. Uchida: Solid State Commun. **63**, 857 (1987)
936. Zhi-xun Shen, J.W. Allen, J.J. Yeh, J.-S.Kang, W. Ellis, W. Spicer, I. Lindau, M.B. Maple, Y.D. Dalichaouch, M.S. Torikachvili, J.Z. Sun, T.H. Geballe: Phys. Rev. B **36**, 8414 (1987)
937. G.A. Kourouklis, A. Jayaraman, W. Weber, J.P. Remeika, G.P. Espinosa, A.S. Cooper, R.G. Maines, Jr.: Phys. Rev. B **36**, 7218 (1987)
938. F. Gervais, P. Echegut, J.M. Bassat, P. Odier: Phys. Rev. B **37**, 9364 (1988)
939. R.J. Birgeneau, C.Y. Chen, D.R. Gabbe, H.P. Jenssen, M.A. Kastner, C.J. Peters, P.J. Picone, Tineke Thio, T.R. Thurston, H.J. Tuller: Phys. Rev. Lett. **59**, 1329 (1987)
940. J.M. Ferreira, B.W. Lee, Y. Dalichaouch, M.S. Torikachvili, K.N. Yang, M.B. Maple: Phys. Rev. B **37**, 1580 (1988)
941. M.E. Reeves, T.A. Friedmann, D.M. Ginsberg: Phys. Rev. B **35**, 7207 (1987)

942. A. Migliori, W.M. Visscher, S.E. Brown, Z. Fisk, S.-W. Cheong, B. Alten, E.T. Ahrens, K.A. Kubat-Martin, J.D. Maynard, Y. Huang, D.R. Kirk, K.A. Gills, H.K. Kim, M.H.W. Chan: Phys. Rev. B **41**, 2098 (1990)
943. M.J. Rosseinsky, K. Prassides, P. Day, A.J. Dianoux: Phys. Rev. B **37**, 2231 (1988)
944. B. Renker, F. Gompf, E. Gering, N. Nucker, D. Ewert, W. Reichardt, H. Rietschel: Z. Phys. B **67**, 15 (1987)
945. W. Reichardt, F. Gompf, L. Pintschovius, N. Pyka, B. Renker, P. Bourges, G. Collin, A.S. Ivanov, N.L. Mitrofanov, A.Yu. Rumiantsev: In *Electron–Phonon Interaction in Oxides Superconductors*, ed. by R. Baquero (World Scientific, Singapore, 1991) p. 1
946. F.W. de Wette, J. Prade, A.D. Kulkarni, U. Schröder, W. Kress: Phys. Rev. B **38**, 6583 (1988)
947. N. Ogita, M. Udagawa, K. Kojima, K. Ohbayashi: J. Phys. Soc. Jpn. **57**, 3932 (1988)
948. G.L. Doll, J.T. Nicholls, M.S. Dresselhaus, A.M. Rao, J.M. Zhang, G.W. Lehman, P.C. Eklund, G. Dresselhaus, A.J. Strauss: Phys. Rev. B **38**, 8850 (1988)
949. K. Strobel, R. Geick: J. Phys. C **9**, 4223 (1976)
950. W. Weber: Phys. Rev. Lett. **58**, 1371 (1987)
951. H. Krakauer, W.E. Pickett, R.E. Cohen: In *Manifestations of the Electron–Phonon Interaction* ed. by R. Baquero (World Scientific, Singapore, 1994) p. 38
952. H. Krakauer, W.E. Pickett, R.E. Cohen: Phys. Rev. B **47**, 1002 (1993)
953. K. Motizuki, M. Shirai: In *Manifestations of the Electron–Phonon Interaction* ed. by R. Baquero (World Scientific, Singapore, 1994) p. 46
954. T. Ekino, J. Akimitsu: Jpn. J. Appl. Phys. Series **7**, 260 (1992)
955. T. Ekino: Physica C **205**, 338 (1993)
956. Q. Huang, J.F. Zasadzinski, N. Tralshawala, K.E. Gray, D.G. Hinks, J.L. Peng, R.L. Greene: Nature **347**, 369 (1990)
957. J.W. Lynn, I.W. Sumarlin, D.A. Neumann, J.J. Rush, J.L. Peng, Z.Y. Li: Phys. Rev. Lett. **66**, 919 (1991)
958. M. Shirai, N. Suzuki, K. Motizuki: Solid State Commun. **73**, 633 (1990)
959. S. Uchida, H. Takagi, K. Kitazawa, S. Tanaka: Jpn. J. Appl. Phys. **26**, L151 (1987)
960. C.W. Chu, P.H. Hor, R.L. Meng, L. Gao, Z.J. Huang: Science **235**, 567 (1987)
961. J.M. Tarascon, L.H. Greene, P. Barboux, W.R. McKinnon, G.W. Hull. T.P. Orlando, K.A. Dekin, S. Foner, E.J. McNiff, Jr.: Phys. Rev. B **36**, 8393 (1987)
962. C. Politis, J. Geerk, M. Dietrich, B. Obst: Z. Phys. B **66**, 141 (1987)
963. K. Kitazawa, M. Sakai, S. Uchida, H. Takagi, K. Kisio, S. Kanbe, S. Tanaka, K.Fueki: Jpn. J. Appl. Phys. **26**, L342 (1987)
964. D.C. Gubser, R.A. Hein, S.H. Lawrence, M.S. Osofsky, D.J. Schrodt, L.E. Toth, S.A. Wolf: Phys. Rev. B **35**, 5350 (1987)
965. N. Ogita, K. Ohbayashi, M. Udagawa, Y. Aoki, Y. Maeno, T. Fujita: Jpn. J. Appl. Phys. **26**, L415 (1987)
966. T. Hasegawa, K. Kishio, M. Aoki, N. Ooba, K. Kitazawa, K. Fueki, S. Uchida, S. Tanaka: Jpn. J. Appl. Phys. **26**, L337 (1987)
967. Y. Saito, T. Noji, A. Endo, N. Matsuzaki, M. Katsumata: Jpn. J. Appl. Phys. **26**, L223 (1987)
968. H. Sawa, S. Suzuki, M. Watanabe, J. Akimitsu, H. Matsubara, H. Watanabe, S. Uchida, K. Kokusho, H. Asano, F. Izumi, E. Takayama-Muromachi: Nature **337**, 347 (1989)

969. Z. Schlesinger, R.T. Collins, M.W. Shafer, E.M. Engler: Phys. Rev. B **36**, 5275 (1987)
970. J.R. Kirtley, C.C. Tsuei, S.I. Park, C.C. Chi, J. Rozen, M.W. Shafer: Phys. Rev. B **35**, 7216 (1987)
971. P.J.M. van Bentum, L.E.C. van de Leemput, L.W.M. Schreurs, P.A.A. Teunissen, H. van Kempen: Phys. Rev. B **36**, 843 (1987)
972. M.E. Hawley, K.E. Gray, D.W. Caponen II, D.G. Hinks: Phys. Rev. B **35**, 7224 (1987)
973. S. Pan, K.W. Ng, A.L. de Lozanne, J.M. Tarascon, L.H. Greene: Phys. Rev. B **35**, 7220 (1987)
974. A.P. Fein, J.R. Kirtley, M.W. Shafer: Phys. Rev. B **37**, 9738 (1988)
975. H. Murakami, S. Ohkuchi, R. Aoki: J. Phys. Soc. Jpn. **63**, 2653 (1994)
976. S.H. Wang, Q. Song, B.P. Clayman, J.L. Peng, L. Zhang, R.N. Shelton: Phys. Rev. Lett. **64**, 1067 (1990)
977. S.M. Anlage, D.H. Wu, J. Mao, S.N. Mao, X.X. Xi, T. Venkatesan, J.L. Peng, R.L. Greene: Phys. Rev. B **50**, 523 (1994)
978. M. Naito, A. Matsuda, K. Kitazawa, S. Kambe, I. Tanaka, H. Kojima: Phys. Rev. B **41**, 4823 (1990)
979. S. Uchida, H. Takagi, S. Tanaka, N. Nakao, N. Miura, K. Kishio, K. Kitazawa, K. Fueki: Jpn. J. Appl. Phys. **26**, L196 (1987)
980. P.J.M. van Bentum, H. van Kempen, L.E.C. van de Leemput, J.A.A.J. Perenboom, L.W.M. Schreurs, P.A.A. Teunissen: Phys. Rev. B **36**, 5279 (1987)
981. D.W. Capone II, D.G. Hinks, J.D. Jorgensen, K. Zhang: Appl. Phys. Lett. **50**, 543 (1987)
982. R.S. List, A.J. Arko, Z. Fisk, S.-W. Cheong, S.D. Conradson, J.D. Thompson, C.B. Pierce, D.E. Peterson, R.J. Bartlett, N.D. Shinn, J.E. Shirber, B.W. Veal, A.P. Paulikas, J.C. Campzano: Phys. Rev. B **38**, 11966 (1988)
983. M.C. Sharbel, C.-H. Park, A. Matsuura, Z.-X. Shen, D.A. Bonn, R. Liang, W.N. Hardy: Phys. Rev. B **55**, 2796 (1997)
984. Q.W. Yan, P.L. Zhang, Z.G. Shen, J.K. Zhao, Y. Ren, Y.N. Wei, T.D. Mao, C.X. Liu, T.S. Ning, K. Sun, Q.S. Yang: Phys. Rev. B **36**, 8810 (1987)
985. J.D. Jorgensen, B.W. Veal, W.K. Kwok, G.W. Crabtree, A. Umezawa, L.J. Nowicki, A.P. Paulikas: Phys. Rev. B **36**, 5731 (1987)
986. R.J. Cava, B. Batlogg, K.M. Rabe, E.A. Rietman, P.K. Gallagher, L.W. Rupp, Jr.: Physica C **156**, 523 (1988)
987. I.-W. Chen, S.J. Keating, C.Y. Keating, X. Wu, J. Xu, P.E. Reyeo-Morel, T.Y. Tien: Solid State Commun. **63**, 997 (1987)
988. M. Tokumoto, H. Ihara, T. Matsubara, M. Hirabayashi, N. Terada, H. Oyanagi, K. Murata, Y. Kimura: Jpn. J. Appl. Phys. **26**, L1565 (1987)
989. J.D. Jorgensen, M.A. Beno, D.G. Hinks, L. Soderholm, K.J. Volin, R.L. Hitterman, J.D. Grace, Ivan K. Schuller: Phys. Rev. B **36**, 3608 (1987)
990. P.P. Freitas, T.S. Plasket: Phys. Rev. B **36**, 5723 (1987)
991. A.T. Fiory, M. Gurvitch, R.J. Cava, G.P. Espinosa: Phys. Rev. B **36**, 7262 (1987)
992. R.W. Cohen, G.D. Cody, J.J. Halloran: Phys. Rev. Lett. **19**, 840 (1967)
993. Y. Tokura, J.B. Torrance, T.C. Huang, A.I. Nazzal: Phys. Rev. B **38**, 7156 (1988)
994. Z.Z. Wang, J. Clayhold, N.P. Ong: Phys. Rev. B **36**, 7222 (1987)
995. R.C. Baetzold: Phys. Rev. B **38**, 11304 (1988)
996. H. Maeda, A. Koizumi, N. Bamba, E. Takayama-Muromachi, F. Izumi, H. Asano, K. Shimizu, H. Moriwaki, H. Matuyama, Y. Kuroda, H. Yamazaki: Physica C **157**, 483 (1989)

997. B. Batlogg, R.J. Cava, A. Jayaraman, R.B. van Dover, G.A. Kourouklis, S. Sunshine, D.W. Murphy, L.W. Rupp, H.S. Chen, A. White, K.T. Short, A.M. Mujsce, E.A. Rietman: Phys. Rev. Lett. **58**, 2333 (1987)
998. K.J. Leary, H-C zur Loye, S.W. Keller, T.A. Faltens, W.K. Ham, J.N. Michaels, A.M. Stacy: Phys. Rev. Lett. **59**, 1236 (1987)
999. L.C. Bourne, A. Zettl, T.W. Barbee III, M.L. Cohen: Phys. Rev. B **36**, 3990 (1987)
1000. H. Katayama-Yoshida, T. Hirooka, A.J. Mascarenhas, Y. Okabe, T. Takahashi, T. Sasaki, A. Ochiai, T. Suzuki, J.I. Pankove, T. Ciszek, S.K. Deb: Jap. J. Appl. Phys. **26**, L2085 (1987)
1001. J.P. Franck, J. Jung, M.A.-K. Mohamed, S. Gygax, G.I. Sproule: Phys. Rev. B **44**, 5318 (1991)
1002. H.J. Bornemann, D. Morris: Phys. Rev. B **44**, 5322 (1991)
1003. D. Zech, H. Keller, K. Conder, E. Kaldis, E. Liarokapis, N. Poulikas, K.A. Müller: Nature **371**, 681 (1994)
1004. K.A. Müller: Z. Phys. B **80**, 193 (1990)
1005. G. Ruani, C. Taliani, M. Muccini, K. Conder, E. Kaldis, H. Keller, D. Zech, K.A. Müller: Physica C **226**, 101 (1994)
1006. G. Zhao, V. Kirtikar, K.K. Singh, A.P.B. Sinha, D.E. Morris, A.V. Inyushkin: Phys. Rev. B **54**, 14956 (1996)
1007. J.P. Carbotte: Rev. Mod. Phys. **62**, 1027 (1990)
1008. A. Oguri: J. Phys. Soc. Jpn. **57**, 2123 (1988)
1009. Y. Yokoya: Phys. Rev. B **56**, 6107 (1997)
1010. A.A. Abrikosov, J.C. Campuzano, K. Gofron: Physica C **214**, 73 (1993)
1011. P.B. Allen, B. Mitrovic: In *Solid State Physics*, Vol. 37, ed. by H. Ehrenreich, F. Seitz, D. Turnbull (Academic Press, New York, 1982) p. 73
1012. L.N. Bulaevskii, O.V. Dolgov, M.O. Ptitsyn: Phys. Rev. B **38**, 11290 (1988)
1013. S.L. Drechsler, N.M. Plakida: Phys. Stat. Sol. (B) **144**, K113 (1987)
1014. L. Pietronero, S. Strassler: Europhys. Lett. **18**, 627 (1996)
1015. C. Grimaldi, L. Pietronero, S. Strassler, C. Grimaldi: Phys. Rev. B 52, 10516 (1995)
1016. D. Zech, K. Conder, H. Keller, E. Kaldis, K.A. Müller: Physica B **219-220**, 136 (1996)
1017. H.-B. Schüttler, C.-H. Pao: Phys. Rev. Lett: **75**, 4504 (1995)
1018. R. Griessen: Phys. Rev. B **36**, 5284 (1987)
1019. A. Driessen, R. Griessen, N. Koeman, E. Salomons, R. Brouwer, D.G. de Groot, K. Heeck, H. Hemmes, J. Rector: Phys. Rev. B **36**, 5602 (1987)
1020. U. Koch, N. Lotter, J. Wittig, W. Assumus, B. Gegenheimer, K. Winzer: Solid State Commun. **67**, 959 (1988)
1021. B. Bucher, J. Karpinski, E. Kaldis, P. Wachter: Physica C **157**, 478 (1989)
1022. F. Marsiglio, R. Akis, J.P. Carbotte: Phys. Rev. B **36**, 5245 (1987)
1023. P. Fulde, J. Keller, G. Zwicknagl: *Solid State Physics*, Vol. 41, ed. by H. Ehrenreich, D. Turnbull (Academic, New York, 1988) p. 1
1024. D. Pines: Physica C **235–240**, 113 (1994)
1025. D.J. Scalapino: Physics Reports **250**, 329 (1995)
1026. T. Dahm, J. Erdmenger, K. Scharnberg, C.T. Rieck: Phys. Rev. B **48**, 3896 (1993)
1027. M. Sigrist, T.M. Rice: Z. Phys. B **68**, 9 (1987)
1028. J.F. Annett: Adv. in Phys. **39**, 83 (1990)
1029. R. Zeyher, M.L. Kulić: Phys. Rev. B **53**, 2850 (1996)
1030. J. Song, J.F. Annett: Phys. Rev. B **51**, 3840 (1995)
1031. S.Y. Savrasov, O.K. Andersen: Phys. Rev. Lett. **77**, 4430 (1996)

1032. H. Kamimura, S. Matsuno, Y. Suwa, H. Ushio: Phys. Rev. Lett. **77**, 723 (1996)
1033. D.A. Wollman, D.J. Van Harkingen, J. Giapintzakis, D.M. Ginsberg: Phys. Rev. Lett. **74**, 797 (1995)
1034. J.R. Kirtley, C.C. Tuei, J.Z. Sun, C.C. Chi, L.S. Yu-Jahnes, A. Gupta, M. Rupp, M.B. Ketchen: Nature **373**, 225 (1995)
1035. L. Alff, H. Takashima, S. Kashiwaya, N. Terada, T. Ito, K. Oka, Y. Tanaka, M. Koyanagi: Physica C **282–287**, 1485 (1997)
1036. S. Kashiwaya, Y. Tanaka, M. Koyanagi, K. Kajimura: Phys. Rev. B **53**, 2667 (1996)
1037. K. Zhang, D.A. Bonn, S. Kamal, R. Liang, D.J. Baar, W.N. Hardy, D. Basov, T. Timusk: Phys. Rev. Lett. **73**, 2484 (1994)
1038. P. Chaudhari, S.Y. Lin: Phys, Rev. Lett. **72**, 1084 (1994)
1039. Ya.G. Ponomarev, B.A. Aminov, M.A. Hein, H. Heinrichs, V.Z. Kresin, G. Müller, H. Piel, K. Rosner, S.V. Tchesnokov, E.B. Tsokur, D. Wehler, K. Winzer, A.V. Yaryagin, K.T. Yusupov: Physica C **243**, 167 (1995)
1040. A.G. Sun, S.H. Han, A.S. Katz, G.A. Gajewski, M.B. Maple, R.C. Dynes: Phys. Rev. B **52**, R15731 (1995)
1041. J. Lesueur, M. Aprili, A. Goulon, T.J. Horton, L. Dumoulin: Phys. Rev. B **55**, R3398 (1997)
1042. T. Imai, T. Shimizu, T. Tsuda, H. Yasuoka, T. Takabatake, Y. Nakazawa, M. Ishikawa: J. Phys. Soc. Jpn. **57**, 1771 (1988)
1043. T. Imai, T. Shimizu, H. Yasuoka, Y. Ueda, K. Kosuge: J. Phys. Soc. Jpn. **57**, L2280 (1988)
1044. P.C. Hammel, M. Takigawa, R.H. Heffner, Z. Fisk, K.C. Ott: Phys. Rev. Lett. **63**, 1992 (1989)
1045. J. van Kranendonk: Physica **XX**, 781 (1954)
1046. H. Yasuoka, T. Shimizu, Y. Ueda, K. Kosuge: J. Phys. Soc. Jpn. **57**, L2659 (1988)
1047. S.E. Barrett, D.J. Durand, C.H. Pennington, C.P. Slichter, T.A. Friedmann, J.P. Rice, D.M. Ginsberg: Phys. Rev. B **41**, 6283 (1990)
1048. L.C. Hebel, C.P. Slichter: Phys. Rev. **113**, 1504 (1959)
1049. K. Kuboki, H. Fukuyama: J. Phys. Soc. Jpn. **58**, L376 (1989)
1050. N. Bulut, D.J. Scalapino: Phys. Rev. Lett. **68**, 706 (1992)
1051. P.B. Allen, D. Rainer: Nature **349**, 396 (1991)
1052. A. Matsushita, T. Oguchi, K. Kimura, T. Matsumoto, T. Hatano, K. Ogawa, S. Takayanagi: Jpn. J. Appl. Phys. **26**, L1953 (1987)
1053. S.E. Inderhees, M.B. Salamon, T.A. Friedmann, D.M. Ginsberg: Phys. Rev. B **36**, 2401 (1987)
1054. A. Junod, A. Bezinge, J. Muller: Physica C **152**, 50 (1988)
1055. T. Atake, Y. Takagi, T. Nakamura, Y. Saito: Phys. Rev. B **37**, 552 (1988)
1056. J.W. Loram, K.A. Mirza, J.R. Cooper, W.Y. Liang: Phys. Rev. Lett. **71**, 1740 (1993)
1057. J.W. Loram, K.A. Mirza, J.R. Cooper, W.Y. Liang, J.M. Wade: J. Supercond. **7**, 243 (1994)
1058. J.W. Loram, K.A. Mirza, J.M. Wade, J.R. Cooper, W.Y. Liang: Physica C **235–240**, 134 (1994)
1059. H.P. Van Der Meulen, J.J. M.Franse, Z. Tarnawski, K. Kadowaki, J.C.P. Klasse, A.A. Menovsky: Physica C **152**, 65 (1988)
1060. J.W. Loram, K.A. Mirza, P.A. Freeman: Physica C **171**, 243 (1990)
1061. J. Heremans, D.T. Morelli, G.W. Smith, S.C. Strite III: Phys. Rev. B **37**, 1604 (1988)

1062. M. Tinkham: *Introduction to Superconductivity* (McGraw-Hill, New York, 1975)
1063. M. Hikita, Y. Tajima, A. Tatsui, Y. Hidaka, T. Iwata, S. Tsurumi: Phys. Rev. B **36**, 7199 (1987)
1064. A. Umezawa, G.W. Crabtree, J.Z. Liu, T.J. Moran, S.K. Malik, L.H. Nunez, W.L. Kwok, C.H. Sowers: Phys. Rev. B **38**, 2843 (1988)
1065. S. Waki, Y. Yamaguchi, K. Oka, H. Unoki, K. Mitsugi: Jpn. J. Appl. Phys. **26**, L2097 (1987)
1066. A. Maeda, T. Yabe, H. Eisaki, K. Uchinokura, S. Tanaka: Jpn. J. Appl. Phys. **26**, L1670 (1987)
1067. T.R. McGuire, T.R. Dinger, P.J.P. Freitas, W.J. Gallagher, T.S. Plaskett, R.L. Sandstrom, T.M. Shaw: Phys. Rev. B **36**, 4032 (1987)
1068. K.W. Kwok, G.W. Crabtree, A. Umezawa, B.W. Veal, J.D. Jorgensen, S.K. Malik, L.J. Nowicki, A.P. Paulikas, L. Nunez: Phys. Rev. B **37**, 106 (1988)
1069. H. Oyanagi, H. Ihara, T. Matsubara, M. Tokumoto, T. Matsushita, M. Hirabayashi, K. Murata, N. Terada, T. Yao, H. Iwasaki, Y. Kimura: Jpn. J. Appl. Phys. **26**, L1561 (1987)
1070. J.S. Moodera, R. Meservey, J.E. Tkaczyk, C.X. Hao, G.A. Gibson, P.M. Tedrow: Phys. Rev. B **37**, 619 (1988)
1071. T. Itoh, K. Takenaka, S. Uchida: Phys. Rev. Lett. **70**, 3995 (1993)
1072. S.-W. Cheong, S.E. Brown, Z. Fisk, R.S. Kwok, J.D. Thompson, E. Zirngiebel, G. Gruner, D.E. Peterson, G.L. Wells, R.B. Schwarz, J.R. Cooper: Phys. Rev. B **36**, 3913 (1987)
1073. J. Clayhold, N.P. Ong, Z.Z. Wang, J.M. Trascon, P. Barboux: Phys. Rev. B **39**, 7324 (1989)
1074. A.T. Fiory, G.S. Grader: Phys. Rev. B **38**, 9198 (1988)
1075. J.M. Harris, Y.F. Yan, N.P. Ong: Phys. Rev. B **46**, 14293 (1992)
1076. P.W. Anderson: Phys. Rev. Lett. **67**, 2092 (1991)
1077. B.P. Stojković, D. Pines: Phys. Rev. B **55**, 8576 (1997)
1078. T. Timusk: *Electron–Phonon Interaction in Oxide Superconductors*, ed. by R. Baquero (World Scientific, Singapore, 1991) p. 16
1079. K. Kanoda, T. Kawagoe, M. Hasumi, T. Takahashi, S. Kagoshima, T. Mizoguchi: J. Phys. Soc. Jpn. **57**, L1554 (1988)
1080. K. Westerholt, H. Bach: Phys. Rev. B **39**, 858 (1989)
1081. C.H. Pennington, D.J. Durand, C.P. Slichter, J.P. Rice, E.D. Bukowski, D.M. Ginsberg: Phys. Rev. B **39**, 274 (1989)
1082. J.M. Tranquada, G. Shirane, B. Keimer, S. Shamoto, M. Sato: Phys. Rev. B **40**, 4503 (1989)
1083. T. Schneider: Z. Phys. B **85**, 187 (1991)
1084. K. Kanoda, T. Takahashi, T. Kawagoe, T. Mizoguchi, S. Kagoshima, M. Hasumi: Jpn. J. Appl. Phys. **26**, L2018 (1987)
1085. W.C. Lee, D.C. Johnston: Phys. Rev. B **41**, 1904 (1990)
1086. N. Nishida, H. Miyatake, D. Shimada, S. Okuma, M. Ishikawa, T. Takabatake, Y. Nakazawa, Y. Kuno, R. Keitel, J.H. Brewer, T.M. Riseman, D.L. Williams, Y. Watanabe, T. Yamazaki, K. Nishiyama, K. Nagamine, E.J. Ansaldo, E. Torikai: J. Phys. Soc. Jpn. **57**, 597 (1988)
1087. J.M. Tranquada, A.H. Moudden, A.I. Goldman, P. Zolliker, D.E. Cox, G. Shirane, S.K. Sinha, D. Vaknin, D.C. Johnston, M.S. Alvarez, A.J. Jacobson, J.T. Lewandowski, J.M. Newsam: Phys. Rev. B **38**, 2477 (1988)
1088. C. Berthier, Y. Berthier, P. Butaud, W.G. Clark, J.A. Gillet, M. Horvatić, P. Segransan, J.Y. Henry: Physica C **185–189**, 1141 (1991)
1089. M. Takigawa, A.P. Reyes, P.C. Hammel, J.D. Thompson, R.H. Heffner, Z. Fisk, K.C. Ott: Phys. Rev. B **43**, 247 (1991)

1090. M. Takigawa, P.C. Hammel, R.H. Heffner, Z. Fisk: Phys. Rev. B **39**, 7371 (1989)
1091. C.P. Slichter: *Principles of Magnetic Resonance*, 3rd. edn., Springer Ser. Solid-State Sci., Vol. 1 (Springer, Berlin, Heidelberg 1989) p. 156
1092. N. Bourges, L.P. Regnault, Y. Sidis, C. Vettier: Phys. Rev. B **53**, 876 (1996)
1093. J. Rossat-Mignod, P. Bourges, F. Onufrieva, L.P. Regnault, J.Y. Henry, P. Burlet, C. Vettier: Physica B **199–200**, 281 (1994)
1094. A.J. Millis, H. Monien, D. Pines: Phys. Rev. B **42**, 167 (1990)
1095. T. Moriya: J. Phys. Soc. Jpn. **18**, 516 (1963)
1096. H.F. Fong, B. Keimer, D. Reznik, D.L. Milius, I. A. Aksay: Phys. Rev. B **54**, 6708 (1996)
1097. C. Hammel, M. Takigawa, R.H. Heffner, Z. Fisk, K.C. Ott: Phys. Rev. Lett. **63**, 1992 (1989)
1098. E.D. Crozier, N. Alberding, K.B. Bauchspiess, A.J. Seary, S. Gygax: Phys. Rev. B **36**, 8288 (1987)
1099. P. Boolchand, R.N. Enzweiler, I. Zitkovsky, J. Wells, W. Bresser, D. McDaniel, R.L. Meng, P.H. Hor, C.W. Chu, C.Y.Huang: Phys. Rev. B **37**, 3766 (1988)
1100. T. Yuen, C.L. Lin, J.E. Crow, G.N. Myer, R.E. Salomon, P. Schlottmann, N. Bykovetz, W.N. Herman: Phys. Rev. B **37**, 3770 (1988)
1101. S. Battacharya, M.J. Higgins, D.C. Johnston, A.J. Jacobson, J.P. Stokes, D.P. Goshorn, J.T. Lewandowski: Phys. Rev. Lett. **60**, 1181 (1988)
1102. J. Schützmann, S. Tajima, S. Miyamoto, Y. Sato, R. Hauff: Phys. Rev. B **52**, 13665 (1995)
1103. C.C. Homes, T. Timusk, D.A. Bonn, R. Liang, W.N. Hardy: Can. J. Phys. **73**, 663 (1995)
1104. D. Mihailovic, T. Mertelj, K.A. Müller: Phys. Rev. B **57**, 6116 (1998)
1105. R.G. Buckley, M.P. Staines, E.M. Haines, M. Ziaei, B.P. Clayman: Physica C **235–240**, 1237 (1994)
1106. J.L. Tallon, J.R. Cooper, P.S.I.P.N. de Silva, G.V.M. Williams, J.W. Loram: Phys. Rev. Lett. **75**, 4114 (1995)
1107. H. Ding, J.C. Campuzano, M.R. Norman, M. Randeria, T. Yokoya, T. Takahashi, T. Takeuchi, T. Mochiku, K. Kadowaki, P. Guptasarma, D.G. Hinks: J. Phys. Chem. Solids **59**, 1888 (1998)
1108. G.V.M. Williams, J.L. Tallon, J.W. Quilty, H.J. Trodahl, N.E. Flower: Phys. Rev. Lett. **80**, 377 (1998)
1109. L.D. Rotter, Z. Schlesinger, R.T. Collins, F. Holtzberg, C. Field, U.W. Welp, G.W. Crabtree, J.Z. Liu, Y. Fang, K.G. Vandervoort, S. Fleshler: Phys. Rev. Lett. **67**, 2741 (1991)
1110. E.T. Heyen, M. Cardona, J. Karpinski, E. Kaldis, S. Rusiecki: Phys. Rev. B **43**, 12958 (1991)
1111. A.P. Litvinchuk, C. Thomsen, M. Cardona: Solid State Commun. **83**, 343 (1992)
1112. J. Rossat-Mignod, L.P. Regnault, C. Vettier, P. Bourges, P. Burlet, J. Bossy, J.Y. Henry, G. Lapertot: Physica C **185–189**, 86 (1991)
1113. Y. Itoh, T. Machi, S. Adachi, A. Fukuoka, K. Tanabe, H. Yasuoka: J. Phys. Soc. Jpn. **67**, 312 (1998)
1114. S. Massida, J. Yu, A.J. Freeman, D.D. Koelling: Phys. Lett. **122**, 198 (1987)
1115. J. Yu, S. Massidda, A.J. Freeman, D.D. Koelling: Phys. Lett. **122**, 203 (1987)
1116. L.F. Mattheiss, D.R. Hamann: Solid State Commun. **63**, 395 (1987)
1117. F. Herman, R.V. Kasowski, W.Y. Hsu: Phys. Rev. B **36**, 6904 (1987)
1118. O.K. Andersen, O. Jepsen, A.I. Liechtenstein, I.I. Mazin: Phys. Rev. B **49**, 4145 (1994)

1119. A.J. Viescas, J.M. Tranquada, A.R. Moodenbaugh, P.D. Johnson: Phys. Rev. B **37**, 3738 (1988)
1120. A.J. Arko, R.S. List, R.J. Bartlett, S.W. Cheong, Z. Fisk, J.D. Thompson, C.G. Olson, A.B. Yang, R. Liu, C. Gu, B.W. Veal, J.Z. Liu, A.P. Paulikas, K. Vandervoort, H. Claus, J.C. Campzano, J.E. Schirber, N.D. Shinn: Phys. Rev. B **40**, 2268 (1989)
1121. C.H. Pennington, D.J. Durand, C.P. Slichter, J.P. Rice, E.D. Bukowski, D.M. Ginsberg: Phys. Rev. B **39**, 2902 (1989)
1122. R. Liu, B.W. Veal, A.P. Paulikas, J.W. Downey, P.J. Kostić, S. Fleshler, U. Welp, C.G. Olson, X. Wu, A.J. Arko, J.J. Joyce: Phys. Rev. B **46**, 11056 (1992)
1123. J.C. Campuzano, G. Jennings, M. Faiz, L. Beaulaigue, B.W. Veal, J.Z. Liu, A.P. Paulikas, K. Vandervoort, H. Claus, R.S. List, A.J. Arko, R.J. Bartlett: Phys. Rev. Lett. **64**, 2308 (1990)
1124. C.M. Fowler, B.L. Freeman, W.L. Hults, J.C. King, F.M. Mueller, J.L. Smith: Phys. Rev. Lett. **68**, 534 (1992)
1125. O.K. Andersen, A.I. Liechtenstein, O. Rodriguez, I.I. Mazin, O. Jepsen, V.P. Antropov, O. Gunnarsson, S. Gopalan: Physica C **185-189**, 147 (1991)
1126. W.E. Pickett, R.E. Cohen, H. Krakauer: Phys. Rev. B **42**, 8764 (1990)
1127. B. Friedl, C. Thomsen, H.-U. Harbermeier, M. Cardona: Solid State Commun. **81**, 989 (1992)
1128. W.E. Pickett: Rev. Mod. Phys. **61**, 433 (1989)
1129. P.B. Kirby, M.R. Harrison, W.G. Freeman, I. Samuel, M.J. Haines: Phys. Rev. B **36**, 8315 (1987)
1130. D.M. Krol, M. Stavola, W. Weber, L.F. Schneemeyer, J.V. Waszczak, S.M. Zahurak, S.G. Kosinski: Phys. Rev. B **36**, 8325 (1987)
1131. A. Yamanaka, F. Minami, K. Watanabe, K. Inoue, S. Takekawa, N. Iyi: Jpn. J. Appl. Phys. **26**, L1404 (1987)
1132. R. Liu, C. Thomsen, W. Kress, M. Cardona, B. Gegenheimer, F.W. de Wette, J. Prade, A.D. Kulkarni, U. Schroder: Phys. Rev. B **37**, 7971 (1988)
1133. R.M. Macfarlane, H.J. Rosen, E.M. Engler, R.D. Jacowitz, V.Y. Lee: Phys. Rev. B **38**, 284 (1988)
1134. L. Mihaly, L. Rosta, G. Coddens, F. Mezei, Gy. Hutiray, G. Kriza, B. Keszei: Phys. Rev. B **36**, 7137 (1987)
1135. J.J. Rhyne, D.A. Neumann, J.A. Gotaas, F. Beech, L.S. Lawrence, S. Wolf, M. Osofsky, D.U. Gubser: Phys. Rev. B **36**, 2294 (1987)
1136. B. Renker, F. Gompf, E. Gering, G. Roth, W. Reichardt, D. Evert, H. Rietschel, H. Mukta: Z. Phys. B **71**, 437 (1988)
1137. L. Pintschovius, N. Pyka, W. Reichardt, A. Yu. Rumiantsev, N. L. Mitrofanov, A.S. Ivanov, G. Collin, P. Bourges: Physica C **185–189**, 156 (1991)
1138. X.Z. Zhou, M. Raudsepp, Q.A. Pankhurst, A.H. Morrish, Y.L. Luo, I. Maartense: Phys. Rev. B **36**, 7230 (1987)
1139. H. You, J.D. Axe, X.B. Kan, S. Hashimoto, S.C. Moss, J.Z. Liu, G.W. Crabtree, D.J. Lam: Phys. Rev. B **38**, 9213 (1988)
1140. R. Zeyher: Z. Phys. B **80**, 187 (1990)
1141. R.E. Cohen, W.E. Pickett, H. Krakauer: Phys. Rev. Lett. **64**, 2575 (1990)
1142. I.I. Mazin, O.K. Andersen, A.I. Liechtenstein, O. Jepsen, V.P. Antropov, S.N. Rashkeev, V.I. Anisimov, J. Zaanen, C.O. Rodgiguez, M. Methfessel: In *Lattice Effects in High-$T_c$ Superconductors*, ed. by Y. Bar Yam, T. Egami, J. Mustre-de Leon, A.R. Bishop (World Scientific, Singapore, 1991) p. 235
1143. G.L. Zhao, J. Callaway: Phys. Rev. B **50**, 9511 (1994)
1144. Y. Shiina, Y.Oi. Nakamura: Solid State Commun. **76**, 1189 (1990)

1145. J.R. Schrieffer: *Theory of Superconductivity* (Addison-Wesley, Redwood City 1964)
1146. L. Pietronero, S. Strässler, C. Grimaldi: Phys. Rev. B **52**, 10516 (1995)
1147. A.R. Bishop, R.L. Martin, K.A. Müller, Z. Tesanovic: Z. Phys. B **76**, 17 (1989)
1148. I. Batistic, A.R. Bishop, R.L. Martin, Z. Tesanovic: Phys. Rev. B **40**, 6896 (1989)
1149. Y. Ito, H. Hasegawa, K. Takagi, K. Miyauchi: Jpn. J. Appl. Phys. **26**, L692 (1987)
1150. J.J. Reilly, M. Suenaga, J.R. Johnson, P. Thompson, A.R. Moodenbaugh: Phys. Rev. B **36**, 5694 (1987)
1151. I. Felner, I. Nomik, Y. Yeshurun: Phys. Rev. B **36**, 3923 (1987)
1152. T. Siegrist, L.F. Schneemeyer, J.V. Waszczak, N.P. Singh, R.L. Opila, B. Batlogg, L.W. Rupp, D.W. Murphy: Phys. Rev. B **36**, 8365 (1987)
1153. E. Takayama-Muromachi, Y. Uchida, K. Kato: Jpn. J. Appl. Phys. **26**, L2087 (1987)
1154. Y. Maeno, T. Nijima, Y. Aoki, M. Kato, K. Hoshino, A. Minami, T. Fujita: Jpn. J. Appl. Phys. **26**, L774 (1987)
1155. M. Hiratani, Y. Ito, K. Miyauchi, T. Kudo: Jpn. J. Appl. Phys. **26**, L1997 (1987)
1156. D.W. Murphy, S. Sunshine, R.B. van Dover, R.J. Cava, B. Batlogg, S.M. Zahurak, L.F. Schneemeyer: Phys. Rev. Lett. **58**, 1888 (1987)
1157. A. Ono, T. Tanaka, H. Nozaki, Y. Ishizawa: Jpn. J. Appl. Phys. **26**, L1687 (1987)
1158. R. Liang, Y. Inaguma, Y. Takagi, T. Nakamura: Jpn. J. Appl. Phys. **26**, L1150 (1987)
1159. A. Manthiram, X.X. Tang, J.B. Goodenough: Phys. Rev. B **37**, 3734 (1988)
1160. M. Oda, T. Murakami, Y. Enomoto, M. Suzuki: Jpn. J. Appl. Phys **26**, L804 (1987)
1161. S. Turumi, M. Hikita, T. Iwata, K. Semba, S. Kurihara: Jpn. J. Appl. Phys. **26**, L856 (1987)
1162. D.W. Murphy, S. Sunshine, R.B. van Dover, R.J. Cava, B. Batlogg, S.M. Zahurak, L.F. Schneemeyer: Phys. Rev. Lett. **58**, 1888 (1987)
1163. O.K. Alexandrov, M. François, T. Graf, K. Yvon: Physica C **172**, 491 (1991)
1164. J.L. Tallon, D.M. Pooke, R.G. Buckley, M.R. Presland, F.J. Blunt: Phys. Rev. B **41**, 7220 (1990)
1165. J.M. Liang, R.S. Liu, Y.T. Huang, S.F. Wu, P.T. Wu, L.J. Chen: Physica C **165**, 347 (1990)
1166. P.J.M. van Bentum, L.E.C. van de Leemput, L.W.M. Schreurs, P.A.A. Teunissen, H. van Kempen: Phys. Rev. B **36**, 843 (1987)
1167. H.F.C. Hoevers, P.J.M. van Bentum, L.E.C. van de Leemput, H. van Kempen, A.J.G. Schellingerhout, D. van der Marel: Physica C **152**, 105 (1988)
1168. A. Barone, A. Di Chihara, G. Peluso, U. Scotti di Uccio, A.M. Cucolo, R. Vaglio, F.C. Matacotta, E.Olzi: Phys. Rev. B **36**, 7121 (1987)
1169. J.S. Tsai, I. Takeuchi, J. Fujita, S. Miura, T. Terashima, Y. Bando, K. Iijima, K. Yamamoto: Physica C **157**, 537 (1989)
1170. M. Sera, S. Shamoto, M. Sato: Solid State Commun. **65**, 997 (1988)
1171. L.N. Bulaevskii, O.V. Dolgov, I.P. Kazakov, S.N. Maksimovskii, M.O. Ptitsyn, V.A. Stepanov, S.I. Vedeneev: Supercond. Sci. Technol. **1**, 205 (1988)
1172. M.A. Izbizky, M. Nunez Regueiro, P. Esquinazi, C. Fainstein: Phys. Rev. B **38**, 9220 (1988)

1173. R.T. Collins, Z. Schlesinger, R.A. Koch, R.B. Laibowitz, T.S. Plaskett, P. Freitas, W.J. Gallagher, R.L. Sandstrom, T.R. Dinger: Phys. Rev. Lett. **59**, 704 (1987)
1174. Z. Schlesinger, R.T. Collins, D.L. Kaiser, F. Holtzberg: Phys. Rev.Lett. **59**, 1958 (1987)
1175. R.T. Collins, Z. Schlesinger, F. Holtzberg, C. Field: Phys. Rev.Lett. **63**, 422 (1989)
1176. E. Seider, M. Bauer, L. Genzel, P. Wyder, A. Jansen, C. Richten: Solid State Commun. **72**, 85 (1989)
1177. T. Ekino, T. Minami, H. Fujii, J. Akimitsu: Physica C **235–240**, 1899 (1994)
1178. E.T. Heyen, R. Liu, C. Thomsen, R. Kremer, M. Cardona: Phys. Rev. B **41**, 11058 (1990)
1179. M. Oda, Y. Hidaka, M. Suzuki, T. Murakami: Phys. Rev. B **38**, 252 (1988)
1180. U. Welp, W.K. Kwok, G.W. Crabtree, K.G. Vandervoort, J.Z. Liu: Phys. Rev. Lett. **62**, 1908 (1989)
1181. L. Krusin-Elbaum, A.P. Malozemoff, Y. Yeshurun, D.C. Cronemeyer, F. Holtzberg: Phys. Rev. B **39**, 2936 (1989)
1182. S.J. Hibble, A.K. Cheetham, A.M. Chippindale, P. Day, J.A. Hriljac: Physica C **156**, 604 (1988)
1183. Y. Koike, T. Nakanomyo, T. Hukase: Jpn. J. Appl. Phys. **27**, L841 (1988)
1184. A. Katsui: Jpn. J. Appl. Phys. **27**, L844 (1988)
1185. L.F. Schneemeyer, R.B. van Dover, S.H. Glarum, S.A, Sunshine, R.M. Fleming, B. Batlogg, T. Siegrist, J.H. Marshall, J.V. Waszczak, L.W. Rupp: Nature **332**, 423 (1988)
1186. A. Ono, S. Sueno, F.P. Okamura: Jpn. J. Appl. Phys. **27**, L786 (1988)
1187. H. Takagi, H. Eisaki, S. Uchida, A. Maeda, S. Tajima, K. Uchinokura, S. Tanaka: Nature **332**, 236 (1988)
1188. S.A, Sunshine, T. Siegrist, L.F. Schneemeyer, D.W. Murphy, R.J. Cava, B. Batlogg, R.B. van Dover, R.M. Fleming, S.H. Glarum, S. Nakahara, R. Farrow, J.J. Krajewski, S.M. Zahurak, J.V. Waszczak, J.H. Marshall, P. Marsh, L.W. Rupp, Jr., W.F. Peck: Phys. Rev. B **38**, 893 (1988)
1189. J.M. Tarascon, Y. le Page, P. Barboux, B.G. Bagley, L.H. Greene, W.R. McKinnon, G.W. Hull, M. Giround, D.M. Hwang: Phys. Rev. B **37**, 9382 (1988)
1190. M. Onoda, A. Yamamoto, E. Takayama-Muromachi, S. Takekawa: Jpn. J. Appl. Phys. **27**, L833 (1988)
1191. T. Kajitani, K. Kusaba, M. Kikuchi, N. Kobayashi, Y. Syono, T.B. Williams, M. Hirabayashi: Jpn. J. Appl. Phys. **27**, L587 (1988)
1192. M.A. Subramanian, J.C. Calabrese, C.C. Torardi, J. Gopalakrishnan, T.R. Askew, R.B. Flippen, K.J. Morrissey, U. Chowdhry, A.W. Sleight: Nature **332**, 420 (1988)
1193. M.A. Subramanian, J.B. Parise, J.C. Calabrese, C.C. Torardi, J. Gopalakrishnan, A.W. Sleight: J. Solid State Chem. **77**, 192 (1988)
1194. D.E. Cox, C.C. Torardi, M.A. Subramanian, A.W. Sleight: Phys. Rev. B **38**, 6624 (1988)
1195. Y. Matsui, H. Maeda, Y. Tanaka, E. Takayama-Muromachi, S. Takekawa, S. Horiuchi: Jpn. J. Appl. Phys. **27**, L827 (1988)
1196. S. Iijima, T. Ichihashi, Y. Shimakawa, T. Manako, Y. Kubo: Jpn. J. Appl. Phys. **27**, L1061 (1988)
1197. Y. le Page, W.R. McKinnon, J.-M. Tarascon, P. Barboux: Phys. Rev. B **40**, 6810 (1989)
1198. P.G. Allen, J. Mustre de Leon, S.D. Conradson, A.R. Bishop: Phys. Rev. B **44**, 9480 (1991)

1199. Shimakawa, Y. Kubo, T. Manako, T. Satoh, S. Iijima, T. Ichihara, H. Igarashi: Physica C **157**, 279 (1989)
1200. D.M. de Leeuw, W.A. Groen, J.C. Jol, H.B. Brom, H.W. Zandbergen: Physica C **166**, 349 (1990)
1201. E. Sonder, B.C. Chakoumakos, B.C. Sales: Phys. Rev. B **40**, 6872 (1989)
1202. T. Watanabe, T. Fujii, A. Matsuda: Phys. Rev. Lett. **79**, 2113 (1997)
1203. G. Triscone, J.-Y. Genoud, T. Graf, A. Junod, J. Muller: Physica C **176**, 247 (1991)
1204. R.G. Buckley, J.L. Tallon, I.W.M. Brown, M.R. Presland, N.E. Fowler, P.W. Gilberd, M. Browden, N.B. Milestone: Physica C **156**, 629 (1988)
1205. A. Maeda, T. Yabe, S. Takebayashi, M. Hase, K. Uchiokura: Phys. Rev. B **41**, 4112 (1990)
1206. T. Tamegai, K. Koga, K. Suzuki, M. Ichihara, F. Sakai, Y. Iye: Jpn. J. Appl. Phys. **28**, L112 (1989)
1207. N. Fukushima, H. Niu, K. Ando: Jpn. J. Appl. Phys. **27**, L790 (1988)
1208. R. Yoshizaki, Y. Saito, Y. Abe, H. Ikeda: Physica C **152**, 408 (1988)
1209. B. Jayaram, P.C. Lanchester, M.T. Weller: Physica C **160**, 17 (1989)
1210. H. Katayama-Yoshida, T. Hirooka, A. Oyamada, Y. Okabe, T. Takahashi, T. Sasaki, A. Ochiai, T. Suzuki, A.J. Mascarenhas, J.I. Pankove, T.F. Ciszek, S.K. Deb, R.B. Goldfarb, Y. Li: Physica C **156**, 481 (1988)
1211. H. Ding, M.R. Norman, J.C. Campuzano, M. Randeria, A.F. Bellman, T. Yokoya, T. Takahashi, T. Mochiku, K. Kadowaki: Phys. Rev. B **54**, R9678 (1996)
1212. Z.-X. Shen, D.S. Dessau, B.O. Wells, D.M. King, W.E. Spicer, A.J. Arko, D. Marshall, L.W. Lombardo, A. Kapitulnik, P. Dickinson, S. Doniach, J. DiCarlo, A.G. Loeser, C.H. Park: Phys. Rev. Lett. **70**, 1553 (1993)
1213. B.O. Wells, Z.-X. Shern, D.S. Dessau, W.E. Spicer, D.B. Mitzi, L. Lambardo, A. Kapitulnik, A.J. Arko: Phys. Rev. B **46**, 11830 (1992)
1214. J. Ma, C. Quitmann, R.J. Kelley, H. Berger, G. Margaritondo, M. Onellion: Science **267**, 862 (1995)
1215. R.J. Kelly, C. Quitmann, M. Onellion, H. Berger, P. Almeras, G. Margaritondo: Science **271**, 1255 (1996)
1216. C. Renner, *Φ*. Fisher: Phys. Rev. B **51**, 9208 (1995)
1217. K. Ichimura, K. Nomura: J. Phys. Soc. Jpn. **62**, 3661 (1993)
1218. B. Barbiellini, *Φ*. Fisher, M. Peter, C. Renner, M. Weger: Physica C **220**, 55 (1994)
1219. J. Kane, K.-W. Ng: Phys. Rev. B **53**, 1 (1996)
1220. A.F. Andreev: Sov. Phys.-JETP **19**, 1228 (1964)
1221. G.E. Blonder, M. Tinkham, T.M. Klapijwk: Phys. Rev. B **25**, 4515 (1982)
1222. G. Varelogiannis, L. Pietronero: Phys. Rev. B **52**, R15753 (1995)
1223. W.L. McMillan, J.M. Rowell: Phys. Rev. Lett. **14**, 108 (1965)
1224. G.M. Eliashberg: Sov. Phys.-JETP **11**, 696 (1960)
1225. R.S. Gonnelli, G.A. Ummarino, V.A. Stepanov: Physica C **275**, 162 (1997)
1226. D. Shimada, Y. Shiina, A. Mottate, Y. Ohyagi, N. Tsuda: Phys. Rev. B **51**, 16495 (1995)
1227. Y. Shiina, D. Shimada, A. Mottate, Y. Ohyagi, N. Tsuda: J. Phys. Soc. Jpn. **64**, 2577 (1995)
1228. S.I. Vedeneev, A.A. Tsvetkov, A.G.M. Jansen, P. Wyder: Physica C **235–240**, 1851 (1994)
1229. R.S. Gonnelli, S.I. Vedeneev, O.V. Dolgov, G.A. Ummarino: Physica C **235–240**, 1861 (1994)
1230. N. Miyakawa, Y. Shiina, T. Kaneko, N. Tsuda: J. Phys. Soc. Jpn. **62**, 2445 (1993)

1231. B. Renker, F. Gompf, D. Evert, P. Adelman, H. Schmidt, E. Gering, H. Hinks: Z. Phys. B **77**, 65 (1989)
1232. D. Shimada, N. Tsuda, U. Paltzer, F.W. de Wette: Physica C **298**, 195 (1998)
1233. D. Shimada, N. Miyakawa, K. Ajiki, K. Hagiwara, N. Tsuda: *Electron–Phonon interaction in Oxide Superconductors*, ed. by R. Baquero (World Scientific, Singapore, 1991) p. 56
1234. R. Koltun, M. Hoffmann, P.C. Splittergerber-Hunnekes, C. Jarchow, G. Gunterodt, V.V. Moschalkov, L.I. Leonyuk: Z. Phys. B **82**, 53 (1991)
1235. Y. de Wilde, N. Miyakawa, P. Guptasarma, M. Iavarone, L. Ozyuzer, J.F. Zasadzinski, P. Romano, D.G. Hinks, C. Kendziora, G.W. Crabtree, K.E. Gray: Phys. Rev. Lett. **80**, 153 (1998)
1236. N. Miyakawa, P. Guptasarma, J.F. Zasadzinski, D.G. Hinks, K.E. Gray: Phys. Rev. Lett. **80**, 157 (1998)
1237. D. Coffey, L. Coffey: Phys. Rev. Lett. **70**, 1529 (1993)
1238. M.R. Norman, H. Ding: Phys. Rev. B **57**, R11089 (1998-II)
1239. C. Renner, B. Revaz, J.-Y. Genoud, K. Kadowaki, $\Phi$. Fisher: Phys. Rev. Lett. **80**, 149 (1998)
1240. M. Oda, K. Hoya, R. Kubota, C. Manabe, N. Momono, T. Nakano, M. Ido: Physica C **281**, 135 (1997)
1241. Y. Ando, G.S. Boebinger, A. Passner, N.L. Wang, C. Geibel, F. Steglich: Phys. Rev. Lett. **77**, 2065 (1996)
1242. Y. Kubo, T. Manako: Physica C **197**, 378 (1992)
1243. D.S. Dessau, Z.-X. Shen, D.M. King, D.S. Marshall, L.W. Lombardo, P.H. Dickinson, A.G. Loeser, J. DiCarlo, C.H. Park, A. Kapitulnik, W.E. Spicer: Phys. Rev. Lett. **71**, 2781 (1993)
1244. A.G. Loeser, D.S. Dessau, Z.-X. Shen: Physica C **263**, 208 (1996)
1245. P. Aebi, J. Osterwalder, P. Schwaller, L. Schlapbach, M. Shimoda, T. Mochiku, K. Kadowaki: Phys. Rev. Lett. **72**, 2757 (1994)
1246. J. Ma, C. Quitmann, R.J. Kelley, P. Alméras, H. Berger, G. Margaritonndo, M. Onellion: Phys. Rev. B **51**, 3832 (1995)
1247. H. Ding, M.R. Norman, T. Yokoya, T. Takeuchi, M. Randeria, J.C. Campuzano, T. Takahashi, T. Mochiku, K. Kadowaki: Phys. Rev. Lett. **78**, 2628 (1997)
1248. S. Massidda, J. Yu, A.J. Freeman: Physica C **152**, 251 (1988)
1249. S. Chakravarty: Phys. Rev. Lett. **74**, 1885 (1995)
1250. D.H. Lu, M. Schmidt, T.R.Cummins, S. Schuppler, F. Lichtenberg, J.G. Bednorz: Phys. Rev. Lett. **76**, 4845 (1996)
1251. H. Ding, T. Yokoya, J.C. Campuzano, T. Takahashi, M. Randeria, M.R. Norman, T. Mochiku, K. Kadowaki, J. Giapintzakis: Nature **382**, 51 (1996)
1252. D.S. Marshall, D.S. Dessau, A.G. Loeser, C.-H. Park, A.Y. Matuura, J.N. Eckstein, I. Bozovic, P. Fournier, A. Kapitulnik, W.E. Spicer, Z.-X. Shen: Phys. Rev. Lett. **76**, 4841 (1996)
1253. Z.-X. Shen, P.J. White, D.L. Feng, C. Kim, G.D. Gu, H. Ikeda, R. Yoshizaki, N. Koshizuka: Science **280**, 259 (1998)
1254. M. Suzuki, S. Karimoto, K. Namekawa: J. Phys. Soc. Jpn. **67**, 732 (1998)
1255. I. Kosztin, Q. Chen, B. Jankó, K. Levin: Phys. Rev. B **58**, R5936 (1998)
1256. U. Paltzer, F.W. de Wette, U. Schröder, E. Rampt: Physica C **301**, 55 (1998)
1257. J. Prade, A.D. Kurkarni, F.W. de Wette, U. Schröder, W. Kress: Phys. Rev. B **39**, 2771 (1989)
1258. A.D. Kurkarni, J. Prade, F.W. de Wette: Phys. Rev. B **40**, 2642 (1989)
1259. A.D. Kurkarni, F.W. de Wette, J. Prade, U. Schröder, W. Kress: Phys. Rev. B **41**, 6409 (1990)
1260. T. Den, A. Yamazaki, J. Akimitsu: Jpn. J. Appl. Phys. **27**, L1620 (1988)

1261. S. Ikeda, H. Ichinose, T. Kimura, T. Matsumoto, H. Maeda, Y. Ishida, K. Ogawa: Jpn. J. Appl. Phys. **27**, L999 (1988)
1262. T. Tamegai, A. Watanabe, K. Koga, I. Oguro, Y. Iye: Jap. J. Appl. Phys. **27**, L1074 (1988)
1263. N. Fukushima, H. Niu, K. Ando: Jpn. J. Appl. Phys. **27**, L1432 (1988)
1264. M.A. Subramanian, A.R. Strzelecki, J. Gopalakrishnan, A.W. Sleight: J. Solid State Chem. **77**, 196 (1988)
1265. G.S. Grader, E.M. Gyorgy, P.K. Gallagher, H.M. O'Bryan, D.W. Johnson, Jr., S. Sunshine, S.M. Zahurak, S. Jin, R.C. Sherwood: Phys. Rev. B **38**, 757 (1988)
1266. P.V.P.S.S. Sastry, I.K. Gopalakrishnan, J.V. Yakhmi, R.M. Iyer: Physica C **157**, 491 (1989)
1267. Y. Ikeda, M. Takano, Z. Hiroi, K. Oda, H. Kitaguchi, J. Takada, Y. Miura, Y. Takeda, O. Yamamoto, H. Mazaki: Jap. J. Appl. Phys. **27**, L2067 (1988)
1268. E.A. Hewat, J.J. Capponi, R.J. Cava, C. Chaillout, M. Marezio, J.L. Tholence: Physica C **157**, 509 (1989)
1269. R.J. Cava, B. Batlogg, J.J. Krajewski, L.W. Rupp, L.F. Schneemeyer, T. Siegrist, R.B. van Dover, P. Marsh, W.F. Peck, Jr., P.K. Gallagher, S.H. Glarum, J.H. Marshall, R.C. Farrow, J.V. Waszczak, R. Hull, P. Trevor: Nature **336**, 211 (1988)
1270. S. Nakajima, M. Kikuchi, Y. Shono, T. Oku, D. Shindo, K. Hiraga, N. Kobayashi, H. Iwasaki, Y. Muto: Physica C **158**, 471 (1989)
1271. H. Ihara, R. Sugise, M. Hirabayashi, N. Terada, M. Jo, K. Hayashi, A. Negishi, M. Tokumoto, Y.Kimura, T. Shimomura: Nature **334**, 510 (1988)
1272. R.M. Hazen, L.W. Finger, R.J. Angel, C.T. Prewitt, N.L. Ross, C.G. Hadidiacos, P.J. Heaney, D.R. Veblen, Z.Z. Sheng, A. El Ali, A.M. Hermann: Phys. Rev. Lett. **60**, 1657 (1988)
1273. S.S.P. Parkin, V.Y. Lee, E.M. Engler, A.I. Nazzal, T.C. Huang, G. Gorman, R. Savoy, R. Beyers: Phys. Rev. Lett. **60**, 2539 (1988)
1274. A.K. Ganguli, K.S. Nanjundaswamy, C.N.R. Rao: Physica C **156**, 788 (1988)
1275. C. Martin, J. Provost, D. Bougault, B. Domenges, C. Michel, M. Hervieu, B. Raveau: Physica C **157**, 460 (1989)
1276. R.S. Liu, Y.T. Huang, W.H. Lee, S.F. Wu, P.T. Wu: Physica C **156**, 791 (1988)
1277. E.A. Hayri, M. Greenblatt: Physica C **156**, 775 (1988)
1278. A. Fukuoka, A. Tokiwa-Yamamoto, M. Itoh, R. Usami, S. Adachi, K. Tanabe: Phys. Rev. B **55**, 6612 (1997)
1279. J.M. Tarascon, Y. Le Page, P. Barvoux, B.G. Bagley, L.H. Greene, W.R. McKinnon, G.W. Hull, M. Giroud, D.M. Hwang: Phys. Rev. B **38**, 8885 (1988)
1280. T. Yoshitake, T. Satoh, Y. Kubo, T. Manako, H. Igarashi: Jpn. J. Appl. Phys. **27**, L1094 (1988)
1281. Y. Hidaka, M. Oda, M. Suzuki, Y. Maeda, Y. Enomoto, T. Murakami: Jpn. J. Appl. Phys. **27**, L538 (1988)
1282. T. Ekino, J. Akimitsu: Phys. Rev. B **40**, 6902 (1989)
1283. H. Ikuta, A. Maeda, K. Uchinokura, S. Tanaka: Jpn. J. Appl. Phys. **27**, L1038 (1988)
1284. S. Vieira, M.A. Ramos, M. Vallet-Regi, J.M. Gonzalez-Calbert: Phys. Rev. B **38**, 9295 (1988)
1285. N. Miyakawa, D. Shimada, T. Kido, N. Tsuda: J. Phys. Soc. Jpn. **58**, 383 (1989)
1286. M. Lee, D.B. Mitzi, A. Kapitulnik, M.R. Beasley: Phys. Rev. B **39**, 801 (1989)

1287. A. Yamanaka, T. Kimura, F. Minami, K. Inoue, S. Takekawa: Jpn. J. Appl. Phys. **27**, L1902 (1988)
1288. D. Kirillov, I. Bozovic, T.H. Geballe, A. Kapitulnik, D.B. Mitzi: Phys. Rev. B **38**, 11955 (1988)
1289. C.G. Olson, R. Liu, D.W. Lynch, B.W. Veal, Y.C. Chang, P.Z. Jiang, J.Z. Liu, A.P. Paulikas, A.J. Arko, R.S. List: Physica C **162–164**, 1697 (1989)
1290. J.-M. Imer, F. Patthey, B. Dardel, W.-D. Schneider, Y. Bauer, Y. Betroff, A. Zettl: Phys. Rev. Lett. **62**, 336 (1989)
1291. R. Escudero, E. Guarner, F. Morales: Physica C **166**, 15 (1990)
1292. N. Miyakawa, D. Shimada, T. Kido, N. Tsuda: J. Phys. Soc. Jpn. **59**, 2473 (1990)
1293. S.I. Vedeneev, V.A. Stepanov: JETP Lett. **49**, 588 (1989)
1294. G. Briceno, A. Zettl: Solid State Commun. **70**, 1055 (1989)
1295. M. Ohuchi, D. Shimada, N. Tsuda: Jpn. J. Appl. Phys. **32**, L251 (1993)
1296. D. Shimada, K. Ajiki, N. Miyakawa, K. Hagiwara, N. Tsuda: Z. Phys. B **85**, 7 (1991)
1297. J. Moreland, D.S. Ginley, E.L. Venturini, B. Morosin: Appl. Phys. Lett. **55**, 1463 (1989)
1298. S. Vieira, J.G. Rodrigo, M.A. Ramos, K.V. Rao, Y. Makino: Phys. Rev. B **40**, 11403 (1989)
1299. M. Suzuki, H. Inoue, L. Rinderer: Phys. Rev. B **41**, 7217 (1990)
1300. N. Miyakawa, J.F. Zasadzinski, L. Ozyuzer, P. Guptasarma, D.G. Hinks, C. Kendziora, K.E. Gray: Phys. Rev. Lett. **83**, 1018 (1999)
1301. Y. Maeno, S. Nishizaki, Z.Q. Mao: J. Supercond. **12** 535 (1999)
1302. D.B. Rogers, R.D. Shannon, C.T. Prewitt, J.L. Gillson, Inorg. Chem. **10**, 723 (1971)
1303. M. Tanaka, M. Hasegawa, H. Takei: J. Phys. Soc. Jpn. **65**, 3973 (1996)

# Index

To avoid excessive length, the subject index does not refer exhaustively to all the various properties of each material mentioned in this book. Thus, the reader is advised to consult the table of contents and/or the respective tables.


1/8 suppression 277, 315
2H-$NbSe_2$ 87

absorption edge 175
ac conductivity 204, 256
adiabatic approximation 11, 62, 64, 68, 89
Anderson localization 15, 152, 153, 197
Anderson's condition 251
Anderson-lattice model 122, 224
Andreev reflection 311
anharmonic Peierls–Hubbard model 100
anharmonic phonon 98, 99, 101, 104
antibonding orbital 9, 165, 166
antiferro-orbital 142
antiferromagnetic correlation 299
antiferromagnetic insulator 72

$BaBiO_3$ 83, 85
band-filling-control metal–insulator transition 144
bandwidth-control metal–insulator transition 144
BCS superconductivity 94, 98, 104, 185, 279
bipolaron 51, 67, 68, 88, 91, 92, 188, 205
BKBO 51, 87, 98, 287
Bloch orbital 132
Bloch–De Dominicis theorem 113
bonding orbital 9, 165, 166
Born–Oppenheimer approximation 11
Bose condensation 93
Bouckaert–Smolchowski–Wigner's symbol 7
breathing mode 51, 84, 286
broken symmetry 62, 74
BSCCO 305
BTK theory 311

catalytic reaction 167
charge density wave, CDW 37, 58, 68, 71, 72, 75, 87, 98, 99, 101, 210
charge ordering 237
charge separation 69
charge stripe 277
charge-transfer energy 134
charge-transfer type insulator 135, 138, 225
cluster model 223
coherence 94, 266, 293
coherence length 185, 295, 310
colossal magnetoresistance 47, 230
compression 250
conductivity 170, 172, 218, 252, 269
configuration interaction, CI 77, 126, 136
$CoPdO_2$ 53
correlation effect 123, 126, 268, 270
corundum structure 44
Coulomb gap 155, 192, 197
Coulomb integral 121
covalency 7, 134
critical field 185, 280, 281, 296, 310
critical separation 36
crystal structure 199, 290, 308
crystal-field splitting 215
Curie–Weiss law 18, 264
cyclotron resonance 163, 169, 302

De Haas–van Alphen oscillation 302

Debye temperature 159, 181, 183, 285, 304
deformation potential approximation 12, 61, 172, 180
density of states, DOS 129, 155, 192, 259, 261, 287
depression in the electronic spectrum 301
$d\epsilon$ orbital 6
$d\epsilon$ conductor 39, 52, 158, 271
$d\gamma$ orbital 6
$d\gamma$ conductor 39, 49, 271
dimensionality 62, 204
direct gap 86, 169
double exchange interaction 18, 48, 143
double well potential 66, 310
$d$–$p$ hybridization 223
drift mobility 161, 173, 181, 193, 218
Drude model 252
$d$-wave order parameter 272, 293, 305, 311
dynamical mean field 150

effective mass 62
$E_g$ symmetry 7
Einstein temperature 193
Einstein's relation 256
elastic constant 263, 285
electrochromic material 53, 195
electron localization 282
electron–electron interaction 10, 57, 67, 70, 77, 107, 183, 184, 212, 248, 282, 305, 314
electron–hole correlation 79
electron–phonon interaction 11, 57, 60, 67, 70, 77, 87, 105, 158, 180, 181, 193, 198, 261, 263, 285, 286, 304, 314
electron–phonon interaction strength 14, 51, 161, 183
Eliashberg equation 105, 305, 313
energy band structure 10, 41, 42, 45, 162, 165, 166, 169, 180, 184, 259, 275, 301
energy gap 72, 130, 138, 223, 259, 279, 293, 312
enhancement of states 13, 62, 161, 183
EPR 207
exchange integral 121
exchange potential 124
exciton 65, 70, 71, 76, 77, 80, 85, 117
extraordinary Hall coefficient 254

$Fe_3O_4$ 151, 243
Fermi liquid theory 146
Fermi surface 160, 260, 301, 315
ferrimagnet 247
ferrite 243
ferro-orbital 142
first-principles band-structure calculations 126
fluctuation 49, 75, 96, 99, 107, 257, 265, 297
Franck–Condon principle 65
Fröhlich Hamiltonian 12, 13
Fröhlich superconductivity 199
frozen-phonon model 286
frustration 250

gas sensor 174
Grassmann algebra 113

Hall coefficient 44, 160, 178, 253, 276, 290, 296
Hall mobility 16, 17, 161, 171, 174, 179, 193, 200, 315
harmonic approximation 99
Hartree–Fock approximation 124, 127, 133
heavy Fermion 150, 191
Hebel–Slichter peak 295
high pressure 271, 278, 293
high-spin state, HS 42, 50
high-$T_c$ superconductor 270
HMMC 74, 117
hole concentration 277, 290
hopping 15, 17, 181
Hubbard model 122
Hubbard–Stratonovitch transformation 96
hybridization 133

incoherent part of the spectral function 150
independent-particle approximation 119
indirect gap 86
inelastic tunneling 313
infrared absorption 284
internal strain 245
inverse adiabatic limit 89
ionic radius 42
isotope effect 98, 100, 101, 104, 106, 182, 240, 263, 279, 291, 300, 310
itineracy 57, 60, 267, 268

Jahn–Teller distortion 231, 240, 276

$K_2NiF_4$ structure 52
Kadowaki–Woods relation 150
Kanamori parameter 127, 131
$\kappa$ 295
Knight shift 202, 294
Kondo effect 11, 144, 282
Koopmans' theorem 125
Korringa relation 298

$La_{1-x}Sr_xMO_3$ 47, 48, 151, 231
$La_{2-x}M_xCuO_4$ 272, 276
large polaron 13, 61, 168, 191, 193
LDA+$U$ method 228, 235
ligand field 5, 128, 129, 222
light absorption 116
light emission spectrum 65
linear combination of atomic orbitals, LCAO 10, 169
linear coupling between electrons and phonons 118
linear lattice relaxation 76
$LiTi_2O_4$ 181
Little-type superconductor 199
$LiV_2O_4$ 181, 190
local density approximation, LDA 120, 126, 227, 235
local lattice distortion 65, 263
logarithmic peak in the density of states 167
logarithmic temperature dependence 282
long-range interaction 61
long-range ordering 251
low-dimensional 55, 71, 199, 209
low-spin state, LS 42, 50
Luttinger's theorem 147

$M_xV_2O_5$ 199
Magnéli phases 45
magnetic polaron 48
magnetic susceptibility 11, 17, 20, 41, 52, 68, 183, 186, 191, 205, 280, 282, 283, 297
magnetoresitance 232
magnon 75
Matthiessen rule 160
McMillan's formula for $T_c$ 183
mean field approximation 119
mean field effect 123
metal–insulator transition, MIT 19, 37, 38, 45, 46, 48, 50, 52, 84, 100, 101, 144, 182, 190, 195, 197, 209, 213, 231, 250, 268, 272
metallic conductor 38
mid-gap absorption 84
mobility 15, 161, 172, 177, 202, 216, 256
mobility edge 15, 153
momentum distribution function 147, 167
Moriya's theory 19, 50, 146, 284
motional narrowing 259, 266
Mott insulator 123, 131
Mott transition 19
Mott–Hubbard splitting 229
Mott–Hubbard type insulator 135, 138
Mulliken's symbol 7
multi-band Hubbard model 122
multiplet correction 130, 137
multiplet theory 129
muon 278, 284, 298
$M_xMoO_3$ 199, 209
$M_xV_2O_5$ 69, 199
$M_xWO_3$ 52, 53, 191

NaCl structure 40
NCCO 287
Néel temperature, $T_N$ 141, 216, 298
negative charge-transfer-energy insulator 135
neutron scattering 265, 284, 298
NMR 207, 256
non-Fermi liquid 315
non-Grassmann path integral theory 107
non-zero density of states 186, 207
nonadiabatic effect 64
nonbonding orbital 165, 166
nonlinear lattice relaxation 76, 80

octahedron 5
one-body Green's function 110
one-dimensional 64, 71, 78, 116, 167, 200, 209
one-dimensional charge density wave 76, 80
one-electron approximation 119, 123
optical conductivity 236, 242, 268
optical phonon 172
orbital diamagnetism 18
orbital ordering 142, 236
order–disorder transformation 248
overlap integral 165
oxide semiconductor 168

$p$–wave order parameter 52, 271

pair-distribution-function analysis, PDF 239
Pauli paramagnetism 18, 45, 161
$PbMo_6S_8$ 87, 281
Peierls instability 36, 72, 77, 80, 209
Peierls–Hubbard model 78, 84, 85
penetration depth 185
percolation model 37, 196
perovskite structure 47
phase diagram 75, 90, 103, 192, 244, 282, 298
phonon 89, 285, 302, 319
phonon density of states 286, 303, 319
phonon softening 285, 286, 290, 299, 304
photoemission 214, 222, 262, 301
plasma frequency 175
polaron 13, 61, 263, 266
positron annihilation 167
pseudogap 296, 300, 314, 315
pyrochlore structure 53

quasi-particle 147

Racah parameter 131
Raman scattering 310
randomness 15, 197, 206
reflectivity 252
$ReO_3$ structure 158
resistivity 14, 20, 68, 233, 245, 255, 296, 315
RKKY interaction 18, 144
$RuO_2$ 45
rutile 45, 169, 177

*s*–*p* conductor 271
*s*–wave order parameter 281, 293
screening of the Coulomb interaction 19, 150, 192, 250
screening of the electron–phonon interaction 198
Seebeck coefficient 16, 188, 219, 241, 254, 300
self-energy 148
self-interaction-corrected LDA, SIC-LDA 126, 227
self-trapped exciton 69, 77
self-trapping 62, 65
short-range interaction 61
short-range ordering 251, 269
single-band Hubbard model 123
single-body problem 60
Slater integral 130
small polaron 15, 168, 179, 180, 202, 268
$SnO_2$ 168
soliton 71, 78, 80, 82
*s*–*p* conductor 39
specific heat 183, 185, 188, 206, 207, 280, 284, 295, 301, 302
spectral function of the electron–electron or –phonon interaction 314
spin correlation length 284
spin density 261
spin density wave, SDW 58, 72, 75, 100, 101, 116
spin glass 206
spin ladder 55, 271
spin Peierls transition 37
spin soliton 82
spin structure 273
spin–lattice relaxation 284, 294, 298
spinel structure 54, 246
$Sr_2RuO_4$ 271, 302
Stokes shift 66
strong–coupling electron–phonon interaction 60, 62, 68, 73, 88, 94, 97, 270
strong–coupling superconductivity 286
superconducting critical temperature 52, 94, 97, 277, 278, 290, 292, 306
superconductivity 87
superexchange interaction 17, 140
superfluid 88, 93, 94
surface states 167
symmetry of the order parameter, see also *d*-, *p*-, and *s*–wave 311

$T_{2g}$ symmetry 7, 163
temperature-independent paramagnetism 18, 187
tetragonal symmetry 9
tetrahedron 5, 246
thermal conductivity 295
thermal expansion 159
thermal expansion coefficient 304
three-dimensional charge density wave 83
tight binding approximation 10, 163
$Ti_4O_7$ 68, 243, 247
$TiO_2$ 168, 177
tolerance factor 48
transfer integral 121
transparent metal 168
triplet superconductivity 52

tunneling 15, 66, 181, 287, 311, 317
two-body problem 67, 69
two-dimensional 64, 88, 167, 198, 211, 282

ultraviolet photoelectron spectroscopy, UPS, ARPES 148, 184, 190, 285, 310, 315

van Hove singularity 302, 315
Van Vleck paramagnetism 18, 162
variable range hopping, VRH 15, 188, 196, 203, 255, 274
Verwey transition 152, 243
$VO_2$ 37, 47
$V_2O_3$ 45

weak–coupling electron–phonon interaction 59, 61, 87, 95
weak–coupling superconductor 181, 186
WHH equation 281, 295
Wigner crystal 152
Wilson ratio 145
$WO_3$ 191

YBCO 288
$YFe_2O_4$ 243, 247, 250

Zaanen–Sawatzky–Allen diagram 138

# Springer Series in Solid-State Sciences

Editors: M. Cardona P. Fulde K. von Klitzing H.-J. Queisser

1 **Principles of Magnetic Resonance** 3rd Edition By C. P. Slichter
2 **Introduction to Solid-State Theory** By O. Madelung
3 **Dynamical Scattering of X-Rays in Crystals** By Z. G. Pinsker
4 **Inelastic Electron Tunneling Spectroscopy** Editor: T. Wolfram
5 **Fundamentals of Crystal Growth I** Macroscopic Equilibrium and Transport Concepts By F. E. Rosenberger
6 **Magnetic Flux Structures in Superconductors** By R. P. Huebener
7 **Green's Functions in Quantum Physics** 2nd Edition By E. N. Economou
8 **Solitons and Condensed Matter Physics** Editors: A. R. Bishop and T. Schneider
9 **Photoferroelectrics** By V. M. Fridkin
10 **Phonon Dispersion Relations in Insulators** By H. Bilz and W. Kress
11 **Electron Transport in Compound Semiconductors** By B. R. Nag
12 **The Physics of Elementary Excitations** By S. Nakajima, Y. Toyozawa, and R. Abe
13 **The Physics of Selenium and Tellurium** Editors: E. Gerlach and P. Grosse
14 **Magnetic Bubble Technology** 2nd Edition By A. H. Eschenfelder
15 **Modern Crystallography I** Fundamentals of Crystals Symmetry, and Methods of Structural Crystallography 2nd Edition By B. K. Vainshtein
16 **Organic Molecular Crystals** Their Electronic States By E. A. Silinsh
17 **The Theory of Magnetism I** Statics and Dynamics By D. C. Mattis
18 **Relaxation of Elementary Excitations** Editors: R. Kubo and E. Hanamura
19 **Solitons** Mathematical Methods for Physicists By. G. Eilenberger
20 **Theory of Nonlinear Lattices** 2nd Edition By M. Toda
21 **Modern Crystallography II** Structure of Crystals 2nd Edition By B. K. Vainshtein, V. L. Indenbom, and V. M. Fridkin
22 **Point Defects in Semiconductors I** Theoretical Aspects By M. Lannoo and J. Bourgoin
23 **Physics in One Dimension** Editors: J. Bernasconi and T. Schneider
24 **Physics in High Magnetics Fields** Editors: S. Chikazumi and N. Miura
25 **Fundamental Physics of Amorphous Semiconductors** Editor: F. Yonezawa
26 **Elastic Media with Microstructure I** One-Dimensional Models By I. A. Kunin
27 **Superconductivity of Transition Metals** Their Alloys and Compounds By S. V. Vonsovsky, Yu. A. Izyumov, and E. Z. Kurmaev
28 **The Structure and Properties of Matter** Editor: T. Matsubara
29 **Electron Correlation and Magnetism in Narrow-Band Systems** Editor: T. Moriya
30 **Statistical Physics I** Equilibrium Statistical Mechanics 2nd Edition By M. Toda, R. Kubo, N. Saito
31 **Statistical Physics II** Nonequilibrium Statistical Mechanics 2nd Edition By R. Kubo, M. Toda, N. Hashitsume
32 **Quantum Theory of Magnetism** 2nd Edition By R. M. White
33 **Mixed Crystals** By A. I. Kitaigorodsky
34 **Phonons: Theory and Experiments I** Lattice Dynamics and Models of Interatomic Forces By P. Brüesch
35 **Point Defects in Semiconductors II** Experimental Aspects By J. Bourgoin and M. Lannoo
36 **Modern Crystallography III** Crystal Growth By A. A. Chernov
37 **Modern Chrystallography IV** Physical Properties of Crystals Editor: L. A. Shuvalov
38 **Physics of Intercalation Compounds** Editors: L. Pietronero and E. Tosatti
39 **Anderson Localization** Editors: Y. Nagaoka and H. Fukuyama
40 **Semiconductor Physics** An Introduction 6th Edition By K. Seeger
41 **The LMTO Method** Muffin-Tin Orbitals and Electronic Structure By H. L. Skriver
42 **Crystal Optics with Spatial Dispersion, and Excitons** 2nd Edition By V. M. Agranovich and V. L. Ginzburg
43 **Structure Analysis of Point Defects in Solids** An Introduction to Multiple Magnetic Resonance Spectroscopy By J.-M. Spaeth, J. R. Niklas, and R. H. Bartram
44 **Elastic Media with Microstructure II** Three-Dimensional Models By I. A. Kunin
45 **Electronic Properties of Doped Semiconductors** By B. I. Shklovskii and A. L. Efros
46 **Topological Disorder in Condensed Matter** Editors: F. Yonezawa and T. Ninomiya

# Springer Series in Solid-State Sciences

Editors: M. Cardona P. Fulde K. von Klitzing H.-J. Queisser

47 **Statics and Dynamics of Nonlinear Systems**
Editors: G. Benedek, H. Bilz, and R. Zeyher

48 **Magnetic Phase Transitions**
Editors: M. Ausloos and R. J. Elliott

49 **Organic Molecular Aggregates**
Electronic Excitation and Interaction Processes
Editors: P. Reineker, H. Haken, and H. C. Wolf

50 **Multiple Diffraction of X-Rays in Crystals**
By Shih-Lin Chang

51 **Phonon Scattering in Condensed Matter**
Editors: W. Eisenmenger, K. Laßmann, and S. Döttinger

52 **Superconductivity in Magnetic and Exotic Materials** Editors: T. Matsubara and A. Kotani

53 **Two-Dimensional Systems, Heterostructures, and Superlattices**
Editors: G. Bauer, F. Kuchar, and H. Heinrich

54 **Magnetic Excitations and Fluctuations**
Editors: S. W. Lovesey, U. Balucani, F. Borsa, and V. Tognetti

55 **The Theory of Magnetism II** Thermodynamics and Statistical Mechanics By D. C. Mattis

56 **Spin Fluctuations in Itinerant Electron Magnetism** By T. Moriya

57 **Polycrystalline Semiconductors**
Physical Properties and Applications
Editor: G. Harbeke

58 **The Recursion Method and Its Applications**
Editors: D. G. Pettifor and D. L. Weaire

59 **Dynamical Processes and Ordering on Solid Surfaces** Editors: A. Yoshimori and M. Tsukada

60 **Excitonic Processes in Solids**
By M. Ueta, H. Kanzaki, K. Kobayashi, Y. Toyozawa, and E. Hanamura

61 **Localization, Interaction, and Transport Phenomena** Editors: B. Kramer, G. Bergmann, and Y. Bruynseraede

62 **Theory of Heavy Fermions and Valence Fluctuations** Editors: T. Kasuya and T. Saso

63 **Electronic Properties of Polymers and Related Compounds**
Editors: H. Kuzmany, M. Mehring, and S. Roth

64 **Symmetries in Physics** Group Theory Applied to Physical Problems 2nd Edition
By W. Ludwig and C. Falter

65 **Phonons: Theory and Experiments II**
Experiments and Interpretation of Experimental Results By P. Brüesch

66 **Phonons: Theory and Experiments III**
Phenomena Related to Phonons
By P. Brüesch

67 **Two-Dimensional Systems: Physics and New Devices**
Editors: G. Bauer, F. Kuchar, and H. Heinrich

68 **Phonon Scattering in Condensed Matter V**
Editors: A. C. Anderson and J. P. Wolfe

69 **Nonlinearity in Condensed Matter**
Editors: A. R. Bishop, D. K. Campbell, P. Kumar, and S. E. Trullinger

70 **From Hamiltonians to Phase Diagrams**
The Electronic and Statistical-Mechanical Theory of sp-Bonded Metals and Alloys By J. Hafner

71 **High Magnetic Fields in Semiconductor Physics**
Editor: G. Landwehr

72 **One-Dimensional Conductors**
By S. Kagoshima, H. Nagasawa, and T. Sambongi

73 **Quantum Solid-State Physics**
Editors: S. V. Vonsovsky and M. I. Katsnelson

74 **Quantum Monte Carlo Methods in Equilibrium and Nonequilibrium Systems** Editor: M. Suzuki

75 **Electronic Structure and Optical Properties of Semiconductors** 2nd Edition
By M. L. Cohen and J. R. Chelikowsky

76 **Electronic Properties of Conjugated Polymers**
Editors: H. Kuzmany, M. Mehring, and S. Roth

77 **Fermi Surface Effects**
Editors: J. Kondo and A. Yoshimori

78 **Group Theory and Its Applications in Physics**
2nd Edition
By T. Inui, Y. Tanabe, and Y. Onodera

79 **Elementary Excitations in Quantum Fluids**
Editors: K. Ohbayashi and M. Watabe

80 **Monte Carlo Simulation in Statistical Physics**
An Introduction 3rd Edition
By K. Binder and D. W. Heermann

81 **Core-Level Spectroscopy in Condensed Systems**
Editors: J. Kanamori and A. Kotani

82 **Photoelectron Spectroscopy**
Principle and Applications 2nd Edition
By S. Hüfner

83 **Physics and Technology of Submicron Structures**
Editors: H. Heinrich, G. Bauer, and F. Kuchar

84 **Beyond the Crystalline State** An Emerging Perspective By G. Venkataraman, D. Sahoo, and V. Balakrishnan

85 **The Quantum Hall Effects**
Fractional and Integral 2nd Edition
By T. Chakraborty and P. Pietiläinen

86 **The Quantum Statistics of Dynamic Processes**
By E. Fick and G. Sauermann

87 **High Magnetic Fields in Semiconductor Physics II**
Transport and Optics Editor: G. Landwehr

88 **Organic Superconductors** 2nd Edition
By T. Ishiguro, K. Yamaji, and G. Saito

89 **Strong Correlation and Superconductivity**
Editors: H. Fukuyama, S. Maekawa, and A. P. Malozemoff

# Springer Series in Solid-State Sciences

Editors: M. Cardona P. Fulde K. von Klitzing H.-J. Queisser

Managing Editor: H. K.V. Lotsch

90 **Earlier and Recent Aspects of Superconductivity**
Editors: J. G. Bednorz and K. A. Müller

91 **Electronic Properties of Conjugated Polymers III** Basic Models and Applications
Editors: H. Kuzmany, M. Mehring, and S. Roth

92 **Physics and Engineering Applications of Magnetism** Editors: Y. Ishikawa and N. Miura

93 **Quasicrystals** Editors: T. Fujiwara and T. Ogawa

94 **Electronic Conduction in Oxides** 2nd Edition
By N. Tsuda, K. Nasu, F. Atsushi, and K. Siratori

95 **Electronic Materials**
A New Era in Materials Science
Editors: J. R. Chelikowsky and A. Franciosi

96 **Electron Liquids** 2nd Edition By A. Isihara

97 **Localization and Confinement of Electrons in Semiconductors**
Editors: F. Kuchar, H. Heinrich, and G. Bauer

98 **Magnetism and the Electronic Structure of Crystals** By V.A. Gubanov, A.I. Liechtenstein, and A.V. Postnikov

99 **Electronic Properties of High-$T_c$ Superconductors and Related Compounds**
Editors: H. Kuzmany, M. Mehring, and J. Fink

100 **Electron Correlations in Molecules and Solids** 3rd Edition By P. Fulde

101 **High Magnetic Fields in Semiconductor Physics III** Quantum Hall Effect, Transport and Optics By G. Landwehr

102 **Conjugated Conducting Polymers**
Editor: H. Kiess

103 **Molecular Dynamics Simulations**
Editor: F. Yonezawa

104 **Products of Random Matrices** in Statistical Physics By A. Crisanti, G. Paladin, and A. Vulpiani

105 **Self-Trapped Excitons**
2nd Edition By K. S. Song and R. T. Williams

106 **Physics of High-Temperature Superconductors**
Editors: S. Maekawa and M. Sato

107 **Electronic Properties of Polymers**
Orientation and Dimensionality of Conjugated Systems Editors: H. Kuzmany, M. Mehring, and S. Roth

108 **Site Symmetry in Crystals**
Theory and Applications 2nd Edition
By R. A. Evarestov and V. P. Smirnov

109 **Transport Phenomena in Mesoscopic Systems** Editors: H. Fukuyama and T. Ando

110 **Superlattices and Other Heterostructures**
Symmetry and Optical Phenomena 2nd Edition
By E. L. Ivchenko and G. E. Pikus

111 **Low-Dimensional Electronic Systems**
New Concepts
Editors: G. Bauer, F. Kuchar, and H. Heinrich

112 **Phonon Scattering in Condensed Matter VII**
Editors: M. Meissner and R. O. Pohl

113 **Electronic Properties of High-$T_c$ Superconductors**
Editors: H. Kuzmany, M. Mehring, and J. Fink

114 **Interatomic Potential and Structural Stability**
Editors: K. Terakura and H. Akai

115 **Ultrafast Spectroscopy of Semiconductors and Semiconductor Nanostructures**
2nd Edition By J. Shah

116 **Electron Spectrum of Gapless Semiconductors**
By J. M. Tsidilkovski

117 **Electronic Properties of Fullerenes**
Editors: H. Kuzmany, J. Fink, M. Mehring, and S. Roth

118 **Correlation Effects in Low-Dimensional Electron Systems**
Editors: A. Okiji and N. Kawakami

119 **Spectroscopy of Mott Insulators and Correlated Metals**
Editors: A. Fujimori and Y. Tokura

120 **Optical Properties of III–V Semiconductors**
The Influence of Multi-Valley Band Structures
By H. Kalt

121 **Elementary Processes in Excitations and Reactions on Solid Surfaces**
Editors: A. Okiji, H. Kasai, and K. Makoshi

122 **Theory of Magnetism**
By K. Yosida

123 **Quantum Kinetics in Transport and Optics of Semiconductors**
By H. Haug and A.-P. Jauho

124 **Relaxations of Excited States and Photo-Induced Structural Phase Transitions**
Editor: K. Nasu

125 **Physics and Chemistry of Transition-Metal Oxides**
Editors: H. Fukuyama and N. Nagaosa

Volumes 1–89 are listed at the end of the book

Printing: Saladruck, Berlin
Binding: Buchbinderei Lüderitz & Bauer, Berlin